T0388041

BIOMASS, BIOFUELS, BIOCHEMICALS

Series Editor

Ashok Pandey

Centre for Innovation and Translational Research,
CSIR-Indian Institute of Toxicology,
Lucknow, India

BIOMASS, BIOFUELS, BIOCHEMICALS

Microbial Fermentation of Biowastes

Edited by

LE ZHANG

NUS Environmental Research Institute, National University of Singapore, Singapore; Energy and Environmental Sustainability for Megacities (E2S2) Phase II, Campus for Research Excellence and Technological Enterprise (CREATE), Singapore

YEN WAH TONG

NUS Environmental Research Institute, National University of Singapore, Singapore; Energy and Environmental Sustainability for Megacities (E2S2) Phase II, Campus for Research Excellence and Technological Enterprise (CREATE), Singapore; Department of Chemical and Biomolecular Engineering, National University of Singapore, Singapore

JINGXIN ZHANG

China-UK Low Carbon College, Shanghai Jiao Tong University, Shanghai, China

ASHOK PANDEY

Centre for Innovation and Translational Research, CSIR-Indian Institute of Toxicology, Lucknow, India

ELSEVIER

Elsevier
Radarweg 29, PO Box 211, 1000 AE Amsterdam, Netherlands
The Boulevard, Langford Lane, Kidlington, Oxford OX5 1GB, United Kingdom
50 Hampshire Street, 5th Floor, Cambridge, MA 02139, United States

Notices

Knowledge and best practice in this field are constantly changing. As new research and experience
broaden our understanding, changes in research methods, professional practices, or medical treatment
may become necessary.

Practitioners and researchers must always rely on their own experience and knowledge in evaluating and
using any information, methods, compounds, or experiments described herein. In using such information or methods they should be mindful of their own safety and the safety of others, including parties for
whom they have a professional responsibility.

To the fullest extent of the law, neither the Publisher nor the authors, contributors, or editors, assume any
liability for any injury and/or damage to persons or property as a matter of products liability, negligence
or otherwise, or from any use or operation of any methods, products, instructions, or ideas contained in
the material herein.

British Library Cataloguing-in-Publication Data
A catalogue record for this book is available from the British Library

Library of Congress Cataloging-in-Publication Data
A catalog record for this book is available from the Library of Congress

ISBN: 978-0-323-90633-3

For Information on all Elsevier publications visit our website at
https://www.elsevier.com/books-and-journals

Publisher: Susan Dennis
Senior Acquisitions Editor: Katie Hammon
Editorial Project Manager: Bernadine A. Miralles
Production Project Manager: Joy Christel Neumarin Honest
    Thangiah
Cover Designer: Greg Harris

Typeset by Aptara, New Delhi, India

Contents

Contributors

Gustavo Amaro Bittencourt   Department of Bioprocess Engineering and Biotechnology, Federal University of Paraná, Centro Politécnico, Curitiba, Paraná, Brazil

Xinhui Bao   State Key Laboratory of Materials-Oriented Chemical Engineering, College of Biotechnology and Pharmaceutical Engineering, Nanjing Tech University, Nanjing, PR China

Maria Giovana Binder Pagnoncelli   Department of Bioprocess Engineering and Biotechnology, Federal University of Paraná, Centro Politécnico, Curitiba, Paraná, Brazil

Yahui Bo   College of Food and Pharmaceutical, Sciences, Ningbo University, Ningbo, Zhejiang, China

Pengfei Cheng   College of Food and Pharmaceutical, Sciences, Ningbo University, Ningbo, Zhejiang, China; Center for Biorefining and Department of Bioproducts and Biosystems, Engineering, University of Minnesota-Twin Cities, Saint Paul, MN, USA

Ruirui Chu   College of Food and Pharmaceutical, Sciences, Ningbo University, Ningbo, Zhejiang, China

Chicaiza-Ortiz Cristhian   China-UK Low Carbon College, Shanghai Jiao Tong University, Shanghai, China; School of Environmental Science and Engineering, Shanghai Jiao Tong University, Shanghai, China; Faculty of Life Sciences, Amazon State University (UEA), Puyo, Pastaza, Ecuador

Yanjun Dai   Energy and Environmental Sustainability for Megacities (E2S2) Phase II, Campus for Research Excellence and Technological Enterprise (CREATE), Singapore; School of Mechanical Engineering, Shanghai Jiao Tong University, Shanghai, China

Júlio César de Carvalho   Department of Bioprocess Engineering and Biotechnology, Federal University of Paraná, Centro Politécnico, Curitiba, Paraná, Brazil

Lingkan Ding   Department of Bioproducts and Biosystems Engineering, University of Minnesota, MN, USA

Weiliang Dong   State Key Laboratory of Materials-Oriented Chemical Engineering, College of Biotechnology and Pharmaceutical Engineering, Nanjing Tech University, Nanjing, PR China; Jiangsu National Synergetic Innovation Center for Advanced Materials, Nanjing Tech University, Nanjing, PR China

Yanzhang Feng   College of Food and Pharmaceutical, Sciences, Ningbo University, Ningbo, Zhejiang, China

Chenjun He   Institute of Ecological and Environmental Sciences, Sichuan Agricultural University, Chengdu, Sichuan, P. R. China; Rural Environment Protection Engineering & Technology Center of Sichuan Province, Sichuan Agricultural University, Chengdu, Sichuan, P. R. China

Bo Hu   Department of Bioproducts and Biosystems Engineering, University of Minnesota, MN, USA

Min Jiang   State Key Laboratory of Materials-Oriented Chemical Engineering, College of Biotechnology and Pharmaceutical Engineering, Nanjing Tech University, Nanjing, PR China; Jiangsu National Synergetic Innovation Center for Advanced Materials, Nanjing Tech University, Nanjing, PR China

Luiz Alberto Junior Letti   Department of Bioprocess Engineering and Biotechnology, Federal University of Paraná, Centro Politécnico, Curitiba, Paraná, Brazil

Xihui Kang   MaREI Centre, Environmental Research Institute, University College Cork, Cork, Ireland; Civil, Structural and Environmental Engineering, School of Engineering and

Architecture, University College Cork, Cork, Ireland

Susan Grace Karp Department of Bioprocess Engineering and Biotechnology, Federal University of Paraná, Centro Politécnico, Curitiba, Paraná, Brazil

Kim Kley Valladares-Diestra Department of Bioprocess Engineering and Biotechnology, Federal University of Paraná, Centro Politécnico, Curitiba, Paraná, Brazil

Jonathan T.E. Lee NUS Environmental Research Institute, National University of Singapore, Singapore; Energy and Environmental Sustainability for Megacities (E2S2) Phase II, Campus for Research Excellence and Technological Enterprise (CREATE), Singapore

Dongyi Li Department of Biology, Institute of Bioresource and Agriculture, Sino-Forest Applied Research Centre for Pearl River Delta Environment, Hong Kong Baptist University, Kowloon Tong, Hong Kong SAR, China

Jingyi Li School of Environmental Science and Engineering, China–America CRC for Environment & Health of Shandong Province, Shandong University, Qingdao, Shandong, China

Wangliang Li CAS Key Laboratory of Green Process and Engineering, Institute of Process Engineering, University of Chinese Academy of Sciences, Beijing, China

Yang Li School of Ocean Science and Technology, Dalian University of Technology, Panjin, Liaoning, China

Nelson Libardi Junior Department of Bioprocess Engineering and Biotechnology, Federal University of Paraná, Centro Politécnico, Curitiba, Paraná, Brazil

Jun Wei Lim NUS Environmental Research Institute, National University of Singapore, Singapore; Energy and Environmental Sustainability for Megacities (E2S2) Phase II, Campus for Research Excellence and Technological Enterprise (CREATE), Singapore

Richen Lin MaREI Centre, Environmental Research Institute, University College Cork, Cork, Ireland; Civil, Structural and Environmental Engineering, School of Engineering and

Architecture, University College Cork, Cork, Ireland

Haojie Liu State Key Laboratory of Materials-Oriented Chemical Engineering, College of Biotechnology and Pharmaceutical Engineering, Nanjing Tech University, Nanjing, PR China

Kai-Chee Loh Energy and Environmental Sustainability for Megacities (E2S2) Phase II, Campus for Research Excellence and Technological Enterprise (CREATE), Singapore; Department of Chemical and Biomolecular Engineering, National University of Singapore, Singapore

Adenise Lorenci Woiciechowski Department of Bioprocess Engineering and Biotechnology, Federal University of Paraná, Centro Politécnico, Curitiba, Paraná, Brazil

Yandu Lu State Key Laboratory of Marine Resource Utilization in South China Sea, Hainan University, Hainan, China

Liwen Luo Department of Biology, Institute of Bioresource and Agriculture, Sino-Forest Applied Research Centre for Pearl River Delta Environment, Hong Kong Baptist University, Kowloon Tong, Hong Kong SAR, China

Tao Luo Biogas Institute of Ministry of Agriculture (BIOMA), Chengdu, P. R. China

Ariane Fátima Murawski de Mello Department of Bioprocess Engineering and Biotechnology, Federal University of Paraná, Centro Politécnico, Curitiba, Paraná, Brazil

Jerry D Murphy MaREI Centre, Environmental Research Institute, University College Cork, Cork, Ireland; Civil, Structural and Environmental Engineering, School of Engineering and Architecture, University College Cork, Cork, Ireland

Nicholas Cheuk Him Ng Department of Biology, Institute of Bioresource and Agriculture, Sino-Forest Applied Research Centre for Pearl River Delta Environment, Hong Kong Baptist University, Kowloon Tong, Hong Kong SAR, China

Qigui Niu School of Environmental Science and Engineering, China–America CRC for Environment & Health of Shandong Province, Shandong University, Qingdao, Shandong, China

Luciana Porto de Souza Vandenberghe Department of Bioprocess Engineering and Biotechnology, Federal University of Paraná, Centro Politécnico, Curitiba, Paraná, Brazil

Xiujuan Qian State Key Laboratory of Materials-Oriented Chemical Engineering, College of Biotechnology and Pharmaceutical Engineering, Nanjing Tech University, Nanjing, PR China

Cristine Rodrigues Department of Bioprocess Engineering and Biotechnology, Federal University of Paraná, Centro Politécnico, Curitiba, Paraná, Brazil

Roger Ruan Center for Biorefining and Department of Bioproducts and Biosystems, Engineering, University of Minnesota-Twin Cities, Saint Paul, MN, USA

Zulma Sarmiento Vásquez Department of Bioprocess Engineering and Biotechnology, Federal University of Paraná, Centro Politécnico, Curitiba, Paraná, Brazil

Fei Shen Institute of Ecological and Environmental Sciences, Sichuan Agricultural University, Chengdu, Sichuan, P. R. China; Rural Environment Protection Engineering & Technology Center of Sichuan Province, Sichuan Agricultural University, Chengdu, Sichuan, P. R. China

Zhiqiang Shi Department of Chemical Engineering, Tianjin University Renai College, Tianjin, China

Carlos Ricardo Soccol Department of Bioprocess Engineering and Biotechnology, Federal University of Paraná, Centro Politécnico, Curitiba, Paraná, Brazil

Liuying Song School of Environmental Science and Engineering, China—America CRC for Environment & Health of Shandong Province, Shandong University, Qingdao, Shandong, China

Bing Song Scion, Te Papa Tipu Innovation Park, Rotorua, New Zealand

Hailin Tian NUS Environmental Research Institute, National University of Singapore, Singapore; Energy and Environmental Sustainability for Megacities (E2S2) Phase II, Campus for Research Excellence and Technological Enterprise (CREATE), Singapore

Yen Wah Tong NUS Environmental Research Institute, National University of Singapore, Singapore; Energy and Environmental Sustainability for Megacities (E2S2) Phase II, Campus for Research Excellence and Technological Enterprise (CREATE), Singapore; Department of Chemical and Biomolecular Engineering, National University of Singapore, Singapore

To-Hung Tsui NUS Environmental Research Institute, National University of Singapore, Singapore; Energy and Environmental Sustainability for Megacities (E2S2) Phase II, Campus for Research Excellence and Technological Enterprise (CREATE), Singapore

David Wall MaREI Centre, Environmental Research Institute, University College Cork, Cork, Ireland; Civil, Structural and Environmental Engineering, School of Engineering and Architecture, University College Cork, Cork, Ireland

Haixia Wang College of Food and Pharmaceutical, Sciences, Ningbo University, Ningbo, Zhejiang, China

Endashaw Workie China-UK Low Carbon College, Shanghai Jiao Tong University, Shanghai, China

Chao Xu Institute of Bast Fiber Crops, Chinese Academy of Agricultural Sciences, Changsha, China

Lijie Xu State Key Laboratory of Materials-Oriented Chemical Engineering, College of Biotechnology and Pharmaceutical Engineering, Nanjing Tech University, Nanjing, PR China

Xiaojun Yan Key Laboratory of Marine Biotechnology of Zhejiang Province, Ningbo University, Ningbo, Zhejiang, China

Miao Yan NUS Environmental Research Institute, National University of Singapore, Singapore; Energy and Environmental Sustainability for Megacities (E2S2) Phase II, Campus for Research Excellence and Technological Enterprise (CREATE), Singapore

Hairong Yuan Centre for Resource and Environmental Research, Beijing University of Chemical Technology, Beijing, P. R. China

Le Zhang NUS Environmental Research Institute, National University of Singapore, Singapore; Energy and Environmental Sustainability for Megacities (E2S2) Phase II, Campus for Research Excellence and Technological Enterprise (CREATE), Singapore

Pengshuai Zhang China-UK Low Carbon College, Shanghai Jiao Tong University, Shanghai, China; School of Environmental Science and Engineering, Shanghai Jiao Tong University, Shanghai, China

Jingxin Zhang China-UK Low Carbon College, Shanghai Jiao Tong University, Shanghai, China

Tengyu Zhang China-UK Low Carbon College, Shanghai Jiao Tong University, Shanghai, China; Department of Chemical and Biological Engineering, University of Sheffield, Sheffield, UK

Jun Zhao Department of Biology, Institute of Bioresource and Agriculture, Sino-Forest Applied Research Centre for Pearl River Delta Environment, Hong Kong Baptist University, Kowloon Tong, Hong Kong SAR, China

Mi Zhou Department of Agricultural, Food and Nutritional Sciences, University of Alberta, Edmonton, AB, Canada

Chengxu Zhou College of Food and Pharmaceutical, Sciences, Ningbo University, Ningbo, Zhejiang, China

Kang Zhou Department of Chemical and Biomolecular Engineering, National University of Singapore, Singapore

Jie Fu J. Zhou Department of Chemical and Biomolecular Engineering, National University of Singapore, Singapore

Jie Zhou State Key Laboratory of Materials-Oriented Chemical Engineering, College of Biotechnology and Pharmaceutical Engineering, Nanjing Tech University, Nanjing, PR China; Jiangsu National Synergetic Innovation Center for Advanced Materials, Nanjing Tech University, Nanjing, PR China

Preface

The book titled **Microbial Fermentation of Biowastes** is a part of the Elsevier comprehensive book series on *BIOMASS, BIOFUELS, BIOCHEMICALS* (Editor-in-Chief: Ashok Pandey). This book intends to cover different aspects of microbial fermentation of biowastes, providing the state-of-art information on various aspects and perspectives for future developments. As modernization, industrialization, and urbanization continue to increase, substantial organic wastes (e.g., food waste, agricultural waste, manure, sludge, wastewater, and plastics, etc.) are being generated, which pose a growing threat to the environment and sustainable development of mankind. To address these challenges, microbial fermentation is seen as a useful technology that can convert the abundant organic wastes to various biofuels and biochemicals in many forms and applications, especially in the global context of the carbon cycle. In the past decade, extensive studies have focused on enhanced microbial fermentation for better biowaste management and higher production of renewable bioenergy or biochemicals. Hitherto, great progress has been achieved in this area via various strategies, including the use of additives, multistage bioreactors, microbial bioaugmentation, genetically engineered microorganisms, codigestion, feedstock pretreatment, enzyme technologies, and hybrid technologies, etc. For instance, enhanced anaerobic digestion is used for biowastes conversion to biomethane and biohydrogen while the enhanced acidogenic fermentation is used for the production of volatile fatty acids from biowastes. In addition, microbial

biodegradation of plastics is increasingly becoming a significant topic in recent years due to the fact that "white pollution" caused by the plastics has become a pressing problem that needs to be solved within the shortest timespan. In recent years, many microbes have been identified as candidates for microbial degradation of plastic debris, but further research remains needed. However, the aforementioned advances have not been consolidated and are quite diverse, and the state-of-the-art has yet to be established.

Motivated by the above gaps, this book aims to provide a solution, using microbial fermentation technology, to address the problem of how to efficiently and economically manage various organic wastes and recycle simultaneously energy/resources from the wastes. To achieve this target, this book summarizes the latest research achievements on the development of various strategies for enhanced microbial fermentation for organic wastes conversion to bioenergy and biochemicals and for the biodegradation of plastic waste. The technical principles and practical applications of the enhanced microbial fermentation processes via various strategies can be a reference for the researchers, engineers, investors, policy makers, and students in the fields of waste management with energy and resource recovery.

The chapter 1 of the book presents an overall introduction to the enhancing strategies for microbial fermentation processes, where particular focus is set on the technical principles of additive strategies, multistage bioreactors, microbial bioaugmentation strategies, genetically engineered microorganisms, codigestion strategies, feedstock pretreatment strategies, enzyme technologies, and hybrid technologies. Chapters 2–6 present progress on the conversion of common wastes such as food waste, agricultural waste, manure, wastewater, and algal residues to bioenergy and biochemicals via enhanced anaerobic digestion. Chapters 7–14 discuss the significant progress achieved on enhancing microbial fermentation via additive strategy, multistage bioreactor strategy, microbial bioaugmentation strategy, genetic engineering approach, codigestion strategy, feedstock pretreatment strategy, enzyme applications, and hybrid technologies. Chapter 15 provides details on the recent advances on enhanced acidogenic fermentation of organic wastes for the production of volatile fatty acids; Chapter 16 presents functional microbial characteristics in acidogenic fermenters of organic wastes for the production of volatile fatty acids. Finally, for the plastic waste management, Chapter 17 presents the recent advances on microbial fermentation for biodegradation and biotransformation of plastics into high value–added chemicals.

The editors gratefully acknowledge the authors for their contributions in this book. Thanks are due to the reviewers who provided valuable suggestions to improve the quality of the chapters. We greatly appreciate Dr Kostas Marinakis, Former Senior Book Acquisition Manager and Ms Katie Hammon, Senior Book Acquisition Manager, Ms Andrea Dulberger, Editorial Project Manager, and others in the Elsevier team associated with this book for their support toward publishing this book.

We are confident that this book would be of great value for the researchers broadly working in the areas of waste to wealth, resource recovery, and production of chemicals and fuels from renewable resources. The book will be of equal value to industry persons as well as policy planners.

Editors
Le Zhang
Yen Wah Tong
Jingxin Zhang
Ashok Pandey

CHAPTER

1

Strategies for enhanced microbial fermentation processes

Le Zhang[a,b], Jonathan T.E. Lee[a,b], Kai-Chee Loh[b,c],
Yanjun Dai[b,d], Yen Wah Tong[a,b,c]

[a]NUS Environmental Research Institute, National University of Singapore, Singapore
[b]Energy and Environmental Sustainability for Megacities (E2S2) Phase II, Campus for Research Excellence and Technological Enterprise (CREATE), Singapore
[c]Department of Chemical and Biomolecular Engineering, National University of Singapore, Singapore
[d]School of Mechanical Engineering, Shanghai Jiao Tong University, Shanghai, China

1.1 Introduction

Ever-increasing global population and anthropogenic activities lead to higher energy demands and rampant environmental deterioration (De Sanctis et al., 2019; Saud et al., 2020). To deal with the latter, the management of various biomass wastes remains a critical challenge (De Schouwer et al., 2019; Foong et al., 2020). Microbial fermentation technologies such as anaerobic digestion (AD) have been extensively adopted to generate bioenergy and biochemicals from diverse biowastes (Chen et al., 2020; Kougias & Angelidaki, 2018; Rawoof et al., 2020). Such biowastes comprise food waste, agricultural waste, horticultural waste, animal manure, wastewater/sludge, and algal residues. After upgrading (Angelidaki et al., 2018), methane-rich biogas derived from AD is frequently converted to heat and electricity using combined heat and power (CHP) units (Akkouche et al., 2020; Di Maria et al., 2019). As a result, methane production efficiency plays an important role in the commercial viability of AD technology in biowaste treatment. Nevertheless, as a multistep (i.e., hydrolysis, acidogenesis, acetogenesis, and methanogenesis) biological process governed by a spectrum of microbiota, AD processes are usually affected by many factors (Chen et al., 2014; Chen et al., 2008; Yenigün & Demirel, 2013), which lead to a relatively low conversion efficiency in some AD plants (Zhang et al., 2019c), particularly for nonhomogeneous substrates. At the heart of AD process, microbial communities are also significantly different for different types of substrates (Fernandez-Bayo et al., 2020; Narihiro & Sekiguchi, 2007; Ting et al., 2020). Indeed,

Biomass, Biofuels, Biochemicals.
DOI: https://doi.org/10.1016/B978-0-323-90633-3.00001-8

the microbial community structures in a given AD system have intrinsic link to bioreactor performance. Hence, it is becoming increasingly essential to comprehensively understand the microbiome in AD systems. Fortunately, bioinformatic analysis of biogas-producing microbial communities in anaerobic bioreactors has become a widely used tool to analyze microbiomes and diverse metabolic pathways (Lim et al., 2020; Zhang et al., 2019b).

In the past decades, many studies have been conducted to develop various strategies to enhance the microbial fermentation of organic matters for higher productivity of biofuels and biochemical; these include additive supplementation (Chiappero et al., 2020; Zhang et al., 2019f), multistage bioreactors (García-Ruíz et al., 2020; Zhang et al., 2020a), microbial bioaugmentation (Lee et al., 2020a; Tsapekos et al., 2017; Yan et al., 2020), genetically engineered microorganisms (Llamas et al., 2020; Sáez-Sáez et al., 2020), codigestion (Gao et al., 2020; Oladejo et al., 2020), feedstock pretreatments (Gunes et al., 2019; Millati et al., 2020), enzyme technologies (Agabo-Garcia et al., 2019; Garcia et al., 2019), and hybrid technologies (Cui et al., 2020; Rezaee et al., 2020). While these studies are diverse, there was not good identification of the state-of-the-art, though slowly appearing in the literature. For instance, additional insights in many new aspects such as nanotechnology, engineered enzymes, multistage anaerobic bioreactor, bioaugmentation via microbial consortia, metagenomics-based data mining, precision fermentation platform via artificial intelligence, and industrial application progress have been reported in recent years.

This book focuses on the most recent progress in various enhancing strategies for microbial fermentation of various biowastes for the production of bioenergy and value-added biochemicals. In this introductory chapter, a brief introduction to the various biowastes and the respective enhancing strategies are the main focus. The advantages and limitations, industrial applications, and future perspectives of various performance-enhancing strategies are also highlighted.

1.2 Characteristics of common biowastes

1.2.1 Food waste

Food waste represents the food discarded during the processes of production, transportation, transaction, processing, and consumption (Tong et al., 2018). Due to the differences in the eating habits of people in various countries and regions, the derived food wastes vary in the major components such as vegetables, meat, rice, eggs, noodles, etc. As such, the heterogeneous compositions of proteins, lipids, carbohydrates, and lignocellulosic components in different food waste result in relatively unstable biodegradation rate and methane production rate. Table 1.1 shows the typical characteristics of common biowastes. The volatile solids (VS) and total solids (TS) contents of food waste are in ranges of approximately 12%–26% and 12%–31%, respectively, which indicates that water accounts for around 69%–88% of the mass in food waste. It does contain a fair amount of proteins, lipids, carbohydrates, and lignocellulosic components that are promising substrates for anaerobic bioreactors to produce high value-added products (Ravindran & Jaiswal, 2016). Notably, the pH value of food waste is rather low (i.e., pH 4.0–5.0) and therefore requires appropriate monitoring and control to maintain an efficient AD process.

TABLE 1.1 Typical characteristics of common biowastes.

Characteristics	Food waste (Chen et al., 2016; Zhang et al., 2020c; Zhou et al., 2015)	Agricultural waste (Hoornweg & Bhada-Tata, 2012; Hu et al., 2015)	Yard waste (Brown & Li, 2013; Yao et al., 2017)	Animal manure (Dechrugsa et al., 2013; Zarkadas et al., 2015; Zhang et al., 2019d)	Waste activated sludge (Ebenezer et al., 2015; Yu et al., 2014; Zhang et al., 2019e)	Algal residues (Suganya et al., 2016; Zhang et al., 2020b)	Units
pH	4.0–5.0	5.6–5.9	6.2–6.8	7.3–7.8	6.5–8.6	-	-
Volatile solids (VS)	11.6–26.3	83.2–88.2	64.9–91.7	5.7–40.7	8.8–12.7	66.6–70.3	%(w.b.)
Total solids (TS)	12.3–30.9	93.7–95.2	93.1–94.3	27.2–84.4	10.8–22.9	87.6–93.0	%(w.b.)
VS/TS	85.3–90.0	89.0–92.7	69.7–97.2	50.0–84.7	55.4–81.5	71.6–80.3	%
Protein	15.32–20.76	-	-	-	32.9–33.8	5.06–71.0	%TS
Lipid	19.62–23.54	-	-	4.94–19.0	-	2.0–40.0	%TS
Carbohydrate	38.63–40.37	-	-	-	-	4.0–57.0	%TS
Cellulose	4.3–9.7	34.9–38.2	24.3–28.9	15.3–25.1	14.7–38.9	-	%TS
Hemicellulose	2.5–3.7	16.7–31.3	9.7–32.1	4.1–20.9	3.8–12.6	-	%TS
Lignin	0.7–1.5	8.4–23.3	7.9–23.0	4.5–19.0	1.2–7.3	-	%TS
Total carbon	31.9–48.3	47.0–49.7	21.3–36.2	39.4–41.5	28.2–34.8	43.42–45.58	%TS
Total nitrogen	2.6–3.2	0.8–1.0	0.8–1.3	1.3–2.3	6.8–6.9	7.26–7.52	%TS
C/N ratio	14.6–18.9	47.0–55.8	15.9–45.3	17.2–32.2	5.0–5.2	5.8–6.3	-
Moisture content	69.1–79.8	6.1–6.5	12.8–13.1	72.5–72.8	98.5–99.8	7.0–12.4	%
Calorific value	14.6–16.0	16.2–19.0	15.1–17.4	16.1–18.8	19.3–23.4	-	MJ/kg-TS

Note: w.b., wet basis.

1.2.2   Agricultural waste and yard waste

Agricultural waste is one of the most abundant biomass resources, and therefore has great potential for the generation of renewable energy and value-added biochemicals (Yu et al., 2019). Essentially, the bulk of agricultural waste is lignocellulosic biomass, which is composed of three polymers (i.e., cellulose, hemicellulose, and lignin) entangled together into a super-structure and is thereby resistant to microbial degradation. Hence, hydrolysis is regarded as the rate-limiting step during AD of lignocellulosic material. Many current studies therefore are focused on enhancing hydrolysis of lignocellulosic biomass through strategies such as microbial bioaugmentation (Lee et al., 2020a; Shetty et al., 2020), feedstock pretreatment (Lee et al., 2020b; Vieira et al., 2020), and enzymatic hydrolysis (da Silva et al., 2020).

Yard waste consists mainly of plant leaves, wood chips, and grasses, which is an inherent component of municipal solid waste, especially in the green cities and towns (Lee et al., 2018). It has been reported that the contents of cellulose, hemicellulose, and lignin of common yard waste are approximately 24%–29%, 10%–32%, and 8%–23%, respectively (Table 1.1). Similar to agricultural waste, a major impediment to industrial AD of yard waste lies in the need for various pretreatments (e.g., enzymatic hydrolysis, physical, chemical, and biological techniques) to facilitate the breakdown of lignocellulosic components so that sugar polymers become bioavailable. Notably, the C:N ratio of yard waste is relatively high (e.g., 55.3), compared to the recommended C:N ratios between 20:1 to 30:1 (Uçkun Kiran et al., 2016). Ergo, yard waste is frequently codigested with other wastes (e.g. food waste) with a lower C:N ratio to balance the nutrition requirement of carbon and nitrogen for microbial growth and metabolic reactions.

1.2.3   Animal manure

Animal manure is mainly derived from animal feces, which usually contain liquid farm manure slurry and a solid farmyard manure. The latter is composed mainly of plant materials such as crop straw that are utilized as bedding materials for various farm animals. Due to the fact that different animals have different digestive systems and diets, the manure derived from them displays diverse qualities, which leads to the necessity of tailored optimization of different AD operating conditions (Li et al., 2015). The digestate of the AD operations using animal manure is usually utilized as fertilizer. Nevertheless, the high abundance of antibiotic resistance genes (ARGs) in animal manure caused by overused antibiotics in livestock feed is a pertinent issue. To mitigate the ARGs issue during AD of manure, the effectiveness of feedstock pretreatment with microwave pretreatment has been evaluated and validated (Zhang et al., 2019d). Another critical issue associated with AD of manure is the inhibition caused by high amounts of ammonia (Fuchs et al., 2018; Sun et al., 2016; Wang et al., 2016). To tackle ammonia inhibition for efficient methane production from manure, several strategies such as substrate dilution (Bujoczek et al., 2000), codigestion (Li et al., 2017), microbial bioaugmentation (Yan et al., 2020), and additives (e.g., biochar (Pan et al., 2019) and trace elements (Molaey et al., 2018)) have been investigated. In addition, animal manure was recently found promising as a feedstock to produce volatile fatty acids (VFAs) using a membrane system (Jomnonkhaow et al., 2020).

1.2.4 Wastewater and waste activated sludge

There has been a paradigm shift from regarding wastewater as a waste requiring treatment, to that of a substrate for recovery of energy and resource in the context of a circular economy (Jiang et al., 2012; Robles et al., 2020). Indeed, wastewater treatment plants in many countries and areas have employed water resource recovery facilities (Diaz-Elsayed et al., 2019; Meena et al., 2019). The traditional wastewater treatment focuses mainly on the pollutant removal and safe discharge of treated wastewater (Kanaujiya et al., 2019), whereas the water resource recovery facilities focus on recycling of nutrient elements (e.g., phosphorus (Chrispim et al., 2019)), energy (Goswami et al., 2019), and clean water (Li et al., 2020) from a wide range of wastewater. Essentially, the organic compounds in wastewater are consumed by microbes in anaerobic reactors. The process efficiency of AD of wastewater can be further enhanced by integrating with other technologies such as microbial electrolysis cells (Hassanein et al., 2017).

Waste activated sludge (WAS) is derived from the secondary aerobic process of municipal wastewater treatment (Shin et al., 2019). WAS commonly contains an array of biodegradable organics, nutrients, pathogens, and heavy metals (Jeong et al., 2019), and therefore is frequently treated by AD for energy recovery prior to disposal (Ruffino et al., 2019). Nevertheless, from the perspective of industrial applications, AD of WAS for biogas production generally presents a relatively low economic feasibility due to the high water content in WAS. To enhance biomethane production rate, high-solids AD of dewatered activated sludge with a TS content of over 15 wt% has been suggested (Guo et al., 2017). Meanwhile, the C:N ratio in WAS is relatively low (e.g., 5.0–5.2), thereby relegating it as unsuitable for an effective anaerobic monodigestion, but a good cosubstrate to anaerobic codigestion. Hitherto, codigestion of WAS with other organic wastes that contain very high C:N ratios such as agricultural waste (Zhang et al., 2019a), yard waste (Lee et al., 2019), and food waste (Du et al., 2020b) have been investigated. In addition, it is noteworthy that sludge drying followed by incineration was recently reported as a sustainable disposal method of excess sludge (Hao et al., 2020). However, more efficient dewatering technologies have to be developed before incineration becomes a feasible alternative option.

1.2.5 Algal residues

Microalgae have been utilized as a potential feedstock for generation of biofuels such as biodiesel (Hossain et al., 2019). However, large amounts of algal residues remain after algae-based biodiesel production (Zhu et al., 2018). Algal residues contain mainly carbohydrate, protein, lipid, and a small amount of hemicellulose, and therefore they can be further converted to value-added products (e.g., biomethane, biohydrogen, bioethanol, and biolipid) using AD or microbial fermentation (Chandra et al., 2019; Kumar et al., 2020; Zhang et al., 2020b). AD of dry algal residues plays an essential role in biorefinery of algal biomass toward zero waste targets. In AD of lipid-extracted algal residues, potential solvent toxicity and unbalanced C:N ratio remain challenges (Bohutskyi et al., 2019). Accordingly, codigestion and feedstock pretreatments (Kumar et al., 2020) could be promising strategies to alleviate solvent inhibition and enhance methane yield.

1.3   Strategies for enhancing microbial fermentation

1.3.1   Additive strategies

The additive strategies for enhancing AD and microbial fermentation processes developed rapidly in the past decade, as there are several advantages such as being able to use infrastructure already in place, easy application, and relatively economical operation costs (Arif et al., 2018; Paritosh et al., 2020; Romero-Güiza et al., 2016; Zhang et al., 2018). These features make additive strategies suitable for potential industrial scale application. Hitherto, various additives have been explored in AD in the form of carbon-rich additives (e.g., activated carbon, biochar and graphene) (Chiappero et al., 2020; Lin et al., 2017; Zhang et al., 2020c), trace metals (Thanh et al., 2016), nanoparticles (Lee & Lee, 2019; Zhu et al., 2020), and zeolite (Montalvo et al., 2020), etc. Carbon-rich additives are capable of simultaneously providing an immobilization substrate for enhanced microbial growth (Fagbohungbe et al., 2016; Weber et al., 1978), promoting direct interspecies electron transfer among microorganisms (Lin et al., 2018; Park et al., 2018), and mitigating acid stress (Luo et al., 2015) and ammonia inhibition (Lü et al., 2016). The technology readiness level for using biochar addition strategy to enhance AD performance was at least 1000 L pilot-scale demonstration, on the way to early commercialization (Hu et al., 2020; Zhang et al., 2020c). Trace metals play an essential role in the growth of anaerobic microorganisms and their metabolic functions, so supplementations of vital trace metals such as Fe, Ni, Co, Mo, Ca, Mg, Mn, W, Se, and Zn have been widely studied (Hijazi et al., 2020; Mancini et al., 2018; Matheri et al., 2016). Nonetheless, in practice they could be added to anaerobic digesters in excessive amounts, leading to potential process inhibition (Thanh et al., 2016). The key research required to avoid potential metal inhibition is an in-depth understanding of chemical speciation and bioavailability of trace metals in anaerobic digesters (Thanh et al., 2016). Bioavailability of trace metals was reported to be increased in anaerobic digesters through appropriate supplementation of metal chelating agents (e.g., nickel-chelator complexes (Zhang et al., 2020d)) and encapsulated metal additive (Zhang et al., 2019f). However, the environmental and human health risk of chelating agents as carriers for metal supplementation greatly restricts their extensive application (Serrano et al., 2017). The two most common nanoparticles added into anaerobic digesters are zero-valent metals (Aguilar-Moreno et al., 2020; Hassanein et al., 2020) and metal oxides (Faisal et al., 2020; Ghofrani-Isfahani et al., 2020). Conductive metal nanoparticles could simultaneously provide key nutrients, aid synthesis of key enzymes and co-enzymes, as well as facilitate interspecies electron transfer in anaerobic digesters. In-depth effects on dominant metabolic pathways of microbial communities, economic feasibility, environmental friendliness, and pilot-scale tests of the aforementioned additives could be areas for further research.

1.3.2   Multistage bioreactors

The idea of enhancing AD via multistage bioreactors stems from the intrinsic feature of AD, namely, multiple sequential steps of hydrolysis, acidogenesis, acetogenesis, and methanogenesis. The different processes are spatially separated in a multistage bioreactor. The most commonly investigated and utilized two-stage bioreactor (Srisowmeya et al., 2020) is such an example. In a two-stage bioreactor, hydrolysis, acidogenesis, and acetogenesis are carried out in the first stage with a pH range of 5.0–7.0, while the methanogenesis is conducted in the second stage with a pH between 7.0 and 8.0 (Li et al., 2018). Compared to the traditional single-stage bioreactor, the two-stage bioreactor is superior due to better control of process parameters, providing better functioning environmental conditions for microbial communities and a resultant improved process performance (De Gioannis et al., 2017; Nguyen et al., 2020). Two-stage bioreactors have been utilized in AD of food waste (Srisowmeya et al., 2020), horticultural waste (Li et al., 2018), vinasse (Fu et al., 2017), sewage sludge and glycerol (Silva et al., 2018), landfill leachate (Begum et al., 2018), vegetable waste (Chatterjee & Mazumder, 2020), organic fraction of municipal solid waste (Lavagnolo et al., 2018), algal residue (Lunprom et al., 2019) for the production of biomethane (García-Ruíz et al., 2020), biohydrogen (Ding et al., 2020), and biohythane (Promnuan et al., 2020). As an enhancement, three-stage anaerobic bioreactors with three separate chambers in a single reactor have been proposed to provide more diverse environment for various functional microorganisms responsible for stabilizing the AD process. Hitherto, three-stage AD of food waste (Zhang et al., 2017b), corn stover (Liu et al., 2019), cornstalks (Cheng & Liu, 2012), tofu wastewater (Vistanti & Malik, 2019), ethanol wastewater (Intanoo et al., 2020), food waste + horse manure (Zhang et al., 2017a), food waste + sludge (Zhang et al., 2019e), food waste + horticultural waste (Zhang et al., 2020a), and fruit waste + vegetable waste (Chatterjee & Mazumder, 2018) has been investigated. Compared to the traditional single-stage bioreactor, the three-stage anaerobic bioreactor can simultaneously allow three available pH windows, including pH ~ 4–5 for substrate hydrolysis, pH ~ 5–7 for acidification, and pH ~ 7–8 for methanogenesis in the first, second, and third stage, respectively. The biggest advantage of a multistage bioreactor is that independent optimization of each process can be conducted without interference from other processes. However, the pH control will further increase the operating cost, which may hinder the large-scale application of multistage bioreactors. Accordingly, a study of the economic feasibility of AD using three-stage bioreactors can be conducted in the near future. Governmental subsidies and technical upgrading (e.g., biochar amendment) could be promising methods to increase the feasibility of the three-stage thermophilic bioreactor at an industrial scale (Zhang et al., 2020a).

1.3.3 Bioaugmentation strategies

Bioaugmentation strategies refer to supplementation of exogenous pure culture, preadapted consortium, or genetically modified microbes harboring specific metabolic activities into the bioreactor to facilitate the microbe-mediated multistep AD process (Herrero & Stuckey, 2015). Hitherto, bioaugmentation strategies with microbial consortia have been successfully utilized to enhance biodegradation of lignocellulosic components (Lee et al., 2020a; Tsapekos et al., 2017), mitigate inhibition caused by organic acids and ammonia (Yan et al., 2020), as well as improve production of biogas, hydrogen, and VFA (Atasoy & Cetecioglu, 2020; Jiang et al., 2020). The main target of bioaugmentation in AD of lignocellulosic biomass is to increase the bottleneck hydrolysis rate of cellulose and hemicellulose. In bioaugmentation operations for mitigating ammonia inhibition or acid inhibition, ammonia-tolerant or acid-tolerant microbial strains are supplemented into the respective anaerobic bioreactors. In addition, methanogens or hydrogen-producing strains can also be bioaugmented to promote methanogenesis or enhance metabolic pathways leading to hydrogen production. Common operational procedures include obtaining microbial cultures, cultivation of microbial cultures, acclimatization of microbes, and bioaugmentation of acclimated microbes. Additionally, long-term storage of acclimated microbes is a key for potential industrial application of bioaugmentation strategies employing microbial consortia. Key findings in bioaugmentation strategies for enhancing AD in recent years (see Chapter 9) are numerous; however, the reported bioaugmentation studies are frequently limited to lab-scale, which remain inadequate for potential industrial application. More large-scale studies should be conducted to assess the technical and economic feasibility.

1.3.4 Genetically engineered microorganisms

A genetically engineered microorganism refers to a microorganism whose genetic material has been altered by genetic engineering techniques or recombinant DNA technologies. Molecular biology techniques have been developed that can overcome the technical limitations of cultivation-based methods and allowed the identification of noncultivable microorganisms involved in AD (Lee et al., 2017b; Lim et al., 2020). These microbial genomics resources have shown great potential for novel insights and new applications in enhanced biofuel productivity. Subsequent to the identification of the predominant bacteria and methanogens (Lim et al., 2018; Zhang et al., 2019b), genetic engineering tools can be utilized to manipulate specific enzymes by altering its DNA sequence for targeted enhancement of biogas production. Notably, the cost effectiveness could be a critical factor to affect the industrial-scale application of the genetic engineering approach. Providentially, the operation costs of genetic engineering and synthetic biology have been decreasing. Thus, the biggest challenge for large-scale application of engineered microorganisms for biowaste fermentation and biofuel production might not be the cost, but the available standardized genetic units and modules in the area of synthetic biology. Furthermore, synthetic biology approaches allows the building of genetically engineered microorganisms with entirely new sequences of DNA, which can make AD more efficient for biogas production. To do so, a considerable understanding of functional enzymes and genes associated with the acetate oxidation, hydrogenotrophic and aceticlastic methanogenesis pathways must be attained. Presently, genetically engineered microorganisms have been utilized in fermentation of organic wastes for the production of biochemicals such as 2-phenylethanol (Wang et al., 2019), single-cell oils (Bandhu et al., 2019), and limonene (Pang et al., 2019).

1.3.5 Codigestion strategies

Codigestion refers to the simultaneous digestion of two or more substrates in the same reactor, a well-established strategy to overcome the limitations of monodigestion (Solé-Bundó et al., 2019). Codigestion possesses advantages such as the balancing of C:N ratio and dilution of inhibitory/toxic compounds in the bioreactor (Brown & Li, 2013). For instance, monodigestion of food waste is challenging due to high C:N ratio and low pH, while the monodigestion of yard waste is slower due to the high lignocellulosic content, and monodigestion of algal residues is associated with the risk of ammonia inhibition and low biodegradability. By combining these biowastes appropriately, codigestion can effectively enhance the rate of substrate digestion, as well as the rate of methane generation. For instance, food waste has been successfully codigested with a wide variety of wastes, including WAS (Du et al., 2020b), animal manure (Oladejo et al., 2020), yard waste (Panigrahi et al., 2020), blackwater (Gao et al., 2020), cymbopogon citratus (Owamah, 2020), rice straw (Kainthola et al., 2020a), algal biomass (Zhang et al., 2020b), etc. Additionally, in some cases, more than two kinds of wastes have been combined for effective codigestion. For instance, a feedstock comprising wheat straw, chicken manure, and sheep manure demonstrated effective AD and enhanced methane production under a fermentative environment of appropriate VFA concentrations and low ammonia (Liu et al., 2015). In addition to AD, the codigestion strategy has also been applied to aerobic fermentation for the production of chemicals such as bioethanol (Lee et al., 2017a), microbial lipids (Chen & Wan, 2017), organic acids (Ong et al., 2019), etc. Codigestion is therefore a promising approach for intensification of production of bioenergy and biochemicals, and could be easily implemented in pilot- and full-scale biorefinery factories.

1.3.6 Feedstock pretreatment strategies

In some recalcitrant substrates, particularly for feedstock with a relatively high lignocellulosic content such as agricultural waste and horticultural waste, pretreatments are essential to improve subsequent fermentation efficiency. In the literature, numerous substrate pretreatment methods have been suggested, including physical, chemical, biological, and combined pretreatments (Atelge et al., 2020; Yu et al., 2019; Zabed et al., 2019; Zhen et al., 2017). Chemical pretreatment can provide better performance in a short period of time, which is a substantial advantage for practical industrial application (Xu et al., 2020). However, the use of strong alkalis or acids could cause potential hazards to the environment. A possible solution is development of more environmentally friendly solvents for chemical pretreatments. In terms of physical pretreatments, hydrothermal pretreatment seems a promising method because it avoids requirement of chemicals (Lee et al., 2020b; Thompson et al., 2020). However, a high energy input is required. Biological pretreatments using microbial species or biological agents have attracted attention as they can avoid the disadvantages of chemicals usage and high energy requirement (Kainthola et al., 2020b). Notwithstanding lower operating costs and being environmentally benign, the main limitation of biological pretreatments would be the relative longer treatment time. A combination of two or more individual pretreatment methods could offer synergistic effects of multiple pretreatment strategies, at the cost of higher pretreatment expenditure (Siami et al., 2020). In the context of the development of a circular economy, the energy performance, economic feasibility, and environmental impact of the aforementioned pretreatment technologies have to be taken into account before adoption.

1.3.7 Application of enzymes

As mentioned previously, hydrolysis is regarded as the rate-limiting step of recalcitrant substrates such as lignocellulosic biomass and sludge (Kainthola et al., 2019; Park et al., 2020). During hydrolysis, complex polymers are converted into corresponding soluble monomers by a suite of extracellular microbial enzymes (Rajin, 2018). Specifically, cellulose polymer is hydrolyzed to cellobiose and glucose monomers, while hemicellulose polymer is hydrolyzed to hexose and pentose monomers. Carbohydrates, lipids, and proteins are hydrolyzed to corresponding monomers, namely, sugars, fatty acids, and amino acids. The concentration of hydrolytic enzymes that are naturally secreted by the microbes in the bioreactor is relatively low. Thus, additional enzymes can be supplemented to facilitate the hydrolysis process of recalcitrant substrates. Essentially, appropriate enzymes as catalysts can decrease activation energy of complex biochemical reactions, thus accelerating the rates of reactions. The advantages of augmentation of enzymes in microbial fermentation processes include high efficiency, high selectivity, environmental friendliness, and operational convenience. A disadvantage of enzyme supplementation is the relatively high cost, which may hinder their industrial application. Thus far, many enzymes have been applied to enhance the performance of the microbial fermentation systems for simultaneous waste degradation and resource recovery (Nigam, 2013; Singh et al., 2016; Singh et al., 2019).

1.3.8 Hybrid technologies

Among the above-mentioned enhancing strategies, various approaches may have different mechanisms by which the enhancement is affected. For instance, some strategies such as pretreatments can enhance the conversion efficiency from the perspective of feedstock properties (Sanchez et al., 2020), while other strategies such as the supplementation of additives (e.g., trace elements) can improve the process from the perspective of nutrition supply (FitzGerald et al., 2019). Strategies such as two-stage (Deng et al., 2019; Liu et al., 2020) and multistage bioreactors (Rabii et al., 2019) can boost the fermentation from the microbial growth and enrichment. By coupling two or more techniques, synergistic effects could be achieved to significantly promote the microbial fermentation processes. So far researchers have sought to combine a variety of the aforementioned strategies. Many of the hybrid technologies or approaches integrate substrate pretreatment with codigestion, additive supplementation, bioreactor optimization, etc. For instance, an increase in methane yield from AD has been reported by integrating alkali-hydrodynamic pretreatment of WAS with two-stage AD process (Grübel & Suschka, 2015); in coupling the codigestion of food waste and animal manure with the operation of a three-stage anaerobic bioreactor (Zhang et al., 2017a); and combining microwave pretreatment with activated carbon supplementation (Zhang et al., 2019d). In addition to AD, cofermentation has also been carried out in two-stage bioreactor systems. Compared with using a single enhancing strategy, hybrid technologies could exhibit higher conversion efficiency and higher productivity. Based on the aforementioned summary, a comprehensive research framework, as shown in Fig. 1.1, is established for enhanced microbial fermentation of common biowastes for production of biofuels and biochemicals. For each strategy, the economic feasibility, environmental sustainability, carbon neutrality, management policies, and system availability should be taken into consideration before implementation.

1.4   Global prospects

Rather than being relegated to the laboratory, several practical industrial applications employing such enhanced microbial fermentation technologies have been operationalized globally. In the biogas industry for example, great efforts have been made by many countries to

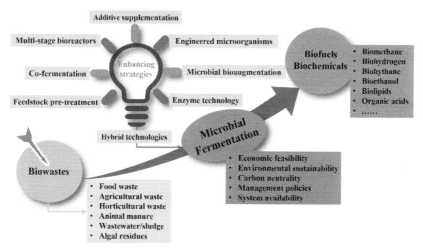

FIG. 1.1 A comprehensive research framework for enhanced microbial fermentation of biowastes for production of biofuels and biochemicals.

research and develop practical applications of biogas, including Germany, Denmark, United Kingdom, Italy, Sweden, United States, France, China, Singapore, etc. Industrial applications of biogas technology in several countries are listed in Table 1.2.

Due to the benefits on reducing the issues of environmental pollution and power energy, many commercial biogas plants have been installed globally. Brazil, Germany, China, the United States, and India are currently leaders of bioenergy production in the world (Tasmaganbetov et al., 2020). As one of the leaders in bioenergy, simultaneous support from different sectors including wastes/water treatment, electricity, agriculture, and natural gas have been given to further decrease technology prices and create a more favorable biogas market in Brazil (Borges et al., 2020). In Germany, the United States, Denmark, and the United Kingdom, at least 9545, 1497, 114, and 265 commissioned AD plants have been installed, respectively (Edwards et al., 2015). Among these countries, the development of Germany's biogas industry has long enjoyed a good reputation, which is due to advanced biogas technology and equipment, strong support from the federal government (e.g., adaptive Renewable Energy Act), and substantial biogas generation and purification companies (Table 1.2). Sustainable development of biogas industry requires adaptive legislation, national regulation, and healthy market order (Thrän et al., 2020). In Germany, over 40,000 new jobs were created and 37,470 m^3/h biogas was produced and utilized in the natural gas pipeline network. Howbeit, some limitations remain, including high investment and operational costs, the complicated structure of the equipment, and the substantial government subsidies needed (Lajdova et al., 2016; Xue et al., 2020). Similarly, the growth of the biogas industry in the United States also requires implementation of some new policies and practices. The US Environmental Protection Agency has adopted biogas from anaerobic digesters as a transportation biofuel under the expanded Renewable Fuel Standard. In addition to regulatory issues, other critical issues in deploying biogas production technology in the United States

TABLE 1.2 Industrial applications of biogas technology in partial countries.

Country	Substrate	Situation	Application	Pros	Cons	Prospects	Reference
Brazil	Municipal waste, vinasse and waste water, agricultural waste	Biogas potential in Brazil has been studied extensively; many emerging biogas innovations are made	Energy supply; transport; natural gas	Supported by renewable fuels law and the National Sanitary Policy	Brazil is far from exploiting its full potential in biogas	Reach 5.3 million cubic meters/day of methane by 2030 to 19.7 million in 2050	(Borges et al., 2020)
Germany	80% corn silage + 15% livestock manure and food industrial waste	> 60 biogas purification plant; capacity: > 37470 m^3/h, about 500 biogas engineering companies; created > 40,000 new jobs	Natural gas pipeline network	Government support by renewable energy law; giving priority to renewable energy in national energy supply; mature biogas technology	Complicated structure of equipment; high investment and operation cost; big government subsidies to initiate	Use biogas replacing 10% natural gas supply by 2030	(Auer et al., 2017; Lajdova et al., 2016)
Denmark	Manure; organic industrial wastes; energy crops such as corn and beets	Rapidly increased centralized biogas plants; biggest biogas plant: 540,000 tons wastes/year; generating 21 million m^3 upgraded biogas annually	Electricity; natural gas grid after biogas upgrading; transport fuel and heating purposes	Supported by waste policies issued by the Danish Energy Agency; Public subsidy support; fruitful circumstances for technical development	Uncertainty about subsidy schemes and energy policies from a new government in the future	To reduce country's CO_2 emission by 39% from 2005 to 2030	(Lybæk & Kjær, 2015)
UK	Food waste, farming waste and wastewater	> 557 plants in live operation; total energy generation of > 10.7 TWh per year; created > 3500 jobs	Electricity generation; waste reducing purpose	Required by certificate of responsibility for renewable obligations	Policy uncertainty	To reduce greenhouse gas emissions by 4%; to create 35,000 new jobs	(Torrijos, 2016)
Italy	Agricultural wastes	Contributed to greening of the gas network	Biogas plant connected to the electricity grid and the gas grid	Supported by National Energy Strategy	Production costs should be further reduced	10 billion m^3 methane rich biogas produced by 2030	(Torrijos, 2016)

Country	Substrate	Situation	Application	Pros	Cons	Prospects	Reference
Southern Sweden	Food industry wastes and agricultural waste	Covered > 10% of county's total energy demand; created > 3300 new jobs	As a fuel in regional transport system, and help reduce greenhouse gases	Supported by regional public policy	Decisive affect from national policy	Fossil free by 2030; to create a biogas market	(Martin & Coenen, 2015)
France	Industrial food wastes and agricultural wastes	> 519 plants producing electricity	> 405 Megawatt (MW) of power generated in total	Biogas market in France continues to develop	-	Cover one-third of its gas requirements with biomethane by 2030	(Torrijos, 2016)
China	Agricultural wastes such as crop straws; animal manure; industrial waste and wastewater; restaurant waste; sludge	Decentralized household-based biogas digesters for scattered rural households; centralized biogas plants for large-scale application	Energy supply; waste and pollution reduction	Substantial supply of feedstock; more commercial-scale biogas projects initiated by companies and the government	High operation cost; Lack of experience in large engineering applications	To produce 3×10^{29} m^3 of biogas by 2030	(Chen & Liu, 2017; Giwa et al., 2020)
Singapore	Food waste; horticulture wastes; animal manure; sewage sludge; brown water	Food waste recycling rate: 18% in 2019; first codigestion plant of sewage sludge and food waste (FW) start in 2015; treating > 40 tons feedstock/day	Powder generation and waste management	Close cooperation between academy and industries; supported by government policy; many related start-up companies in waste management area	More efforts needed in garbage source separation; more industrial applications of various enhancing strategies	Goal: > 400 tons/day of food waste; to create a "zero-waste" country	(Ng et al., 2017)

are economic and technical issues. Specifically, the challenges include the low energy content of the biogas, the relatively low biogas generation rate, and the relatively high costs in biogas upgrading (Shen et al., 2015). In Denmark, public subsidies and waste recycling policies issued by the Danish Energy Agency promoted rapid installation of centralized biogas plants, which convert industrial organic wastes and energy crops into biogas. Accordingly, over 540,000 tons wastes in Denmark were anaerobically treated to generate more than 21 million m^3 upgraded biogas annually, which would be utilized for power (e.g., natural gas grid, transport fuel and electricity production), heat, and industrial applications. Unfortunately, due to the uncertainty in future subsidy schemes and energy policies, biogas in Denmark could be a limited resource dependent on the structure of the agricultural sector (Korberg et al., 2020).

In Italy, Southern Sweden, and France, biogas industries based off industrial food wastes and agricultural wastes contributed to creating new jobs, a greening of the natural gas network, reducing organic wastes, and producing electricity (Torrijos, 2016). However, biogas plant in southern Italy was reported to contribute to NO, SO_2, and accumulated nitrate, therefore biogas upgrading before its utilization should be conducted (Merico et al., 2020). In China, one of main objectives of China's energy sector in terms of bionatural gas yield is to produce 3×10^{29} m^3 of biogas by 2030 (Giwa et al., 2020). Hitherto, over 30,000 medium- and large-sized biogas plants treating agricultural wastes have been built in China, especially the rural areas (Chen et al., 2012). The current disadvantages of China's biogas industry include insufficient standardization, lack of maintenance in pilot-scale engineering plants, underdeveloped biogas upgrading technology, and the need for more private investment and government financial support (Wang et al., 2020; Zheng et al., 2020). Therefore, more technological breakthroughs are required in China before a stable biogas market can material-ize. In addition, ongoing foreign collaborations (e.g., Sino–German cooperation and Sino–Dutch cooperation) can also help improve China's biogas industry (Giwa et al., 2020). Singapore's organic waste streams are mainly food waste, horticultural waste, sludge, waste-water, and animal manure. Over 40 tons of waste in codigestion biogas plant can be anaero-bically treated daily for biogas production and subsequent power generation (Ng et al., 2017). Due to the advantages of strong government support and close collaboration between indus-tries and academia, several larger biogas plants with capacity of 400 tons/day of food waste have been built in Singapore in 2020. One current challenge is for efficient waste separation at source. For instance, quick separation of plastic waste from the heterogeneous organic waste streams is a critical technical issue because of contamination. Alternatively, the use of biodegradable plastics should be advocated and practical technologies for plastic biodegra-dation should be developed. Meanwhile, more industrial applications of various strategies such as feedstock pretreatment and additives supplementation should be tested at the industrial scale.

Among the aforementioned enhancing strategies for AD, only some techniques have been widely employed, including pretreatments, codigestion, parameter optimization, and biore-actor optimization. Globally, it is challenging to accurately estimate the percentage of the existing biogas plants that utilize enhancing strategies; however, it is believed to be increasing progressively due to the multiple benefits of waste reduction, environmental protection, and energy supply. The codigestion strategy has been widely used in biogas industry globally. For instance, codigestion of wastewater with food waste is being investigated in Singapore's commercial scale waste treatment systems. In addition to AD, some enhancing strategies such as two-stage fermentation have shown the potential in aerobic fermentation of organic wastes for production of chemicals (Du et al., 2020a). Many strategies for enhancing microbial fermentation have been introduced in this chapter; however, a large proportion is not adopted on a commercial scale. In the short term, pretreatment technology such as hydrothermal pretreatment (Carrere et al., 2016) holds great investment potential, if the equipment with an affordable cost and high efficiency is available. In the medium term, promising additives (e.g., biochar, activated carbon, and trace elements) for enhancing microbial fermentation might be worth the investment. Nonetheless, most of the current studies on the additive strategies are carried out at lab scale; the economic feasibility of using additives in commercial-scale systems remains to be explored. On the longer term, genetically engineered microorganisms might be attractive to the investors. The derived omics technologies for data mining and applications of synthetic biology in AD and microbial fermentation also hold great potential to revolutionize the biogas industry (Jouzani & Sharafi, 2018; Lim et al., 2020). Overall, more research efforts are required to investigate the effects of scale-up for many of the enhancement techniques before they can be effectively employed in industrial installations.

1.5 Conclusions and perspectives

Microbial fermentation technology of biowastes plays a vital role in the development of a circular economy as it can simultaneously achieve waste management and resource recovery. However, the microbial fermentation process can be affected by a variety of factors stemming from feedstock, operating conditions, microorganism, bioreactor, and nutrition supply etcetera. The product yields and process stability have to be improved to advance the related technologies to practical application. In this chapter, a brief introduction of common strategies for enhancing microbial fermentation for conversion of biowastes into biofuels and biochemical has been provided. Numerous studies have demonstrated the technical feasibility of many popular enhancement strategies, including additive supplementation, multistage bioreactors, microbial bioaugmentation, utilization of genetically engineered microorganisms, codigestion, feedstock pretreatment, and enzyme technologies. However, each approach has its own pros and cons and therefore hybrid technologies can be a viable alternative to harness the capabilities of the individual methods. Cofermentation or codigestion of two or more biowastes holds great potential in full-scale installations, and cheaper yet more efficient microbial agents should also be developed for biological pretreatments. Trace elements and nanoparticles are promising additives to enhance microbial fermentation at present, while multistage bioreactor strategy, which allows functional segregation and microbial enrichment in different spatial positions in one reactor, should be further investigated for other aspects such as life cycle analysis. The scaling-up of any enhancing strategies requires that the associated economic sustainability, feasibility, carbon neutrality, suitability for government policies, and system availability also be taken into consideration. Technical advances in microbial community analysis will facilitate the current understanding on intrinsic mechanisms of each enhancing strategy, while modern biomolecular engineering techniques such as metagenomics-based sequencing have revealed key enzymes and genes associated with products generation. Upon in-depth understanding of the predominant microbes in the bioreactors, bioaugmentation employing microbial consortia can be utilized to further improve the related fermentation processes. Further development of biogas industries also requires adaptive policy regulation for technological innovation and a more active biogas market.

Acknowledgments

This project was funded by the National Research Foundation, Prime Minister's Office, Singapore under its Campus for Research Excellence and Technological Enterprise (CREATE) Program.

References

Agabo-Garcia, C., Pérez, M., Rodríguez-Morgado, B., Parrado, J., Solera, R., 2019. Biomethane production improvement by enzymatic pre-treatments and enhancers of sewage sludge anaerobic digestion. Fuel 255, 115713.

Aguilar-Moreno, G.S., Navarro-Cerón, E., Velázquez-Hernández, A., Hernández-Eugenio, G., Aguilar-Méndez, M.Á., Espinosa-Solares, T., 2020. Enhancing methane yield of chicken litter in anaerobic digestion using magnetite nanoparticles. Renew. Energy 147, 204–213.

Akkouche, N., Loubar, K., Nepveu, F., Kadi, M.E.A., Tazerout, M., 2020. Micro-combined heat and power using dual fuel engine and biogas from discontinuous anaerobic digestion. Energy Convers. Manage. 205, 112407.

Angelidaki, I., Treu, L., Tsapekos, P., Luo, G., Campanaro, S., Wenzel, H., Kougias, P.G., 2018. Biogas upgrading and utilization: current status and perspectives. Biotechnol. Adv. 36 (2), 452–466.

Arif, S., Liaquat, R., Adil, M., 2018. Applications of materials as additives in anaerobic digestion technology. Renew. Sust. Energ. Rev. 97, 354–366.

Atasoy, M., Cetecioglu, Z., 2020. Butyric acid dominant volatile fatty acids production: bio-augmentation of mixed culture fermentation by *Clostridium butyricum*. J. Environ. Chem. Eng. 8 (6), 104496.

Atelge, M.R., Atabani, A.E., Banu, J.R., Krisa, D., Kaya, M., Eskicioglu, C., Kumar, G., Lee, C., Yildiz, Y.Ş., Unalan, S., Mohanasundaram, R., Duman, F., 2020. A critical review of pretreatment technologies to enhance anaerobic digestion and energy recovery. Fuel 270, 117494.

Auer, A., Vande Burgt, N.H., Abram, F., Barry, G., Fenton, O., Markey, B.K., Nolan, S., Richards, K., Bolton, D., De Waal, T., 2017. Agricultural anaerobic digestion power plants in Ireland and Germany: policy and practice. J. Sci. Food Agric. 97 (3), 719–723.

Bandhu, S., Bansal, N., Dasgupta, D., Junghare, V., Sidana, A., Kalyan, G., Hazra, S., Ghosh, D., 2019. Overproduction of single cell oil from xylose rich sugarcane bagasse hydrolysate by an engineered oleaginous yeast *Rhodotorula mucilaginosa* IIPL32. Fuel 254, 115653.

Begum, S., Anupoju, G.R., Sridhar, S., Bhargava, S.K., Jegatheesan, V., Eshtiaghi, N., 2018. Evaluation of single and two stage anaerobic digestion of landfill leachate: Effect of pH and initial organic loading rate on volatile fatty acid (VFA) and biogas production. Bioresour. Technol. 251, 364–373.

Bohutskyi, P., Phan, D., Spierling, R.E., Kopachevsky, A.M., Bouwer, E.J., Lundquist, T.J., Betenbaugh, M.J., 2019. Production of lipid-containing algal-bacterial polyculture in wastewater and biomethanation of lipid extracted residues: enhancing methane yield through hydrothermal pretreatment and relieving solvent toxicity through co-digestion. Sci. Total Environ. 653, 1377–1394.

Borges, C.P., Sobczak, J.C., Silberg, T.R., Uriona-Maldonado, M., Vaz, C.R., 2020. A systems modeling approach to estimate biogas potential from biomass sources in Brazil. Renew. Sust. Energ. Rev., 110518.

Brown, D., Li, Y., 2013. Solid state anaerobic co-digestion of yard waste and food waste for biogas production. Bioresour. Technol. 127, 275–280.

Bujoczek, G., Oleszkiewicz, J., Sparling, R., Cenkowski, S., 2000. High solid anaerobic digestion of chicken manure. J. Agric. Eng. Res. 76 (1), 51–60.

Carrere, H., Antonopoulou, G., Affes, R., Passos, F., Battimelli, A., Lyberatos, G., Ferrer, I., 2016. Review of feedstock pretreatment strategies for improved anaerobic digestion: from lab-scale research to full-scale application. Bioresour. Technol. 199, 386–397.

Chandra, R., Iqbal, H.M., Vishal, G., Lee, H.-S., Nagra, S., 2019. Algal biorefinery: a sustainable approach to valorize algal-based biomass towards multiple product recovery. Bioresour. Technol. 278, 346–359.

Chatterjee, B., Mazumder, D., 2018. Performance evaluation of three-stage anaerobic digestion (AD) for stabilization of fruit and vegetable waste (FVW). J. Indian Chem. Soc. 95, 65–80.

Chatterjee, B., Mazumder, D., 2020. New approach of characterizing fruit and vegetable waste (FVW) to ascertain its biological stabilization via two-stage anaerobic digestion (AD). Biomass Bioenergy 139, 105594.

Chen, H., Wang, W., Xue, L., Chen, C., Liu, G., Zhang, R., 2016. Effects of ammonia on anaerobic digestion of food waste: process performance and microbial community. Energy Fuels 30 (7), 5749–5757.

Chen, J.L., Ortiz, R., Steele, T.W., Stuckey, D.C., 2014. Toxicants inhibiting anaerobic digestion: a review. Biotechnol. Adv. 32 (8), 1523–1534.

Chen, L., Zhao, L., Ren, C., Wang, F., 2012. The progress and prospects of rural biogas production in China. Energy Policy 51, 58–63.

Chen, Q., Liu, T., 2017. Biogas system in rural China: upgrading from decentralized to centralized? Renew. Sust. Energy Rev. 78, 933–944.

Chen, Y., Cheng, J.J., Creamer, K.S., 2008. Inhibition of anaerobic digestion process: a review. Bioresour. Technol. 99 (10), 4044–4064.

Chen, Y-d, Yang, Z., Ren, N-q, Ho, S.-H., 2020. Optimizing the production of short and medium chain fatty acids (SCFAs and MCFAs) from waste activated sludge using different alkyl polyglucose surfactants, through bacterial metabolic analysis. J. Hazard. Mater. 384, 121384.

Chen, Z., Wan, C., 2017. Co-fermentation of lignocellulose-based glucose and inhibitory compounds for lipid synthesis by *Rhodococcus jostii* RHA1. Process Biochem. 57, 159–166.

Cheng, X.-Y., Liu, C.-Z., 2012. Enhanced coproduction of hydrogen and methane from cornstalks by a three-stage anaerobic fermentation process integrated with alkaline hydrolysis. Bioresour. Technol. 104, 373–379.

Chiappero, M., Norouzi, O., Hu, M., Demichelis, F., Berruti, F., Di Maria, F., Mašek, O., Fiore, S., 2020. Review of biochar role as additive in anaerobic digestion processes. Renew. Sust. Energy Rev. 131, 110037.

Chrispim, M.C., Scholz, M., Nolasco, M.A., 2019. Phosphorus recovery from municipal wastewater treatment: critical review of challenges and opportunities for developing countries. J. Environ. Manage. 248, 109268.

Cui, M.-H., Sangeetha, T., Gao, L., Wang, A.-J., 2020. Hydrodynamics of up-flow hybrid anaerobic digestion reactors with built-in bioelectrochemical system. J. Hazard. Mater. 382, 121046.

da Silva, A.S.A., Espinheira, R.P., Teixeira, R.S.S., de Souza, M.F., Ferreira-Leitão, V., Bon, E.P., 2020. Constraints and advances in high-solids enzymatic hydrolysis of lignocellulosic biomass: a critical review. Biotechnol. Biofuels 13 (1), 1–28.

De Gioannis, G., Muntoni, A., Polettini, A., Pomi, R., Spiga, D., 2017. Energy recovery from one-and two-stage anaerobic digestion of food waste. Waste Manag 68, 595–602.

De Sanctis, M., Chimienti, S., Pastore, C., Piergrossi, V., Di Iaconi, C., 2019. Energy efficiency improvement of thermal hydrolysis and anaerobic digestion of *Posidonia oceanica* residues. Appl. Energy 252, 113457.

De Schouwer, F., Claes, L., Vandekerkhove, A., Verduyckt, J., De Vos, D.E., 2019. Protein-rich biomass waste as a resource for future biorefineries: state of the art, challenges, and opportunities. ChemSusChem 12 (7), 1272–1303.

Dechrugsa, S., Kantachote, D., Chaiprapat, S., 2013. Effects of inoculum to substrate ratio, substrate mix ratio and inoculum source on batch co-digestion of grass and pig manure. Bioresour. Technol. 146, 101–108.

Deng, C., Lin, R., Cheng, J., Murphy, J.D., 2019. Can acid pre-treatment enhance biohydrogen and biomethane production from grass silage in single-stage and two-stage fermentation processes? Energy Convers. Manage. 195, 738–747.

Di Maria, F., Sisani, F., Norouzi, O., Mersky, R.L., 2019. The effectiveness of anaerobic digestion of bio-waste in replacing primary energies: an EU28 case study. Renew. Sust. Energy Rev. 108, 347–354.

Diaz-Elsayed, N., Rezaei, N., Guo, T., Mohebbi, S., Zhang, Q., 2019. Wastewater-based resource recovery technologies across scale: a review. Resour. Conserv. Recycl. 145, 94–112.

Ding, L., Cheng, J., Lin, R., Deng, C., Zhou, J., Murphy, J.D., 2020. Improving biohydrogen and biomethane co-production via two-stage dark fermentation and anaerobic digestion of the pretreated seaweed *Laminaria digitata*. J. Clean. Prod. 251, 119666.

Du, C., Li, Y., Zong, H., Yuan, T., Yuan, W., Jiang, Y., 2020a. Production of bioethanol and xylitol from non-detoxified corn cob via a two-stage fermentation strategy. Bioresour. Technol. 310, 123427.

Du, M., Liu, X., Wang, D., Yang, Q., Duan, A., Chen, H., Liu, Y., Wang, Q., Ni, B.-J., 2020b. Understanding the fate and impact of capsaicin in anaerobic co-digestion of food waste and waste activated sludge. Water Res. 188, 116539.

Ebenezer, A.V., Arulazhagan, P., Kumar, S.A., Yeom, I.-T., Banu, J.R., 2015. Effect of deflocculation on the efficiency of low-energy microwave pretreatment and anaerobic biodegradation of waste activated sludge. Appl. Energy 145, 104–110.

Edwards, J., Othman, M., Burn, S., 2015. A review of policy drivers and barriers for the use of anaerobic digestion in Europe, the United States and Australia. Renew. Sust. Energy Rev. 52, 815–828.

Fagbohungbe, M.O., Herbert, B.M., Hurst, L., Li, H., Usmani, S.Q., Semple, K.T., 2016. Impact of biochar on the anaerobic digestion of citrus peel waste. Bioresour. Technol. 216, 142–149.

Faisal, S., Salama, E.-S., Malik, K., Lee, S-h, Li, X., 2020. Anaerobic digestion of cabbage and cauliflower biowaste: Impact of iron oxide nanoparticles (IONPs) on biomethane and microbial communities alteration. Bioresour. Technol. Rep. 12, 100567.

Fernandez-Bayo, J.D., Simmons, C.W., VanderGheynst, J.S., 2020. Characterization of digestate microbial community structure following thermophilic anaerobic digestion with varying levels of green and food wastes. J. Ind. Microbiol. Biotechnol. 47, 1031–1044.

FitzGerald, J.A., Wall, D.M., Jackson, S.A., Murphy, J.D., Dobson, A.D., 2019. Trace element supplementation is associated with increases in fermenting bacteria in biogas mono-digestion of grass silage. Renew. Energy 138, 980–986.

Foong, S.Y., Liew, R.K., Yang, Y., Cheng, Y.W., Yek, P.N.Y., Mahari, W.A.W., Lee, X.Y., Han, C.S., Vo, D.-V.N., Van Le, Q., 2020. Valorization of biomass waste to engineered activated biochar by microwave pyrolysis: progress, challenges, and future directions. Chem. Eng. J. 389, 124401.

Fu, S.-F., Xu, X.-H., Dai, M., Yuan, X.-Z., Guo, R.-B., 2017. Hydrogen and methane production from vinasse using two-stage anaerobic digestion. Process Saf. Environ. Prot. 107, 81–86.

Fuchs, W., Wang, X., Gabauer, W., Ortner, M., Li, Z., 2018. Tackling ammonia inhibition for efficient biogas production from chicken manure: status and technical trends in Europe and China. Renew. Sust. Energy Rev. 97, 186–199.

Gao, M., Zhang, L., Liu, Y., 2020. High-loading food waste and blackwater anaerobic co-digestion: Maximizing bioenergy recovery. Chem. Eng. J. 394, 124911.

Garcia, N.H., Benedetti, M., Bolzonella, D., 2019. Effects of enzymes addition on biogas production from anaerobic digestion of agricultural biomasses. Waste Biomass Valoriz. 10 (12), 3711–3722.

García-Ruíz, M.J., Castellano-Hinojosa, A., Armato, C., González-Martínez, A., González-López, J., Osorio, F., 2020. Biogas production and microbial community structure in a stable-stage of a two-stage anaerobic digester. AIChE J. 66 (2), e16807.

Ghofrani-Isfahani, P., Baniamerian, H., Tsapekos, P., Alvarado-Morales, M., Kasama, T., Shahrokhi, M., Vossoughi, M., Angelidaki, I., 2020. Effect of metal oxide based TiO_2 nanoparticles on anaerobic digestion process of ligno-cellulosic substrate. Energy 191, 116580.

Giwa, A.S., Ali, N., Ahmad, I., Asif, M., Guo, R.-B., Li, F.-L., Lu, M., 2020. Prospects of China's biogas: fundamentals, challenges and considerations. Energy Rep. 6, 2973–2987.

Goswami, L., Kumar, R.V., Pakshirajan, K., Pugazhenthi, G., 2019. A novel integrated biodegradation—microfiltration system for sustainable wastewater treatment and energy recovery. J. Hazard. Mater. 365, 707–715.

Grübel, K., Suschka, J., 2015. Hybrid alkali-hydrodynamic disintegration of waste-activated sludge before two-stage anaerobic digestion process. Environ. Sci. Pollut. Res. 22 (10), 7258–7270.

Gunes, B., Stokes, J., Davis, P., Connolly, C., Lawler, J., 2019. Pre-treatments to enhance biogas yield and quality from anaerobic digestion of whiskey distillery and brewery wastes: a review. Renew. Sust. Energy Rev. 113, 109281.

Guo, H., Du, L., Liang, J., Yang, Z., Cui, G., Zhang, K., 2017. Influence of alkaline-thermal pretreatment on high-solids anaerobic digestion of dewatered activated sludge. BioResources 12 (1), 195–210.

Hao, X., Chen, Q., van Loosdrecht, M.C., Li, J., Jiang, H., 2020. Sustainable disposal of excess sludge: incineration without anaerobic digestion. Water Res. 170, 115298.

Hassanein, A., Keller, E., Lansing, S., 2020. Effect of metal nanoparticles in anaerobic digestion production and plant uptake from effluent fertilizer. Bioresour. Technol. 321, 124455.

Hassanein, A., Witarsa, F., Guo, X., Yong, L., Lansing, S., Qiu, L., 2017. Next generation digestion: complementing anaerobic digestion (AD) with a novel microbial electrolysis cell (MEC) design. Int. J. Hydrog. Energy 42 (48), 28681–28689.

Herrero, M., Stuckey, D.C., 2015. Bioaugmentation and its application in wastewater treatment: a review. Chemosphere 140, 119–128.

Hijazi, O., Abdelsalam, E., Samer, M., Amer, B., Yacoub, I., Moselhy, M., Attia, Y., Bernhardt, H., 2020. Environmental impacts concerning the addition of trace metals in the process of biogas production from anaerobic digestion of slurry. J. Clean. Prod. 243, 118593.

Hoornweg, D., Bhada-Tata, P., 2012. What a waste: a global review of solid waste management. Urban Development & Local Government Unit, World Bank, Washington DC, USA.

Hossain, N., Mahlia, T., Saidur, R., 2019. Latest development in microalgae-biofuel production with nano-additives. Biotechnol. Biofuels 12 (1), 125.

Hu, Q., Jung, J., Chen, D., Leong, K., Song, S., Li, F., Mohan, B., Yao, Z., Prabhakar, A.K., Lin, X.H., 2020. Biochar industry to circular economy. Sci. Total Environ. 757, 143820.

Hu, Y., Pang, Y., Yuan, H., Zou, D., Liu, Y., Zhu, B., Chufo, W.A., Jaffar, M., Li, X., 2015. Promoting anaerobic biogasification of corn stover through biological pretreatment by liquid fraction of digestate (LFD). Bioresour. Technol. 175, 167–173.

Intanoo, P., Watcharanurak, T., Chavadej, S., 2020. Evolution of methane and hydrogen from ethanol wastewater with maximization of energy yield by three-stage anaerobic sequencing batch reactor system. Int. J. Hydrog. Energy 45, 9469–9483.

Jeong, S.Y., Chang, S.W., Ngo, H.H., Guo, W., Nghiem, L.D., Banu, J.R., Jeon, B.-H., Nguyen, D.D., 2019. Influence of thermal hydrolysis pretreatment on physicochemical properties and anaerobic biodegradability of waste activated sludge with different solids content. Waste Manag. 85, 214–221.

Jiang, J., Li, L., Li, Y., He, Y., Wang, C., Sun, Y., 2020. Bioaugmentation to enhance anaerobic digestion of food waste: dosage, frequency and economic analysis. Bioresour. Technol. 307, 123256.

Jiang, Y., Marang, L., Tamis, J., van Loosdrecht, M.C., Dijkman, H., Kleerebezem, R., 2012. Waste to resource: converting paper mill wastewater to bioplastic. Water Res. 46 (17), 5517–5530.

Jomnonkhaow, U., Uwineza, C., Mahboubi, A., Wainaina, S., Reungsang, A., Taherzadeh, M.J., 2020. Membrane bioreactor-assisted volatile fatty acids production and in situ recovery from cow manure. Bioresour. Technol. 321, 124456.

Jouzani, G.S., Sharafi, R., 2018. New "omics" technologies and biogas production. In: Tabatabaei, M., Ghanavati, H. (Eds.), Biogas. Springer, pp. 419–436.

Kainthola, J., Kalamdhad, A.S., Goud, V.V., 2019. A review on enhanced biogas production from anaerobic digestion of lignocellulosic biomass by different enhancement techniques. Process Biochem. 84, 81–90.

Kainthola, J., Kalamdhad, A.S., Goud, V.V., 2020a. Optimization of process parameters for accelerated methane yield from anaerobic co-digestion of rice straw and food waste. Renew. Energy 149, 1352–1359.

Kainthola, J., Podder, A., Fechner, M., Goel, R., 2020b. An overview of fungal pretreatment processes for anaerobic digestion: applications, bottlenecks and future needs. Bioresour. Technol. 321 (33), 124397.

Kanaujiya, D.K., Paul, T., Sinharoy, A., Pakshirajan, K., 2019. Biological treatment processes for the removal of organic micropollutants from wastewater: a review. Curr. Pollut. Rep. 5 (3), 112–128.

Korberg, A.D., Skov, I.R., Mathiesen, B.V., 2020. The role of biogas and biogas-derived fuels in a 100% renewable energy system in Denmark. Energy 199, 117426.

Kougias, P.G., Angelidaki, I., 2018. Biogas and its opportunities—a review. Front. Env. Sci. Eng. 12, 1–12.

Kumar, M., Sun, Y., Rathour, R., Pandey, A., Thakur, I.S., Tsang, D.C.W., 2020. Algae as potential feedstock for the production of biofuels and value-added products: opportunities and challenges. Sci. Total Environ. 716, 137116.

Lajdova, Z., Lajda, J., Bielik, P., 2016. The impact of the biogas industry on agricultural sector in Germany. Agric. Econ. 62 (1), 1–8.

Lavagnolo, M.C., Girotto, F., Rafieenia, R., Danieli, L., Alibardi, L., 2018. Two-stage anaerobic digestion of the organic fraction of municipal solid waste—effects of process conditions during batch tests. Renew. Energy 126, 14–20.

Lee, C.-R., Sung, B.H., Lim, K.-M., Kim, M.-J., Sohn, M.J., Bae, J.-H., Sohn, J.-H., 2017a. Co-fermentation using recombinant *Saccharomyces cerevisiae* yeast strains hyper-secreting different cellulases for the production of cellulosic bioethanol. Sci. Rep. 7 (1), 1–14.

Lee, E., Bittencourt, P., Casimir, L., Jimenez, E., Wang, M., Zhang, Q., Ergas, S.J., 2019. Biogas production from high solids anaerobic co-digestion of food waste, yard waste and waste activated sludge. Waste Manag 95, 432–439.

Lee, J.T.E., Ee, A.W.L., Tong, Y.W., 2018. Environmental impact comparison of four options to treat the cellulosic fraction of municipal solid waste (CF-MSW) in green megacities. Waste Manag. 78, 677–685.

Lee, J.T.E., He, J.Z., Tong, Y.W., 2017b. Acclimatization of a mixed-animal manure inoculum to the anaerobic digestion of *Axonopus compressus* as characterized by DGGE and Illumina MiSeq. Bioresour. Technol. 245, 1148–1154.

Lee, J.T.E., Khan, M.U., Tian, H., Ee, A.W.L., Lim, E.Y., Dai, Y., Tong, Y.W., Ahring, B.K., 2020b. Improving methane yield of oil palm empty fruit bunches by wet oxidation pretreatment: mesophilic and thermophilic anaerobic digestion conditions and the associated global warming potential effects. Energy Convers. Manage. 225, 113438.

Lee, J.T.E., Wang, Q., Lim, E.Y., Liu, Z., He, J.Z., Tong, Y.W., 2020a. Optimization of bioaugmentation of the anaerobic digestion of *Axonopus compressus* cowgrass for the production of biomethane. J. Clean. Prod. 258, 120932.

Lee, Y.-J., Lee, D.-J., 2019. Impact of adding metal nanoparticles on anaerobic digestion performance—a review. Bioresour. Technol. 292, 121926.

Li, K., Liu, R., Sun, C., 2015. Comparison of anaerobic digestion characteristics and kinetics of four livestock manures with different substrate concentrations. Bioresour. Technol. 198, 133–140.

Li, L., Shi, W., Yu, S., 2020. Research on forward osmosis membrane technology still needs improvement in water recovery and wastewater treatment. Water 12 (1), 107.

Li, R., Duan, N., Zhang, Y., Liu, Z., Li, B., Zhang, D., Dong, T., 2017. Anaerobic co-digestion of chicken manure and microalgae *Chlorella* sp.: methane potential, microbial diversity and synergistic impact evaluation. Waste Manag 68, 120–127.

Li, W., Loh, K.-C., Zhang, J., Tong, Y.W., Dai, Y., 2018. Two-stage anaerobic digestion of food waste and horticultural waste in high-solid system. Appl. Energy 209, 400–408.

Lim, J.W., Ge, T., Tong, Y.W., 2018. Monitoring of microbial communities in anaerobic digestion sludge for biogas optimisation. Waste Manag. 71, 334–341.

Li, Y., Khanal, S.K., Lim, J.W., Park, T., Tong, Y.W., Yu, Z., 2020. The microbiome driving anaerobic digestion and microbial analysis Li, Y., Khanal, S.K. In: Li, Y., Khanal, S.K. (Eds.), Advances in Bioenergy, 5. Elsevier, pp. 1–61.

Lin, R., Cheng, J., Ding, L., Murphy, J.D., 2018. Improved efficiency of anaerobic digestion through direct interspecies electron transfer at mesophilic and thermophilic temperature ranges. Chem. Eng. J. 350, 681–691.

Lin, R., Cheng, J., Zhang, J., Zhou, J., Cen, K., Murphy, J.D., 2017. Boosting biomethane yield and production rate with graphene: the potential of direct interspecies electron transfer in anaerobic digestion. Bioresour. Technol. 239, 345–352.

Liu, L., Zhang, T., Wan, H., Chen, Y., Wang, X., Yang, G., Ren, G., 2015. Anaerobic co-digestion of animal manure and wheat straw for optimized biogas production by the addition of magnetite and zeolite. Energy Convers. Manage. 97, 132–139.

Liu, W., Mao, W., Zhang, C., Lu, X., Xiao, X., Zhao, Z., Lin, J., 2020. Co-fermentation of a sugar mixture for microbial lipid production in a two-stage fermentation mode under non-sterile conditions. Sustain. Energ. Fuels 4 (5), 2380–2385.

Liu, Y., Wachemo, A.C., Yuan, H., Li, X., 2019. Anaerobic digestion performance and microbial community structure of corn stover in three-stage continuously stirred tank reactors. Bioresour. Technol. 287, 121339.

Llamas, M., Magdalena, J.A., González-Fernández, C., Tomás-Pejó, E., 2020. Volatile fatty acids as novel building blocks for oil-based chemistry via oleaginous yeast fermentation. Biotechnol. Bioeng. 117 (1), 238–250.

Lü, F., Luo, C., Shao, L., He, P., 2016. Biochar alleviates combined stress of ammonium and acids by firstly enriching *Methanosaeta* and then *Methanosarcina*. Water Res. 90, 34–43.

Lunprom, S., Phanduang, O., Salakkam, A., Liao, Q., Imai, T., Reungsang, A., 2019. Bio-hythane production from residual biomass of *Chlorella* sp. biomass through a two-stage anaerobic digestion. Int. J. Hydrog. Energy 44 (6), 3339–3346.

Luo, C., Lü, F., Shao, L., He, P., 2015. Application of eco-compatible biochar in anaerobic digestion to relieve acid stress and promote the selective colonization of functional microbes. Water Res. 68, 710–718.

Lybæk, R., Kjær, T., 2015. Municipalities as facilitators, regulators and energy consumers for enhancing the dissemination of biogas technology in Denmark. Int. J. Sustain. Energy Planning Manag. 8, 17–30.

Mancini, G., Papirio, S., Riccardelli, G., Lens, P.N., Esposito, G., 2018. Trace elements dosing and alkaline pretreatment in the anaerobic digestion of rice straw. Bioresour. Technol. 247, 897–903.

Martin, H., Coenen, L., 2015. Institutional context and cluster emergence: the biogas industry in Southern Sweden. Eur. Plan. Stud. 23 (10), 2009–2027.

Matheri, A.N., Belaid, M., Seodigeng, T., Ngila, J.C., 2016. The role of trace elements on anaerobic co-digestion in biogas production, Proc. World Congress on Engineering. London, UK.

Meena, R.A.A., Kannah, R.Y., Sindhu, J., Ragavi, J., Kumar, G., Gunasekaran, M., Banu, J.R., 2019. Trends and resource recovery in biological wastewater treatment system. Bioresour. Technol. Rep. 7, 100235.

Merico, E., Grasso, F.M., Cesari, D., Decesari, S., Belosi, F., Manarini, F., De Nuntiis, P., Rinaldi, M., Gambaro, A., Morabito, E., Contini, D., 2020. Characterisation of atmospheric pollution near an industrial site with a biogas production and combustion plant in southern Italy. Sci. Total Environ. 717, 137220.

Millati, R., Wikandari, R., Ariyanto, T., Putri, R.U., Taherzadeh, M.J., 2020. Pretreatment technologies for anaerobic digestion of lignocelluloses and toxic feedstocks. Bioresour. Technol. 304, 122998.

Molaey, R., Bayrakdar, A., Sürmeli, R.Ö., Çalli, B., 2018. Anaerobic digestion of chicken manure: mitigating process inhibition at high ammonia concentrations by selenium supplementation. Biomass Bioenergy 108, 439–446.

Montalvo, S., Huiliñir, C., Borja, R., Sánchez, E., Herrmann, C., 2020. Application of zeolites for biological treatment processes of solid wastes and wastewaters—a review. Bioresour. Technol. 301, 122808.

Narihiro, T., Sekiguchi, Y., 2007. Microbial communities in anaerobic digestion processes for waste and wastewater treatment: a microbiological update. Curr. Opin. Biotech. 18 (3), 273–278.

Ng, B.J.H., Mao, Y., Chen, C.-L., Rajagopal, R., Wang, J.-Y., 2017. Municipal food waste management in Singapore: practices, challenges and recommendations. J. Mater. Cycles Waste Manag. 19 (1), 560–569.

Nguyen, P.-D., Tran, N.-S.T., Nguyen, T.-T., Dang, B.-T., Le, M.-T.T., Bui, X.-T., Mukai, F., Kobayashi, H., Ngo, H.H., 2020. Long-term operation of the pilot scale two-stage anaerobic digestion of municipal biowaste in Ho Chi Minh City. Sci. Total Environ. 766, 142562.

Nigam, P.S., 2013. Microbial enzymes with special characteristics for biotechnological applications. Biomolecules 3 (3), 597–611.

Oladejo, O.S., Dahunsi, S.O., Adesulu-Dahunsi, A.T., Ojo, S.O., Lawal, A.I., Idowu, E.O., Olanipekun, A.A., Ibikunle, R.A., Osueke, C.O., Ajayi, O.E., 2020. Energy generation from anaerobic co-digestion of food waste, cow dung and piggery dung. Bioresour. Technol. 313, 123694.

Ong, K.L., Li, C., Li, X., Zhang, Y., Xu, J., Lin, C.S.K., 2019. Co-fermentation of glucose and xylose from sugarcane bagasse into succinic acid by Yarrowia lipolytica. Biochem. Eng. J. 148, 108–115.

Owamah, H., 2020. Biogas yield assessment from the anaerobic co-digestion of food waste and cymbopogon citratus. J. Mater. Cycles Waste Manag. 22 (6), 2012–2019.

Pan, J., Ma, J., Zhai, L., Liu, H., 2019. Enhanced methane production and syntrophic connection between microorganisms during semi-continuous anaerobic digestion of chicken manure by adding biochar. J. Clean. Prod. 240, 118178.

Pang, Y., Zhao, Y., Li, S., Zhao, Y., Li, J., Hu, Z., Zhang, C., Xiao, D., Yu, A., 2019. Engineering the oleaginous yeast Yarrowia lipolytica to produce limonene from waste cooking oil. Biotechnol. Biofuels 12 (1), 241.

Panigrahi, S., Sharma, H.B., Dubey, B.K., 2020. Anaerobic co-digestion of food waste with pretreated yard waste: a comparative study of methane production, kinetic modeling and energy balance. J. Clean. Prod. 243, 118480.

Paritosh, K., Yadav, M., Chawade, A., Sahoo, D., Kesharwani, N., Pareek, N., Vivekanand, V., 2020. Additives as a support structure for specific biochemical activity boosts in anaerobic digestion: a review. Front. Energy Res. 8, 88.

Park, J.-H., Kang, H.-J., Park, K.-H., Park, H.-D., 2018. Direct interspecies electron transfer via conductive materials: a perspective for anaerobic digestion applications. Bioresour. Technol. 254, 300–311.

Park, M., Kim, N., Jung, S., Jeong, T.-Y., Park, D., 2020. Optimization and comparison of methane production and residual characteristics in mesophilic anaerobic digestion of sewage sludge by hydrothermal treatment. Chemosphere 264, 128516.

Promnuan, K., Higuchi, T., Imai, T., Kongjan, P., Reungsang, A., Sompong, O., 2020. Simultaneous biohythane production and sulfate removal from rubber sheet wastewater by two-stage anaerobic digestion. Int. J. Hydrog. Energy 45 (1), 263–274.

Rabii, A., Aldin, S., Dahman, Y., Elbeshbishy, E., 2019. A review on anaerobic co-digestion with a focus on the microbial populations and the effect of multi-stage digester configuration. Energies 12 (6), 1106.

Rajin, M., 2018. A current review on the application of enzymes in anaerobic digestion. In: Horan, N., Yaser, A.Z., Wid, N. (Eds.), Anaerobic Digestion Processes. Springer, Cham, pp. 55–70.

Ravindran, R., Jaiswal, A.K., 2016. A comprehensive review on pre-treatment strategy for lignocellulosic food industry waste: challenges and opportunities. Bioresour. Technol. 199, 92–102.

Rawoof, S.A.A., Kumar, P.S., Vo, D.-V.N., Devaraj, K., Mani, Y., Devaraj, T., Subramanian, S., 2020. Production of optically pure lactic acid by microbial fermentation: a review. Environ. Chem. Lett. 19 (2), 1–18.

Rezaee, M., Gitipour, S., Sarrafzadeh, M.-H., 2020. Different pathways to integrate anaerobic digestion and thermochemical processes: moving toward the circular economy concept. Environ. Energy Eco. Res. 4 (1), 57–67.

Robles, I., O'Dwyer, E., Guo, M., 2020. Waste-to-Resource value chain optimisation: combining spatial, chemical and technoeconomic aspects. Water Res. 178, 115842.

Romero-Güiza, M., Vila, J., Mata-Alvarez, J., Chimenos, J., Astals, S., 2016. The role of additives on anaerobic digestion: a review. Renew. Sust. Energy Rev. 58, 1486–1499.

Ruffino, B., Cerutti, A., Campo, G., Scibilia, G., Lorenzi, E., Zanetti, M., 2019. Improvement of energy recovery from the digestion of waste activated sludge (WAS) through intermediate treatments: the effect of the hydraulic retention time (HRT) of the first-stage digestion. Appl. Energy 240, 191–204.

Sáez-Sáez, J., Wang, G., Marella, E.R., Sudarsan, S., Pastor, M.C., Borodina, I., 2020. Engineering the oleaginous yeast *Yarrowia lipolytica* for high-level resveratrol production. Metab. Eng. 62, 51–61.

Sanchez, N., Ruiz, R., Plazas, A., Vasquez, J., Cobo, M., 2020. Effect of pretreatment on the ethanol and fusel alcohol production during fermentation of sugarcane press-mud. Biochem. Eng. J. 161, 107668.

Saud, S., Chen, S., Haseeb, A., Sumayya, 2020. The role of financial development and globalization in the environment: accounting ecological footprint indicators for selected one-belt-one-road initiative countries. J. Clean. Prod. 250, 119518.

Serrano, A., Pinto-Ibieta, F., Braga, A., Jeison, D., Borja, R., Fermoso, F.G., 2017. Risks of using EDTA as an agent for trace metals dosing in anaerobic digestion of olive mill solid waste. Environ. Technol. 38 (24), 3137–3144.

Shen, Y., Linville, J.L., Urgun-Demirtas, M., Mintz, M.M., Snyder, S.W., 2015. An overview of biogas production and utilization at full-scale wastewater treatment plants (WWTPs) in the United States: challenges and opportunities towards energy-neutral WWTPs. Renew. Sust. Energ. Rev. 50, 346–362.

Shetty, D., Joshi, A., Dagar, S.S., Ks hirsagar, P., Dhakephalkar, P.K., 2020. Bioaugmentation of anaerobic fungus *Orpinomyces joyonii* boosts sustainable biomethanation of rice straw without pretreatment. Biomass Bioenergy 138, 105546.

Shin, J., Cho, S.-K., Lee, J., Hwang, K., Chung, J.W., Jang, H.-N., Shin, S.G., 2019. Performance and microbial community dynamics in anaerobic digestion of waste activated sludge: impact of immigration. Energies 12 (3), 573.

Siami, S., Aminzadeh, B., Karimi, R., Hallaji, S.M., 2020. Process optimization and effect of thermal, alkaline, H_2O_2 oxidation and combination pretreatment of sewage sludge on solubilization and anaerobic digestion. BMC Biotechnol. 20, 1–12.

Silva, F.M., Mahler, C.F., Oliveira, L.B., Bassin, J.P., 2018. Hydrogen and methane production in a two-stage anaerobic digestion system by co-digestion of food waste, sewage sludge and glycerol. Waste Manag. 76, 339–349.

Singh, R., Kumar, M., Mittal, A., Mehta, P.K., 2016. Microbial enzymes: industrial progress in 21st century. 3 Biotech 6 (2), 174.

Singh, R.S., Singh, T., Pandey, A., 2019. Microbial enzymes—an overview. In: Singh, R.S., Singhania, R.R., Pandey, A., Larroche, C. (Eds.), Advances in Enzyme Technology. Elsevier, pp. 1–40.

Solé-Bundó, M., Passos, F., Romero-Güiza, M.S., Ferrer, I., Astals, S., 2019. Co-digestion strategies to enhance microalgae anaerobic digestion: a review. Renew. Sust. Energy Rev. 112, 471–482.

Srisowmeya, G., Chakravarthy, M., Devi, G.N., 2020. Critical considerations in two-stage anaerobic digestion of food waste–a review. Renew. Sust. Energy Rev. 119, 109587.

Suganya, T., Varman, M., Masjuki, H., Renganathan, S., 2016. Macroalgae and microalgae as a potential source for commercial applications along with biofuels production: a biorefinery approach. Renew. Sust. Energy Rev. 55, 909–941.

Sun, C., Cao, W., Banks, C.J., Heaven, S., Liu, R., 2016. Biogas production from undiluted chicken manure and maize silage: a study of ammonia inhibition in high solids anaerobic digestion. Bioresour. Technol. 218, 1215–1223.

Tasmaganbetov, A.B., Ataniyazov, Z., Basshieva, Z., Muhammedov, A.U., Yessengeldina, A., 2020. World practice of using biogas as alternative energy. Int. J. Energy Econ. Policy 10 (5), 348.

Thanh, P.M., Ketheesan, B., Yan, Z., Stuckey, D., 2016. Trace metal speciation and bioavailability in anaerobic digestion: a review. Biotechnol. Adv. 34 (2), 122–136.

Thompson, T.M., Young, B.R., Baroutian, S., 2020. Efficiency of hydrothermal pretreatment on the anaerobic digestion of pelagic Sargassum for biogas and fertiliser recovery. Fuel 279, 118527.

Thrän, D., Schaubach, K., Majer, S., Horschig, T., 2020. Governance of sustainability in the German biogas sector—adaptive management of the Renewable Energy Act between agriculture and the energy sector. Energy Sustain. Soc. 10 (1), 3.

Ting, H.N.J., Lin, L., Cruz, R.B., Chowdhury, B., Karidio, I., Zaman, H., Dhar, B.R., 2020. Transitions of microbial communities in the solid and liquid phases during high-solids anaerobic digestion of organic fraction of municipal solid waste. Bioresour. Technol. 317, 123951.

Tong, H., Shen, Y., Zhang, J., Wang, C.-H., Ge, T.S., Tong, Y.W., 2018. A comparative life cycle assessment on four waste-to-energy scenarios for food waste generated in eateries. Appl. Energy 225, 1143–1157.

Torrijos, M., 2016. State of development of biogas production in Europe. Procedia Environ. Sci. 35, 881–889.

Tsapekos, P., Kougias, P., Vasileiou, S., Treu, L., Campanaro, S., Lyberatos, G., Angelidaki, I., 2017. Bioaugmentation with hydrolytic microbes to improve the anaerobic biodegradability of lignocellulosic agricultural residues. Bioresour. Technol. 234, 350–359.

Uçkun Kiran, E., Stamatelatou, K., Antonopoulou, G., Lyberatos, G., 2016. Production of biogas via anaerobic digestion. In: Luque, R., Lin, C.S.K., Wilson, K., Clark, J. (Eds.), Handbook of Biofuels Production: Processes and Technologies. Woodhead Publishing, Cambridge, pp. 259–301.

Vieira, S., Barros, M.V., Sydney, A.C.N., Piekarski, C.M., de Francisco, A.C., de Souza Vandenberghe, L.P., Sydney, E.B., 2020. Sustainability of sugarcane lignocellulosic biomass pretreatment for the production of bioethanol. Bioresour. Technol. 299, 122635.

Vistanti, H., Malik, R.A., 2019. Enhanced performance of multi-stage anaerobic digestion of tofu wastewater: role of recirculation. Jurnal Riset Teknologi Pencegahan Pencemaran Industri 10 (1), 29–37.

Wang, H., Zhang, Y., Angelidaki, I., 2016. Ammonia inhibition on hydrogen enriched anaerobic digestion of manure under mesophilic and thermophilic conditions. Water Res. 105, 314–319.

Wang, X., Yan, R., Zhao, Y., Cheng, S., Han, Y., Yang, S., Cai, D., Mang, H.-P., Li, Z., 2020. Biogas standard system in China. Renew. Energy 157, 1265–1273.

Wang, Y., Zhang, H., Lu, X., Zong, H., Zhuge, B., 2019. Advances in 2-phenylethanol production from engineered microorganisms. Biotechnol. Adv. 37 (3), 403–409.

Weber, J., Walter, Pirbazari, M., Melson, G., 1978. Biological growth on activated carbon: an investigation by scanning electron microscopy. Environ. Sci. Technol. 12 (7), 817–819.

Xu, H., Li, Y., Hua, D., Zhao, Y., Mu, H., Chen, H., Chen, G., 2020. Enhancing the anaerobic digestion of corn stover by chemical pretreatment with the black liquor from the paper industry. Bioresour. Technol. 306, 123090.

Xue, S., Song, J., Wang, X., Shang, Z., Sheng, C., Li, C., Zhu, Y., Liu, J., 2020. A systematic comparison of biogas development and related policies between China and Europe and corresponding insights. Renew. Sust. Energy Rev. 117, 109474.

Yan, M., Treu, L., Campanaro, S., Tian, H., Zhu, X., Khoshnevisan, B., Tsapekos, P., Angelidaki, I., Fotidis, I.A., 2020. Effect of ammonia on anaerobic digestion of municipal solid waste: inhibitory performance, bioaugmentation and microbiome functional reconstruction. Chem. Eng. J. 401, 126159.

Yao, Z., Li, W., Kan, X., Dai, Y., Tong, Y.W., Wang, C.-H., 2017. Anaerobic digestion and gasification hybrid system for potential energy recovery from yard waste and woody biomass. Energy 124, 133–145.

Yenigün, O., Demirel, B., 2013. Ammonia inhibition in anaerobic digestion: a review. Process Biochem. 48 (5-6), 901–911.

Yu, B., Xu, J., Yuan, H., Lou, Z., Lin, J., Zhu, N., 2014. Enhancement of anaerobic digestion of waste activated sludge by electrochemical pretreatment. Fuel 130, 279–285.

Yu, Q., Liu, R., Li, K., Ma, R., 2019. A review of crop straw pretreatment methods for biogas production by anaerobic digestion in China. Renew. Sust. Energy Rev. 107, 51–58.

Zabed, H.M., Akter, S., Yun, J., Zhang, G., Awad, F.N., Qi, X., Sahu, J.N., 2019. Recent advances in biological pretreatment of microalgae and lignocellulosic biomass for biofuel production. Renew. Sust. Energy Rev. 105, 105–128.

Zarkadas, I.S., Sofikiti, A.S., Voudrias, E.A., Pilidis, G.A., 2015. Thermophilic anaerobic digestion of pasteurised food wastes and dairy cattle manure in batch and large volume laboratory digesters: focusing on mixing ratios. Renew. Energy 80, 432–440.

Zhang, J., Loh, K.-C., Lee, J., Wang, C.-H., Dai, Y., Tong, Y.W., 2017a. Three-stage anaerobic co-digestion of food waste and horse manure. Sci. Rep. 7 (1), 1269.

Zhang, J., Loh, K.-C., Li, W., Lim, J.W., Dai, Y., Tong, Y.W., 2017b. Three-stage anaerobic digester for food waste. Appl. Energy 194, 287–295.

Zhang, J., Luo, W., Wang, Y., Li, G., Liu, Y., Gong, X., 2019a. Anaerobic cultivation of waste activated sludge to inoculate solid state anaerobic co-digestion of agricultural wastes: effects of different cultivated periods. Bioresour. Technol. 294, 122078.

Zhang, L., Kuroki, A., Loh, K.-C., Seok, J.K., Dai, Y., Tong, Y.W., 2020a. Highly efficient anaerobic co-digestion of food waste and horticultural waste using a three-stage thermophilic bioreactor: performance evaluation, microbial community analysis, and energy balance assessment. Energy. Convers. Manage. 223, 113290.

Zhang, L., Li, F., Kuroki, A., Loh, K.-C., Wang, C.-H., Dai, Y., Tong, Y.W., 2020b. Methane yield enhancement of mesophilic and thermophilic anaerobic co-digestion of algal biomass and food waste using algal biochar: semi-continuous operation and microbial community analysis. Bioresour. Technol. 302, 122892.

Zhang, L., Lim, E.Y., Loh, K.-C., Ok, Y.S., Lee, J.T., Shen, Y., Wang, C.-H., Dai, Y., Tong, Y.W., 2020c. Biochar enhanced thermophilic anaerobic digestion of food waste: focusing on biochar particle size, microbial community analysis and pilot-scale application. Energ. Convers. Manage. 209, 112654.

Zhang, L., Loh, K.-C., Lim, J.W., Zhang, J., 2019b. Bioinformatics analysis of metagenomics data of biogas-producing microbial communities in anaerobic digesters: a review. Renew. Sust. Energy Rev. 100, 110–126.

Zhang, L., Loh, K.-C., Zhang, J., 2019c. Enhanced biogas production from anaerobic digestion of solid organic wastes: current status and prospects. Bioresour. Technol. Rep. 5, 280–296.

Zhang, L., Loh, K.-C., Zhang, J., 2019d. Jointly reducing antibiotic resistance genes and improving methane yield in anaerobic digestion of chicken manure by feedstock microwave pretreatment and activated carbon supplementation. Chem. Eng. J. 372, 815–824.

Zhang, L., Loh, K.-C., Zhang, J., Mao, L., Tong, Y.W., Wang, C.-H., Dai, Y., 2019e. Three-stage anaerobic co-digestion of food waste and waste activated sludge: identifying bacterial and methanogenic archaeal communities and their correlations with performance parameters. Bioresour. Technol. 285, 121333.

Zhang, L., Zhang, J., Loh, K.-C., 2018. Activated carbon enhanced anaerobic digestion of food waste-laboratory-scale and pilot-scale operation. Waste Manag. 75, 270–279.

Zhang, L., Zhang, J., Loh, K.-C., 2019f. Enhanced food waste anaerobic digestion: an encapsulated metal additive for shear stress-based controlled release. J. Clean. Prod. 235, 85–95.

Zhang, M., Fan, Z., Hu, Z., Luo, X., 2020d. Enhanced anaerobic digestion with the addition of chelator-nickel complexes to improve nickel bioavailability. Sci. Total Environ. 759, 143458.

Zhen, G., Lu, X., Kato, H., Zhao, Y., Li, Y.-Y., 2017. Overview of pretreatment strategies for enhancing sewage sludge disintegration and subsequent anaerobic digestion: current advances, full-scale application and future perspectives. Renew. Sust. Energy Rev. 69, 559–577.

Zheng, L., Chen, J., Zhao, M., Cheng, S., Wang, L., Mang, H.-P., Li, Z., 2020. What could China give to and take from other countries in terms of the development of the biogas industry? Sustainability 12 (4), 1490.

Zhou, Q., Yuan, H., Liu, Y., Zou, D., Zhu, B., Chufo, W.A., Jaffar, M., Li, X., 2015. Using feature objects aided strategy to evaluate the biomethane production of food waste and corn stalk anaerobic co-digestion. Bioresour. Technol. 179, 611–614.

Zhu, L., Li, Z., Hiltunen, E., 2018. Theoretical assessment of biomethane production from algal residues after biodiesel production. Wires Energy Environ. 7 (1), e273.

Zhu, X., Blanco, E., Bhatti, M., Borrion, A., 2020. Impact of metallic nanoparticles on anaerobic digestion: a systematic review. Sci. Total Environ. 757, 143747.

Conversion of food waste to bioenergy and biochemicals *via* anaerobic digestion

Liwen Luo[a], Nicholas Cheuk Him Ng[a], Jun Zhao[a], Dongyi Li[a], Zhiqiang Shi[b], Mi Zhou[c]

[a]Department of Biology, Institute of Bioresource and Agriculture, Sino-Forest Applied Research Centre for Pearl River Delta Environment, Hong Kong Baptist University, Kowloon Tong, Hong Kong SAR, China, [b]Department of Chemical Engineering, Tianjin University Renai College, Tianjin, China, [c]Department of Agricultural, Food and Nutritional Sciences, University of Alberta, Edmonton, AB, Canada

2.1 Introduction

Around 2 billion tons of municipal solid waste is produced every year in the whole world, 30%–50% of which belongs to the organic biodegradable waste such as food waste. Food waste might be generated from the food production section, storage and processing section, and consuming process, it has been defined as "the decrease in quantity or quality of food" (Delgado et al., 2021). A great deal of food waste production has a serious impact on the environment such as odor emission, greenhouse gas emission, and water contamination (Nordahl et al., 2020). Therefore, an appropriate scheme regarding food waste management is needed to reduce the problems related to food waste disposal and to recover bioresource in terms of bioenergy and biochemicals.

Food waste can be collected from households and restaurants and its composition may vary depending on the time, the locations, and the diet habits (Parizeau et al., 2015). However, in general, food waste is mainly composed of easily degradable carbohydrate (50%–60%, dry weight basis), protein (15%–5%, dry weight basis), lipids (13%–30%, dry weight basis), and traces in inorganic fractions (Negri et al., 2020). Table 2.1 shows the food waste characteristics sourcing from different countries in the literature. Thus, food waste can be converted into different biochemicals and energies by suitable bioprocess (Dahiya & Joseph, 2015). Converting food waste into biochemicals is an alternative resource recovery method instead of fossil-based

TABLE 2.1   Characteristics of food waste from the literature. Adopted and modified from Luo et al. (2019).

Source	pH	Carbohy-drate (g/L)	Protein (g/L)	Lipid (g/L)	TS	VS	Region	Reference
A canteen of Hangzhou Di-anzi University	N/A	42.7 ± 0.8 (g/100g)	11.2 ± 0.4 (g/100g)	6.7 ± 0.8 (g/100g)	19.6 ± 1.2 (g/100g)	17.8 ± 0.9 (g/100g)	China	(Han et al., 2015)
A mess of the Beijing Univer-sity of Chemical Technology	5.2 ± 0.3	N/A	N/A	N/A	18.5 ± 0.1%	17.0 ± 0.1%	China	(Zhang et al., 2013)
A restaurant of Myongji University	6.5	N/A	14.98	23.3	181 (g/L)	171 (g/L)	Korea	(Zhang et al., 2011)
A food waste–recycling plant in Kwangyang	3.68 ± 0.23	17.5 ± 4.9	21.7 ± 2.4	17.7 ± 5.0	79.5 ± 8.7 (g/L)	63.4 ± 8.3 (g/L)	Korea	(Shin et al., 2010)
Food waste	3.9 ± 0.1	271 ± 57.5 (TCOD, total chemi-cal oxygen demand)	0.504 ± 0.153 (Ammonia)	N/A	17.8 ± 1.2%	16.1 ± 1.2%	Norway	(Zamanza-deh et al., 2016)
The local Environmental Protection Ad-ministration	4.5 ± 0.2	143 ± 50		24 ± 11	266 ± 68 (g/L)	256± 66 (g/L)	Taiwan	(Wang et al., 2010)
A restaurant	6.5	N/A	7.3 ± 12.2	9.4 ± 11.9	61 ± 32 (g/L)	56 ± 33 (g/L)	Egypt	(Tawfik et al., 2011)
Synthetic food waste Kitchen waste	N/A	N/A	N/A	N/A	39.5%	N/A	Hong Kong	(Xu et al., 2011)
Resume of food waste charac-teristics from 102 samples throughout the world between 2001 and 2014	5–6	21% of dry matter	26% of dry matter	28% of dry mat-ter	N/A	94% of dry matter	N/A	(Fisgativa et al., 2016)

chemical production. Food waste served as a substrate in anaerobic digestion process for methane generation is highly developed by academics and industries (Xu et al., 2017). Recently, more research interests regarding food waste bioconversion have been expanded to the production of commodity chemicals (Bilal & Iqbal, 2019), liquid fuel (Antonopoulou et al., 2020), hydrogen (Kiran et al., 2014; Pagliaccia et al., 2016), and bioelectricity (Kiran et al., 2014). As reported by Lim and Wang (2013), some pretreatment approaches involving physical, chemical, enzymatic methods have been suggested to accelerate food waste hydrolysis and thus enhance methane gas production and biochemicals generation. Besides the methanogenic process, acidogenic process, bioelectrochemical configuration, and biodiesel generation are attracting a great deal of interests. As compared to the chemical approaches, waste-to-energy conversion through biological technologies requires lower chemical consumption and energy investment. Energy and resources recovery from food waste via biological way has been proposed as a cost-effective and environment friendly technology for source recovery of food waste. Combining these bioprocesses to achieve food waste utilization in a sustainable manner can give a new insight for food waste management. Therefore, this chapter aims to comprehensively discuss the conversion of food waste into bioenergy and biochemicals using anaerobic digestion principles. To do so, this chapter starts from anaerobic digestion process, moves to the possible products generated, and then introduces available technologies. The relating content offers a full attempt to picture different futuristic technologies and their potential integration for developing biocircular economy from food waste.

2.2 Conversion of food waste to bioenergy and biochemicals

Anaerobic digestion is a bioprocess for treating organic waste without the oxygen participation throughout involving four stages: hydrolysis, acidogenesis, acetogenesis, and methanogenesis (as shown in Fig. 2.1) (Khalid et al., 2011; Luo et al., 2019). Throughout the anaerobic biodegradation process, metabolic products such as hydrogen, methane, volatile fatty acid (VFAs), and ethanol could be considered to recover and utilized into various application such as electricity production (Fasahati et al., 2017), bioplastic synthesis (Abdel-Rahman & Sonomoto, 2016), microbial fuel/electrolysis cells (Zheng et al., 2017), and biodiesel production (Selvakumar & Sivashanmugam, 2017). Thus, no matter biogas fuel or high valuable products, anaerobic digestion process is identified as the most developing technology for valorization of food waste toward effective biogas and valuable biochemicals production

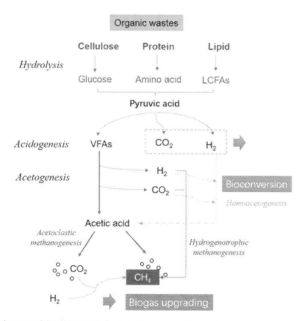

FIG. 2.1   Bioprocess of anaerobic digestion for treating organic waste.

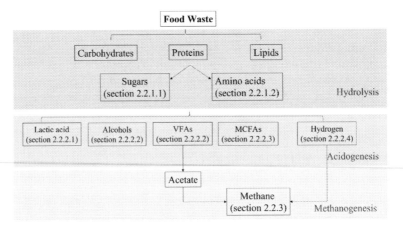

FIG. 2.2   **Conversion of food waste to bioenergy and biochemicals during anaerobic digestion.**

(Fig. 2.2). As mentioned above, four anaerobic steps are contributed by the action of different microorganisms involved: hydrolytic bacteria, acidogenic bacteria, hydrogen producing-consuming bacteria, and methanogenic archaea (Luo et al., 2019). Correspondingly, different products are generated in different stages. In general, sugar and amino acid are dominant in hydrolysis, VFAs and hydrogens are produced in acidogenesis, acetate and hydrogen are the main products in acetogenesis, while the methane gas is the final product for methanogenesis. Therefore, we will introduce what kinds of bioproducts can be potentially enriched and recovered from food waste through anaerobic digestion technology in this section.

2.2.1   Bioproducts generated during hydrolysis

2.2.1.1   *Sugars*

Sugars are widely consumed in food industry for its sweetness and high energy, and they are commonly produced by the hydrolytic step of carbohydrates (Rajaeifar et al., 2019). Monosaccharides, such as galactose and glucose, are known as simple sugars consisting of only a basic unit of carbohydrate. These simple sugars are synthesized by depolymerization of complex carbohydrates. For instance, glucose is synthesized by breaking down cellulose (Huang & Fu, 2013). Generating monosaccharides (e.g., galactose and glucose) from food wastes and unconsumed beverage has become a hot topic in recent years (Fischer & Bipp, 2005; Sainio et al., 2013; Scharf et al., 2020). For example, more than 0.35 g glucose can be recovered from per gram of food waste residual using anaerobic fermentation (Fischer & Bipp, 2005). In addition, Zhang et al. (2020) optimized condition of two-stage enzymatic hydrolysis and eventually obtained the highest reducing sugar concentration of 204 g/L from food waste.

Enzymatic anaerobic digestion of organic waste with high carbohydrate content, followed by depolymerization of complex, unused carbohydrates (e.g. cellulose, pectin and lignin), has been gaining attention to increase sugar production (Hafid et al., 2015). Nevertheless, the biggest issue regarding sugar production is the segregation of different sugars. Success in extracting these reducing sugars from anaerobic digestion effluent in a large scale may lead to big market opportunities. Simple sugars, such as fructose, can be worth up to US$1100 per ton, while xylose is priced up to US$2500 per ton (Bhaumik & Dhepe, 2015). The high market value of these sugars corresponds closely to their potential of being ideal substrates for a variety of products such as alcohol, dietary fiber for food industry, medicines for pharmaceutical industry, or even valuable chemicals such as hydroxymethylfurfural which has a market price of US$300,000–350,000. To maximize the sugar production from enzymatic saccharification, it is more practical to separate the hydrolysis process from the fermentation process, as they have a different set of optimum conditions (Jiang et al., 2014).

2.2.1.2 Amino acids

Proteins are constituted by one or several peptides, which are basically chains of amino acids connected by peptide bonds (Gerardi, 2003). After the enzymatic cleavage of these bonds, microorganisms degrade proteins into amino acids, which can be utilized to produce organic acids. On the other hand, amino acids can also be repackaged into cell proteinaceous matter by microbes given that adequate energy is provided in the form of carbohydrates (Duong et al., 2019). Alanine, cysteine and histidine, leucine and tyrosine are some examples of amino acids recovered from organic waste.

Little is known about the status of amino acid in anaerobic digestion, yet hydrolysis is believed by many to be the rate-determining step of protein degradation, rather than amino acid fermentation (Fache et al., 2015). Nevertheless, there was a contradictory result found by Duong et al. (2019) who have identified about 20 amino acids during the first 8 h of the gelatin digestion, indicating amino acids accumulation. After all, amino acids are further broken down by the metabolic pathways below: Strickland reaction (for pairs of amino acids), oxidative deamination, and reductive deamination of single amino acids (Shen et al., 2017).

With the aim to synthesize amino acids, Strickland reaction and deamination conducted mainly by *Clostridium* sp. should be avoided (Park et al., 2014). The increase in pH attributed to the protein digestion plays a pivotal role in buffering the acidic condition created by VFA buildup; however extra attention should be paid to ammonia toxicity in the meantime (Ma et al., 2018). To better control the desired reaction, genetically modified *Clostridium* strains have the potential to be incorporated into the digester.

Amino acids production from renewable substrates is less popular. Still, there are researchers who worked on the microbial transformation of amino acid, and the other investigated the chemocatalytic approach (Deng et al., 2018). Cellulose and hemicellulose are the target substrates in most of these studies (Dusselier et al., 2013). One of the breakthroughs of this area would be the successful glutamate production by *Corynebacterium glutamicum*, after which other amino acids can be produced by anaerobic fermentation (Vassilev et al., 2018).

2.2.2 Bioproducts generated during acidogenesis

2.2.2.1 Lactic acid

Lactic acid is a particularly important intermediate during food waste acidogenesis when operation pH is lower around 3.2–4.5 (Hunter et al., 2020). *Lactobacillus* is one of the main functioning bacteria for lactate fermentation pathway (Luo & Wong, 2019). According to the theoretical production reaction, lactate fermentation can be divided as homolactic and heterolactic fermentation pathway, in which one mole glucose is converted a two-mole lactate and one-mole lactate associated with ethanol/acetate production, respectively (Zhou et al., 2018). As reported, the global demand for lactate was approximately close to half million in 2016

(Hatti-Kaul et al., 2016). Lactate might be widely used as acidulant, preservative, flavor additive in chemical and pharmaceutical synthesis and food industry (Vijayakumar et al., 2008).

2.2.2.2 Volatile fatty acids (VFAs) and alcohols

VFAs, including acetate, propionate, butyrate, and valerate are the essential precursors for the productions of bioplastic, biodiesel, and biofertilizer (Atasoy et al., 2020). In general, VFA production comes from petroleum-based industry, while these processes are associated with energy consumption and negative environmental impacts. In this case, bioprocess for VFA production is attracting more interest. When considering VFA production during anaerobic digestion process, there is no need to control sterile fermentation and thus a complex waste such as food waste is a suitable and cost-effective substrate (Yin et al., 2016). For anaerobic digestion process, there are several anaerobic fermentation pathways correlating the VFA production such as acetate-ethanol fermentation, propionate fermentation, butyrate fermentation, and mix-acids fermentation pathways (Table 2.2). Corresponding metabolites

TABLE 2.2 Metabolic pathway in AD at different steps.

Acidogenesis		
Acetate-ethanol type	$\Delta G_0'$ **(kJ/mol)**	**Eq.**
$C_6H_{12}O_6 + 2H_2O \rightarrow 2CH_3COOH + 2CO_2 + 4H_2$	−135.6	(1)
$C_6H_{12}O_6 + 4H_2O + 2NAD^+ \rightarrow 2CH_3COO^- + 2HCO_{3^-} + 2NADH + 2H_2 + 6H^+$	+215.7	(2)
$C_6H_{12}O_6 + 2H_2O + 2NADH \rightarrow 2CH_3CH2OH + 2HCO_3^- + 2NAD + 2H_2$	+234.8	(3)
Propionate type		
$C_6H_{12}O_6 \rightarrow CH_3COO^- + CH_3CH_2COO^- + CO_2 + H_2 + 2H^+$	+287.0	(4)
Butyrate type		
$C_6H_{12}O_6 \rightarrow CH_3CH_2CH_2COOH + 2CO_2 + 2H_2$	−257.1	(5)
$C_6H_{12}O_6 + 2H_2O \rightarrow CH_3CH_2CH_2COO^- + 2HCO_{3^-} + 2H_2 + 3H^+$	−261.5	(6)
Acetogenesis		
Propionate: $CH_3CH_2COOH + 2H_2O \rightarrow CH_3COOH + 3H_2 + CO_2$	+76.2	(7)
Butyrate: $CH_3CH_2CH_2COOH + 2H_2O \rightarrow 2CH_3COOH + 2H_2$	+48.4	(8)
Lactate: $CH_3CHOHCOOH + 2H_2O \rightarrow CH_3COOH + HCO_3^- + 2H_2$	−4.2	(9)
Ethanol: $CH_3CH_2OH + H_2O \rightarrow CH_3COOH + 2H_2$	+9.6	(10)
Syntrophic acetate oxidizing reaction (SAO): $CH_3COOH + 2H_2O \rightarrow 2CO_2 + 4H_2$	+104.6	(11)
Homoacetogenesis		
Autotrophic: $4H_2 + 2CO_2 \rightarrow CH_3COOH + 2H_2O$	−104.6	(12)
Heterotrophic: $C_6H_{12}O_6 \rightarrow 3CH_3COO^- + 3H^+$	−310.9	(13)
Methanogenesis		
Acetoclastic methanogenesis: $CH_3COOH \rightarrow CH_4 + CO_2$	−31.0	(14)
Hydrogenotrophic methanogenesis: $4H_2 + CO_2 \rightarrow CH_4 + 2H_2O$	−135.0	(15)

in acetate-, propionate-, and butyrate-fermentation pathways account for more than 60% of total metabolic products in each mentioned metabolic pathway (Zhou et al., 2018). Individual acids would be favorably generated in these fermentation pathways. However, all VFAs will occur in mixed fermentation pathways. Fig. 2.1 presents the metabolic network during anaerobic digestion process. Therein, pyruvate and acetyl-CoA are the critical points, which are considered to be two key links between metabolite pathways and their products (Luo et al., 2019). During operation, the VFA production is mainly manipulated by the organic loading rate, the operational pH, and microbial diversity. Among them, pH is identified as the critical parameter to regulate the anaerobic fermentation pathway (Mohd-Zaki et al., 2016). For example, butyrate was the dominant metabolic product at a pH ranging of 4.5–4.7; when pH increased to 6, the butyrate fermentation was changed into mixed acids fermentation (Huang et al., 2016). In addition, alcohols production also can be found in acidogenesis stage through solventogenic pathway (Li et al., 2020). The typical pathway is named as acetone–butanol–ethanol fermentation, in which glucose will be degraded into acetyl-CoA first and transformed into acetoacetyl-CoA and eventually formed into end-point products such as acetone, butanol, or ethanol (Han et al., 2013). Therefore, it is possible to produce and recover specific VFAs and alcohols from food waste during anaerobic fermentation process if we can regulate the metabolic pathways in appropriate ways.

2.2.2.3 *Medium-chain fatty acids*

As reported by Reddy et al. (2018), medium-chain fatty acids (MCFA) production can be processed through biological process in anaerobic digestion system. Therein, reversing β-oxidation pathway was demonstrated as the effective functioning pathway for MCFA synthesis. Unlike the conversion of VFAs into methane (CH_4), the MCFAs' synthesis, in which two carbon atoms could be used for elongation process, could potentially reduce carbon loss so as to improve carbon conversion efficiency during anaerobic process. Sufficient carbon source, proper reducing agents, and electron donors are required in the pathway of reverse β-oxidation. Electron donors, such as lactate, hydrogen gas, and ethanol could also be used in addition to the simple carbon source. For instance, ethanol can be served as one electron donor when synthesizing caproic acid (Cavalcante et al., 2017). MCFAs are more competitive than short-chain VFAs (C_2–C_4), as MCFAs have more carbon atoms (C_6-C_{10}) which offer lower ratio of oxygen/carbon and greater hydrophobic effect. In this sense, the separation of MCFAs from anaerobic fermentation would be possible and easier (Wu et al., 2020). Furthermore, as compared to ethanol and VFAs, lower solubility and higher energy value will be provided by longer hydrocarbon chain of MCFAs as well. For example, the heating values of ethanol and caproic acid are 1319 kJ/mol and 3452 kJ/mol, respectively, indicating the greatly lower energy intensity in short-chain organics (Grootscholten et al., 2013). In terms of the application of MCFAs, it can be used for a great deal of industries such as biodiesel production, antimicrobial technology, and flavoring intermediates. Therefore, developing MCFAs biosynthesis technology using food waste would not only provide a new direction for waste management but also obtain high valuable products, which would probably contribute to biocircular economy (Dahiya et al., 2018; Menon & Lyng, 2021).

2.2.2.4 *Hydrogen*

Hydrogen gas (H_2) is a renewable energy carrier, and plausible synthesis methods had garnered worldwide research interest recently. Biological production methods include photosynthesis, photofermentation, microbial electrolysis cell, and anaerobic fermentation, or a combination of these methods (Chen et al., 2016; Nagarajan et al., 2021; Shin & Youn, 2005; Srirangan et al., 2011). Hydrogen-producing bacteria take part in some biological hydrogen gas production (Wang & Wan, 2008). There was a wide array of wastes being tested for biological hydrogen production, such as vegetable-predominant market waste, food waste, wastewater, and municipal sewage. Some pilot studies were performed to investigate the feasibility of using food waste as the feedstock (Jarunglumlert et al., 2018; Lee & Chung, 2010). The main reason is that it mostly consists of highly degradable organic matter and can be continuously supplied at low cost making the process both technically efficient and economically viable. Pretreatment, pH, temperature, hydraulic retention time (HRT), cosubstrate application, reactor set-up, and feedstock are all essential factors manipulating hydrogen production in anaerobic digestion (Li & Fang, 2007). Introducing microbial consortia to apply with food waste in hydrogen production has various benefits as the process can be operated flexibly under nonsterile environments with easy manipulation and up-scaling possibility (Favaro et al., 2013; Jo et al., 2007). Also, the substrate with high carbon content and production of diverse products with appreciable market values are also competitive features. Combination of anaerobic fermentation and photofermentation is plausible because metabolites created in the former can be fed into the latter for further transformation (Su et al., 2009). This integration can be implemented in both two-phase and single-phase set-up each with respective advantages. In contrast to photofermentation, microbe-assisted conversion of biomass to hydrogen without the presence of light is termed as dark fermentation. Theoretically, hydrogen production is maximized when dark fermentation and photofermentation are combined. With the potential of food waste feeding into biological hydrogen production, more pilot plants are anticipated to be built to further investigate hydrogen gas as a new green fuel for transportation.

2.2.3   Potential bioenergy generated from methanogenesis

2.2.3.1 *Methane*

Biogas production from food waste has been intensively researched, as it is the direct product of anaerobic digestion (Bong et al., 2018; Chiu & Lo, 2016; Yang et al., 2019). Within the biogas, methane is the major product. Besides, the aqueous solubility of methane is very low that it inherently segregates, and less effort is needed for extraction. Therefore, it is a popular alternative to fossil fuels as a renewable energy source. In Europe alone, the forecast market value for biogas in 2025 is expected to be the double of 2018 at US\$6 billion (Hunter et al., 2021). Also, anaerobic digestion contributed to over 70% of the market revenue showcasing a solid economic incentive for biogas production through anaerobic digestion (Liu et al., 2012). Nevertheless, biogas has only 45%–65% methane in normal operation. The higher the methane proportion in the biogas, the higher the energy generated. Hence, the increase in methane recovery is the objective of many research studies (Campuzano & Gonzalez-Martinez, 2016; Heydari & Peyvandi, 2020). However, there are a lot of viable metabolic pathways for methane production, and optimization varies in every case. Numerous parameters have been studied, for example, reaction conditions (pH, temperature, C:N ratio, HRT, and loading rate), microbial consortia addition, and types of feedstocks (Luo et al., 2019; Xu et al., 2017). Methanogenesis, the last stage of anaerobic digestion, is the most important stage of methane production, which is mainly governed by methanogens (Luo et al., 2016). These microbes are slow growing and condition sensitive making this final step as the rate-limiting process.

2.2.3.2 *Hythane*

Although hydrogen and methane can be recovered from anaerobic system as an ideal energy source, individual limitations of hydrogen and methane also are reported in previous studies (Rena et al., 2020). Hythane, mixing methane and hydrogen with the ratio of 1 to 4, is recognized as the suitable biofuel due to clear and high calorific value (Rena et al., 2020). Food waste has been used as the feedstock for hythane production in various studies ranging from lab scale to pilot scale (Cavinato et al., 2012; Kumara et al., 2019). In general, hythane can be achieved in two-phase anaerobic digestion system. For example, Cavinato et al., (2012) reported that the hydrogen and methane yields obtained in acidogenic and methanogenic phases were 66.7 L/kg VS and 720 L/kg VS (volatile solids), respectively, which was argued as the typical composition for hythane. In addition, when using food waste as a substrate, the hythane production yield was increased from 144, 156, 159 to 163 L with the increasing of food waste loading from 60, 70, 90 to100 gCOD/L, respectively (Dahiya et al., 2018). It indicated that food waste can be acted as a potential suitable substrate for hythane production, which might be easily commercialized using two-phase anaerobic digestion scheme.

2.2.3.3 *Anaerobic configurations for bioenergy production*

Single-phase anaerobic digestion

For methane production, the configuration is identified as an essential factor for affecting the methanogenic efficiency. Single-phase anaerobic digestion system is a simple design, in which all above anaerobic reaction steps simultaneously occur. Up to now, over 90% of full-scale anaerobic plants apply single-phase anaerobic system to treat organic waste. Single-phase anaerobic system can be operated into wet or dry condition. The organic loading rate of 5 gVS/L/d was the general full-scale value in practice for wet condition that rarely exceeded 10 gVS/L/d. In general, the maximum OLR (organic loading rate) is around 15 gVS/L/d, but the degradation efficiency would be highly limited once reached this value (Scherer et al., 2000). Thus, many studies reported the single-phase anaerobic digestion process when OLR was lower than 15 gVS/L/d. It suggested that the OLR of 7.7 ± 6.4 gVS/L/d associated with methane production yield of 414 ± 31 L/kgVS was the general condition for single-phase anaerobic digestion when using food waste as the feedstock (Negri et al., 2020). As compared to the wet condition, dry condition for single-phase anaerobic system contains higher solids content and lower water content but finally provides lower methane yield and volatile solids removal efficiency mainly due to the VFAs accumulation (Luo et al., 2019). As revealed by Xu et al. (2020), single-phase anaerobic reactor treating food waste was easily unstable because of the VFA limitation and lower pH, resulting in the lower even the failure of biogas production. Thus, how to ensure the operational stability is the key concerning single-phase anaerobic digestion treating food waste.

Two-phase anaerobic digestion

Differ to a single-phase anaerobic digestion system, a two-phase anaerobic system separates hydrolytic-acidogenic and methanogenic phase into two individual reactors. In this system, acidogenic bacteria and hydrogen-producing bacteria are dominant microorganisms for VFAs and hydrogen production at the first phase to break down the complex organics into simple molecular components; acetogenic and methanogenic microorganisms are enriched in the second phase to effectively convert VFAs into acetate and hydrogen and thus facilitating the methane production. Because of the separated configuration, two-phase anaerobic digestion system is proposed with several benefits including: (1) achieving the multiple bioenergy in terms of hydrogen and methane, (2) gaining better stable operating condition during anaerobic degradation, (3) having better manipulation of acidogenic fermentation and methanogenic process, (4) increasing the whole energy conversion efficiency from food waste (Zhou et al., 2018).

It is interesting that there is no clear information regarding the increasing of overall energy production in two-phase anaerobic digestion system from the literature. Even through, most studies have illustrated that the energy conversion efficiency increased around 20%–60% in separated anaerobic system as compared to that in single-phase anaerobic system (Luo & Angelidaki, 2012). However, there was no significant difference between single- and two-phase anaerobic systems in lab-scale methane production study, which was reported by Schievano et al. (2012). Furthermore, Negri et al. (2020) reviewed different food waste degradation studies and conclude the fact that, on the average, there was no statistic difference in energy production when we compared single- and two-phase anaerobic digestion approaches. Taking capital cost and energy yield into consideration, it needs to gain other valuable biochemicals from two-phase anaerobic approach to increase the bioeconomic value.

Electro-fermentation

When we introduce electrochemical configuration into regular anaerobic fermentation process, this integrated novel system is named as electro-fermentation (Moscoviz et al., 2016). During electro-fermentation process, microbial fermentation would be interfered by electricity stimulation with electrodes. The electrodes can play a role in electron transfer and also can serve as one kind of resource balancing the fermentative process. Thus, both regular fermentation kinetics and pathways would probably be changed in novel electro-fermentation system (Nikhil et al., 2016; Schievano et al., 2016). The electrode-related rates could be improved to apply these electrode-based electron acceptors due to the elimination of redox limitations, so that it facilitates the targeting reactions to improve the generation of specific products with higher conversion efficiencies. Although electro-fermentation belongs to a newly developing technology, it could be recognized as a potential approach to converting organics into valuable chemicals considering its principles. Therein, once applying external potential (negative or positive), the appearance of electricity stimulation would cause an interaction effect between microorganisms and electrode on solid–liquid interfaces affecting the metabolic rates in electro-fermentation process (Nikhil et al., 2016). Considering either hydrogen evolution reaction or oxygen–reduction reaction, how to utilize food waste effectively and decrease electrochemical losses regarding to electrochemistry of substrate-electron binding should be conscientiously studied in next. As compared to regular electrodes, the specific electrodes with modifications on the carbon surface, which may affect selectivity, economic viability, and sensitivity, need to be researched more (Kumar et al., 2017). Electrode potential is a crucial parameter to evaluate the efficiency of bioelectrochemical system during electro-fermentation process and thus provide a guide to reduce the electro-chemical losses and harvest maximum energy. In general, most bacteria have negative charges, adding an extra positive potential on anode could effectively enhance the electrostatic interaction on the biofilm between microbes and electrodes (Luo et al., 2016; Luo et al., 2018). While adding negative potential on cathode, the reducing reaction will be facilitated. The conversion of food waste into valuable products will be accelerated and enhanced by the microbial interactions and operational conditions during electro-fermentation process (Venkata Mohan et al., 2014). The potential of electro-fermentation to valorize food waste is depicted by the energy productions in terms of bioelectricity, biofuels including hydrogen and methane, and other metabolic intermediates such as lactate and VFAs from solid-state food waste when applying bioelectro-fermentation process. Therefore, the synergistic effects of integrating electrochemistry and regular bioprocess allow the waste management in a green and sustainable manner. And thus, the electro-fermentation approach attracts more and more interests from academia and industry.

2.3 Integrated bioprocesses with food waste anaerobic digestion for bioresource recovery

Different efficiencies produce various effluents with distinct properties generated from different anaerobic steps. For food waste valorization, these effluents coming from hydrolytic-acidogenic stage or methanogenic stage contains a great deal of metabolic organics, which can be used as substrates to generate various products such as biofuels, bioplastics, and biofertilizers. To improve the economic benefits, it is necessary to integrate different bioprocesses to achieve multiple products and enhance the operating efficiency. Therefore, several integrated bioprocess models were concluded and discussed here.

2.3.1 Production of polyhydroxyalkanoates (PHAs) from food waste after anaerobic digestion

PHAs are typical biopolymers, which can be internally synthesized by various microorganisms. Under stressful or nutrient-limiting conditions, PHAs can be generated in the internal cell to conserve energy and carbon for growth. The advancement of sustainable and renewable biopolymers is required to meet the environmental and economic needs (Yadav et al., 2020). The bioplastics have developed greatly because of their biodegradability and thermoplastic properties similar to plastics derived from petroleum. Their applications are diverse, ranging from paints to implants, from packaging to tissue engineering, from adhesives to chiral chemicals for pharmaceutical industry.

In the production of PHA, biowastes such as dyeing wastewater, bakery waste, sugar cane waste, and food waste could be acted as substrates (Amulya et al., 2016; Haque et al., 2016; Serafim et al., 2008). It has been reported that acidogenic fermentation can be integrated with PHA production (Reddy & Mohan, 2012). For example, when the effluent generated from food waste acidogenic fermentation was applied into PHA production with a multistage mode, the PHA productivity obtained was 23.7% (Amulya et al., 2014). In fermented food waste, PHA yield is 39.6%, while for unfermented food waste it is 35.6%. It was mainly attributed to the higher available VFA in the acidogenic effluents after fermentation. Higher process efficiency can be obtained in the integrated system that combined bioplastics production and food waste acidogenesis compared to the sole food waste degradation process (Reddy & Mohan, 2012). Thus, it is feasible to apply food waste or its fermentation effluent in bioplastics production, which would further expand the economic value of food waste after anaerobic technology.

2.3.2 Bio-based electricity generation

Microbial fuel cell (MFC), converting biowaste into electricity using microorganisms, is an environmental technique with promising application potential (Zeppilli et al., 2019). The microorganisms acting as a catalyst in the anode produce reductants (protons and electrons), which facilitate the degradation of organic substrate via electrochemical oxidation. In the cathode, the electron acceptor works as an electron driving force and promotes electrons transfer process throughout the external circuit, resulting in the electricity production. In MFC, anode and cathode are separated by a membrane that only allows the proton, that is, H^+ transferred into the cathode (Zeppilli et al., 2019). However, organic stream such as food waste placed in the cathode chamber participates in reductive reactions and transferring H^+ and e^- (Srikanth & Mohan, 2012). Pant et al. (2013) established two-step process where integrated fermentation and MFC convert food waste in sewage into VFAs and electricity simultaneously. Another innovative food waste resourceful approach was proposed by Xin et al. (2018). In this study, ultrafast pretreatment was used to hydrolyze food waste first, and then solid and liquid are separated for biofertilizer after composting and electricity production using MFC, respectively. Thus, combining anaerobic technology with MFC would simultaneously produce electricity and convert waste.

2.3.3 Biofertilizer produced in anaerobic digestion treating food waste

Liquid and solid digestates could be collected after anaerobic digestion, which are recognized as biofertilizer due to their high N and P concentration (Hassaneen et al., 2020; Tallou et al., 2020). The application of anaerobic effluents as biofertilizer is a promising and environmental approach regarding the combination of waste control and utilization. Biofertilizer is a cost-effective and renewable nutrient source for plant growth. In the effluent from food waste acidogenesis, it involves various microorganisms such as *Pseudomonas, Bacteroides, Penicillium, Bacillus, Klebsiella, Aspergillus, Shigella,* and *Salmonella* which is identified as probiotics so that the food waste effluent can be applied to the production of biofertilizers (Soares-Silva, 2016). Among them, *Pseudomonas* and *Bacillus* are considered as soluble phosphate biofertilizers, while *Clostridium* and *Klebsiella* are identified as nitrogen-fixing biofertilizers (Alfa et al., 2014). These probiotics can be used to accelerate the microbial metabolism in soil, and release the available nutrients to plants. Apart from the microorganisms, a large number of organic compounds and biomass contained in the liquid effluent suggested that food waste effluent can act as fertilizer to enhance the nutrient concentration of soil that is, fertility. Comparing to chemical fertilizers, it would be better to fertilize crops with effluent due to the ricing available nutrients. Generation of biofertilizers from food waste may be helpful to divert the negative carbon footprint from greenhouse gas emission to organic fertilizer.

2.4   Challenges and limitations in bioconversion of food waste to bioresources *via* anaerobic digestion

The increasing demand for energy and materials requires the development of sustainable technology, in particular the bioprocesses. Anaerobic digestion can be identified as the promising biotechnology for valorization biowaste value and contributing sustainable economy. Along with the anaerobic bioprocesses such as hydrolysis, acidogenesis, and methanogenesis, sugars, carboxylic acids, biopolymers, electricity, biofertilizers, and biofuels can be potentially recovered from food waste. Several studies have made efforts to expand the anaerobic digestion applications treating food waste in practical; however, rare studies achieve the great implement due to some challenges correlated to anaerobic digestion process. Some identified challenges for expanding anaerobic digestion development treating food waste are as follows.

The foremost obstacle that hinders the anaerobic digestion development is the complicated reaction pathways that undergo within the digesters. This obstacle is attributed mainly to the high microbial diversity inside the reactors, which makes the effort of elucidating all the metabolic pathways extremely difficult. The present studies mostly focus on how a single strain participates in a specific stage of anaerobic digestion, which possibly shed light on the understanding of specific mechanisms. However, the complex interactions of different microbial communities, as a whole, contribute to the whole process. The possible solution to this challenge necessitates the characterization of microorganisms combined with the close monitoring of the intermediate production dynamics throughout the process.

The unstable composition of food waste is another huge challenge. Its composition varies upon different sources (household, restaurant, agricultural), regions, and even seasonal food supply. Studies conducted using a local food waste source can only be limited to local application. Characterization of the substrate composition may provide a better understanding and lead to a more extensive use of the result. Knowing every specific component of the collected food waste is important for creating linkage between components and metabolic pathways, and in turn, the dynamic change of intermediates production. Studying the effect of functional groups or bonds composition in specific food waste components may also be relevant.

Another hurdle that needs to be overcome would be the potential presence of inhibitory effects. The groups of microorganisms involved in different stages have different optimal operational conditions. Research for the advancement in reactor set-up such as two-phase anaerobic digestion and integration of compatible processes can potentially increase the substrate utilization efficiencies and minimize the major inhibition.

Biological challenges aside, some intermediates with increasing concentration are found to pose significant antagonistic effect on the downstream processes. Close monitoring on the intermediate accumulation in the system is needed. The current studies mainly focus on the analysis of VFA, pH, and gaseous products. However, many intermediates are in low level and are difficult to detect using the present technology. More sensitive and sophisticated measuring techniques should be in place to allow low concentration detection. This presents a research gap in the analytical technology.

On the other hand, the segregation of specific intermediates from the system is another research direction that promotes easier detection and further utilization of pure anaerobic products. As the composition of anaerobic digestion products is highly heterogeneous, extraction of pure constituents is extremely hard. Gaseous products with low density, such as VFA and biogas are relatively easy to separate as compared to those in a liquid state. Being the precursors of various marketable chemicals, terpenes are volatile in nature and can also be collected with the help of absorbents or biofilters.

Multiple approaches have been explored with regard to different targeted molecules for isolation. For phenolic compounds, an array of techniques has been investigated including ultrasonic-assisted, resin adsorption, microwave-assisted extraction, supercritical fluid extraction, Soxhlet, and accelerated solvent extraction. When comes to alcohol and sugars segregation, distillation and membrane filtering are widely used methods, respectively. The carbon backbone of short-chain fatty acids can be elongated to form MCFA that are poorly soluble in water allowing easier separation. The development of isolation and extraction methods is essential for maximizing the potential of food waste in anaerobic digestion system, like those seen in traditional refinery. Nevertheless, some of these methods are developed and operated at the expense of environment and capital. Therefore, thorough and tailor-made consideration is necessary.

In spite of these obstacles, enhancing the understanding of intermediates during anaerobic digestion gives chances for valorization and system optimization. Intermediates such as terpenes and SCCA (short chain carboxylic acid) are both profitable and relatively easy for isolation. These compounds provide an additional source of income which can potentially make the anaerobic digestion process more economically sensible. On the other hand, intermediates with low economic value and extractability may still be worth attention. Tracking of these compounds may contribute to the mapping of the metabolic pathways in AD, which in turns indicates the digestor health. Advancement in this area will probably aim at specific types of food waste with stable composition, where the effect due to the varying substrate composition is minimized and product quality can be controlled easier.

Although there are myriads of research worked on the standard anaerobic intermediates mentioned in this chapter, these compounds still have the potential to further enrich the current anaerobic digestion knowledge. For instance, the current research on amino acids is often contradictory in their results, which is likely due to the structural differences among the wide array of amino acid. More in-depth investigation on the mechanism is required to prevent over generalization on the amino acid pathways in anaerobic digestion process. In addition, there is a research gap in examining the effect of phytochemicals on anaerobic degradation. A large proportion of food waste is from plant sources, making research in this area reasonable.

Last but not least, research from other fields may provide new idea for improving anaerobic digestion performance. For example, a biological study aiming at understanding the condition inside ruminants' digestive system may shed light on the design and operational condition of the digester. Multidisciplinary collaboration increases the chance of innovation.

2.5   Conclusions and perspectives

The ever-growing demand for resources and energy has been the reason for the gradual transition from fossil-based economy to sustainable green bioeconomy. Biocircular economy suggests heavy reliance on green substrates that can produce green products having multiple implications in an array of sectors including science, management, and engineering. In this case, food waste with continuous and considerable supply presents to be an ideal feedstock for bioeconomy. To be environment friendly, avoidable food waste can be minimized by elimination, reusing, and reducing. The remaining part and the inconsumable food waste require a management strategy. Anaerobic digestion technology that involves conversion of biomass to energy and bio-based products is a possible solution. The possible contribution processes include hydrolysis, acidogenesis, acetogenesis, methanogenesis. A wide range of products, such as green fuels (biohydrogen, methane, ethanol), marketable compounds (sugars, VFAs), electricity, biofertilizer (food waste digestate, compost), livestock feed, and so on, can be well produced at each discussed step in anaerobic digestion system (as seen in Section 2.3). The integration of these processes (as introduced in Section 2.4) can better make use of various reaction intermediates and digestate generated during different processes. A case in point is the combination of dark fermentation, anaerobic digestion, and MFCs, in which biohydrogen, methane, bioelectricity, and biofertilizer can be coproduced. The diverse bioproducts can be utilized in various situations and pave the way to a more robust and prosperous development of bioeconomy. By the advancement of anaerobic digestion technology, food waste with a huge amount of highly degradable organic matter possesses an indispensable potential to be recycled. Resource recovery from food waste through anaerobic digestion process is expected to reduce the reliance on oil and petroleum refinery, so as to relieve the negative impacts brought by the latter such as global warming, resource scarcity, pollution, and loss of ecosystem services.

References

Abdel-Rahman, M.A., Sonomoto, K., 2016. Opportunities to overcome the current limitations and challenges for efficient microbial production of optically pure lactic acid. J. Biotechnol. 236, 176–192.

Alfa, M., Adie, D., Igboro, S., Oranusi, U., Dahunsi, S., Akali, D., 2014. Assessment of biofertilizer quality and health implications of anaerobic digestion effluent of cow dung and chicken droppings. Renew. Energ. 63, 681–686.

Amulya, K., Reddy, M.V., Mohan, S.V., 2014. Acidogenic spent wash valorization through polyhydroxyalkanoate (PHA) synthesis coupled with fermentative biohydrogen production. Bioresource Technol. 158, 336–342.

Amulya, K., Dahiya, S., Mohan, S.V., 2016. Building a bio-based economy through waste remediation: innovation towards sustainable future. In: Bioremediation and bioeconomy. Elsevier, pp. 497–521.

Antonopoulou, G., Alexandropoulou, M., Ntaikou, I., Lyberatos, G., 2020. From waste to fuel: Energy recovery from household food waste via its bioconversion to energy carriers based on microbiological processes. Sci. Total Environ. 732, 139230.

Atasoy, M., Eyice, O., Cetecioglu, Z., 2020. A comprehensive study of volatile fatty acids production from batch reactor to anaerobic sequencing batch reactor by using cheese processing wastewater. Bioresource Technol., 311.

Bhaumik, P., Dhepe, P.L., 2015. Conversion of biomass into sugars. In: Murzin, D., Simakova, O. (Eds.), Biomass sugars for non-fuel applications. RSC Green Chemistry, pp. 1–53.

Bilal, M., Iqbal, H.M., 2019. Sustainable bioconversion of food waste into high-value products by immobilized enzymes to meet bio-economy challenges and opportunities–A review. Food Res. Int., 123, 226–240.

Bong, C.P.C., Lim, L.Y., Lee, C.T., Klemes, J.J., Ho, C.S., Ho, W.S., 2018. The characterisation and treatment of food waste for improvement of biogas production during anaerobic digestion - A review. J. Clean Prod. 172, 1545–1558.

Campuzano, R., Gonzalez-Martinez, S., 2016. Characteristics of the organic fraction of municipal solid waste and methane production: A review. Waste Manage. 54, 3–12.

Cavalcante, W.D., Leitao, R.C., Gehring, T.A., Angenent, L.T., Santaella, S.T., 2017. Anaerobic fermentation for n-caproic acid production: A review. Process Biochem. 54, 106–119.

Cavinato, C., Giuliano, A., Bolzonella, D., Pavan, P., Cecchi, F., 2012. Bio-hythane production from food waste by dark fermentation coupled with anaerobic digestion process: A long-term pilot scale experience. Int. J. Hydrogen Energ. 37 (15), 11549–11555.

Chen, Y., Chen, M., Shen, N., Zeng, R.J., 2016. H-2 production by the thermoelectric microconverter coupled with microbial electrolysis cell. Int. J. Hydrogen Energ. 41 (48), 22760–22768.

Chiu, S.L.H., Lo, I.M.C., 2016. Reviewing the anaerobic digestion and co-digestion process of food waste from the perspectives on biogas production performance and environmental impacts. Environ. Sci. Pollut. R. 23 (24), 24435–24450.

Dahiya, S., Joseph, J., 2015. High rate biomethanation technology for solid waste management and rapid biogas production: an emphasis on reactor design parameters. Bioresource Technol. 188, 73–78.

Dahiya, S., Kumar, A.N., Sravan, J.S., Chatterjee, S., Sarkar, O., Mohan, S.V., 2018. Food waste biorefinery: Sustainable strategy for circular bioeconomy. Bioresource Technol. 248, 2–12.

Delgado, L., Schuster, M., Torero, M., 2021. Quantity and quality food losses across the value Chain: A Comparative analysis. Food Policy, 98, 101958.

Deng, W., Wang, Y., Zhang, S., Gupta, K.M., Hülsey, M.J., Asakura, H., Liu, L., Han, Y., Karp, E.M., Beckham, G.T., 2018. Catalytic amino acid production from biomass-derived intermediates. P. Natl. Acad. Sci. 115 (20), 5093–5098.

Duong, T.H., Grolle, K., Nga, T.T.V., Zeeman, G., Temmink, H., Van Eekert, M., 2019. Protein hydrolysis and fermentation under methanogenic and acidifying conditions. Biotechnol. Biofuels 12 (1), 1–10.

Dusselier, M., Van Wouwe, P., Dewaele, A., Makshina, E., Sels, B.F., 2013. Lactic acid as a platform chemical in the biobased economy: the role of chemocatalysis. Energ. Environ. Sci. 6 (5), 1415–1442.

Fache, M., Boutevin, B., Caillol, S., 2015. Vanillin, a key-intermediate of biobased polymers. Eur. Polym. J. 68, 488–502.

Fasahati, P., Saffron, C.M., Woo, H.C., Liu, J.J., 2017. Potential of brown algae for sustainable electricity production through anaerobic digestion. Energ. Convers. Manage. 135, 297–307.

Favaro, L., Alibardi, L., Lavagnolo, M.C., Casella, S., Basaglia, M., 2013. Effects of inoculum and indigenous microflora on hydrogen production from the organic fraction of municipal solid waste. Int. J. Hydrogen Energ. 38 (27), 11774–11779.

Fischer, K., Bipp, H.-P., 2005. Generation of organic acids and monosaccharides by hydrolytic and oxidative transformation of food processing residues. Bioresource Technol. 96 (7), 831–842.

Gerardi, M.H., 2003. The microbiology of anaerobic digesters. John Wiley & Sons.

Grootscholten, T.I.M., dal Borgo, F.K., Hamelers, H.V.M., Buisman, C.J.N., 2013. Promoting chain elongation in mixed culture acidification reactors by addition of ethanol. Biomass Bioenerg. 48, 10–16.

Hafid, H.S., Shah, U.K.M., Baharudin, A.S., 2015. Enhanced fermentable sugar production from kitchen waste using various pretreatments. J. Environ. Manage. 156, 290–298.

Han, S.H., Cho, D.H., Kim, Y.H., Shin, S.J., 2013. Biobutanol production from 2-year-old willow biomass by acid hydrolysis and acetone-butanol-ethanol fermentation. Energy 61, 13–17.

Haque, M.A., Kachrimanidou, V., Koutinas, A., Lin, C.S.K., 2016. Valorization of bakery waste for biocolorant and enzyme production by Monascus purpureus. J. Biotechnol. 231, 55–64.

Hassaneen, F.Y., Abdallah, M.S., Ahmed, N., Taha, M.M., Abd ElAziz, S.M.M., El-Mokhtar, M.A., Badary, M.S., Allam, N.K., 2020. Innovative nanocomposite formulations for enhancing biogas and biofertilizers production from anaerobic digestion of organic waste. Bioresource Technol. 309, 123350.

Hatti-Kaul, R., Mamo, G., Mattiasson, B., 2016. Anaerobes in Biotechnology Preface. Anaerobes in Biotechnology 156, V–VI.

Heydari, A., Peyvandi, K., 2020. Study of biosurfactant effects on methane recovery from gas hydrate by CO2 replacement and depressurization. Fuel, 272.

Huang, Y.-B., Fu, Y., 2013. Hydrolysis of cellulose to glucose by solid acid catalysts. Green Chem. 15 (5), 1095–1111.

Huang, H.N., Chen, Y.G., Zheng, X., Su, Y.L., Wan, R., Yang, S.Y., 2016. Distribution of tetracycline resistance genes in anaerobic treatment of waste sludge: The role of pH in regulating tetracycline resistant bacteria and horizontal gene transfer. Bioresource Technol. 218, 1284–1289.

Hunter, S.M., Blanco, E., Borrion, A., 2021. Expanding the anaerobic digestion map: A review of intermediates in the digestion of food waste. Sci. Total Environ., 767.

Hunter, S.M., Blanco, E., Borrion, A., 2020. Expanding the anaerobic digestion map: A review of intermediates in the digestion of food waste. Sci. Total Environ., 144265.

Jarunglumlert, T., Prommuak, C., Putmai, N., Pavasant, P., 2018. Scaling-up bio-hydrogen production from food waste: Feasibilies and challenges. Int. J. Hydrogen Energ. 43 (2), 634–648.

Jiang, J., Gong, C., Wang, J., Tian, S., Zhang, Y., 2014. Effects of ultrasound pre-treatment on the amount of dissolved organic matter extracted from food waste. Bioresource Technol. 155, 266–271.

Jo, J.H., Jeon, C.O., Lee, D.S., Park, J.M., 2007. Process stability and microbial community structure in anaerobic hydrogen-producing microflora from food waste containing kimchi. J. Biotechnol. 131 (3), 300–308.

Khalid, A., Arshad, M., Anjum, M., Mahmood, T., Dawson, L., 2011. The anaerobic digestion of solid organic waste. Waste Manage. 31 (8), 1737–1744.

Kiran, E.U., Trzcinski, A.P., Ng, W.J., Liu, Y., 2014. Bioconversion of food waste to energy: A review Fuel 134, 389–399.

Kumar, G., Saratale, R.G., Kadier, A., Sivagurunathan, P., Zhen, G.Y., Kim, S.H., Saratale, G.D., 2017. A review on bio-electrochemical systems (BESs) for the syngas and value added biochemicals production. Chemosphere 177, 84–92.

Kumara, C.P., Rena, Meenakshi, A., Khapre, A.S., Kumar, S., Anshul, A., Singh, L., Kim, S.H., Lee, B.D., Kumar, R., 2019. Bio-Hythane production from organic fraction of municipal solid waste in single and two stage anaerobic digestion processes. Bioresource Technol., 294.

Lee, Y.W., Chung, J., 2010. Bioproduction of hydrogen from food waste by pilot-scale combined hydrogen/methane fermentation. Int. J. Hydrogen Energ. 35 (21), 11746–11755.

Li, C.L., Fang, H.H.P., 2007. Fermentative hydrogen production from wastewater and solid wastes by mixed cultures. Crit. Rev. Env. Sci. Tec. 37 (1), 1–39.

Li, S.B., Huang, L., Ke, C.Z., Pang, Z.W., Liu, L.M., 2020. Pathway dissection, regulation, engineering and application: lessons learned from biobutanol production by solventogenic clostridia. Biotechnol. Biofuels 13 (1).

Lim, J.W., Wang, J.Y., 2013. Enhanced hydrolysis and methane yield by applying microaeration pretreatment to the anaerobic co-digestion of brown water and food waste. Waste Manage. 33 (4), 813–819.

Liu, X., Gao, X.B., Wang, W., Zheng, L., Zhou, Y.J., Sun, Y.F., 2012. Pilot-scale anaerobic co-digestion of municipal biomass waste: Focusing on biogas production and GHG reduction. Renew. Energ. 44, 463–468.

Luo, G., Angelidaki, I., 2012. Integrated biogas upgrading and hydrogen utilization in an anaerobic reactor containing enriched hydrogenotrophic methanogenic culture. Biotechnol. Bioeng. 109 (11), 2729–2736.

Luo, L., Xu, S., Selvam, A., Wong, J.W., 2016. Assistant role of bioelectrode on methanogenic reactor under ammonia stress. Bioresource Technol. 217, 72–81.

Luo, L., Kaur, G., Wong, J.W., 2019. A mini-review on the metabolic pathways of food waste two-phase anaerobic digestion system. Waste Manage. Res. 37 (4), 333–346.

Luo, L., Wong, J.W., 2019. Enhanced food waste degradation in integrated two-phase anaerobic digestion: effect of leachate recirculation ratio. Bioresource Technol. 291, 121813.

Ma, H., Guo, Y., Qin, Y., Li, Y.-Y., 2018. Nutrient recovery technologies integrated with energy recovery by waste biomass anaerobic digestion. Bioresour. Technol. 269, 520–531.

Menon, A., Lyng, J.G., 2021. Circular bioeconomy solutions: driving anaerobic digestion of waste streams towards production of high value medium chain fatty acids. Rev. Environ. Sci. Biotechnol. 20 (1), 189–208.

Mohd-Zaki, Z., Bastidas-Oyanedel, J.R., Lu, Y., Hoelzle, R., Pratt, S., Slater, F.R., Batstone, D.J., 2016. Influence of pH regulation mode in glucose fermentation on product selection and process stability. Microorganisms4, 2.

Moscoviz, R., Toledo-Alarcon, J., Trably, E., Bernet, N., 2016. Electro-fermentation: how to drive fermentation using electrochemical systems. Trends Biotechnol. 34 (11), 856–865.

Nagarajan, D., Dong, C.-D., Chen, C.-Y., Lee, D.-J., Chang, J.-S., 2021. Biohydrogen production from microalgae—major bottlenecks and future research perspectives. Biotechnol. J. 2000124.

Negri, C., Ricci, M., Zilio, M., D'Imporzano, G., Qiao, W., Dong, R., Adani, F., 2020. Anaerobic digestion of food waste for bio-energy production in China and Southeast Asia: a review. Renew. Sustain. Energy Rev. 133, 110138.

Nikhil, G.N., Yeruva, D.K., Mohan, S.V., Swamy, Y.V., 2016. Assessing potential cathodes for resource recovery through wastewater treatment and salinity removal using non-buffered microbial electrochemical systems. Bioresour. Technol. 215, 247–253.

Nordahl, S.L., Devkota, J.P., Amirebrahimi, J., Smith, S.J., Breunig, H.M., Preble, C.V., Satchwell, A.J., Jin, L., Brown, N.J., Kirchstetter, T.W., 2020. Life-cycle greenhouse gas emissions and human health trade-offs of organic waste management strategies. Environ. Sci. Technol. 54 (15), 9200–9209.

Pagliaccia, P., Gallipoli, A., Gianico, A., Montecchio, D., Braguglia, C., 2016. Single stage anaerobic bioconversion of food waste in mono and co-digestion with olive husks: impact of thermal pretreatment on hydrogen and methane production. Int. J. Hydrog. Energy 41 (2), 905–915.

Pant, D., Arslan, D., Van Bogaert, G., Gallego, Y.A., De Wever, H., Diels, L., Vanbroekhoven, K., 2013. Integrated conversion of food waste diluted with sewage into volatile fatty acids through fermentation and electricity through a fuel cell. Environ. Technol. 34 (13–14), 1935–1945.

Parizeau, K., von Massow, M., Martin, R., 2015. Household-level dynamics of food waste production and related beliefs, attitudes, and behaviours in Guelph, Ontario. Waste Manage. 35, 207–217.

Park, J., Park, S., Kim, M., 2014. Anaerobic degradation of amino acids generated from the hydrolysis of sewage sludge. Environ. Technol. 35 (9), 1133–1139.

Rajaeifar, M.A., Hemayati, S.S., Tabatabaei, M., Aghbashlo, M., Mahmoudi, S.B., 2019. A review on beet sugar industry with a focus on implementation of waste-to-energy strategy for power supply. Renew. Sustain. Energy Rev. 103, 423–442.

Reddy, M.V., Mohan, S.V., 2012. Influence of aerobic and anoxic microenvironments on polyhydroxyalkanoates (PHA) production from food waste and acidogenic effluents using aerobic consortia. Bioresour. Technol. 103 (1), 313–321.

Reddy, M.V., Mohan, S.V., Chang, Y.C., 2018. Medium-chain fatty acids (MCFA) production through anaerobic fermentation using *Clostridium kluyveri*: effect of ethanol and acetate. Appl. Biochem. Biotechnol. 185 (3), 594–605.

Rena, B., Zacharia, K.M., Yadav, S., Machhirake, N.P., Kim, S.H., Lee, B.D., Jeong, H., Singh, L., Kumar, S., Kumar, R., 2020. Bio-hydrogen and bio-methane potential analysis for production of bio-hythane using various agricultural residues. Bioresour. Technol. 309.

Sainio, T., Kallioinen, M., Nakari, O., Mänttäri, M., 2013. Production and recovery of monosaccharides from lignocellulose hot water extracts in a pulp mill biorefinery. Bioresour. Technol. 135, 730–737.

Scharf, R., Wang, R., Maycock, J., Ho, P., Chen, S., Orfila, C., 2020. Valorisation of potato (Solanum tuberosum) peel waste: extraction of fibre, monosaccharides and uronic acids. Waste Biomass Valori 11 (5), 2123–2128.

Scherer, P.A., Vollmer, G.R., Fakhouri, T., Martensen, S., 2000. Development of a methanogenic process to degrade exhaustively the organic fraction of municipal "grey waste" under thermophilic and hyperthermophilic conditions. Water Sci. Technol. 41 (3), 83–91.

Schievano, A., Sciarria, T.P., Vanbroekhoven, K., De Wever, H., Puig, S., Andersen, S.J., Rabaey, K., Pant, D., 2016. Electro-fermentation—merging electrochemistry with fermentation in industrial applications. Trends Biotechnol. 34 (11), 866–878.

Schievano, A., Tenca, A., Scaglia, B., Merlino, G., Rizzi, A., Daffonchio, D., Oberti, R., Adani, F., 2012. Two-stage vs single-stage thermophilic anaerobic digestion: comparison of energy production and biodegradation efficiencies. Environ. Sci. Technol. 46 (15), 8502–8510.

Selvakumar, P., Sivashanmugam, P., 2017. Optimization of lipase production from organic solid waste by anaerobic digestion and its application in biodiesel production. Fuel Process. Technol. 165, 1–8.

Serafim, L.S., Lemos, P.C., Albuquerque, M.G., Reis, M.A., 2008. Strategies for PHA production by mixed cultures and renewable waste materials. Appl. Microbiol. Biotechnol. 81 (4), 615–628.

Shen, D., Yin, J., Yu, X., Wang, M., Long, Y., Shentu, J., Chen, T., 2017. Acidogenic fermentation characteristics of different types of protein-rich substrates in food waste to produce volatile fatty acids. Bioresour. Technol. 227, 125–132.

Shin, H.S., Youn, J.H., 2005. Conversion of food waste into hydrogen by thermophilic acidogenesis. Biodegradation 16 (1), 33–44.

Shin, S.G., Han, G., Lim, J., Lee, C., Hwang, S., 2010. A comprehensive microbial insight into two-stage anaerobic digestion of food waste-recycling wastewater. Water Res. 44 (17), 4838–4849.

Soares-Silva, I., 2016. The role of the gut microbiome on chronic kidney disease. Adv. Appl. Microbiol. 96, 65–94.

Srikanth, S., Mohan, S.V., 2012. Influence of terminal electron acceptor availability to the anodic oxidation on the electrogenic activity of microbial fuel cell (MFC). Bioresour. Technol. 123, 480–487.

Srirangan, K., Pyne, M.E., Chou, C.P., 2011. Biochemical and genetic engineering strategies to enhance hydrogen production in photosynthetic algae and cyanobacteria. Bioresour. Technol. 102 (18), 8589–8604.

Su, H.B., Cheng, J., Zhou, J.H., Song, W.L., Cen, K.F., 2009. Improving hydrogen production from cassava starch by combination of dark and photo fermentation. Int. J. Hydrog. Energy 34 (4), 1780–1786.

Tallou, A., Salcedo, F.P., Haouas, A., Jamali, M.Y., Atif, K., Aziz, F., Amir, S., 2020. Assessment of biogas and biofertilizer produced from anaerobic co-digestion of olive mill wastewater with municipal wastewater and cow dung. Environ. Technol. Innov. 20, 101152.

Tawfik, A., Salem, A., El-Qelish, M., 2011. Two stage anaerobic baffled reactors for bio-hydrogen production from municipal food waste. Bioresour. Technol. 102 (18), 8723–8726.

Vassilev, I., Gießelmann, G., Schwechheimer, S.K., Wittmann, C., Virdis, B., Krömer, J.O., 2018. Anodic electrofermentation: anaerobic production of L-Lysine by recombinant *Corynebacterium glutamicum*. Biotechnol. Bioeng. 115 (6), 1499–1508.

Venkata Mohan, S., Velvizhi, G., Vamshi Krishna, K., Lenin Babu, M., 2014. Microbial catalyzed electrochemical systems: a bio-factory with multi-facet applications. Bioresour. Technol. 165, 355–364.

Vijayakumar, J., Aravindan, R., Viruthagiri, T., 2008. Recent trends in the production, purification and application of lactic acid. Chem. Biochem. Eng. Q. 22 (2), 245–264.

Wang, J.L., Wan, W., 2008. Comparison of different pretreatment methods for enriching hydrogen-producing bacteria from digested sludge. Int. J. Hydrog. Energy 33 (12), 2934–2941.

Wang, Y.-H., Li, S.-L., Chen, I.-C., Tseng, I.-C., Cheng, S.-S., 2010. A study of the process control and hydrolytic characteristics in a thermophilic hydrogen fermentor fed with starch-rich kitchen waste by using molecular-biological methods and amylase assay. Int. J. Hydrog. Energy 35 (23), 13004–13012.

Wu, S.L., Wei, W., Sun, J., Xu, Q.X., Dai, X.H., Ni, B.J., 2020. Medium-chain fatty acids and long-chain alcohols production from waste activated sludge via two-stage anaerobic fermentation. Water Res. 186.

Xin, X., Ma, Y., Liu, Y., 2018. Electric energy production from food waste: microbial fuel cells versus anaerobic digestion. Bioresource technol 255, 281–287.

Xu, S., Luo, L., Selvam, A., Wong, J., 2017. Strategies to increase energy recovery from phase-separated anaerobic digestion of organic solid waste. In: Current Developments in Biotechnology and Bioengineering. Elsevier, Amsterdam, pp. 113–134.

Xu, S.Y., Lam, H.P., Karthikeyan, O.P., Wong, J.W., 2011. Optimization of food waste hydrolysis in leach bed coupled with methanogenic reactor: effect of pH and bulking agent. Bioresour. Technol. 102 (4), 3702–3708.

Xu, S.Y., Wang, C.Y., Sun, Y.Y., Luo, L.W., Wong, J.W.C., 2020. Assessing the stability of co-digesting sewage sludge with pig manure under different mixing ratios. Waste Manage. 114, 299–306.

Yadav, B., Pandey, A., Kumar, L.R., Tyagi, R.D., 2020. Bioconversion of waste (water)/residues to bioplastics—a circular bioeconomy approach. Bioresour. Technol. 298, 122584.

Yang, Y., Bao, W.Q., Xie, G.H., 2019. Estimate of restaurant food waste and its biogas production potential in China. J. Clean. Prod. 211, 309–320.

Yin, J., Yu, X., Wang, K., Shen, D., 2016. Acidogenic fermentation of the main substrates of food waste to produce volatile fatty acids. Int. J. Hydrog. Energy 41 (46), 21713–21720.

Zamanzadeh, M., Hagen, L.H., Svensson, K., Linjordet, R., Horn, S.J., 2016. Anaerobic digestion of food waste—effect of recirculation and temperature on performance and microbiology. Water Res. 96, 246–254.

Zeppilli, M., Paiano, P., Villano, M., Majone, M., 2019. Anodic vs cathodic potentiostatic control of a methane producing microbial electrolysis cell aimed at biogas upgrading. Biochem. Eng. J. 152, 107393.

Zhang, C., Kang, X., Wang, F., Tian, Y., Liu, T., Su, Y., Qian, T., Zhang, Y., 2020. Valorization of food waste for cost-effective reducing sugar recovery in a two-stage enzymatic hydrolysis platform. Energy, 208, 118379.

Zhang, C., Xiao, G., Peng, L., Su, H., Tan, T., 2013. The anaerobic co-digestion of food waste and cattle manure. Bioresour. Technol. 129, 170–176.

Zhang, L., Lee, Y.-W., Jahng, D., 2011. Anaerobic co-digestion of food waste and piggery wastewater: focusing on the role of trace elements. Bioresour. Technol. 102 (8), 5048–5059.

Zheng, X., Chen, Y., Wang, X., Wu, J., 2017. Using mixed sludge-derived short-chain fatty acids enhances power generation of microbial fuel cells. Energy Procedia 105, 1282–1288.

Zhou, M.M., Yan, B.H., Wong, J.W.C., Zhang, Y., 2018. Enhanced volatile fatty acids production from anaerobic fermentation of food waste: a mini-review focusing on acidogenic metabolic pathways. Bioresour. Technol. 248, 68–78.

Conversion of agricultural wastes to bioenergy and biochemicals *via* anaerobic digestion

Chenjun He[a,b], *Tao Luo*[c], *Hairong Yuan*[d], *Fei Shen*[a,b]

[a]Institute of Ecological and Environmental Sciences, Sichuan Agricultural University, Chengdu, Sichuan, P. R. China, [b]Rural Environment Protection Engineering & Technology Center of Sichuan Province, Sichuan Agricultural University, Chengdu, Sichuan, P. R. China, [c]Biogas Institute of Ministry of Agriculture (BIOMA), Chengdu, P. R. China, [d]Centre for Resource and Environmental Research, Beijing University of Chemical Technology, Beijing, P. R. China

3.1 Introduction

Agricultural wastes are generally defined based on the sector classification of society, and they are widely regarded as unwanted wastes, especially the organic wastes, derived from various agricultural activities. Thus, the agricultural wastes can be widely included as the agricultural residues from crop production, the livestock manures, and the wastes from agro-based industries or the input agro-chemical products (Dai et al., 2018; Yan et al., 2021). As well-known, the latter three types of wastes have the properties of pollution and mainly relate to environmental issues and food security; therefore, these wastes are currently well controlled from the levels of environmental consciousness to the environmental policies, even the legislations; the corresponding technologies for treating or valorizing these wastes are relatively more mature to satisfy the application scenarios. Although the pollutant intensity of agricultural residues is not stronger than that of the other wastes, disposal of them in some inappropriate ways still can trigger some environmental issues, for example, air pollution by directly incinerating them in field, and water eutrophication (Cuéllar and Webber, 2008; Sun et al., 2020; Yan et al., 2021). Besides, agricultural residues from the crop production mainly occupy the whole wastes from various agricultural activities, and are characterized by wide distribution, renewability, and availability (Dai et al., 2018; Xiao et al., 2014). Therefore, the agricultural residues,

Biomass, Biofuels, Biochemicals.
DOI: https://doi.org/10.1016/B978-0-323-90633-3.00007-9

especially the lignocellulosic residues that display a distinguishable resource property, are attracting more and more attentions on their valorization.

Facing the increasingly deteriorated environment and the increasingly depleted fossil energy sources, the agricultural residues have become one of the most favorable alternatives to the conventional energy sources, and efficient energy conversion technologies from agricultural residues are receiving much more attention (Du et al., 2019; Wang et al., 2015; Wang et al., 2018). As restricted by various factors, such as the technical maturity, the resource distribution, the input costs, and the supportive policies, anaerobic digestion for biogas (biomethane or biohydrogen) displays great potentials in application among various bioenergy conversation technologies using agricultural residues. In detail, anaerobic digestion provides the technical advantages over other bioenergy types due to its energy efficiency and environmental friendliness; it also offers numerous application advantages in the sense that they do not need tedious and expensive maintenance steps, and are adaptable to the climatic conditions of most countries (Bharathiraja et al., 2018; Paudel et al., 2017; Shang et al., 2019). Besides, anaerobic digestion of agricultural wastes generally releases methane-rich biogas, an excellent gaseous fuel, especially after being purified as the bionatural gas. Moreover, the contained nutrients and residual organic matters can be recycled via soil application of the digested residues (Bharathiraja et al., 2018).

Although anaerobic digestion of agricultural residues describes a beautiful scenario, and seems to be promising, it is still facing many challenges on supporting the whole chain to be efficient, profitable, and sustainable, when the large-scale application was attempted (Fig. 3.1).

The first big challenge comes from the feedstock supply because agricultural residues are commonly characterized by very low bulk density, resulting in the hardly acceptable costs in the logistics stream including the collection, transportation, and storage (Wang et al., 2016). The second challenge is greatly related to the pretreatment stream. Unlike other nonlignocellulosic biomass, agricultural residues contain relatively high lignin content (15%–25%, dry weight), and the lingo-carbohydrate complexes from strong bonds are resistant to microbial

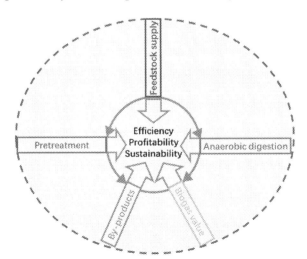

FIG. 3.1   The potential challenges for anaerobic digestion of agricultural residues in application scenario.

attack, resulting in difficult degradation, poor flowability, and accumulation of substrates (Ahmad et al., 2018; Paudel et al., 2017). Furthermore, the characteristic floating of agricultural residues accelerates the scum formation without interruption during the whole anaerobic digestion process. As a result, the floating layer of agricultural residues on the liquid surface cannot be digested completely, and the scum also prevents methane releasing (Pan et al., 2019; Singh et al., 2019). To substantially improve the utilization efficiency of agricultural residues for bioenergy via anaerobic digestion, a proper pretreatment for this type substrate is a strongly suggested step. Anaerobic digestion is the core stream of the conversion of agricultural residues into bioenergy. The poor digestion environment resulting in low biogas yield is the main problem of digestion stream, which can be regarded as another challenge to the efficiency and profitability. Besides, the end products, especially the byproduct (the digested residues) suffers the low quality, high pollution risk, and potential variability, which are practically challenging the sustainability of the whole chain of anaerobic digestion (Alengebawy et al., 2021; Wang et al., 2019).

In this context, the current chapter makes some technical summaries for the above-mentioned challenges according to the previous work in our group, and some meaningful investigations from other researchers. Hopefully, these summaries could offer some useful references for the in-depth investigations and possible applications in future.

3.2   Biomass densification to promote feedstock supply efficiency

As a Chinese saying that "One cannot make bricks without straw," a smooth feedstock supply chain is critically important to all kinds of sectors, which determines the feasibility of whole process. The same principle also goes for, even decides the bioenergy sector from agricultural residues by anaerobic digestion. Certainly, the feedstock supply chain for biogas from agricultural residues is mainly related to many aspects in practice, for example, the feedstock types and availability, the geographical features, climates, infrastructure constructions, policies, etc., which decide an important parameter of operational radius (Ghiani et al., 2014; Thiriet et al., 2020; Zhao et al., 2020). Currently, many studies on the feedstock supply of agricultural residues for biogas focus on the planning management strategies to optimize the possible operational radius. It is undeniable thatvt the measures from planning and management offer the maximal possibility to enlarge the operational radius, but the technical development to reduce the costs in feedstock supply chain, which also can be a potential path from other aspect for the promotion of operational radius or reduction of system inputs. Herein, we share our work from the technical aspect to intensify the efficiency of feedstock supply chain using the densification technique.

3.2.1   Technical feasibility of applying biomass densification for anaerobic digestion

The biomass densification technology was initially developed for the solid fuel production from lignocellulose, and the densification systems for biomass have been adapted from other highly efficient processing industries such as feed, food, and pharmacy including pellet mill, briquette press, screw extruder, tabletizer, and agglomerator. The very low bulk density mainly controls the supply costs of agricultural residues in logistics stream, especially, the streams of transportation and storage. It can be imaged that the biomass densification can significantly promote the bulk density of agricultural residues, and it will be possible to promote the efficiency of these streams (Li et al., 2014; Miranda et al., 2018; Mostafa et al., 2019). It is well known that the binding mechanism of densification undergoes the formation of solid bridges, which relates to the chemical reactions including the sintering solidification, the hardening of melted substances, and the crystallization of soluble materials (Miranda et al., 2018; Wang et al., 2016). Thus, the possibility of biochemical conversion of biomass-densified products should be prenatally investigated thoroughly before the densification can be employed in the anaerobic digestion.

3.2.1.1 Does anaerobic digestion of densified biomass require a recrushing step?

The extremely compact bonding in the densified substrates with high density and hardness may greatly reduce their sensitivity to the pretreatment; thereby they should be recrushed prior to for the following anaerobic digestion to increase their surface. However, the autoswelling of the densified biomass, including the pellets and the briquettes, will happen when they are in the high moisture conditions (Li et al., 2014; Tooyserkani et al., 2013). Thus, densified biomass, as the solid fuels, generally should consider their moisture before being packaged for goods. Taking advantage of this property, it will offer a possible to directly load the densified biomass for anaerobic digestion or the pretreatment step. Hereby, the water-involved pretreatment (hydrothermal pretreatment and diluted-NaOH pretreatment) was attempted to show the possibility. Their autoswelling behaviors of densified biomass in the water or the NaOH solution before the pretreatment are plotted in Fig. 3.2.

As two typical densified formations for agricultural residual, the briquettes achieved the maximum water or NaOH solution absorption (the completely-swollen state) within 24 h; and it only took about 1.0 h to achieve the completely swollen state for the pellets in water

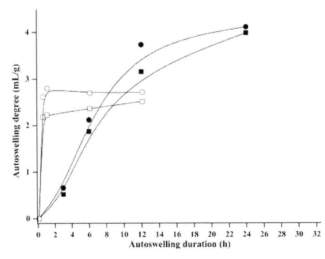

FIG. 3.2   The auto-swelling behaviors of densified biomass (○ Pellets in NaOH solution; □ Pellets in water; ● Briquettes in NaOH solution; ■ Briquettes in water; cited from: Li et al., 2014).

or the NaOH solution. Of course, the compression severity for briquettes was generally higher than that of pellets, and enhanced the solid bridge or the interparticle attraction forces, which will be responsible for this difference (Bajwa et al., 2018). As a result, the autoswelling nature of the typical densified agricultural residues can decompress themselves in the water, which will not require an extra step to recrush the densified biomass and save the energy input for size-reduction in application. Moreover, it will be suitable for the following water-involved pretreatment prior to anaerobic digestion.

3.2.1.2 Is biomass densification negative to the efficiency of anaerobic digestion?

According to the evaluation on the cumulative biogas production (Fig. 3.3A) at solid loading of 30 g total solids (TS)/kg, the pelleted cover stover biogas production was 349.9 mL/g volatile solids (VS), and significantly higher than that of the corresponding unpelleted corn stover (278.1 mL/g VS). Moreover, the briquetted corn stover also exhibited a slightly higher cumulative biogas production comparing with the unbriquetted corn stover. These results indicated that the biomass densification had no negative effects on the biogas production. Instead, approximated 25.8% and 0.6% improvements on the cumulative biogas production were observed in the group of pellets and briquettes, respectively. Even though the densification of briquettes was attempted on anaerobic digestion with higher solid loadings (Fig. 3.3B), the improvement was intensified as the organic loading was increased, especially when the higher organic loading of 80 g TS/kg was employed, an 11.7% improvement on biogas production could be observed. According to Fig. 3.3C, the improvement on biogas production may relate to the higher bulk density of briquettes, which may potentially retain more free water in the anaerobic digestion system comparing the unbriquetted biomass, which can facilitate mass transfer and improve the accessibility of microorganisms and enzymes to substrates.

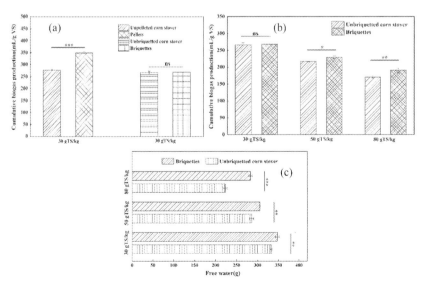

FIG. 3.3 The free water content of briquettes and unbriquetted corn stover at organic loading of 30, 50, and 80 g TS/kg. ***$P < 0.001$ ($n = 3$); **$P < 0.01$($n = 3$);**$P < 0.05$; (cited from Wang et al., 2016).

Based on these observations, biomass densification can be potentially employed for anaerobic digestion of lignocellulosic biomass for technically alleviating the low efficiency on logistics.

3.2.2 Economic feasibility of biomass densification for anaerobic digestion

In detail, the logistics stream for feedstock of straw biogas plants mainly includes mechanical collection, transportation, storage, rubbing (or smashing), and feeding for anaerobic digestion. Besides higher costs on transportation, the seasonal collection of agricultural residues requires a large amount of area for long-term storage with a great potential risk of fire; thereby the cost for storage is relatively higher in practice. Technically, biomass densification can offer lower transportation efficiency, greatly reduce the space requirements on storage, and display positive effects on digestion efficiency; however, the densification itself is the high energy–dependent process. Thus, the economic feasibility of biomass densification on anaerobic digestion should be evaluated for the potential application.

Accordingly, briquette technique was involved in the anaerobic digestion of agricultural residues for a large scale of biogas plant in Henan Province, China. The mass flow-rate of the feedstock rubbing biomass or briquetting biomass was about 120 t/d, and the effective volume of the digester was 3000 m^3. TS loading was fixed at about 8%, and mesophilic condition (35 °C) was employed for anaerobic digestion. The involved briquette technique was compared with the unbriquetted one based on two processing routes in Fig. 3.4.

Emergy analysis was employed to evaluate these two models, considering their emergy investment and emergy recovery (benefits) by establishing long-term sustainability and measuring contributions to the environmental stress. Results showed the methane yield in Model I was 66.74% higher than that of Model II. Two models required almost the same emergy investment input, while Model I obtained a greater quantity of net emergy (16.5% higher) in comparison with Model II. The net emergy yield ratio (biogas only) of Model I and Model II were 1.67 and 0.99, respectively, suggesting less market competitiveness for commercial operations with Model II. The investment in the storage phase dramatically was reduced by the involvement of densification technique in terms of straw yard costs due to some compulsive requirements on the construction area and the fire management. Consequently, the logistic costs of Model I could be reduced to approximately US$34,514 annually for the whole system operation in the investigated scale here. Based on these results, the involvement of biomass densification into the current biogas plant by anaerobic digestion of agricultural residues basically displays the economic feasibility and to be more profitable in application.

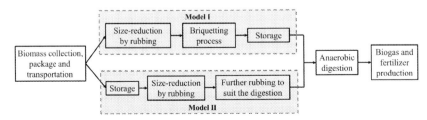

FIG. 3.4 The logistics streams with densification (Model I) and without densification (Model II).

3.3 Pretreatment of agricultural residues for anaerobic digestion

As is well-known, hydrolysis is the first step of anaerobic digestion, which is the process of converting insoluble organic matter such as cellulose, fats, and proteins into sugars, fatty acids, and other substances (Bharathiraja et al., 2018). However, the refractory structure and complex chemical composition of lignocellulosic greatly hindered the hydrolysis of microorganisms, which further affected the subsequent conversion by anaerobic microorganisms (Wang et al., 2018). Therefore, hydrolysis is considered as an important step in anaerobic digestion for methane production, and is assumed as the rate-limiting step especially for substrate-like lignocellulose (Phuttaro et al., 2019). As the typical lignocellulosic biomass, agricultural residues should be pretreated to promote their accessibility to microorganisms and enzymes thereby pretreatment has been realized to be a strongly suggested step prior to its degradation for anaerobic digestion (Ahmad et al., 2018; Mirmohamadsadeghia et al., 2021). As displayed in Fig. 3.5, the biomass pretreatment methods are mainly introduced from the sector of lignocellulosic ethanol and are generally classified into chemical, physical, biological pretreatment methods, or their combination (Ahmad et al., 2018; Karimi and Taherzadeh, 2016; Liu et al., 2020). Physical pretreatment mainly employs the mechanical functions to reduce particle size and crystallinity of fiber, thus increases the surface availability; this method is efficient substantially but highly energy-dependent (Theuretzbacher et al., 2015), thereby it is not suggested to use the physical pretreatment only. Chemical pretreatment is in virtue of the input chemicals, such as bases, oxidants, and acids, substantial efficiency also can be achieved, but suffers some problems such as higher cost of chemical input, carbon loss, inhibitors, and hard valorization of digested residues (Jung et al., 2013; Mehta et al., 2020; Schroyen et al., 2014). Biological pretreatment uses microorganisms or enzymes to decompose lignocellulose into simpler and more biodegradable molecules such as sugar, starch, and pectin. Thus, a shorter digestion time, improved total solid reduction, and biogas conversion rate could be observed after this pretreatment; therefore, can be regarded as a totally green way. However, a lower efficiency is the main embarrassing issues. Moreover, the microorganisms or preparation of enzymes themselves required carbon sources, which will increase the carbon loss in anaerobic digestion for the aimed bioenergy or overload the carbon input for

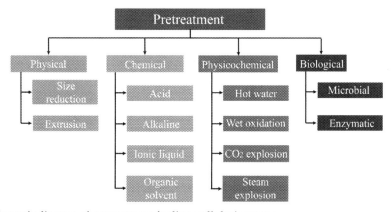

FIG. 3.5   Schematic diagram of pretreatment for lignocellulosic wastes.

the enzyme production (Abraham et al., 2019). Obviously, some relative suitable techniques should be carefully selected for the anaerobic digestion of agricultural residues, although the mentioned pretreatment techniques are basically investigated in the research work.

Hydrothermal pretreatment, involved in lignocellulosic feedstock and water only, has been widely accepted as a green technology without potential chemical consumption and potential pollution (Xiang et al., 2021). Typically, it can remove most of hemicellulose and part of lignin in biomass via degrading them into soluble fractions, and loosen the recalcitrant structure as well. Therefore, hydrothermal pretreatment has been widely applied for facilitating biofuels production, especially bioethanol, from lignocellulosic feedstocks (Mahmoodi et al., 2018; Rezania et al., 2020). In this context, hydrothermal method was directly adopted from producing bioethanol to anaerobic digestion for biogas, because considerable studies generally believed the achieved improvement on digestion performances was mainly attributed to the structure-breaking and the promoted accessibility of solid substrate after pretreatment (Shang et al., 2019; Wang et al., 2018; Zhou et al., 2017). However, most of these investigations were subjected to solid–liquid separation, and only the obtained insoluble fraction was generally utilized for digestion (Wang et al., 2018). This will be bound to reduce the biogas yield once the derived soluble fraction from hydrothermal pretreatment could not be digested together. Herein, we proposed the conception of whole slurry from hydrothermal pretreatment for anaerobic digestion.

3.3.1 Hydrothermal pretreatment of lignocellulosic biomass for anaerobic digestion

As displayed in Fig. 3.6, although the employed agricultural residues of rice straw can be well deconstructed by hydrothermal pretreatment, especially at higher pretreatment severities (Fig. 3.6A–C, and E), its anaerobic digestion in whole slurry is not improved greatly, and

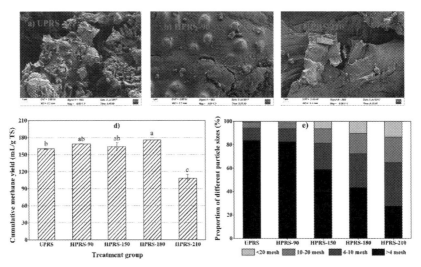

FIG. 3.6 Scanning electron microscope (SEM) morphology, size distribution, and anaerobic digestion performances of UPRS and HPRS. UPRS refers to the unpretrated rice straw; HPRS-# refers to the pretreated rice straw at the number-corresponding temperature (Cited from Wang et al., 2018).

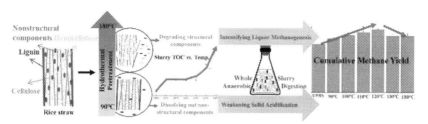

FIG. 3.7 The summary why low severity hydrothermal pretreatment on lignocellulose improves anaerobic digestion performances (cited from Xiang et al., 2021).

only maximum 3% of biogas yield is promoted. Instead, serious inhibitions happened at extra higher pretreatment temperatures, and 30% biogas yield reduction appears at 210 °C (Fig. 3.6D). Technically, hydrothermal pretreatment, especially at higher temperatures (>180 °C), is not recommended for biomethane production; however, lower temperatures for hydrothermal pretreatment are potentially suggested as the prior step for anaerobic digestion (Wang et al., 2018).

Based on the results above, hydrothermal pretreatment was beneficial to improve anaerobic digestion performance of rice straw in whole slurry. However, the relatively lower temperatures (100–120 °C) for the pretreatment displayed a better performance on biomethane production according to further selection. This finding offers an advantage of hydrothermal pretreatment in application for a lower energy input to maintain very high temperature. But, why can low severity hydrothermal pretreatment on lignocellulose improve anaerobic digestion performances? As summarized in Fig. 3.7 (Xiang et al., 2021), the hydrothermal pretreatment released the soluble fractions from rice straw into liquor, which shortened the acidification process by acidogens, and achieved more direct methanogenesis. However, this more direct methanogenesis weakened the acidification of solid fractions, which consequently weakened the performances of anaerobic digestion in whole slurry, especially the obtained slurry at higher pretreatment severities.

3.3.2 Net energy of hydrothermal pretreatment at lower temperature for anaerobic digestion

Although low temperature hydrothermal pretreatment could be considered as an effective method to improve biogas production, further investigations are required on its application scenario. According to results above, the pretreatment temperature of 100 °C was selected for further evaluation on the net energy production in application to check the feasibility for practical operation. Accordingly, the density of whole slurry was regarded as 1.0 kg/L for the calculation of net energy production, annual average atmospheric temperature was assumed as 20 °C, and heat capacity (equivalent to that of water) was presumed to be 4.18 J/(g•°C). The heat loss during preheating, inefficient heating, or waste-heat utilization was not included, thereby it was deemed as an ideal-case scenario. Besides, the retted agricultural residue of rice straw was employed as the control without any extra heat supply.

Table 3.1 shows the energy requirements for retting rice straw and hydrothermal pretreatment on rice straw; it could be indicated that very high temperature is not suitable for

TABLE 3.1   Energy balance of retting and hydrothermal pretreatment (cited from Luo et al., 2019).

Pretreatments	Energy required for (rice straw) RS heating (MJ/ton)	Energy output (MJ/ton)	Net energy (MJ/ton)
Retting	0	3349.5	3349.5
100 °C	−334.4	4114.1	3779.7
180 °C	−668.8	3450.0	−2481.2

commercial operation regardless of the biogas yield, net energy, and the convenience for continuous operation. Pretreatment using lower than 100 °C could achieve 12.8% higher net biogas production from retted rice straw. For the application scenario, a lower temperature hydrothermal pretreatment might be a feasible option under certain working conditions as follows: (1) mesophilic or thermophilic fermentation can be conducted with preheated feed-stock (whole slurry) to maintain digestion at a stable state; (2) agricultural residues can be heated with less pretreated water; (3) excess heat captured from other processes, such as solar energy or heat from combination heat and power system, could be used to heat pretreated water (>40 °C); and (4) some waste heat could be reused for reducing biogas demand.

As the surplus biogas failed to meet the whole slurry heating requirement, the net energy production appeared to be lower than that naturally retted rice straw in the control. Never-theless, this method could still be a feasible approach for mesophilic or thermophilic fermen-tation. Besides, better net energy production could be achieved in practical applications with an optimal design of the integral anaerobic digestion system or extra accessible heat source.

3.4   Enhancement techniques for anaerobic digestion

As a biological process, anaerobic digestion is to convert various organic substrates into biogas by microorganisms in the oxygen-free conditions (Wang et al., 2017). This process comprises of four stages, namely hydrolysis, acidogenesis, acetogenesis, and methanogene-sis, which are carried out by different set of microorganisms (i.e., acidogens, methanogens etc.). (Bharathiraja et al., 2018; He et al., 2017; Mirmohamadsadeghia et al., 2021). Anaerobic digestion of this type substrate also can be regulated by various operational parameters, such as temperature, pH, C/N ratio, alkalinity, organic loading rate (OLR), hydraulic retention time (HRT), and concentration of volatile fatty acids (Mao et al., 2015). Besides, the afore-mentioned pretreatment for agricultural residues, many investigations have been carried out to enhance the digestion performances from the following three aspects: (1) promotion of inoculation efficiency includes inoculum selection and optimization, leachate recirculation; (2) modulation of the operating parameters, such as temperature, pH and buffer capacity, OLR, HRT, mixing/agitation condition, C/N ratio, supplementation trace elements, section of accelerator, and removal of the potential inhibitors or toxicants (Mao et al., 2015; Wang et al., 2017); and (3) the codigestion with other substrates. For the application scenario, we considered to enhance the digestion performances from the supplementation of some cheaper additives or the wastes as accelerators, improving mixing efficiency, and design of the digesters.

3.4.1 Waste bottom ash of biomass power generation as an accelerator

During the biomass incineration for power generation, most of the ash fraction in biomass went into the bottom ash, and the mineral elements, especially the alkaline-earth metals, were mainly converted into oxides at high temperatures, which featured the bottom ash by alkaline pH (10.7 by average) (Vassilev et al., 2013). Besides, the inherent minerals in biomass will be greatly remained in the bottom ash after incineration, which made the bottom ash characterized by abundant minerals (Vassilev et al., 2010).

As a solid waste from biomass incineration for power generation, bottom ash was featured by alkaline pH of 10.5. At least 35 mineral elements can be detected in the bottom ash, in which the macroelements of microorganism growth, such as, P, S, K, Ca, and Fe, were determined as 3122 mg/kg, 13488 mg/kg, 36312 mg/kg, 42312 mg/kg, and 8447 mg/kg, respectively. In addition, some trace elements, such as Cu, Zn, Mn, Mo, Co, Ni, Sn, Se, Cr, W, and V, can also be captured. When bottom ash loading was promoted in batch anaerobic digestion of rice straw, the biomethane yield was improved by approximately 25% at the bottom ash loading of 0.37% comparing with 7.4% in the group of 0.12% NaOH (equivalent alkalinity with 0.37% bottom ash), suggesting that the containing mineral elements, as a nutrient provider, were responsible for the improvement (Fig. 3.8). However, excessive loadings were negative to biomethane production, resulting in the yield reduction of 3.6%–13.4%. As rice straw was soaked in the extracting solution of bottom ash at a prolonged duration of 2 days to 7 days, methane yield was correspondingly improved by 25.0% and 28.3%, respectively, which was 37.7% and 44.4% higher than the groups without bottom ash involvement. These results suggested that bottom ash provided base materials for alkali pretreatment on rice straw. As a summary, pretreating rice straw with appropriate bottom ash loading of 0.37% for 7 days was recommended in application, by which bottom ash substantially bifunctioned with alkali pretreatment and nutrient supplementation to improve anaerobic digestion performances (Xiao et al., 2019).

Obviously, the bottom ash residues from biomass power generation can be an accelerator to improve the anaerobic digestion of agricultural residues. Moreover, for the application, anaerobic digestion is facing another challenge on the efficient treatment or valorization of the digested residues. Considering the digestion residues commonly utilized as fertilizer and

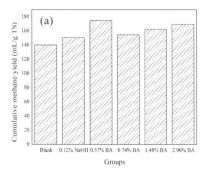
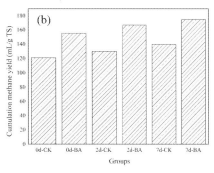

FIG. 3.8 **(A) Cumulative methane yield by adding with different loadings to check the ash residues as a nutrient provider; (B) Cumulative methane yield from the immersed rice straw for different duration to check function of a base provider for pretreatment (cited from Xiao et al., 2019).**

FIG. 3.9 Conceptual graph of a "win-win" strategy for valorizing the residual ash waste from biomass power generation and promoting methane production in anaerobic digestion (cited from Xiao et al., 2021).

residual ash mainly derived from combustion of agricultural residues, it can be easily understood that the larger amounts of nutrient elements in residual ash can greatly improve the fertility quality of the digestion residues, which can well recycle the minerals from soil to soil. As reported, residual ash supplementation in anaerobic digestion can promote the digestate dewaterability (Wei et al., 2020), which will benefit the solid/liquid separation of digestate, widening its valorization as soil fertilizer or facilitating the treatment of wastewater. Although some heavy metals were inevitably concentrated in residual ash because most organic fractions were removed by the incineration; their concentrations were at extremely low levels in the digestate, even at 5% residual ash loading. As compared with the standard of the Anaerobic Digested Fertilizer (NYT2596-2014) in China, the introduced heavy metals from 5% residual ash supplementation are all lower than the threshold values; moreover, some of these heavy metals are microelements and essential to crop growth.

It offers a new integration concept to bridge the agriculture waste and other organic wastes, such municipal solid wastes (MSW), food wastes, residual sludge, by which more biomethane can be yielded for bioenergy, and the power generation from biomass incineration can be operated cleaner. Besides, the fertilizer-grade digestion residues can be obtained for soil application, and the minerals can be recycled from soil to soil (Fig. 3.9). From the perspective of industrial symbiosis, the current work describes a "win-win" scenario between two bioenergy forms, in which the pollution risk and disposal cost of residual ash waste in biomass power generation can be greatly reduced, and the energy recovery from anaerobic digestion can be promoted significantly. Based on this "win-win" scenario, the sustainability and economy of these two bioenergy industries will be enhanced greatly.

3.4.2 Improving the mixing performances of anaerobic digestion of agricultural residues

Agricultural residues are characterized by low density, high water absorbability, and poor fluidity, resulting in the serious nonhomogenization, poor heat and mass transfer in anaerobic digestion. Besides, the characteristic floating of agricultural residues accelerates scum formation during the entire anaerobic digestion process. The floating layer of substrate on the liquid surface cannot be completely digested, and physically hinders methane release. The mixing therefore is more important for agricultural residue digestion than for other traditional substrates, such as food wastes, sludges. Thus, the optimized mixing parameters for the traditionally investigated substrates could not be completely applicable for the

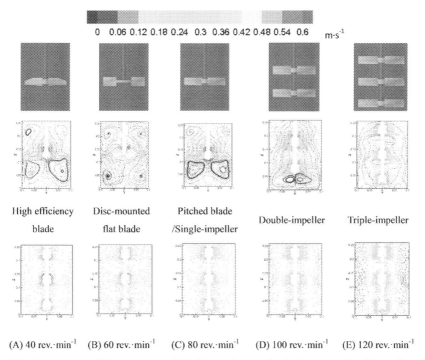

0   0.06 0.12 0.18 0.24 0.3 0.36 0.42 0.48 0.54 0.6   m·s⁻¹

High efficiency blade   Disc-mounted flat blade   Pitched blade /Single-impeller   Double-impeller   Triple-impeller

(A) 40 rev.·min⁻¹   (B) 60 rev.·min⁻¹   (C) 80 rev.·min⁻¹   (D) 100 rev.·min⁻¹   (E) 120 rev.·min⁻¹

FIG. 3.10   Velocity vectors at different types of blade\number, and agitating rates (cited from Shen et al., 2013).

agricultural residues. Hence, the mixing performances for anaerobic digestion of agricultural residues should be reconsidered. Currently, most investigations on the mixing efficiency are based on the lab-scale, and it is very hard to smoothly enlarge to the application scale because the amplification effects of biochemical reactors. Thus, a simple and valid method is necessarily required in application. Herein, the blade type and impeller number, two important parameters in the completely stirred tank reactor (CSTR) for anaerobic digestion of agricultural residues, are considered for the optimization. As for the optimization techniques, the computational fluid dynamics (CFD) is a well-established tool for numerical analysis, which offers many advantages over actual experiments. It can give a complete description for different flow patterns, and is widely used to examine different design scenarios in a short time with a relatively low cost (de Oliveira Marum et al., 2021; Shen et al., 2013; Wutz et al., 2020). In this context, the CFD simulation was employed to investigate the mixing performances and to determine the suitable stirring parameters for efficient biogas production from rice straw (Fig. 3.10). The results indicated that the mixing performances could be improved by the triple-impellers with a pitched blade, and a complete mixing was easily achieved at the stirring rate of 80 rev/min compared with 20–60 rev/min. However, the mixing could not be significantly improved when the stirring rate was further increased from 80 to 160 rev/min. Approximately 36% cumulative biogas yield could be promoted by the optimized parameters. The employed CFD technique could provide some useful guides for the design and operation of biogas plants using agricultural residues as substrates.

Besides, the apparent viscosities of lignocellulosic biomass slurries in the anaerobic digestion would be undoubtedly increased when the biomass TS was in higher level. Correspondingly, the challenges on mixing substrates with the high TS during the anaerobic digestion will be greatly strengthened, and the corresponding high energy consumption on the mixing should give a special consideration. Based on previous work, the decrease of particle sizes from 20 mesh to 80 mesh would reduce apparent viscosity of the employed corn stover and red-oak sawdust during the saccharification for ethanol production resulting in more than 50% improvement on enzymatic hydrolysis (Dasari and Eric Berson, 2007; Viamajala et al., 2009). Besides size reduction, the temperature during the digestion also closely related to the flow behaviors of lignocellulosic biomass slurries. When the manure was digested in temperature range of 30–60 °C, the typical rheological properties of consistency coefficient and apparent viscosity displayed the negative correlations with temperature (El-Mashad et al., 2005; Liu et al., 2019). In this context, improving the rheological properties of the employed lignocellulosic biomass can be a possible way to promote the anaerobic digestion performances at high TS, and to optimize the energy inputs for the mixing (Tian et al., 2014). To clarify agitation energy-reduction potential in anaerobic digestion at high TS by improving rheological properties, size reduction and temperature increase were investigated (Table 3.2). Results indicated the slurry of agricultural residues of corn stover exhibited a typical pseudoplastic flow at TS of 4.23%–7.32%. At low TS of 4.23%, rheological properties were not obviously affected by particle size and temperature. However, when TS in slurry was increased to 7.32%, there was 10.37% shear stress reduction by size-reduction from 20 to 80 mesh, and 11.73% reduction of shear stress by temperature increase from 25 to 55 °C. When the variations of power consumption by TS increase (PTS) was employed for evaluating energy-reduction potential, the maximum of 9.2% PTS reduction was achieved by size reduction from 20 to 80 mesh at 35 °C. Moreover, PTS reduction of 10.3%/10°C was achieved at 20 mesh compared with 9.0%/10 °C at 80 mesh.

TABLE 3.2　The agitation energy-reduction potential by size reduction and temperature increase (cited from Tian et al., 2014).

T/° C	Particle size	TS/%(w/w)	P/watt	P_{TS}/ watt	Particle size	T/° C	TS/%(w/w)	P/watt	P_{TS}/ watt
25	20 mesh	4.23	1.02	1.07	20 mesh	25	4.23	1.02	
		7.32	4.32				7.32	4.32	1.07
	80 mesh	4.23	1.01	0.98		35	4.23	0.94	
		7.32	4.05				7.32	3.86	0.94
35	20 mesh	4.23	0.94	0.94		55	4.23	0.88	
		7.32	3.86				7.32	3.16	0.74
	80 mesh	4.23	0.92	0.86	80 mesh	25	4.23	1.01	
		7.32	3.57				7.32	4.05	0.98
55	20 mesh	4.23	0.88	0.74		35	4.23	0.92	
		7.32	3.16				7.32	3.57	0.86
	80 mesh	4.23	0.92	0.72		55	4.23	0.92	
		7.32	3.14				7.32	3.14	0.72

3.4.3 Digester design to promote the anaerobic digestion of agricultural residues

The CSTR is the widely employed model reactor for anaerobic digestion of agricultural residues in the lab investigations or the large-scale digester (Kress et al., 2018). Generally, the substrates in the CSTR can be well mixed by top-mounted/side-mounted mechanical stirring equipment, by which the digestion duration (HRT) can be greatly shortened and the digestion efficiency can be promoted accordingly. Meanwhile, the mechanical agitator can overcome the crusting layer of agricultural residues during the anaerobic digestion, which can guarantee the continuous operation in application (Liu et al., 2019). Although the formation of crusting layer could be well solved by the agitation, the particle sizes and substrate loadings are generally required lower than 10 mm and 4%–6% (TS) in practice. Besides, the CSTR also suffers some practical challenges such as the relatively lower substrate loading rate, the relatively higher cost input, and operation fee from the agitation (Kress et al., 2018). Besides optimization on the mixing performances and size reduction of the loaded substrates, developing a novel digester is another way to improve the efficiency of anaerobic digestion of agricultural residues. Currently, the agricultural mechanization has been greatly promoted, by which the residues from the main crops can be directly chopped into pieces 30–50 mm in length during their harvesting process in field. The chopped agricultural residues were directly employed for anaerobic digestion, which would dramatically improve the economic benefits because of the decrease on the cost of size reduction (Luo et al., 2015).

Herein, a novel downward plug-flow anaerobic digester (DPAD) was designed for the chopped agricultural residues in Fig. 3.11. The working zone of the DPAD is separated into

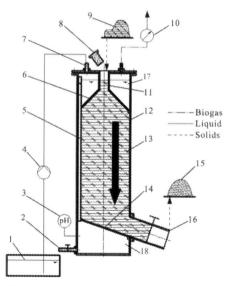

FIG. 3.11 **Schematic diagram of the DPAD.** (1) Buffer tank (temporary storage), (2) liquid outlet, (3) pH sensor, (4) recirculated pump (liquid recirculating and tap water adding), (5) connecting pipe (facilitating liquid permeation and biogas releasing), (6) upper sieve (three-phase separator), (7) liquid inlet, (8) gravity-driver (maintain the entire SSB below the liquid surface), (9) POM (new organic material), (10) gas meter, (11) feed pipe, (12) DPAD, (13) SSB (solid-state bed), (14) lower sieve (solid/liquid separation), (15) sludge (solid residue), (16) sludge outlet, (17) upper liquid zone, (18) lower liquid zone (cited from Luo et al., 2015).

three sections: a lower liquid zone, an upper liquid zone, and the solid-state bed (SSB) in the middle. The SSB is composed of two kinds of sieves. The upper sieve serves as a three-phase separator and keeps the SSB below the liquid surface. The lower sieve serves to separate solids and liquids in the digestate. The gravity driver, which is made of steel, is used to prevent the new particle organic material (POM) from flowing out through the feed pipe, and to maintain the entire SSB below the liquid surface throughout the digestion. The connecting pipe, at the highest point above the liquid surface, maintains a pressure difference between the lower liquid zone and the upper liquid zone to facilitate liquid permeation. Moreover, the connecting pipe also allows for the release of biogas produced in the lower part of the SSB and the lower liquid zone. The sludge outlet for solid residue is located at the lowest point of the lower sieve.

A lab-scale of 70 L DPAD was made to check it operation performances (Fig. 3.12). The lab batch digester of CSTR was employed to compare with DPAD. It could be found that the digestion efficiency of DPAD could be promoted by 21.1% based on the final cumulative methane yield (40 days). This could be attributed to the substrate, which can be kept below the liquid surface without the floating, which will be beneficial for the microbial community

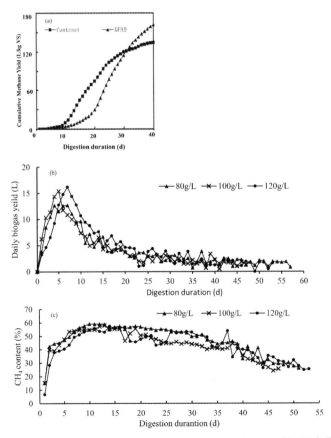

FIG. 3.12   The digestion performances of DPAD: (A) Cumulative methane yield, (B) daily biogas yield, (C) CH$_4$ content.

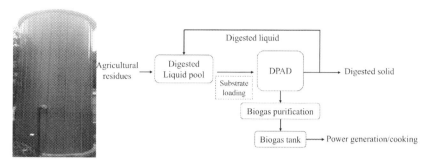

FIG. 3.13   Hundred cubic meters of the chopped straw biogas engineering process flow chart.

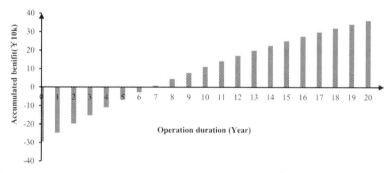

FIG. 3.14   The accumulative benefit versus operation duration of 100 m³ DPAD.

to adhere to the substrate, and facilitate the degradation; meanwhile, feeding and recirculation on the 20th day can improve the distribution of substrates and microorganisms and strengthen the buffering capacity. No significant differences can be observed from the daily biogas yield, as well as methane content, from the increased TS loading from 80 g/L to 120 g/L, suggesting the DPAD can work under the relatively higher substrate loading comparing with the traditional CSTR. These results indicated that the anaerobic digestion of agricultural residues using the newly designed reactor of DPAD could achieve superior digestion performances and work well at higher substrate loadings.

A 100 m³ DPAD was built for anaerobically digestion the rice straw (Fig. 3.13), the similar digestion performances could be achieved as displayed in the 70 L digester. As displayed in Fig. 3.14, the economic feasibility indicated the pay-off period for this project was 6.75 years, meaning this scale DPAD can start to be profitable from the year of 7. The pay-off period was shortened by 1.5–2.0 years in contrast with the same scale CSTR. This comparison suggested that the DPAD displayed a better application feasibility comparing with tradition digester of CSTR, and can be recommended for agricultural residues in application.

3.5   Utilization of the anaerobic digestion products

Anaerobic digestion of agricultural residues, as the other organic substrates, can yield the biogas and digested residues; they greatly relate to the profitability of whole process of anaerobic digestion. Moreover, the sustainability will be also affected if the product could not be profitably consumed. Especially, the digested residues have some severe health risks, odor, environmental pollution, and visual problems if they could not be utilized or treated in proper ways. Thereby, more work should focus on the utilizations of digestion products.

3.5.1   Utilization of biogas and upgradation

Biogas, as the main products from anaerobic digestion, contains 35%–75% methane (CH_4), 25%–65% carbon dioxide (CO_2), and some trace gas such as hydrogen sulfide (H_2S), hydrogen (H_2), and their contents are decided by the employed substrates for anaerobic digestion. Currently, the methane-rich biogas generally has two basic end uses on the production of heat or steam, and electricity generation, which are the widely employed ways in application scenario. Recently, biogas cleanup and upgrading have been the important tendencies to increase the heating value of biogas, and to meet the compatibility in various gas appliances (engines, boilers, fuel cells, vehicles, etc.).

Besides the traditional energy utilizations, biogas is cheap and a potential renewable carbon source for upgraded energy products or chemicals. Moreover, its usage addresses the imperative need for attaining sustainable development and eco-friendly production of the upgraded energy and added-value chemicals (Kapoor et al., 2020; Rafiee et al., 2021). The purified biogas with methane can be reformed to be synthesis gas ($H_2 + CO$) or H_2 production (Kapoor et al., 2020). Practically, any composition (poor, equimolar, or rich in CH_4) of the purified biogas itself is a suitable feed for the methane reforming. The so-called syngas from the methane reforming can be further synthesized to liquid energy carriers via the Fischer–Tropsch synthesis path (Ghorbani et al., 2021; Li et al., 2020). An alternative and more attractive route to biomethane valorization would be its utilization for ethylene (C_2H_4) production by means of the one-step oxidative coupling of methane reaction (Chukeaw et al., 2019). As for the CO_2 product from biogas upgrading process, some other carbonaceous products, such as CO, methanol, dimethyl ether, higher alcohols\hydrocarbons\formic species can be produced via catalytic-chemical, photochemical, and electrochemical technologies or their combinations (Li et al., 2020). However, the mentioned paths for the biogas upgrading still require more work from the aspects of lab-investigation, pilot, demonstration, and commercialization application.

3.5.2   Valorization of the digested residues by producing biochar

Although anaerobically digesting agricultural residues for biogas production, as an effective method, has been gradually popularized, the disposal of digested residues has become an urgent issue to be solved especially with the increased scale of anaerobic digestion. According to the practical experiences, the solid fraction in the digested residues of agricultural residues displayed more rapid sedimentation velocity comparing with other substrates, such as animal manures, food wastes. This property facilitated the solid/liquid separation. Thus, the separated liquid can be directly employed for integration of water and fertilizer via irrigation for crop production. Considering the potential advantage of biochar products on controlling the environmental pollution, the separated solid is attempted to prepare biochar to seek a marketization potential product. As 500 °C was selected for preparing biochar from the anaerobically digested rice straw, the derived biochar displayed a comparable adsorption capacity on the heavy metal pollutants, for example, Pb(II), Cd(II), and the adsorption to the heavy metal ions by this type biochar was dominated via forming silicates precipitation and complexing with carboxylate groups due to the relatively higher mineral content and oxygen-containing functional groups (Li et al., 2018). Besides, other types of pollutants, such as dyes, organic pollutants, and crop nutrients, such as N, P, also can be efficiently captured the biochar from the digested residues (Sun et al., 2013), displaying the broad application in environment management. This will offer a new path of digested residues from the anaerobic digestion of agricultural residues.

3.6 Conclusions and perspectives

This chapter mainly focuses on the whole chain of anaerobic digestion technique including the feedstock supply, the substrate pretreatment, the enhancements of the digestion efficiency, and the valorizations of digested residues.

In the stream of feedstock supply, technically, biomass densification is a feasible way to be employed in anaerobic digestion of agricultural residues to promote the whole logistics efficiency. Due to autoswelling nature, densified agricultural residues can be directly loaded for digestion without recrushing for size reduction. The densified agricultural residues digestion performances are not negatively affected by the densification, instead, some improvements on the methane yield could be observed, moreover the densified agricultural residues can well suspend in the digester comparing with the undensified biomass, which also reduce the formation of the crusting layer and its negative effects on digestion efficiency. Besides, the involvement of biomass densification also promotes the economic efficiency, which improved the sustainability of the whole process of anaerobic digestion using agricultural residues for bioenergy.

In the stream of pretreatment, hydrothermal pretreatment can be an optional technique for the anaerobic digestion. Especially, the relatively lower temperature for the hydrothermal pretreatment on the agricultural displays a better digestion performance and net energy output, thereby is recommended for the application. Moreover, the whole slurry digestion of hydrothermally pretreated substrates can reduce the carbon loss, which can maximally utilize carbon to the bioenergy of biomethane. In addition, the low-temperature hydrothermal pretreatment with only water involvement can be regarded as a green method, which will not introduce the undesirable chemicals, which will be safe to the following soil application of digested slurry.

In the stream of digestion stream, the biomass ash could be a cheaper accelerator to promote the digestion performances due to the bifunction of supplementing minerals for the anaerobes and base provider for pretreatment. It will be a "win-win" strategy to employ the bottom ash into anaerobic digestion, by which the potential pollution of bottom ash residues could be solved, and the digestion performances can be promoted as well. The CFD techniques will be a convenient and reliable way to optimization of the mixing performances in the application of anaerobic digestion of agricultural residues. The improvement on the rheological properties by modulating substrate size and digestion temperature will benefit the energy input for the mixing in the CSTR. As a newly developed digester, DPAD displays the advantages on the digestion performance and the introduced substrate loading, as well as the relatively lower operation fee, thereby can be regarded as a dedicated reactor for the lignocellulosic biomass.

For the stream of byproduct valorization, the functional biochar preparation from the separated solid from digested residues can be a feasible way to broaden its valorization in the pollution control, which can be organized as the strategy of waste control by waste.

Thus, although the anaerobic digestion of agricultural residues for bioenergy is a technically feasible and widely recommended in application, it is still facing many difficulties that are lowering the whole chain efficiency and frustrate the investor's enthusiasms. For these facing difficulties, the solution way will be a system engineering, and the improvements on a single technology or in a single stream could not be enough efficient to support the whole chain, although they have some positive functions in the whole sector and can be partially consulted for application. Therefore, how to integrate the advantages of current various technologies as a model process will be strongly required in the future work. More investigations should be focused on the systematic analysis and optimization on an integrated process in considering the technical feasibility efficiency, economic profitability, and environmental sustainability. The life cycle assessment is the recommended technique for this issue.

Besides the anaerobic digestion itself for the valorization of agricultural residues for bioenergy, it is deserved to perform more investigations on the other bioenergy products, such as hydrogen from direction digestion, biodiesel from the digested residues, and biochemicals (fatty acids). In addition, a bigger scale integration anaerobic digestion with other bioenergy techniques, for example, bioethanol, pyrolysis, carbonization, combustion, which will probably achieve "win-win", "win-win-win" and even "multiwin" strategies.

References

Abraham, A., Mathew, A.K., Park, H., Choi, O., Sindhu, R., Parameswaran, B., Pandey, A., Park, J., Sang, B., 2019. Pretreatment strategies for enhanced biogas production from lignocellulosic biomass. Bioresour. Technol. 301, 122725.

Ahmad, F., Silva, E.L., Varesche, M.B.A., 2018. Hydrothermal processing of biomass for anaerobic digestion—a review. Renew. Sustain. Energy Rev. 98, 108–124.

Alengebawy, A., Jin, K., Ran, Y., Peng, J., Zhang, X., Ai, P., 2021. Advanced pre-treatment of stripped biogas slurry by polyaluminum chloride coagulation and biochar adsorption coupled with ceramic membrane filtration. Chemosphere 267, 129197.

Bajwa, D.S., Peterson, T., Sharma, N., Shojaeiarani, J., Bajwa, S.G., 2018. A review of densified solid biomass for energy production. Renew. Sustain. Energy Rev. 96, 296–305.

Bharathiraja, B., Sudharsana, T., Jayamuthunagai, J., Praveenkumar, R., Chozhavendhan, S., Iyyappan, J., 2018. Biogas production–a review on composition, fuel properties, feed stock and principles of anaerobic digestion. Renew. Sustain. Energy Rev. 90, 570–582.

Chukeaw, T., Sringam, S., Chareonpanich, M., Seubsai, A., 2019. Screening of single and binary catalysts for oxidative coupling of methane to value-added chemicals. Mol. Catal. 470, 40–47.

Cuéllar, A.D., Webber, M.E., 2008. Cow power: the energy and emissions benefits of converting manure to biogas. Environ. Res. Lett. 3, 034002.

Dai, Y., Sun, Q., Wang, W., Lu, L., Liu, M., Li, J., Yang, S., Sun, Y., Zhang, K., Xu, J., Zheng, W., Hu, Z., Yang, Y., Gao, Y., Chen, Y., Zhang, X., Gao, F., Zhang, Y., 2018. Utilizations of agricultural waste as adsorbent for the removal of contaminants: a review. Chemosphere 211, 235–253.

Dasari, R.K., Eric Berson, R., 2007. The effect of particle size on hydrolysis reaction rates and rheological properties in cellulosic slurries. Appl. Biochem. Biotechnol. 137–140, 289–299.

de Oliveira Marum, V.J., Reis, L.B., Maffei, F.S., Ranjbarzadeh, S., Korkischko, I., Meneghini, J.R., 2021. Performance analysis of a water ejector using computational fluid dynamics (CFD) simulations and mathematical modeling. Energy 220, 119779.

Du, J., Qian, Y., Xi, Y., Lü, X., 2019. Hydrothermal and alkaline thermal pretreatment at mild temperature in solid state for physicochemical properties and biogas production from anaerobic digestion of rice straw. Renew. Energy. 139, 261–267.

El-Mashad, H.M., Van Loon, W.K.P., Zeeman, G., Bot, G.P.A., 2005. Rheological properties of dairy cattle manure. Bioresour. Technol. 96, 531–535.

Ghiani, G., Laganà, D., Manni, E., Musmanno, R., Vigo, D., 2014. Operations research in solid waste management: a survey of strategic and tactical issues. Comput. Oper. Res. 44, 22–32.

Ghorbani, B., Ebrahimi, A., Rooholamini, S., Ziabasharhagh, M., 2021. Integrated Fischer-Tropsch synthesis process with hydrogen liquefaction cycle. J. Clean. Prod. 283, 124592.

He, L., Huang, H., Zhang, Z., Lei, Z., Lin, B., 2017. Energy recovery from rice straw through hydrothermal pretreatment and subsequent biomethane production. Energy Fuels 31, 10850–10857.

Jung, Y.H., Kim, I.J., Kim, H.K., Kim, K.H., 2013. Dilute acid pretreatment of lignocellulose for whole slurry ethanol fermentation. Bioresour. Technol. 132, 109–114.

Kapoor, R., Ghosh, P., Tyagi, B., Vijay, V.K., Vijay, V., Thakur, I.S., Kamyab, H., Nguyen, D.D., Kumar, A., 2020. Advances in biogas valorization and utilization systems: a comprehensive review. J. Clean. Prod. 273, 123052.

Karimi, K., Taherzadeh, M.J., 2016. A critical review of analytical methods in pretreatment of lignocelluloses: composition, imaging, and crystallinity. Bioresour. Technol. 200, 1008–1018.

Kress, P., Nägele, H.J., Oechsner, H., Ruile, S., 2018. Effect of agitation time on nutrient distribution in full-scale CSTR biogas digesters. Bioresour. Technol. 247, 1–6.

Li, J., Shen, F., Yang, G., Zhang, Y., Deng, S., Zhang, J., Zeng, Y., Luo, T., Mei, Z., 2018. Valorizing rice straw and its anaerobically digested residues for biochar to remove Pb(II) from aqueous solution. Int. J. Polym. Sci. 2018, 1–11.

Li, Y., Li, X., Shen, F., Wang, Z., Yang, G., Lin, L., Zhang, Y., Zeng, Y., Deng, S., 2014. Responses of biomass briquetting and pelleting to water-involved pretreatments and subsequent enzymatic hydrolysis. Bioresour. Technol. 151, 54–62.

Li, Z., Lin, Q., Li, M., Cao, J., Liu, F., Pan, H., Wang, Z., Kawi, S., 2020. Recent advances in process and catalyst for CO_2 reforming of methane. Renew. Sustain. Energy Rev. 134, 110312.

Liu, W., Wu, R., Hu, Y., Ren, Q., Hou, Q., Ni, Y., 2020. Improving enzymatic hydrolysis of mechanically refined poplar branches with assistance of hydrothermal and Fenton pretreatment. Bioresour. Technol. 316, 123920.

Liu, Y., Chen, J., Song, J., Hai, Z., Lu, X., Ji, X., Wang, C., 2019. Adjusting the rheological properties of corn-straw slurry to reduce the agitation power consumption in anaerobic digestion. Bioresour. Technol. 272, 360–369.

Luo, T., Huang, H., Mei, Z., Shen, F., Ge, Y., Hu, G., Meng, X., 2019. Hydrothermal pretreatment of rice straw at relatively lower temperature to improve biogas production via anaerobic digestion. Chin. Chem. Lett. 30, 1219–1223.

Luo, T., Long, Y., Li, J., Meng, X., Mei, Z., Long, E., Dai, B., 2015. Performance of a novel downward plug-flow anaerobic digester for methane production from chopped straw. BioResources 10, 943–955.

Mahmoodi, P., Karimi, K., Taherzadeh, M.J., 2018. Hydrothermal processing as pretreatment for efficient production of ethanol and biogas from municipal solid waste. Bioresour. Technol. 261, 166–175.

Mao, C., Feng, Y., Wang, X., Ren, G., 2015. Review on research achievements of biogas from anaerobic digestion. Renew. Sustain. Energy Rev. 45, 540–555.

Mehta, S., Jha, S., Liang, H., 2020. Lignocellulose materials for supercapacitor and battery electrodes: a review. Renew. Sustain. Energy Rev. 134, 110345.

Miranda, M.T., Sepúlveda, F.J., Arranz, J.I., Montero, I., Rojas, C.V., 2018. Analysis of pelletizing from corn cob waste. J. Environ. Manage. 228, 303–311.

Mirmohamadsadeghia, S., Karimiab, K., Azarbaijani, R., Yeganeh, L.P., Angelidaki, I., Nizami, A.S., Bhatf, R., Dashora, K., KumarVijay, V., Aghbashlo, M., Kumar Gupta, V., Tabatabaei, M., 2021. Pretreatment of lignocelluloses for enhanced biogas production: a review on influencing mechanisms and the importance of microbial diversity. Renew. Sustain. Energy Rev. 135, 110173.

Mostafa, M.E., Hu, S., Wang, Y., Su, S., Hu, X., Elsayed, S.A., 2019. The significance of pelletization operating conditions : an analysis of physical and mechanical characteristics as well as energy consumption of biomass pellets. Renew. Sustain. Energy Rev. 105, 332–348.

Pan, S., Wen, C., Liu, Q., Chi, Y., Mi, H., Li, Z., Du, L., Huang, R., Wei, Y., 2019. A novel hydraulic biogas digester controlling the scum formation in batch and semi-continuous tests using banana stems. Bioresour. Technol. 286, 121372.

Paudel, S.R., Banjara, S.P., Choi, O.K., Park, K.Y., Kim, Y.M., Lee, J.W., 2017. Pretreatment of agricultural biomass for anaerobic digestion: current state and challenges. Bioresour. Technol. 245, 1194–1205.

Phuttaro, C., Sawatdeenarunat, C., Surendra, K.C., Boonsawang, P., Chaiprapat, S., Khanal, S.K., 2019. Anaerobic digestion of hydrothermally-pretreated lignocellulosic biomass: influence of pretreatment temperatures, inhibitors and soluble organics on methane yield. Bioresour. Technol. 284, 128–138.

Rafiee, A., Khalilpour, K.R., Prest, J., Skryabin, I., 2021. Biogas as an energy vector. Biomass Bioenergy 144, 105935.

Rezania, S., Oryani, B., Cho, J., Talaiekhozani, A., Sabbagh, F., Hashemi, B., Rupani, F.P., Mohammadi, A.A., 2020. Different pretreatment technologies of lignocellulosic biomass for bioethanol production: an overview. Energy 199, 117457.

Schroyen, M., Vervaeren, H., Hulle, S., Raes, K., 2014. Impact of enzymatic pretreatment on corn stover degradation and biogas production. Bioresour. Technol. 173, 59–66.

Shang, G., Zhang, C., Wang, F., Qiu, L., Guo, X., Xu, F., 2019. Liquid hot water pretreatment to enhance the anaerobic digestion of wheat straw—effects of temperature and retention time. Environ. Sci. Pollut. Res. 26 (28), 29424–29434.

Shen, F., Tian, L., Yuan, H., Pang, Y., Chen, S., Zou, D., Zhu, B., Liu, Y., Li, X., 2013. Improving the mixing performances of rice straw anaerobic digestion for higher biogas production by computational fluid dynamics (CFD) simulation. Appl. Biochem. Biotechnol. 171, 626–642.

Singh, B., Szamosi, Z., Siménfalvi, Z., 2019. State of the art on mixing in an anaerobic digester: a review. Renew. Energy 141, 922–936.

Sun, L., Wan, S., Luo, W., 2013. Biochars prepared from anaerobic digestion residue, palm bark, and eucalyptus for adsorption of cationic methylene blue dye: characterization, equilibrium, and kinetic studies. Bioresour. Technol. 140, 406–413.

Sun, M., Xu, X., Wang, C., Bai, Y., Fu, C., Zhang, L., Fu, R., Wang, Y., 2020. Environmental burdens of the comprehensive utilization of straw : wheat straw utilization from a life-cycle perspective. J. Clean. Prod. 259, 120702.

Theuretzbacher, F., Lizasoain, J., Lefever, C., Saylor, M.K., Enguidanos, R., Weran, N., Gronauer, A., Bauer, A., 2015. Steam explosion pretreatment of wheat straw to improve methane yields: investigation of the degradation kinetics of structural compounds during anaerobic digestion. Bioresour. Technol. 179, 299–305.

Thiriet, P., Bioteau, T., Tremier, A., 2020. Optimization method to construct micro-anaerobic digesters networks for decentralized biowaste treatment in urban and peri-urban areas. J. Clean. Prod. 243, 118478.

Tian, L., Shen, F., Yuan, H., Zou, D., Liu, Y., Zhu, B., Li, X., 2014. Reducing agitation energy-consumption by improving rheological properties of corn stover substrate in anaerobic digestion. Bioresour. Technol. 168, 86–91.

Tooyserkani, Z., Kumar, L., Sokhansanj, S., Saddler, J., Bi, X.T., Lim, C.J., Lau, A., Melin, S., 2013. SO₂-catalyzed steam pretreatment enhances the strength and stability of softwood pellets. Bioresour. Technol. 130, 59–68.

Vassilev, S.V., Baxter, D., Andersen, L.K., Vassileva, C.G., 2010. An overview of the chemical composition of biomass. Fuel 89, 913–933.

Vassilev, S.V., Baxter, D., Andersen, L.K., Vassileva, C.G., 2013. An overview of the composition and application of biomass ash. Part 1. Phase-mineral and chemical composition and classification. Fuel 105, 40–76.

Viamajala, S., McMillan, J.D., Schell, D.J., Elander, R.T., 2009. Rheology of corn stover slurries at high solids concentrations—effects of saccharification and particle size. Bioresour. Technol. 100, 925–934.

Wang, C., Wang, J., Wu, W., Qian, J., Song, S., Yue, Z., 2019. Feasibility of activated carbon derived from anaerobic digester residues for supercapacitors. J. Power Sour. 412, 683–688.

Wang, D., Ai, J., Shen, F., Zhang, Y., Deng, S., Zhang, J., Zeng, Y., Song, C., 2017. Improving anaerobic digestion of easy-acidification substrates by promoting buffering capacity using biochar derived from vermicompost. Bioresour. Technol. 227, 286–296.

Wang, D., Huang, H., Shen, F., Yang, G., Zhang, Y., Deng, S., Zhang, J., Zeng, Y., Hu, Y., 2016. Effects of biomass densification on anaerobic digestion for biogas production. RSC Adv. 6, 91748–91755.

Wang, D., Shen, F., Yang, G., Zhang, Y., Deng, S., Zhang, J., Zeng, Y., Luo, T., Mei, Z., 2018. Can hydrothermal pretreatment improve anaerobic digestion for biogas from lignocellulosic biomass? Bioresour. Technol. 249, 117–124.

Wang, Q., Hu, J., Shen, F., Mei, Z., Yang, G., Zhang, Y., Hu, Y., 2015. Pretreating wheat straw by the concentrated phosphoric acid plus hydrogen peroxide (PHP): investigations on pretreatment conditions and structure changes. Bioresour. Technol. 199, 245–257.

Wei, W., Liu, X., Wu, L., Wang, D., Bao, T., Ni, B.J., 2020. Sludge incineration bottom ash enhances anaerobic digestion of primary sludge toward highly efficient sludge anaerobic codigestion. ACS Sustain. Chem. Eng. 8, 3005–3012.

Wutz, J., Waterkotte, B., Heitmann, K., Wucherpfennig, T., 2020. Computational fluid dynamics (CFD) as a tool for industrial UF/DF tank optimization. Biochem. Eng. J. 160, 107617.

Xiang, C., Tian, D., Hu, J., Huang, M., Shen, F., Zhang, Y., Yang, G., Zeng, Y., Deng, S., 2021. Why can hydrothermally pretreating lignocellulose in low severities improve anaerobic digestion performances? Sci. Total Environ. 752, 141929.

Xiao, Q., Chen, W., Tian, D., Shen, F., Hu, J., Long, L., Zeng, Y., Yang, G., Deng, S., 2019. Integrating the bottom ash residue from biomass power generation into anaerobic digestion to improve biogas production from lignocellulosic biomass. Energy Fuels 34 (2), 1101–1110.

Xiao, Q., Hu, J., Huang, M., Shen, F., Tian, D., Zeng, Y., Jang, M.-.K., 2021. Valorizing the waste bottom ash for improving anaerobic digestion performances towards a "win-win" strategy between biomass power generation and biomethane production. J. Clean. Prod. 295, 126508.

Xiao, X., Bian, J., Li, M., Xu, H., Xiao, B., Sun, R., 2014. Enhanced enzymatic hydrolysis of bamboo (*Dendrocalamus giganteus* Munro) culm by hydrothermal pretreatment. Bioresour. Technol. 159, 41–47.

Yan, B., Yan, J., Li, Y., Qin, Y., Yang, L., 2021. Spatial distribution of biogas potential, utilization ratio and development potential of biogas from agricultural waste in China. J. Clean. Prod. 292, 126077.

Zhao, X., Joseph, B., Kuhn, J., Ozcan, S., 2020. Biogas reforming to syngas: a review. iScience 23, 101082.

Zhou, X., Li, Q., Zhang, Y., Gu, Y., 2017. Effect of hydrothermal pretreatment on Miscanthus anaerobic digestion. Bioresour. Technol. 224, 721–726.

Conversion of manure to bioenergy and biochemicals via anaerobic digestion

Qigui Niu*, Liuying Song*, Jingyi Li*

School of Environmental Science and Engineering, China–America CRC for Environment &
Health of Shandong Province, Shandong University, Qingdao, Shandong, China
*These authors contributed equal to this chapter.

4.1 Introduction

The livestock and poultry breeding industry has become an important pillar industry all over the world's rural economy. As the number of global livestock and poultry increased, so did the total amount of manure (Quaik et al., 2020). Regionally, the annual production of manure in the 27 European member states was estimated to be about 1.4 billion tons (Gurmessa et al., 2020). However, in the United States, over 1 billion tons of manure are produced annually. Similarly, China has generated about 3.97×10^9 tons of animal manure each year in recent decades (Peidong et al., 2009). Moreover, the increased intensive poultry breeding industries have produced a large number of solid and liquid waste, especially chicken manure (CM) that generates resource with a high total nitrogen content as high as 4.03×10^8 tons every year that is of great significance to energy production (Zhang et al., 2014).

Land use was the main traditional way to deal with this huge manure. However, the land applications of livestock and poultry manures were also the main way of heavy metal pollution in soils, especially the abundant Zn and Cu, principally contributed by livestock and poultry manures (Liu et al., 2020). Moreover, manure management is an important issue due to the pathogenic microbes, such as the livestock *Cryptosporidium* spp. loads to land on a global scale (Vermeulen et al., 2017). Thus, to reduce the risk of the emergence and spread of antibiotic-resistant microorganisms resulting from antibiotic residues, it is important to properly dispose of animal waste (cow, swine, sheep, and poultry) before land applications, with the purpose of minimization the residues of antibiotics and other veterinary drugs.

Anaerobic digestion (AD) technology is commonly applied in the disposal of solid wastes and wastewater from municipal, agricultural, and industrial sources with the sustainable bioenergy production (Cheng et al., 2021). It is an effective technology for waste management, pollution mitigation, and greenhouse gas emissions reduction with renewable biogas

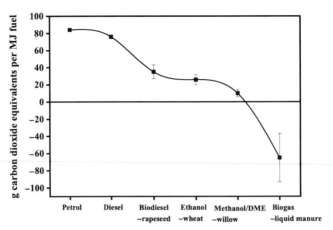

FIG. 4.1   Comparison of carbon emission reduction through anaerobic digestion.

utilization (Li et al., 2017a). With the post purification, the biogas with high biomethane (>90%) can be applied for combined heat and power plants with heat and electricity serving the society by state grid. In the context of carbon neutrality, the benefits of AD for manure treatment become important (Borjesson & Mattiasson, 2008) (Fig. 4.1). Besides the energy production, Tian et al. (2016) also revealed that antibiotic resistance genes and their horizontal and vertical transfer were significantly decreased after high-temperature digestion while using one-step temperature increase from mesophilic to thermophilic AD (Tian et al., 2016).

Besides the energy generation, nowadays, more values from AD become a hot topic in the current sustainable development. More and more researchers have focused on the biochemicals via AD. Conversion of manure not only bioenergy but also biochemicals via AD is a new direction for engineering application. This chapter mainly focused on the AD of manure with biogas generation and valuable material generation.

4.2   Biogas and biomethane production from manures through AD

During AD process (hydrolysis→acidogenesis→acetogenesis→methanogenesis), methanogens as the last electron-acceptors play a key role in the methanogenesis step converting intermediates (acetate, H_2/CO_2, methanol, etc.) from the fermentation process into methane. Consequently, the methanogenic pathway can be divided into four routes including hydrogenotrophic methanogenetic route ($4H_2 + CO_2 \rightarrow CH_4 + 2H_2O$), acetoclastic methanogenetic route ($CH_3COOH \rightarrow CH_4 + CO_2$), methylotrophic methanogenetic route ($CH_3OH + H_2 \rightarrow CH_4 + H_2O$), and electron-dependent methanogenetic route ($8H^+ + 8e^- + CO_2 \rightarrow CH_4 + 2H_2O$) (Lü et al., 2020).

To date, over 110 species of methanogens are reported all over the world's different habitats classified into six orders. The dominant methanogens commonly reported in the engineering application are acetoclastic (*Methanosarcina, Methanothrix, Methanosarcina*), hydrogenotrophic (*Methanococcus, Methanobacterium, Methanothermobacter, Methanocaldococcus, Methanomicrobium*),

TABLE 4.1 Kinetic constant of acidogenesis and methanogenesis (Niu & Li, 2016).

Kinetic constant	Hydrolysis and acidogenesis			Methanogenesis			
	Cellulose	Starch	Glucose	Acetic acid	Propionic acid	Butyric acid	Mixed acid
μ_{max}/d	0.66	12.9	15.6	0.26	0.28	0.18	0.414
v_{max}/d	1.11	37.5	66.2	5.03	6.46	7.96	17.1
SRT_{min}/d	1.52	0.08	0.06	3.85	5.38	5.58	2.42

μ_{max} is the maximum specific appreciation rate, v_{max} is the maximum substrate utilization rate, SRT_{min} is to maintain the minimum sludge residence time of the bacteria.

and methylotrophic methanogens (*Methanococcoides* and *Methanobacterium*) (Liu & Whitman, 2008). The kinetic constant of acidogenesis and methanogenesis is showed in Table 4.1 while treating with the typical substrate of cellulose, starch, and glucose, respectively (Niu & Li, 2016). The mean appreciation time of acetoclastic *Methanosaeta* spp. needs 1–2 days, while hydrogenotrophic *Methanobacterium* spp. only needs 2–4 h. Thus hydrogenotrophic methanogens are more compacted than acetoclastic methanogens under shock loading conditions.

Generally speaking, the methane generation of each organic waste can be predictively calculated based on the stoichiometry equation. The elemental composition of the raw CM was analyzed by dry matter. The stoichiometric biochemical reaction formula, $C_nH_aO_bN_c$ + $(n - 0.25a - 0.5b + 1.75c) H_2O \rightarrow (0.5n + 0.125a - 0.25b - 0.375c) CH_4 + (0.5n - 0.125a + 0.25b - 0.625c) CO_2 + cNH_4 + cHCO_3^-$ was used to describe the element balance and biogas conversion efficiency (Niu et al., 2014). The experience value and stoichiometric formula were identified for the four kinds of typical manure as described in Table 4.2.

TABLE 4.2 The experience value of biogas (Niu & Li, 2016) and stoichiometry formula for the generation.

Raw material	Biogas (m^3/kg)	CH_4 (%)	CO_2 (%)	Methane (m^3/kg)	OLR_{max}	Stoichiometry molecule formula	Reference
Cow manure	0.26–0.28	50–60	34–38	0.14	4	$C_{22}H_{29}O_{10}N + 6.5H_2O \rightarrow$ $9.25CH_4 + 6.75CO_2 + NH_4^+ +$ HCO_3^-	(Niu & Li, 2016)
Sheep manure	0.22–0.24	40–50	37.6	0.1	–	$C_{33.7}H_{54.0}O_{24.0}NS_{0.3} + 33.9H_2O \rightarrow$ $30.0CH_4 + 15.5CO_2 + NH_4^+ +$ $HCO_3^- + 0.3H_2S$	(Erdogdu et al., 2018)
Poultry manure	0.4–0.6	50–72	30–50	0.27	3	$C_{7.5}H_{12.4}O_{4.8}NS_{0.13} + 4.15H_2O \rightarrow$ $3.7CH_4 + 2.8CO_2 + NH_4^+ +$ $HCO_3^- + 0.13H_2S$	(Niu & Li, 2016)
Swine manure	0.22–0.73	50–60	40–50	–	5	$C_{10.3}H_{16.4}O_{5.2}NS_{0.1} + 10.55H_2O \rightarrow$ $6.28CH_4 + 3.78CO_2 + NH_4^+ +$ $HCO_3^- + 0.1H_2S$	(Sadaka & Devender, 2015)

TABLE 4.3   The characteristic of different manures.

Manure	Carbon (%)	Nitrogen (%)	Hydrogen (%)	Oxygen (%)	Sulfur (%)	Reference
Chicken manure	39.3 ± 0.61	4.4 ± 0.33	5.4 ± 0.46	34.1	0 ± 0	(Wang et al., 2017)
Chicken manure	35.2 ± 0.45	4.83 ± 0.05	5.44 ± 0.24	30.12 ± 0.18	0.84 ± 0.1	(Niu et al., 2014)
Cow manure	34.37 ± 5.45	1.46 ± 0.06	5.01 ± 0.66	58.86	0.3 ± 0.02	(Zhang et al., 2016)
Cow manure	36.81	2.3	4.7	24.67	0.36	(Li et al., 2015)
Cow manure	41.8	2.75	6.11	32.99	0.35	(Li et al., 2015)
Swine manure	46.61	0.16	6.6	46.49	0.14	(Zhang et al., 2019a)
Swine manure	39.4	2.3	NA	NA	NA	(Ye et al., 2013)
Swine manure	44.48 ± 0.12	4.18 ± 0.06	6.19 ± 0.06	44.28	0.87 ± 0.11	(Tian et al., 2015)
Sheep manure	29.94	3.67	3.36	NA	NA	(Boostani et al., 2019)
Sheep manure	23.8	1.12	NA	NA	NA	(Bustamante et al., 2011)

However, till now, AD still faces many challenges while dealing with manure that contains complex components and especially lower C/N compared to other organic waste (Table 4.3). Generally, the maximum outputs of biogas and methane can be achieved when the C/N is around 25 (Piatek et al., 2016). However, the C/N of livestock manure ranges from 9 to 16. As is a well-known inhibitor of AD, excess ammonia can cause accumulation of volatile fatty acid (VFA) and interference with microbial metabolism during AD, and ultimately leading to inefficient AD. The low CH_4 production, poor synergistic metabolism, and weak bioutilization make the AD systems unstable, especially for CM digestion. Therefore, researchers aimed at improving the efficiency and stability of AD are still in progress.

4.2.1   Ammonia-tolerant digestion of low C/N manures

Livestock manure has a lower C/N than other organic waste, especially of CM owning much higher nitrogen content than other manure (cow manure, swine manure, and sheep manure (SM)). Excessive ammonia produced by hydrolysis has a poisonous and repressive effect on the anaerobic microbial activity and leads to poor system stability. The core inhibitor is the free ammonia (FA) with an underlying mechanism of permeate through the cell membranes freely. The ionized FA inside the cell makes the pH imbalance inside and outside (Fig. 4.2). The transportation of the materials and enzyme activity decreased to a low level under the imbalanced pH condition (Yan et al., 2019). As previously reported, an FA concentration of 700–1100 mg/L had inhibitory effects on a variety of substrates (Hansen et al., 1998). Some attempts have been made to alleviate ammonia inhibition with low total solid (TS) digestion (0.5%–3%); however, the large amount of wastewater produced from low TS digestion means higher costs for subsequent treatment.

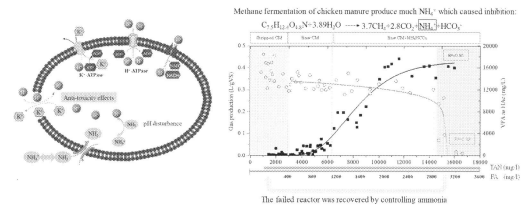

FIG. 4.2   **Ammonia inhibition mechanism and the characteristic of suppression recovery (Niu et al., 2013).**

In our previous research (Niu et al., 2013), a continuously stirred tank reactor (CSTR) with a volume of 12 L was used to carry out long-term AD for CM involving 10% TS (higher TS content than traditional wet digestion) to investigate the stability, inhibition effect, and recovery rate of the inhibited reactor. Besides, the methane conversion efficiency and VFA accumulation as ammonia changes and methanogenesis, acidogenesis, and hydrolysis were investigated both with and without inhibition. The result showed that it was feasible to conduct mesophilic fermentation with a high solid of CM at total ammonia nitrogen (TAN) lower than 5000 mg/L. However, VFA was accumulated responding to ammonia with TAN varying from 5000 mg/L to 10,000 mg/L, whose interactive inhibition with TAN led to the process failure. Surprisingly, the ammonia inhibition threshold was extended to 15,000 mg/L through long-term domestication and successfully recovered a seriously inhibited system with a TAN of 16,000 mg/L by a fluxing and then washing strategy (Fig. 4.2).

As reported above, higher ammonia can lead to the accumulation of intermediate products in AD, leading to synergetic inhibition with excessive VFA (Shi et al., 2017). Moreover, propionate had a stronger inhibition of methane production at thermophilic conditions, especially in the substrate concentration of more than 5000 mg chemical oxygen demand (COD)/L. The largest specific maximum methanogenic activity decreased from 0.029 to 0.022 g COD_{CH4}/g volatile solid (VS)/d. In addition, the phenomenon that microorganisms cannot adapt to thermophilic propionate degradation conditions (at a maximum lag time of 12 days) showed vast substrate inhibition. As an intermediate product of valerate degradation, the degradation rate of propionate was significantly lower than that of valerate (0.036 g COD/d vs. 0.104 g COD/d). Therefore, propionate greatly limited the degradation rate of valerate under thermophilic AD.

4.2.2   Codigestion of manures to improve methane production

To expand the applications of AD to larger areas and improve digestion efficiency, codigestion was usually applied for the manure digestion. Anaerobic codigestion is cooperative digestion of two or more biodegradable substrates, which is more efficient and productive than the digestion of a single substrate. Synergistic digestion or codigestion of multibiodegradable substrates maximizes the production of biogas because the mixing of substrates results in higher biogas production per unit mass. The combination of multiple substrates not only optimizes operational parameters such as the C/N ratio but also enhances the mutualism of microorganisms by enriching both trace and macronutrients. This helps the microbes work together in "syntrophic interaction" (one's products can be the other's substrates), which caused the system to have a more comprehensive AD pathway. Therefore, multiple substrates could enhance the AD process in biogas plants, as they can optimize the C/N ratio and provide essential nutrients required.

4.2.2.1 Codigestion of different kinds of manure

Our previous work innovatively proposed the codigestion of SM and CM (Song et al., 2019a), which was not only conducive to controlling the inhibition threshold of FA and TAN but also beneficial to save the cost of storage and transportation. The optimal value of VS ratio of SM to CM ($Rs/c = 0.4$) was attained. The highest biomethane production was 219.67 mL/gVSadd and biomethane production rate was 0.378 mL/gVSadd/h (Fig. 4.3). Similarly, the optimal codigestion ratio of *Chlorella* sp. and CM was 2:8, and the biogas production increased by 76.86% (Li et al., 2017b). Moreover, it was reported that the production of biomethane

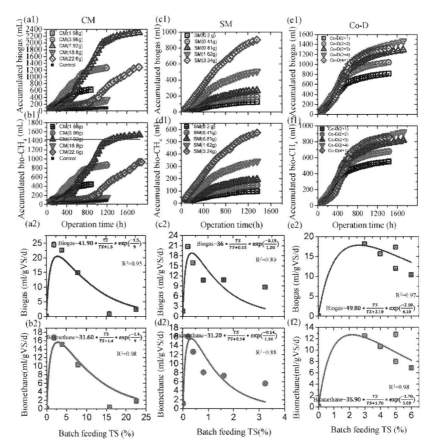

FIG. 4.3   The biogas and biomethane production for CM, SM and codigestion (Song et al., 2019a).

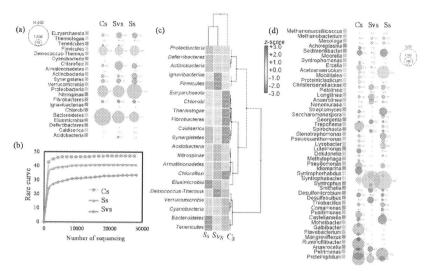

FIG. 4.4   Microbial community comparison of mono- and codigestion (Song et al., 2019b).

increased by 93.0% during the codigestion of CM and agricultural waste (Abouelenien et al., 2014). Furthermore, the acetoclastic *Methanosaeta* was predominated in the codigestion system sample with occupation 92.07% in the Archaea community.

The high diversity of hydrogenotrophic methanogens and acetoclastic methanogens during the codigestion increased the stability of the process (Fig. 4.4). Similar research about codigestion of SM and cow manure indicated that *Methanoculleus* and *Candidatus Cloacimonas* were key microbes (Li et al., 2020). As a syntrophic bacterium, *C. Cloacimonas* played a key role in acidogenesis and its syntrophic capacity was related to the production of methane from hydrogen-supported methane. The hydrogenotrophic *Methanoculleus* (coenzyme-B sulfoethylthiotransferase) was associated with an increased methane yield.

This study also compared the monodigestion and codigestion of CM and SM from the kinetic investigation, autofluorescence analysis and microbial community evaluation (Song et al., 2019a). The kinetic results showed that the Gompertz parameters ($R^2 > 0.98$) were better than the improved Aiba model ($R^2 > 0.88$), which indicated the optimal conditions of monodigestion and codigestion. In particular, the modified Haldane equation was used to simulate the kinetics of different TS for each test with R^2 values ranging from 0.92 to 0.99 (Table 4.4). The improved Gompertz model was used in this study to fit the biomethane and biogas-producing performance under anaerobic conditions (Song et al., 2019c). The model can be written as

$$P(t) = P_0 \times \exp\left[-\exp\left(\frac{K_{max} \times e \times (A-t)}{P_0} + 1\right)\right] \qquad (4.1)$$

where P refers to gas/methane yield at time t, mL/g VS sludge; P_0 is the maximum gas potential, mL/g VS sludge; K_{max} is the maximum gas production rate, mL/gVS/h; A refers to the lag phase of gas production, h; t is the reaction time of the experiment, h; and e is a constant.

TABLE 4.4   Kinetic parameters comparison of mono and codigestion.

Mixture rate		TS	VS	R^2	P	SD	K	SD	A	SD	$t \times (h)$ P_{max}	K(mL/gTS/d)	SD	$K_{Gampotiz}$
SM biogas		0.20	0.16	1.00	132.19	1.82	0.14	0.00	5.85	8.57	17	20.78	0.57	0.157183
		0.41	0.33	1.00	194.66	2.52	0.22	0.00	23.73	8.11	64	15.92	0.42	0.08177
		0.81	0.66	1.00	277.29	3.61	0.29	0.01	41.75	7.30	165	10.75	0.25	0.038782
		1.62	1.31	1.00	517.81	5.48	0.59	0.01	66.04	6.19	113	10.83	0.23	0.020909
		3.24	2.62	1.00	954.08	15.16	0.95	0.02	59.47	7.80	589	8.69	0.21	0.009104
SM biomethane		0.20	0.16	0.99	101.42	1.67	0.11	0.00	−22.29	11.15	64	16.09	0.56	0.158675
		0.41	0.33	0.99	148.70	2.18	0.17	0.00	−2.37	10.24	64	12.66	0.43	0.085171
		0.81	0.66	1.00	196.14	2.62	0.22	0.01	13.81	8.65	64	8.08	0.23	0.041219
		1.62	1.31	1.00	345.02	3.93	0.40	0.01	34.12	7.31	64	7.34	0.18	0.021275
		3.24	2.62	1.00	596.37	9.42	0.61	0.01	27.25	8.78	64	5.60	0.15	0.009385
Co-digestion	2 + 1	3.00	1.80	1.00	790.36	4.16	1.37	0.02	80.88	4.50	382	18.23	0.37	0.017083
	2 + 2	4.00	2.60	1.00	1017.15	6.65	1.70	0.03	87.89	5.46	382	15.68	0.37	0.012372
	2 + 3	5.00	3.60	1.00	1233.06	9.78	1.81	0.04	78.46	7.40	436	12.06	0.35	0.008691
	2 + 4	6.00	4.20	1.00	1435.33	10.58	1.82	0.03	101.47	6.17	476	10.40	0.23	0.006265
	4 + 1	5.00	2.80	1.00	1348.35	14.91	2.02	0.05	165.34	7.69	496	17.35	0.88	0.008895
Co-digestion	2 + 1	3.00	1.80	1.00	536.36	2.80	0.94	0.02	87.53	4.45	382	12.55	0.25	0.017337
	2 + 2	4.00	2.60	1.00	661.61	4.68	1.15	0.02	100.72	5.91	382	10.63	0.28	0.01289
	2 + 3	5.00	3.60	1.00	776.41	6.12	1.20	0.03	88.96	7.42	436	7.99	0.24	0.009145
	2 + 4	6.00	4.20	1.00	900.98	6.80	1.20	0.02	117.55	6.38	476	6.88	0.16	0.006595
	4 + 1	5.00	2.80	1.00	867.84	9.81	1.49	0.05	225.24	7.87	549	12.78	0.75	0.010185

Mixture rate	TS	VS	R^2	P	SD	K	SD	A	SD	t × (h) P_{max}	K(mL/gTS/d)	SD	$K_{Gampotiz}$
CM	1.98	1.03	0.99	683.89	17.62	1.05	0.03	22.67	8.45	380	24.45	0.94	0.02292
	3.96	2.06	1.00	1311.25	10.20	1.94	0.03	82.85	4.56	430	22.65	0.42	0.011074
	7.92	4.11	0.99	2480.58	41.07	2.56	0.07	309.74	10.23	800	14.93	0.47	0.003858
	15.84	8.23	0.98	368.23	16.38	0.31	0.01	3.56	14.49	708	0.91	0.04	0.001586
	22.63	11.75	0.98	3495.89	655.00	1.19	0.10	685.40	74.49	1400	2.42	0.25	0.000444
	Control		0.98	95.26	2.87	0.10	0.00	−23.44	15.08				
CM	1.98	1.03	0.99	490.76	14.45	0.72	0.02	16.96	9.21	380	16.74	0.68	0.021862
	3.96	2.06	1.00	909.21	8.06	1.30	0.02	87.61	5.00	430	15.13	0.30	0.010666
	7.92	4.11	0.99	1642.14	29.13	1.77	0.05	344.64	10.95	800	10.34	0.37	0.004036
	15.84	8.23	0.98	189.21	16.76	0.12	0.00	−53.92	20.16	708	0.36	0.02	0.001212
	22.63	11.75	0.98	2588.99	487.09	0.88	0.07	779.54	74.74	1400	1.80	0.19	0.000447
	Control		0.98	95.26	2.87	0.10	0.00	−23.44	15.08				

The optimized codigestion of CM and SM contributed to balanced metabolism and improved biomethane yield (Fig. 4.4). Codigestion was beneficial to improving biogas production, economic benefits derived from the sharing of equipment, easier handling of mixed wastes, and a synergistic effect with the added nutrients supporting microbial growth. In the process of codigestion, it is important to select the optimal ratio of codigestion substrate for promoting positive and/or synergistic interactions between feedstocks, avoid inhibition, and maximize methane production. Interaction (synergistic or antagonistic) effects were optimized by identifying the different substrate mixtures with the maximum biomethane yield. These positive synergies can be attributed to a variety of factors, including balanced nutrient composition, aroused synergistic effects of microorganisms, an associated increase in buffering capacity, and a reduced effect of toxic compounds on the digestion process (Zheng et al., 2015).

The seed sludge (S_{seed}) for inoculation, the starving sludge (S_{starv}), the codigestion sludge (S_{CO}) with $R_{s/c}$ of 3.2, the monodigestion of CM (S_{CM}), and the monodigestion of SM (S_{SM}) were compared and flocked. Especially, there were some significant differences at the genus level (Fig. 4.5). The increase in methanogenic activity in the codigestion system was attributed to the increase of acetoclastic methanogens and the stable abundance of hydrogenotrophic methanogens.

Moreover, at the genus level, the samples were analyzed by PCoA (Fig. 4.6A). The samples of S_{CO} and S_{Starv} are significantly different from the other samples with significant differences in the microbial community. The main differences of functional genus abundance and potential dominated metabolic pathways of microbial communities during monodigestion and codigestion were showed in Fig. 4.6B.

The chosen metabolism KEGG Orthology (KOs) was divided into two groups with main groups of methane metabolism, nitrogen metabolism, starch and sucrose metabolism, and NADH/NAD$^+$-related metabolism. Furthermore, the coenzyme F_{420}-related KOs reflected the methanogens metabolism of hydrogenotrophic with more than tenfold higher in S_{CO} than in S_{SM} and S_{CM} suggesting that the codigestion has more complicated methanogenesis pathways than monodigestion. Interestingly, the person correlation parameters indicated that the coenzyme F_{420}-related and N acetyl–related KOs were closely corrected with methane production, which was verified in the processing activity in methanogenesis of hydrogenotrophic and acetoclastic. Therefore, this study provides new insights for evaluating the effects

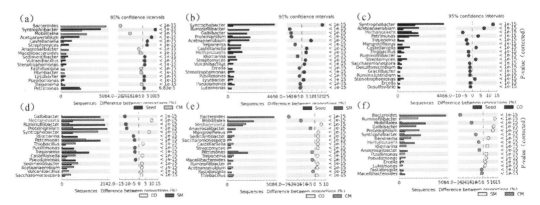

FIG. 4.5   The functional genus comparison of each mono- and codigestion (Song et al., 2019a).

(a) PCoA analysis

(b) The functional genes involved

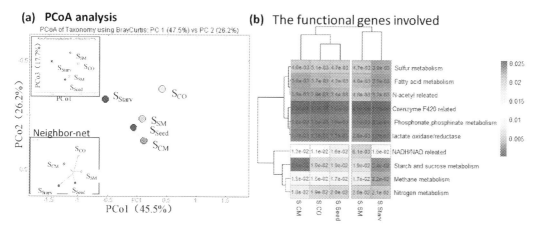

FIG. 4.6 The PCoA analysis and the functional genes in mono- and codigestion (Song et al., 2019a).

of different substrate diets on the functional metabolic pathway, given the feed strategies' efficiency of the digestion control. However, for further study of the digestion and symbiosis of complex organic matter, further mechanistic studies are needed for the metabolic dynamics and interaction pathways of the microbial community. These studies concluded that the optimized codigestion process had a powerful synergistic effect on the production of biomethane.

4.2.2.2 Cardboard and CM codigestion

According to the report "Development Status and Trend of Green Packaging for Express Delivery in China in 2017," the large number of cartons used in express service was about 3.7 billion in China. The growth of population and social economics would make the packaged carton keep increasing. The main problem we face currently is that the carton cannot be recycled but be thrown into the dumpster along with the other household wastes. Consequently, such discarded cardboard was hard to reuse (Li et al., 2019). In the codigestion condition, following the addition of the carton, biomethane production was gradually decreased from 341.01 to 246.81 mL/gVS. However, from the perspective of weighted specific methane yield, the VS ratios of 35:65 for anaerobic codigestion of CM and cartons achieved a promising improvement in all CM and carton codigestion and monodigestion. Moreover, the synergistic effect of codigestion significantly controlled VFAs production and improved the buffer capacity. The appropriate C/N and the optimal COD/NH_4^+ in cosubstances were achieved for better degradation (Hassan et al., 2016; Zheng et al., 2021). Meanwhile, higher cellulase activity (+31.6%) was observed to exhibit higher hydrolysis capability. For acidogenesis, more extracellular polymeric substances were secreted for electron transfer enhancement, which was instrumental in the high metabolic activity in codigestion (Xiao et al., 2017). Further, codigestion achieved optimal pH value leading to a lower Gibbs free energy for acetate production (Zhao et al., 2020). Taking sorghum stem and pig manure codigestion as an example, via the buffering effect of the produced ammonia, cow manure helps to maintain a stable pH value during AD. In addition, high alkalinity of cow manure can avoid acidification of the system to some extent, enabling the pH values to maintain at around 7.0 (Zhang et al., 2016). Moreover, the higher abundance of hydrogen-producing bacteria and methanogens could enhance the VFAs degradability, which eliminated acetate accumulation and inhibition (Cazier et al., 2019). As for CM monodigestion, the high VFAs concentration could inhibit the acetoclastic methanogens (Li et al., 2017b). However, during codigestion, the hydrogenotrophic pathway and the acetoclastic pathway were more active and resulted in higher gas production (Capson-Tojo et al., 2018).

4.2.2.3 *Enteromorpha and CM codigestion*

In recent years, serious eutrophication of seawater has induced *Enteromorpha prolifera* green tide causing a serious economic loss globally (Zhao et al., 2020). Moreover, from May to July 2008, more than 1 million tons of drifting *Enteromorpha* (EN) had caused the world's largest green tide in the Yellow Sea of China. Codigestion proved an efficient reutilization and disposal of EN. In CM and EN codigestion system, the biomethane production was increasing with CM addition. A 1.43%–4.11% improvement of biomethane yield was observed in CM and EN codigestion, and the highest biomethane production improvement was obtained with CM:EN = 2:1. The antagonistic effect for methane production was observed when the EN and CM are mixed in 1:1 and 3:1 ratios. A similar antagonistic effect was observed in swine slurries and grass silage anaerobic codigestion (Himanshu et al., 2018). Actually, the endogenous salinity of EN might contribute to the antagonistic effect. A previous study reported that the low salinity would stimulate the activity of dehydrogenase and coenzyme F420 for higher biomethane production, in which high salinity would induce salinity stress and lower biomethane production (Zhang et al., 2020). There were two gas production peaks in EN and CM codigestion biomethane production rate curve. The existence of easy-degradable matter and the hard-degradable matter would cause the two gas production peaks (Li et al., 2017b). Compared with monodigestion, EN codigestion systems exhibited the first peaks with 0.948–1.422 mL/gVS/h and 23.5%–49.9% improvements.

Codigestion significantly enhanced the cellulase activity in EN and CM codigestion with a 234% enhancement. Moreover, codigestion induced the succession of microbial communities. As typical acidogenic bacteria with the capacity to degrade cellulose, the abundance of *Bacteroidetes* gradually increased with EN increasing (Wei et al., 2020). Meanwhile, the highest abundance of *Firmicutes* was achieved in codigestion (CM: EN = 2:1), which had a high activity for cellulose and protein degradation. Some scholars have found that the *Firmicutes* to *Bacteroidetes* ratio (F/B) might have potential utility to elevate process stability (Chen et al., 2016). With CM content increasing, the F/B ratio showed an increasing trend, which indicated that codigestion was conducive to system stability. As for the most dominant methanogens genus, *Methanosaeta* and *Methanobacterium* showed significant dynamic changes in codigestion systems. *Methanosaeta* had higher resistance to the endogenous salinity of EN, which exhibited a higher abundance in EN monodigestion (Zhang et al., 2020).

4.2.2.4 *Codigestion of green waste and chicken manure*

A recent study reported that the total potential bioenergy generated by green waste (GW) was about 260 petajoules in China (Shi et al., 2013). However, most of the GW was landfilled or unscientifically dumped, which caused various environmental problems. In CM and GW codigestion system, a 3.83%–7.74% improvement of biomethane yield was observed, and the highest biomethane production improvement was obtained with CM:GW = 2:1 (7.62%). The lower biogas production in EN monodigestion and GW monodigestion might be decided by high cellulose content (Zhao et al., 2020). The cellulosic biomass had a significant negative correlation to biogas production. The complex lignocellulosic structure and composition limited hydrolysis and efficient methanogenesis (Lü et al., 2020). As for the biogas production rate curve, two gas production peaks were observed. In the codigestion of GW and CM, the biomethane production rate of the first and second peaks were 0.963–1.277 mL/gVS/h and 0.739–1.070 mL/gVS/h, respectively. The cellulase activity analysis showed a 30.1% enhancement in GW codigestion systems.

The performance of animal manure with different organic waste is compared in the codigestion condition (Table 4.5). Chuenchart et al. (2020) found that a high temperature of anaerobic codigestion could achieve efficient energy recovery and waste treatment by accelerating and enhancing synergy (Chuenchart et al., 2020). To overcome the challenges of process inhibition, codigestion of CM and food waste (FW) of 10% TSs was investigated within an 87 L of bench-scale anaerobic CSTR. Fluchtige organische sauren (FOS)/totales anorganisches carbonat (TAC) (the ratio of fluchtige organische sauren to totales anorganisches carbonat) indicates the stability of the anaerobic system and is a preventive parameter for system failure. Compared to the monodigestion (the maximum value of about 1.6), FOS/TAC values were within the optimal range (0.3–0.4) during the codigestion period, except for operational problems with the system. This indicated that the organic loading rate (OLR) capacity of the anaerobic codigestion system was higher due to the synergistic effect of microorganisms, buffering capacity, and the adjustment of C/N in the feedstock on the nutrient balance. Besides, the percentage increases of specific methane yields in codigestion over monodigestion were 33.2%, 10.4%, 12.1%, and 89.9% at OLRs of 1, 2, 3, and 4 kg VS/m^3/d, respectively, as a result of the improved stability of codigestion.

4.2.3 Enhancement of biogas production through AD

Several conventional methods have been applied to diminish the extent of instability caused by substrate in AD process. Recently, supplementation of various additives for AD has received increasing attention to reinforce reactor performance in terms of system stability, biogas production, and treatment capacity (Jang et al., 2018).

Biochar, a carbon material from the pyrolysis of biomass or wastes, has been used to enhance biogas production in the AD due to its ability including promoting biofilm formation, mitigating ammonia, and acid inhibition (Sunyoto et al., 2016). As a porous and biostable material, biochar always served as a soil amendment and it can be assumed that porous biochar with a relatively high surface area provides a habitat for the adhesion and growth of microbes (Luo et al., 2015). Furthermore, it is speculated that biochar benefits stimulating microbial activity and promoting direct interspecies electron transfer (DIET) among the bacteria and methanogens. For example, the enrichment of electroactive *Anaerolineaceae* and *Methanosaeta* (typical microbes for DIET) with biochar (BC) added brought about the occurrence of syntrophic degradation from butyrate to acetate (Wang et al., 2018). The previous study has investigated the effects of conductive materials on AD and concluded that DIET promoted the methanogenic significantly with the CH$_4$ production rate increased by 33% (Viggi et al., 2014). In the research of Pan et al. (2019), biochar was added to the semicontinuous AD reactor of CM with different OLRs (0.625, 3.125, and 6.25 g VS/L/d). The results

TABLE 4.5   Co-digestion performance of manure with different organic waste.

Reactor	Feed stock	C/N ratio or composition	T (°C)	Digestion Time (d)	TS (%)	VS	BioCH$_4$ production (L/gVS$_{add}$)	Reference
Batch	CM, CS	CM:CS = 1:1.4 C/N = 20:1	37	30	/	3g/L	0.28	(Li et al., 2014)
Batch	DM, CM, WS	DM/CM = 40.3:59.7 C/N = 27.2:1	35	30	/	/	0.25	(Wang et al., 2012)
Batch	CM, AP	CM: AP = 2:1 C/N = 18.5:1	37	70	/	32 g/L	0.08	(Li et al., 2018a)
Batch	TCM, AWS	TCM/AWS = 7:3	55	62	10	/	0.7	(Abouelenien et al., 2014)
Batch	CM, ChS	CM: ChR = 8:2	35	35	/	40 g/L	0.17	(Li et al., 2017b)
Batch	CM, TO-WS	C/N = 20:1	37	42	5	/	0.39	(Hassan et al., 2016)
Batch	CM, SM	CM:SM = 2:5	35	60	6.45	3.95 g	0.22	(Song et al., 2019b)
Batch	CS, CM	CS:CM = 3:1	37	30	/	3 g/L	0.22	(Li et al., 2013)
Batch	CM, CD, Mol	CM:CD:Mol = 1:1:1	35	32	34 g/L	17.2	0.26	(Misi & Forster, 2001)
Batch	DM, CS, TR	DM:CS:TR = 32:48:20	35	35	29	/	0.35	(Li et al., 2018b)
Batch	CM, CB	CM: CB = 70:30 C/N = 15.5:1	37	27	/	15 g/L	0.35	(Zhao et al., 2021)
Batch	CD, FW	CD: FW = 35: 65 C/N = 18.5: 1	38	60	/	15 g/L	0.39	(Zhang et al., 2021)
Batch	CD, CS	CD: CS = 35: 65 C/N = 25.2	38	60	/	15 g/L	0.30	(Zhang et al., 2021)
Batch	PM, RS	C/N = 24.8	35	20	15	/	0.27	(Paranhos et al., 2020)
Batch	PM, CC	C/N = 18.3	35	20	15	/	0.29	(Paranhos et al., 2020)
Batch	PM, PS	C/N = 16.6	35	20	15	/	0.16	(Paranhos et al., 2020)
Batch	PM, SW	C/N = 17.6	35	20	15	/	0.26	(Paranhos et al., 2020)

CM, chicken manure; CD, cow dung; PM, swine manure; CS, corn stover; DM, dairy manure; WS, wheat straw; AP, apple pulp; TCM, treated chicken manure; ChS, Chlorella sp.; TO-WS, thermo-oxidative cleaved wheat straw; SM, sheep manure; AWS, agricultural wastes; Mol, molasses; TR, tomato residues; CB, cardboard; FW, food waste; RS, rice straw; CC, corn cob; PS, peanut shell; SW, sawdust indicated that methane yields were increased by 33%, 36%, and 32% with adding biochar at the three different OLR, respectively. It was also confirmed that biochar facilitates the DIET between fermenting bacteria and electrophilic methanogens to achieve the most efficient digestion strategy. The study also revealed that biochar application can improve the acidogenesis and methanogenesis efficiency of macromolecule organics for energy recovery, and stimulate nitrogen removal by activating *Epsilonproteobacteria* (typically denitrifying bacteria). Yang et al. studied the impacts of biochar on methanogenesis during AD of swine manure. They found that by adding an optimal biochar dosage of 5%–10%, methane production was significantly improved by 25%, which was ascribed to the enhancement of DIET. Biochar addition alleviated the need for cytochrome-c as the components of interspecies electron connection and enriched the microorganisms involved in DIET. It was reported that *Defluviitoga, Thermovirga,* and *Cloacibacillus* held a great potential to take part in DIET with *Methanosaeta* (Yang et al., 2021).

Zero-valent iron (ZVI), a promising reductant, has been widely applied in studies for strengthening AD performance. ZVI addition could not only decrease the oxidation–reduction potential to build a favorable habitat for microbes but also contribute electrons and provide trace elements for microbes to enhance their enzymatic activity. Moreover, ZVI could act as a conductive platform to promote electron transfer among anaerobes. Owing to the benefits above, ZVI has been used in the anaerobic biotreatment of animal manure, leachate, waste sludge, FW, etc. In a previous study, the hydrolysis acidification of municipal sludge was significantly stimulated, and biogas yield was increased by 15.70% (Zhang et al., 2019b). Moreover, ZVI was recommended to promote butyrate biodegradation and methanogenic activity to avoid excessive acidification at high OLR. In the study of high solid AD of swine manure (Meng et al., 2020), the potential of dosing ZVI was added to enhance methanogenic capacity via batch tests under mesophilic conditions. ZVI conduce to the anaerobic conversion of propionate to acetate and H_2 and alleviates ammonia inhibition (total ammonia was over 5.0 g-N/L) which was a bottleneck problem in the digestion of nitrogen-rich substances. The methane yield was increased by 22.2% at ≥160 mM ZVI dosage and the duration of high solid AD was further shortened by 50.6% at ≥320 mM ZVI dosage. The improvement of methanogenesis was attributed to the full utilization of propionate and the expedited decomposition of posterior-biodegradable organic matter which was associated with ZVI added. Results of sequencing and qPCR (mcrA) showed that ZVI might stimulate the growth of *Methanosarcina* (from 27.9% to 78.3%) and *Syntrophomonas* (0.5% to 3.7%) which possibly participate in DIET to enhance propionate biodegradation. Besides, even at FA almost 1.0 g-N/L, the major methanogenesis might still be in the effective acetoclastic pathway, as syntrophic acetate oxidizing bacteria reduced to almost absent at 320 mM ZVI dosage. Thus, dosing ZVI could enhance the high solid AD under TAN inhibition and a higher dosage was needed to resist FA inhibition. Liang et al. have applied the single and combined action of microscale ZVI and magnetite to the AD of diluted swine manure (Liang et al., 2017). After 30 days, the cumulative methane production from the AD of swine manure ranged from 246.9 to 334.5 mL/gVS added. The results of the first-order kinetic model indicated that the addition of ZVI and/or magnetite increased the biogas production potential, rather than the biogas production rate constant. Furthermore, the endproduct of ZVI anaerobic corrosion, magnetite might benefit to promote methane production through DIET in the AD process.

4.3   Additional value from AD

4.3.1   Biohydrogen production from manure through AD

Biohydrogen, as a clean renewable resource, has a high calorific value in the global energy supply with only water generated after utilization, which has the potential to replace fossil fuels. Generally, most biohydrogen generation uses municipal solid wastes as substrates, such as FWs mixture with vegetables, grains and meats, organic waste from a tofu plant, rice winery wastewater, sweet potato starch residues, paper mill wastes, and wheat starch coproduct. However, the huge and annually increased amount of animal manure provided an almost inexhaustible candidate for renewable biomass for producing biohydrogen through dark fermentation. The feasibility of swine manure digestion used for biohydrogen generation has been investigated, and the codigestion of swine manure with glucose in a semicontinuously fed process was conducted under different hydraulic retention times (HRT) (Zhu et al., 2009). The process obtained a maximum biogas production (21.4–38.3 L/d) and H_2 content (35.8–37.6% v/v) is obtained with a significant 18.7×10^{-3} g H_2/gVS conversion ratio at the best HRT of 16h (pH of 5) (Zhu et al., 2009).

4.3.2   Bioethanol production through AD

Recently, the bioethanol production from manure was also reported, especially from dairy manure, which was rich in hemicellulose, cellulose, and nitrogen source. After optimizing the NaOH pretreatment, dairy manure was used for bioethanol generation, and the maximum of 10.55 g/L ethanol was achieved at the optimization conditions (21.14 g/L of glucose and 9.48 g/L of xylose) without an additional nitrogen source for balancing the C/N ratio (You et al., 2017). This successful ethanol fermentation generated 71.91% ethanol from dairy manure. This result demonstrated the economic benefits of dairy manure treatment. While using the saccharification production obtained by acid pretreatment (3.5% H_2SO_4, 121 °C, 30 min), the digestion process obtained the maximum ethanol of 56.32 mg/g dry cattle manure, 12.69 mg/g poultry manure and 27.98 mg/g swine manure with free yeast cocultures (Bona et al., 2018).

4.3.3   Biohythane production through AD

Generally, the calorific value of methane is 50.07 MJ/Kg, while the calorific value of hydrogen is 143 MJ/Kg. Biohythane constituted by methane and hydrogen (5%–25% v/v) has better combustion performances with high calorific value and low emission of carbon dioxide. Biohythane fermentation is becoming a desirable technology for manure treatment in recent years.

    Separate methane fermentation reaction (2673 kJ): $C_6H_{12}O_6 \rightarrow 3CH_4 + 3CO_2$
    Two-stage fermentation of hydrogen and methane (2926 kJ):
    Hydrogen fermentation: $C_6H_{12}O_6 + 2H_2O \rightarrow 2CH_3COOH + 2CO_2 + 4H_2$
    Methane fermentation from acetic acid: $2CH_3COOH \rightarrow 2CH_4 + 2CO_2$ (Vermeulen et al., 2017)
    Overall reaction: $C_6H_{12}O_6 + 2H_2O \rightarrow 2CH_4 + 4CO_2 + 4H_2$

Annually, 3.8 billion tons of manure and 900 million tons of straw are generated in China. These two typical agricultural wastes are the ideal materials for biohythane fermentation. While using straw and manure as a cosubstrate, the highest biohydrogen yield of 16.68 ± 1.88 mL/gVS was obtained, meanwhile, the subsequent maximum biomethane production of 197.73 ± 11.77 mL/gVS was achieved with hydrogen to methane ratio of 8.44% (v/v) (Chen et al., 2021). Biohythane digestion is usually conducted in single-stage and two-stage reactors. While, the two-stage biohythane digestion is preferred as its short time duration of 13 days to 18 days with high effective energy recovery (67.70%) via degradable organic waste (Lay et al., 2020). Till now the most reported microbe were two groups: the acidogenic community including *Caldicellulosiruptor* sp., *Thermotoga* sp., *Enterobacter* sp., *Clostridium* sp. and *Thermoanaero bacterium* sp., while the methanogenesis community includes the predominate methanogens of *Methanosarcina* and *Methanoculleus* (Lay et al., 2020).

4.3.4   Volatile fatty acids production through AD

Green chemical production from organic waste is currently one of the key topics of the circular economy, especially from manure digestion. The intermediate products of AD could be useful materials and green chemicals for other industrial utilization. Among the intermediate products, the VFAs were the most promising products in the AD while feeding with CM. More importantly, as a chemical platform for developing a wide spectrum of essential products, VFAs could be used as primary materials for polymers, alcohols, olefins, and ketones production, which are currently derived from fossil-based resources (Chen et al., 2013). The optimized operational conditions for achieving the high VFAs in the AD process from CM were reported (Yin et al., 2021). The results showed that VFAs generation from CM digestion owning much higher value than just conversion to the conventional biogas. The highest net VFAs yields could increase up to 0.53 gVFA/gVS, which could be achieved by heat shocking the inoculum and CM. The acetate, propionate, and butyrate were generated in the AD during angiogenesis of CM. The process control indicators of alkaline, initial pH, CM characteristic and the heat-shocking as well as the inoculum have been concluded to influent the VFAs accumulation (Yin et al., 2021).

4.3.5   Lactic acid production through AD

Lactic acid (LA), an intermediate metabolite of AD, has high added values, which can be used as the raw material for food, medicine, leather, and biodegradable plastics production. Previous studies reported that LA could be produced from the AD of sole manure, but the yield is relatively low. The highest LA yield is 92.96±42.4 mg COD /gVS$_{added}$ during the digestion of sole swine manure (Cao et al., 2020). Similarly, in the sole swine manure digestion, the LA concentration is 195.52 mg COD/gVS$_{added}$ with the inoculation of LA bacteria (Lian et al., 2020). As an effective enhance method, codigestion has been applied to improve the LA production. Previous tests showed that adding apple waste or potato waste significantly promoted the LA accumulation from the AD of swine manure. The optimal mixing ratio of apple waste or potato waste to swine manure of 75:25 (VS$_{added}$) was achieved with the maximum concentrations of LA of 690.32 and 222.77 mg COD/gVS$_{added}$, which were around 3.53 and 1.14 times of that in the swine manure monodigestion (Lian et al., 2020).

Moreover, the codigestion of swine manure and corn silage also promoted the LA production with a 216.90 mg COD/gVS_{added}, which was 2.33 times higher than monoswine manure digestion (Cao et al., 2020).

4.3.6 Long-chain fatty acids production through AD

Codigestion of manure with lipid-rich wastes appears as a robust process technology that can increase biogas production during AD, accompanied by the production of long-chain fatty acids (LCFA). According to the previous study, the highest concentration of total LCFA (including palmitic acid, stearic acid, and oleic acid) is 1761.4 mg/kgLM during the codigestion of liquid dairy manure (LM) and 3% decanter sludge (Pitk et al., 2014). In the study above, the higher LCFA concentration was obtained with the value of 3256.3 mg/kgLM in the codigestion of LM and technical fat from a slaughterhouse. Besides, oily waste was added to the anaerobic digester fed with cow manure (CM) and FW, which stimulated the LCFA production. Neves et al. reported that only the palmitic acid (9 ± 3 gCOD/kgTS) and stearic acid (3 ± 1 gCOD/kgTS) were detected in the digestion system with CM and FW, while the highest total LCFA (palmitic acid, myristic acid, oleic acid, and stearic acid) content (375.2 gCOD/kgTS) was obtained with the addition of oily waste (18.0 gCOD/L) (Neves et al., 2009).

4.4   Conclusions and perspectives

The livestock and poultry breeding industry has become an important pillar industry all over the world's economy making a huge manure production. Meanwhile, bioenergy and green chemicals production from organic waste is currently one of the key topics of the circular economy, especially from the manure through AD. The four steps of AD process (hydrolysis →acidogenesis→acetogenesis→ methanogenesis) can be inhibited by excessive ammonia and VFA concentration. Ammonia-tolerant digestion can benefit the mono manure digestion, while the codigestion with different organic waste with C/N regulation proved a synthetic efficiency for methane production.

Even as an ancient technology, AD is facing new opportunities in the context of global advocacy for carbon nowadays. Moreover, the additional value of biochemicals such as bioethanol, biohydrogen, LA as well as long/short chain fatty acids, and even other higher value-added products such as biodegradable plastics generated from AD will attract more and more attention in the future.

References

Abouelenien, F., Namba, Y., Kosseva, M.R., Nishio, N., Nakashimada, Y., 2014. Enhancement of methane production from co-digestion of chicken manure with agricultural wastes. Bioresour. Technol. 159, 80–87.
Bona, D., Vecchiet, A., Pin, M., Fornasier, F., Mondini, C., Guzzon, R., Silvestri, S., 2018. The biorefinery concept applied to bioethanol and biomethane production from manure. Waste Biomass Valori 9 (11), 2133–2143.
Boostani, H.R., Najafi-Ghiri, M., Hardie, A.G., Khalili, D., 2019. Comparison of Pb stabilization in a contaminated calcareous soil by application of vermicompost and sheep manure and their biochars produced at two temperatures. Appl. Geochem. 102, 121–128.

Borjesson, P., Mattiasson, B., 2008. Biogas as a resource-efficient vehicle fuel. Trends Biotechnol. 26 (1), 7–13.

Bustamante, M.A., Said-Pullicino, D., Agulló, E., Andreu, J., Paredes, C., Moral, R., 2011. Application of winery and distillery waste composts to a Jumilla (SE Spain) vineyard: effects on the characteristics of a calcareous sandy-loam soil. Agric., Ecosyst. Environ. 140 (1), 80–87.

Cao, Q., Zhang, W., Zheng, Y., Lian, T., Dong, H., 2020. Production of short-chain carboxylic acids by co-digestion of swine manure and corn silage: effect of carbon-nitrogen ratio. Trans. ASABE 63 (2), 445–454.

Capson-Tojo, G., Trably, E., Rouez, M., Crest, M., Bernet, N., Steyer, J.-P., Delgenès, J.-P., Escudié, R., 2018. Methanosarcina plays a main role during methanogenesis of high-solids food waste and cardboard. Waste Manage 76, 423–430.

Cazier, E.A., Trably, E., Steyer, J.-P., Escudie, R., 2019. Reversibility of hydrolysis inhibition at high hydrogen partial pressure in dry anaerobic digestion processes fed with wheat straw and inoculated with anaerobic granular sludge. Waste Manage 85, 498–505.

Chen, H., Huang, R., Wu, J., Zhang, W.Z., Han, Y.P., Xiao, B.Y., Wang, D.B., Zhou, Y.Y., Liu, B., Yu, G.L., 2021. Bio-hythane production and microbial characteristics of two alternating mesophilic and thermophilic two-stage anaerobic co-digesters fed with rice straw and pig manure. Bioresour. Technol. 320, 124303.

Chen, H., Meng, H., Nie, Z., Zhang, M., 2013. Polyhydroxyalkanoate production from fermented volatile fatty acids: effect of pH and feeding regimes. Bioresour. Technol. 128, 533–538.

Chen, S., Cheng, H., Wyckoff, K.N., He, Q., 2016. Linkages of Firmicutes and Bacteroidetes populations to methanogenic process performance. J. Ind. Microbiol. Biotechnol. 43 (6), 771–781.

Cheng, H., Li, Y., Hu, Y., Guo, G., Cong, M., Xiao, B., Li, Y.-Y., 2021. Bioenergy recovery from methanogenic co-digestion of food waste and sewage sludge by a high-solid anaerobic membrane bioreactor (AnMBR): mass balance and energy potential. Bioresour. Technol. 326, 124754.

Chuenchart, W., Logan, M., Leelayouthayotin, C., Visvanathan, C., 2020. Enhancement of food waste thermophilic anaerobic digestion through synergistic effect with chicken manure. Biomass Bioenergy 136, 105541.

Erdogdu, A.E., Polat, R., Ozbay, G., 2018. Pyrolysis of goat manure to produce bio-oil. Eng. Sci. Technol. 22 (2), 452–457.

Gurmessa, B., Pedretti, E.F., Cocco, S., Cardelli, V., Corti, G., 2020. Manure anaerobic digestion effects and the role of pre- and post-treatments on veterinary antibiotics and antibiotic resistance genes removal efficiency. Sci. Total Environ. 721, 137532.

Hansen, K.H., Angelidaki, I., Ahring, B.K., 1998. Anaerobic digestion of swine manure: inhibition by ammonia. Water Res. 32 (1), 5–12.

Hassan, M., Ding, W., Shi, Z., Zhao, S., 2016. Methane enhancement through co-digestion of chicken manure and thermo-oxidative cleaved wheat straw with waste activated sludge: a C/N optimization case. Bioresour. Technol. 211, 534–541.

Himanshu, H., Murphy, J.D., Grant, J., O'Kiely, P., 2018. Antagonistic effects on biogas and methane output when co-digesting cattle and pig slurries with grass silage in in vitro batch anaerobic digestion. Biomass Bioenergy 109, 190–198.

Jang, H.M., Choi, Y.K., Kan, E.S., 2018. Effects of dairy manure-derived biochar on psychrophilic, mesophilic and thermophilic anaerobic digestions of dairy manure. Bioresour. Technol. 250, 927–931.

Lay, C.H., Kumar, G., Mudhoo, A., Lin, C.Y., Leu, H.J., Shobana, S., Nguyen, M.L.T., 2020. Recent trends and prospects in biohythane research: an overview. Int. J. Hydrog. Energy 45 (10), 5864–5873.

Li, D., Liu, S., Mi, L., Li, Z., Yuan, Y., Yan, Z., Liu, X., 2015. Effects of feedstock ratio and organic loading rate on the anaerobic mesophilic co-digestion of rice straw and cow manure. Bioresour. Technol. 189, 319–326.

Li, D., Song, L., Fang, H., Li, P., Teng, Y., Li, Y.-Y., Liu, R., Niu, Q., 2019. Accelerated bio-methane production rate in thermophilic digestion of cardboard with appropriate biochar: dose-response kinetic assays, hybrid synergistic mechanism, and microbial networks analysis. Bioresour. Technol. 290, 121782.

Li, J., Sun, S., Yan, P., Fang, L., Yu, Y., Xiang, Y., Wang, D., Gong, Y., Gong, Y., Zhang, Z., 2017a. Microbial communities in the functional areas of a biofilm reactor with anaerobic-aerobic process for oily wastewater treatment. Bioresour. Technol. 238, 7–15.

Li, K., Liu, R., Cui, S., Yu, Q., Ma, R., 2018a. Anaerobic co-digestion of animal manures with corn stover or apple pulp for enhanced biogas production. Renew. Energy 118, 335–342.

Li, R.R., Duan, N., Zhang, Y.H., Liu, Z.D., Li, B.M., Zhang, D.M., Dong, T.L., 2017b. Anaerobic co-digestion of chicken manure and microalgae *Chlorella* sp.: methane potential, microbial diversity and synergistic impact evaluation. Waste Manage. 68, 120–127.

Li, Y., Achinas, S., Zhao, J., Geurkink, B., Krooneman, J., Euverink, G.J.W., 2020. Co-digestion of cow and sheep manure: performance evaluation and relative microbial activity. Renew. Energ. 153, 553–563.

Li, Y., Lu, J., Xu, F., Li, Y., Li, D., Wang, G., Li, S., Zhang, H., Wu, Y., Shah, A., Li, G., 2018b. Reactor performance and economic evaluation of anaerobic co-digestion of dairy manure with corn stover and tomato residues under liquid, hemi-solid, and solid state conditions. Bioresour. Technol. 270, 103–112.

Li, Y., Zhang, R., Chen, C., Liu, G., He, Y., Liu, X., 2013. Biogas production from co-digestion of corn stover and chicken manure under anaerobic wet, hemi-solid, and solid state conditions. Bioresour. Technol. 149, 406–412.

Li, Y., Zhang, R., He, Y., Zhang, C., Liu, X., Chen, C., Liu, G., 2014. Anaerobic co-digestion of chicken manure and corn stover in batch and continuously stirred tank reactor (CSTR). Bioresour. Technol. 156, 342–347.

Lian, T., Zhang, W., Cao, Q., Wang, S., Dong, H., 2020. Enhanced lactic acid production from the anaerobic co-digestion of swine manure with apple or potato waste via ratio adjustment. Bioresour. Technol. 318, 124237.

Liang, Y.G., Li, X.J., Zhang, J., Zhang, L.G., Cheng, B.J., 2017. Effect of microscale ZVI/magnetite on methane production and bioavailability of heavy metals during anaerobic digestion of diluted pig manure. Environ. Sci. Pollut. Res. 24 (13), 12328–12337.

Liu, W.R., Zeng, D., She, L., Su, W.X., He, D.C., Wu, G.Y., Ma, X.R., Jiang, S., Jiang, C.H., Ying, G.G., 2020. Comparisons of pollution characteristics, emission situations, and mass loads for heavy metals in the manures of different livestock and poultry in China. Sci. Total Environ. 734, 139023.

Liu, Y.C., Whitman, W.B., 2008. Metabolic, phylogenetic, and ecological diversity of the methanogenic archaea. Ann. NY Acad. Sci. 1125, 171–189.

Lü, C., Shen, Y., Li, C., Zhu, N., Yuan, H., 2020. Redox-active biochar and conductive graphite stimulate methanogenic metabolism in anaerobic digestion of waste-activated sludge: beyond direct interspecies electron transfer. ACS Sustain. Chem. Eng. 8 (33), 12626–12636.

Luo, C.H., Lu, F., Shao, L.M., He, P.J., 2015. Application of eco-compatible biochar in anaerobic digestion to relieve acid stress and promote the selective colonization of functional microbes. Water Res. 68, 710–718.

Meng, X., Sui, Q., Liu, J., Yu, D., Wang, Y., Wei, Y., 2020. Relieving ammonia inhibition by zero-valent iron (ZVI) dosing to enhance methanogenesis in the high solid anaerobic digestion of swine manure. Waste Manage. 118, 452–462.

Misi, S.N., Forster, C.F., 2001. Batch co-digestion of multi-component agro-wastes. Bioresour. Technol. 80 (1), 19–28.

Neves, L., Oliveira, R., Alves, M.M., 2009. Fate of LCFA in the co-digestion of cow manure, food waste and discontinuous addition of oil. Water Res. 43 (20), 5142–5150.

Niu, Q., Hojo, T., Qiao, W., Qiang, H., Li, Y.-Y., 2014. Characterization of methanogenesis, acidogenesis and hydrolysis in thermophilic methane fermentation of chicken manure. Chem. Eng. J. 244, 587–596.

Niu, Q., Li, Y.-Y., 2016. Recycling of Livestock Manure into Bioenergy. Springer, *Singapore*.

Niu, Q., Qiao, W., Qiang, H., Hojo, T., Li, Y.-Y., 2013. Mesophilic methane fermentation of chicken manure at a wide range of ammonia concentration: stability, inhibition and recovery. Bioresour. Technol. 137, 358–367.

Pan, J.T., Ma, J.Y., Zhai, L.M., Liu, H.B., 2019. Enhanced methane production and syntrophic connection between microorganisms during semi-continuous anaerobic digestion of chicken manure by adding biochar. J. Clean Prod. 240, 118178.

Paranhos, A.G.d.O., Adarme, O.F.H., Barreto, G.F., Silva, SdQ, Aquino, SFd, 2020. Methane production by co-digestion of poultry manure and lignocellulosic biomass: kinetic and energy assessment. Bioresour. Technol. 300, 122588.

Peidong, Z., Yanli, Y., Yongsheng, T., Xutong, Y., Yongkai, Z., Yonghong, Z., Lisheng, W., 2009. Bioenergy industries development in China: dilemma and solution. Renew. Sustain. Energ. Rev. 13 (9), 2571–2579.

Piatek, M., Lisowski, A., Kasprzycka, A., Lisowska, B., 2016. The dynamics of an anaerobic digestion of crop substrates with an unfavourable carbon to nitrogen ratio. Bioresour. Technol. 216, 607–612.

Pitk, P., Palatsi, J., Kaparaju, P., Fernandez, B., Vilu, R., 2014. Mesophilic co-digestion of dairy manure and lipid rich solid slaughterhouse wastes: process efficiency, limitations and floating granules formation. Bioresour. Technol. 166, 168–177.

Quaik, S., Embrandiri, A., Ravindran, B., Hossain, K., Al-Dhabi, N.A., Arasu, M.V., Ignacimuthu, S., Ismail, N., 2020. Veterinary antibiotics in animal manure and manure laden soil: scenario and challenges in Asian countries. J. King Saud Uni–v. Sci. 32 (2), 1300–1305.

Sadaka, S., Devender, K.V., 2015. Evaluation of chemically coagulated swine manure solids as value-added products. Sustain. Bioenergy Syst. 5 (4), 136–150.

Shi, X., Lin, J., Zuo, J., Li, P., Li, X., Guo, X., 2017. Effects of free ammonia on volatile fatty acid accumulation and process performance in the anaerobic digestion of two typical bio-wastes. J. Environ. Sci.-China 55, 49–57.

Shi, Y., Ge, Y., Chang, J., Shao, H., Tang, Y., 2013. Garden waste biomass for renewable and sustainable energy production in China: potential, challenges and development. Renew. Sust. Energ. Rev. 22, 432–437.

Song, L., Li, D., Cao, X., Tang, Y., Liu, R., Niu, Q., Li, Y.-Y., 2019a. Optimizing biomethane production of mesophilic chicken manure and sheep manure digestion: mono-digestion and co-digestion kinetic investigation, autofluorescence analysis and microbial community assessment. J. Environ. Manage. 237, 103–113.

Song, L., Li, D., Fang, H., Cao, X., Liu, R., Niu, Q., Li, Y.-Y., 2019b. Revealing the correlation of biomethane generation, DOM fluorescence, and microbial community in the mesophilic co-digestion of chicken manure and sheep manure at different mixture ratio. Environ. Sci. Pollut. Res. 26 (19), 19411–19424.

Song, L., Yong, S., Li, D., Liu, R., Niu, Q., 2019c. The auto fluorescence characteristics, specific activity, and microbial community structure in batch tests of mono-chicken manure digestion. Waste Manage 83, 57–67.

Sunyoto, N.M.S., Zhu, M.M., Zhang, Z.Z., Zhang, D.K., 2016. Effect of biochar addition on hydrogen and methane production in two-phase anaerobic digestion of aqueous carbohydrates food waste. Bioresour. Technol. 219, 29–36.

Tian, H., Duan, N., Lin, C., Li, X., Zhong, M., 2015. Anaerobic co-digestion of kitchen waste and pig manure with different mixing ratios. J. Biosci. Bioeng. 120 (1), 51–57.

Tian, Z., Zhang, Y., Yu, B., Yang, M., 2016. Changes of resistome, mobilome and potential hosts of antibiotic resistance genes during the transformation of anaerobic digestion from mesophilic to thermophilic. Water Res. 98, 261–269.

Vermeulen, L.C., Benders, J., Medema, G., Hofstra, N., 2017. Global Cryptosporidium loads from livestock manure. Environ. Sci. Technol. 51 (15), 8663–8671.

Viggi, C.C., Rossetti, S., Fazi, S., Paiano, P., Majone, M., Aulenta, F., 2014. Magnetite particles triggering a faster and more robust syntrophic pathway of methanogenic propionate degradation. Environ. Sci. Technol. 48 (13), 7536–7543.

Wang, D., Ai, J., Shen, F., Yang, G., Zhang, Y., Deng, S., Zhang, J., Zeng, Y., Song, C., 2017. Improving anaerobic digestion of easy-acidification substrates by promoting buffering capacity using biochar derived from vermicompost. Bioresour. Technol. 227, 286–296.

Wang, G.J., Li, Q., Gao, X., Wang, X.C.C., 2018. Synergetic promotion of syntrophic methane production from anaerobic digestion of complex organic wastes by biochar: performance and associated mechanisms. Bioresour. Technol. 250, 812–820.

Wang, X., Yang, G., Feng, Y., Ren, G., Han, X., 2012. Optimizing feeding composition and carbon–nitrogen ratios for improved methane yield during anaerobic co-digestion of dairy, chicken manure and wheat straw. Bioresour. Technol. 120, 78–83.

Wei, Y., Wachemo, A.C., Yuan, H., Li, X., 2020. Enhanced hydrolysis and acidification strategy for efficient co-digestion of pretreated corn stover with chicken manure: digestion performance and microbial community structure. Sci. Total Environ. 720, 137401.

Xiao, Y., Zhang, E., Zhang, J., Dai, Y., Yang, Z., Christensen, H.E.M., Ulstrup, J., Zhao, F., 2017. Extracellular polymeric substances are transient media for microbial extracellular electron transfer. Sci. Adv. 3 (7), e1700623.

Yan, W., Lu, D., Liu, J., Zhou, Y., 2019. The interactive effects of ammonia and carbon nanotube on anaerobic digestion. Chem. Eng. J. 372, 332–340.

Yang, S., Chen, Z.Q., Wen, Q.X., 2021. Impacts of biochar on anaerobic digestion of swine manure: methanogenesis and antibiotic resistance genes dissemination. Bioresour. Technol. 324, 124679.

Ye, J., Li, D., Sun, Y., Wang, G., Yuan, Z., Zhen, F., Wang, Y., 2013. Improved biogas production from rice straw by co-digestion with kitchen waste and pig manure. Waste Manage 33 (12), 2653–2658.

Yin, D-m, Mahboubi, A., Wainaina, S., Qiao, W., Taherzadeh, M.J., 2021. The effect of mono- and multiple fermentation parameters on volatile fatty acids (VFAs) production from chicken manure via anaerobic digestion. Bioresour. Technol. 330, 124992.

You, Y., Liu, S., Wu, B., Wang, Y.W., Zhu, Q.L., Qin, H., Tan, F.R., Ruan, Z.Y., Ma, K.D., Dai, L.C., Zhang, M., Hu, G.Q., He, M.X., 2017. Bio-ethanol production by *Zymomonas mobilis* using pretreated dairy manure as a carbon and nitrogen source. RSC Adv. 7 (7), 3768–3779.

Zhang, J., Zhang, R., He, Q., Ji, B., Wang, H., Yang, K., 2020. Adaptation to salinity: response of biogas production and microbial communities in anaerobic digestion of kitchen waste to salinity stress. J. Biosci. Bioeng. 130 (2), 173–178.

Zhang, T., Yang, Y., Liu, L., Han, Y., Ren, G., Yang, G., 2014. Improved biogas production from chicken manure anaerobic digestion using cereal residues as co-substrates. Energy Fuels 28 (4), 2490–2495.

Zhang, W., Wang, X., Xing, W., Li, R., Yang, T., Yao, N., Lv, D., 2021. Links between synergistic effects and microbial community characteristics of anaerobic co-digestion of food waste, cattle manure and corn straw. Bioresour. Technol. 329, 124919.

Zhang, X., Mao, X., Pi, L., Wu, T., Hu, Y., 2019a. Adsorptive and capacitive properties of the activated carbons derived from pig manure residues. J. Environ. Chem. Eng. 7 (3), 103066.

Zhang, Y., Yang, Z., Xu, R., Xiang, Y., Jia, M., Hu, J., Zheng, Y., Xiong, W., Cao, J., 2019b. Enhanced mesophilic anaerobic digestion of waste sludge with the iron nanoparticles addition and kinetic analysis. Sci. Total Environ. 683, 124–133.

Zhang, Z., Zhang, G., Li, W., Li, C., Xu, G., 2016. Enhanced biogas production from sorghum stem by co-digestion with cow manure. Int. J. Hydrog. Energy 41 (21), 9153–9158.

Zhao, S., Chen, W., Luo, W., Fang, H., Lv, H., Liu, R., Niu, Q., 2021. Anaerobic co-digestion of chicken manure and cardboard waste: focusing on methane production, microbial community analysis and energy evaluation. Bioresour. Technol. 321, 124429.

Zhao, S., Li, P., Fang, H., Song, L., Li, D., Liu, R., Niu, Q., 2020. Enhancement methane fermentation of *Enteromorpha prolifera* waste by *Saccharomyces cerevisiae*: batch kinetic investigation, dissolved organic matter characterization, and synergistic mechanism. Environ. Sci. Pollut. Res. 27 (14), 16254–16267.

Zheng, Z., Cai, Y., Zhang, Y., Zhao, Y., Gao, Y., Cui, Z., Hu, Y., Wang, X., 2021. The effects of C/N (10–25) on the relationship of substrates, metabolites, and microorganisms in "inhibited steady-state" of anaerobic digestion. Water Res. 188, 116466.

Zheng, Z., Liu, J., Yuan, X., Wang, X., Zhu, W., Yang, F., Cui, Z., 2015. Effect of dairy manure to switchgrass co-digestion ratio on methane production and the bacterial community in batch anaerobic digestion. Appl. Energy 151, 249–257.

Zhu, J., Li, Y., Wu, X., Miller, C., Chen, P., Ruan, R., 2009. Swine manure fermentation for hydrogen production. Bioresour. Technol. 100 (22), 5472–5477.

CHAPTER

5

Conversion of wastewater to bioenergy and biochemicals *via* anaerobic digestion

Yang Li

School of Ocean Science and Technology, Dalian University of Technology, Panjin, Liaoning, China

5.1   Introduction

Wastewaters are the flow of used water from community or productive processes, which contain domestic wastewater, industrial wastewater, agriculture wastewater, and also the runoff rainwater (Zhen et al., 2019). The discharge of inefficiently treated wastewaters will cause water quality deterioration of ocean and river (Chan et al., 2009). To combat this increasing problem on aquatic environment, strict regulation on wastewaters discharge is being implemented by various governmental bodies (Chan et al., 2009). The biological oxygen demand (BOD) and chemical oxygen demand (COD) are the major goal parameters of wastewaters treatment (Verhoeven & Meuleman, 1999; Di Lorenzo et al., 2009). BOD reflects the content of biodegradable organic matter in the water body and COD reflects the degree of pollution of the reducing substances in water. Actually, the most common reducing substances are organic matters.

Organic matters in wastewaters include granular organics and dissolved organics. Specifically, granular organics consist of living organisms (such as zooplankton, phytoplankton, and zoogloeal) and abiotic organic particles (self-settlement with time). The dissolved organics consist of lipids, proteins, carbohydrates, vitamins, humus, etc. Nowadays, there are a wide range of physiochemical and biological technologies used for organic wastewaters treatment in the world (Demirel et al., 2005). However, as the reagent costs are high and the soluble COD removal is poor in physiochemical treatment processes, biological processes are usually preferred (Vidal et al., 2000; Demirel et al., 2005). Among biological treatment processes, aerobic activated sludge and anaerobic treatment are commonly employed for organic wastewaters treatment (Demirel et al., 2005). Anaerobic treatment of wastewaters has recently gained worldwide attention due to its simplicity, low construction costs, low land requirements, plain operation and maintenance, low sludge production, low energy requirements, and high energy production capacity in the form of biogas compared to aerobic treatment (Appels et al., 2011).

Anaerobic digestion (AD) has been the process most often used to stabilize wastewaters since the early 20th century and remains in development at present. With a series of synergistic biological degradation processes performed by microbial consortia in the absence of oxygen, organics are converted to methane and carbon dioxide in the AD process. In this respect, intensive studies have been conducted in the past few decades, and the various "green technologies" have been extensively explored (Angenent et al., 2004; Hahn-Hägerdal et al., 2006; Li & Yu, 2011). The typical configurations of AD for the anaerobic treatment of organic wastewaters are up-flow anaerobic sludge blanket (UASB) reactor, expanded granular sludge bed reactor (EGSB), anaerobic sequencing batch reactor (ASBR), anaerobic membrane bioreactor, and continuously stirred tank reactor (CSTR). Via these anaerobic treatment processes, the wastewaters are potential commodities from which bioenergy and biochemicals may be produced. The typical bioenergy recovered are methane and hydrogen, the production of which could replace fossil fuels, reduce carbon emission, and slow global warming (Angenent et al., 2004). The typical valuable biochemicals are bioethanol, polyhydroxyalkanoates (PHAs), biopesticides, and bioflocculant, the production of which provides a great commercial value. This chapter mainly focuses on some mature and novel technical summaries for the bioenergy and biochemicals production during anaerobic treatment of organic wastewaters.

5.2 Bioenergy production

The universal energy demand is increasing immensely owing to fossil fuel depletion (Banu et al., 2020). "World energy outlook 2019" published by US Energy Information Administration reported that the world energy demand is predicted to increase by 50% from 2018 to 2050, among which the renewable energy will increase the most quickly (The International Energy Agency). However, the fossil fuels are not considered to be renewable and therefore, their utilization is limited to some extent (Banu et al., 2020). The increase in global energy demand and decrease in fossil fuel have urged to find alternative renewable and carbon-neutral energy sources (Banu et al., 2020). Wastewaters, generated during the processes of industry, agriculture, or people's activities, represent a typical renewable source of energy in consideration of high contents of nutrients and readily biodegradable organic substances (Zhen et al., 2019). Achieving resource utilization and safe disposal of wastewaters are major issues related to the environmental safety and the healthy development of human society (Zhen et al., 2019). Therefore, considerable interest has been attracted concerning the reuse of organic wastewaters to produce bioenergy (Qiao et al., 2013; Zhen et al., 2019). AD is currently one of the most attractive technologies both to recover energy and resources and to treat the environmental pollutants from organic wastewaters through bioconversion processes (Appels et al., 2008; Rittmann, 2008).

Fig. 5.1 defines the recognized three-step AD converting the complex organic matters in wastewaters to useful bioenergy outlets: breaking down the complex organics or macromolecules to simple chemical forms that can be taken up by the microorganisms, fermentation to simpler products, and the final stabilization by specialized microorganisms to an

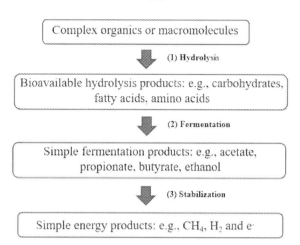

FIG. 5.1 **The three steps for converting complex organics in wastewaters to useful energy outputs.** The stabilization step is what determines the form of bioenergy that is produced: CH_4, H_2, or electricity.

energy-rich form (Rittmann, 2008). The last step—generation of an energy-rich form—is the inherent advantage of microbial energy conversion systems, because it avoids the large energy cost for extracting the bioenergy from water. The most common energy outputs from organics in wastewaters during AD (Fig. 5.1) are methane gas (CH_4), and hydrogen gas (H_2) (Rittmann, 2008).

5.2.1 Methane production

The process of anaerobic biotreatment involves the conversion of organic compounds in an oxygen-free environment by the consortia of bacteria to produce a useful energy carrier gas called biogas (Rashama et al., 2019), which consists mainly of 55%–70% (v/v) methane having an energy value of 35.8 kJ/L under standard conditions (Montiel Corona & Razo-Flores, 2018; Park et al., 2018). During the last decades, global production of biogas and biomethane from AD of biomass resources has increased substantially by a global volume of 59 billion m^3 for biogas, equaling to 35 billion m^3 of methane from 2000 to 2014 (Weiland, 2010; Ramm et al., 2019). Further growth of biogas production is expected to reach the set goals on decarbonization with the overall objective of limiting the increase of global average temperature to below 2 °C above preindustrial levels as stated at the United Nations Framework Convention on Climate Change (UNFCCC) 21st Conference of the Parties (COP21) in 2015 (Ramm et al., 2019). Biogas production is most established in the United States and Europe. Besides, there is also a high demand and potential for biogas serving as an alternative clean energy source in Asia, Africa, and Latin America (Weiland, 2010; Ramm et al., 2019).

Methanogenic AD is a classical anaerobic bioconversion process that has been practiced for over a century and used in full-scale facilities worldwide to produce biogas (Angenent et al., 2004; Li & Yu, 2011). The food web of AD is reasonably well understood (Fig. 5.2) (Angenent et al., 2004), which is a complicated process that involves a mixture of consortia of microorganisms (Harper & Pohland, 1986; Schnürer et al., 1999; Li & Yu, 2011). The whole methane fermentation is divided up into four phases: hydrolysis, acidogenesis, acetogenesis,

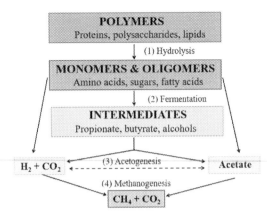

FIG. 5.2   **Intricate food web of methanogenic anaerobic digestion.** Several trophic groups of microorganisms work together to convert complex organic material into methane and carbon dioxide.

and methanogenesis. The individual degradation steps are carried out by different consortia of microorganisms, which partly stand in syntrophic interrelation and place different requirements on the environment (Angelidaki et al., 1993):

1. *Hydrolysis*: Hydrolyzing and fermenting microorganisms such as *Bacteriocides, Clostridia, Bifidobacteria, Streptococci,* and *Enterobacteriaceae* initially attack on large organic molecules to convert them into smaller, simple ones, for example, carbohydrates to sugars, proteins to amino acids, and lipids to fatty acids.
2. *Acidogenesis*: Acidogenic bacteria convert simple organic molecules to mainly acetate and hydrogen and varying amounts of volatile fatty acids (VFAs) such as propionate and butyrate.
3. *Acetogenesis*: The volatile fatty acids are converted into acetate and hydrogen by obligate hydrogen-producing acetogenic bacteria.
4. *Methanogenesis*: Methanogenic bacteria use acetate or hydrogen to form methane and carbon dioxide.

An important breakthrough was made in the methanogenic AD about 30 years ago, with the development of the UASB reactor (Lettinga et al., 1980; Angenent et al., 2004), which efficiently retains the complex microbial consortium without the need for immobilization on a carrier material by the formation of biological granules with good setting characteristics (Harper & Pohland, 1986; Schnürer et al., 1999; Li & Yu, 2011). The mean sludge residence time of UASBs is much longer than the hydraulic retention time (HRT), due to this self-immobilization process (Angenent et al., 2004). The performance of UASBs depends on the mean sludge retention time and reactor volume depends on the HRT; therefore, UASBs can efficiently convert organic compounds in wastewaters into methane in small "high-rate" reactors (Angenent et al., 2004). Approximately 60% of the thousands of anaerobic full-scale treatment facilities worldwide are known based on the UASB design concept, treating a diverse range of industrial wastewaters (Jantsch et al., 2002; Karim & Gupta, 2003; Angenent et al., 2004). A significant limitation of UASB reactors is the interference of suspended solids in the incoming wastewater with granulation and reactor performance (Kalogo & Verstraete,

1999; Angenent et al., 2004). Hence, other high-rate systems, such as ASBR, were developed to better handle high-suspended solids in wastewaters. However, there seems to be limitations inherently insurmountable by ASBRs, as they are one-stage bioreactors. The fast-growing acidogens and prevailing acidification usually lead to organic acid accumulation, which further suppresses the activities of acetogens and methanogens (Li & Yu, 2011). As a likely solution, two-stage bioconversion processes gained increasing popularity, and opened up the possibility for a fundamental breakthrough of the above barriers (Li & Yu, 2011). Through a properly designed two-stage bioconversion process, the organic compounds that are not directly utilizable by the functional microorganisms can be stepwisely converted to valuable products, usually at higher conversion efficiency than one-stage bioconversion processes (Demirel & Yenigün, 2002). The concept of two-stage AD has now been widely accepted and successfully demonstrated, and the two-stage bioconversion processes for methane production from wastewaters has been a common practice worldwide (Kim et al., 2004; Shin et al., 2010). Nevertheless, the existing two-stage bioconversion systems mostly still suffer from low conversion efficiency for wastewaters treatment due to several remaining limitations. First of all, the presence of refractory organics in wastewaters may constitute a big challenge for the conversion, and this barrier is actually encountered in all bioconversion processes (Angenent et al., 2004; Gavrilescu & Chisti, 2005; Li & Yu, 2011). Furthermore, the separation of hydrolysis/acidogenesis and methanogenesis could also negatively affect the syntrophic association and interspecies hydrogen transfer (IHT) between acidogens/acetogens and methanogens (Li & Yu, 2011). These factors should be comprehensively considered when designing two-stage bioconversion processes for methane production. Notwithstanding the many remaining limitations, two-stage bioconversion processes undoubtedly present a promising and versatile strategy that is readily applicable to almost all bioprocesses, thus embracing numerous possibilities for future application (Li & Yu, 2011).

Although the structure of methanogenic anaerobic reactors is improved, slow reactions, odor production, the requirement of a buffer for maintaining neutral pH, and unsuitability for low-strength wastewater are always the disadvantages of all AD processes (Park et al., 2018). Many studies have focused on overcoming the slow reaction of AD. The majority have attempted to optimize operational conditions (e.g., temperature, pH, organic loading, and retention time). Catalysts such as Fe, Ni, Zn, Mn, Cu, Co, and Mo were also used to accelerate AD (Choong et al., 2016; Park et al., 2018). Nevertheless, above methods could not fundamentally solve the problem of AD's slow reaction. The mechanism of AD is actually the electrons and energy transfer between various microorganisms by cooperative interactions (Lee et al., 2012; Park et al., 2018).

Hydrogen is regarded as a diffusive electron carrier during the process of electron transfer from fatty acids or alcohols to carbon dioxide or methane. As hydrogen is produced by acidogenic/acetogenic bacteria and utilized by hydrogenotrophic methanogens, this process is an interspecies electron transfer (IET) (Li & Fang, 2007; Park et al., 2018). The production of hydrogen from fatty acids or alcohols is only thermodynamically feasible (i.e., $\Delta G < 0$) when hydrogen concentrations are very low ($H_2 < 10^{-4}$ atm). This condition is achieved by the consumption of hydrogen by hydrogenotrophic methanogens (Li et al., 2018; Park et al., 2018). Therefore, syntrophy between hydrogen-producing bacteria and hydrogenotrophic methanogens is essential for their survival (Cord-Ruwisch et al., 1998; Park et al., 2018). In addition, low concentrations of hydrogen limit the methane formation rate during AD (Boone et al., 1989;

Park et al., 2018). Thus, IET via hydrogen is assumed to be a "bottleneck" in methane formation (Kato et al., 2012; Park et al., 2018).

In 2010, a new electron transfer approach without mediating diffusive electron carriers was proposed as direct interspecies electron transport (DIET) (Summers et al., 2010). The discovery that methanogens (*Methanothrix* and *Methanosarcina*) can accept electrons via DIET for the reduction of carbon dioxide into methane in defined cocultures provides some novel methods to accelerate anaerobic methane production (Summers et al., 2010; Rotaru et al., 2014; Zhao et al., 2018). Compared with the traditional syntrophic metabolism mode of IHT, the advantages of DIET are obvious. On the one hand, DIET can proceed via electrically conductive pili or outer surface c-type cytochromes (Summers et al., 2010), or a combination of biological and abiological electron transfer components, such as conductive materials (Fig. 5.3) (Liu et al., 2012; Chen et al., 2014; Chen et al., 2015; Liu et al., 2015), no longer requiring the hydrogen as the essential electron carrier as well as overcoming the thermodynamic limitation of hydrogen production (Stams & Plugge, 2009). On the other hand, DIET can provide more rapid electron transfer than IHT (44.9×10^3 e$^-$/cps vs 5.24×10^3 e$^-$/cps) (Storck et al., 2016), supporting *Methanothrix* and *Methanosarcina* species to grow fast. Therefore, establishing the DIET-based syntrophic metabolism during the AD process is expected to achieve a better methanogenesis.

As the DIET-based syntrophy, such as *Geobacter* species, is usually not abundant in anaerobic digesters, conductive materials were provided as the electrical conduit for interspecies electron exchange to help enrich them (Yang et al., 2017; Peng et al., 2018; Zhao et al., 2018).

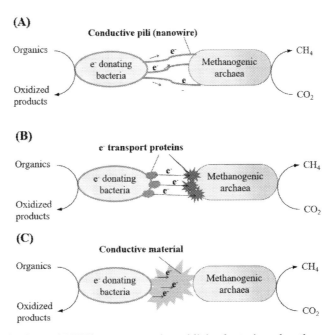

FIG. 5.3   **Three mechanisms of DIET between organics-oxidizing bacteria and methanogenic archaea.**
(A) DIET via conductive pili. (B) DIET via membrane-bound electron transport proteins. (C) DIET via abiotic conductive materials. This figure is a modification of a schematic presented in a previous paper.

Surprisingly, most of the studies on DIET via conductive materials have found reduced lag times for initiating methane production, enhanced methane production rates, high methane yields, and resistance to inhibitory conditions (Park et al., 2018). These findings suggest that conventional AD can be significantly improved by supplementing with conductive materials (Park et al., 2018). By supplementing with conductive materials, methane formation rates increased by 79%–300% and methane yields increased by 100%–178% (Park et al., 2018). As Cruz Viggi et al. (2014) reported in their theoretical calculations, the maximum electron carrier flux via magnetite was 10^6 higher than that via hydrogen.

AD on DIET via conductive materials has great practical potential as this technology can solve some of the disadvantages of AD such as the long lag time, slow methane formation rate, low methane yield, and vulnerability to unfavorable conditions (Park et al., 2018). However, it will take time for this technology to be deployed in full-scale AD plants (Park et al., 2018). There are several technical difficulties to be overcome before practical application: (1) During DIET, the electron-donating bacteria prefer to use fatty acids or alcohols, but no other complex organics. One possible approach is to use a reactor that can generate fatty acids and alcohols for selectively enriching electron-donating bacteria, such as *Geobacter* species, before the methanogenic reactor (Park et al., 2018). This concept was pioneered by Zhao et al. (2017), who successfully demonstrated DIET via granular active carbon (GAC) in a methanogenic reactor by placing an acidogenic reactor before it to treat complex organic waste. (2) Continuous addition of conductive materials into the digester would be necessary; otherwise, the conductive materials would be washed out from the digester. Maintaining conductive materials in the digester is a promising method (Park et al., 2018). This can be achieved by filling biomass-support media fabricated from conductive materials in the AD reactors. In this respect, the media should be fabricated to facilitate efficient attachment of the relevant microorganisms, then the methanogenic process would be accelerated (Park et al., 2018).

The finding that promoting DIET improves methane yields and increases resilience against instabilities has suggested that re-engineering anaerobic digesters to favor DIET will enhance the effectiveness of this important bioenergy strategy (Zhao et al., 2020). To date, investigations on the impact of amendments of conductive materials on DIET have primarily been empirical (Zhao et al., 2020). Furthermore, improved engineering of AD, based on existing information, should proceed in parallel with the suggested microbiological studies with a focus on technologies that are sustainable and economically feasible for long-term operation (Zhao et al., 2020). A holistic approach that considers physical/chemical modifications to favor DIET-based communities as well as the deployment of conductive materials is likely to be most effective (Zhao et al., 2020). This is an exciting time in AD. Modern tolls of microbiology and engineering show promise for advancing the understanding of the function of this early bioenergy technology and improving its effectiveness to enhance the sustainable conversion of wastewaters to bioenergy (Zhao et al., 2020).

5.2.2 Biohydrogen production

H_2 is considered as a clean fuel due to its combustion with oxygen, generating only water and energy as subproducts, which is another low-solubility gas that can be a naturally separating electron and energy output from anaerobic microbial systems (Fig. 5.1) (Rittmann, 2008; Menezes et al., 2019). H_2 has approximately the same heat of combustion as CH_4

($\triangle H^0 = 237$ kJ/mol or 119 kJ/e$^-$ eq), and it can be combusted to gain the energy value similar to CH_4 (Rittmann, 2008). In 2025, it is expected that the contribution of hydrogen to the energy market will reach 10% of the total energy demand (Argun et al., 2008; Menezes et al., 2019). H_2 has the very large added benefit that it can be used as the fuel for a conventional fuel cell, where electricity can be produced without combustion, producing pollution-free electrical energy with an efficiency at least 50% higher than that of a combustion-steam-turbine approach (Rittmann, 2008). In addition to its applicability as an energy source, H_2 is used as a raw material for manufacturing electronic devices, chemical products, hydrogenated fats, fertilizers, oils in the food industry, rocket fuel, refrigeration fluid, steel, desulphurization, and refining gasoline (Menezes et al., 2019). Today, H_2 gas production is a large and mature industry (about 108 m^3/year in the United States), but virtually hydrogen production mainly comes from reforming fossil fuels that contribute to CO_2 emissions (Kapdan & Kargi, 2006; Rittmann, 2008; Ball & Wietschel, 2009; Menezes et al., 2019). Having a biomass source for H_2 would be a giant boon for renewable, carbon-free energy (Rittmann, 2008).

A variety of biohydrogen production processes have been developed, such as dark fermentation, photofermentation and biophotolysis to directly recover hydrogen from wastewaters (Yu & Fang, 2002; Kapdan & Kargi, 2006; Das & Veziroglu, 2008; Fang et al., 2009; Li & Yu, 2011). The most common approach for converting biomass to H_2 is a dark fermentation process that is essentially a truncated version of methanogenesis (Ren et al., 2006; Vatsala et al., 2008; Banu et al., 2020). Dark fermentation process involves biological conversion of organic compounds to biohydrogen in the absence of light and is mediated by anaerobic microbes (Banu et al., 2020). Naturally, the hydrolytic bacteria first utilize the organics from substrate into simpler ones. Obligatory hydrogen-generating acidogens utilized hydrolyzed products to produce acid and hydrogen (Banu et al., 2020). The substrates rich in organic compounds are highly potent for enhanced biohydrogen production via dark fermentation. Utilization of inexpensive and plentifully obtainable feedstock is considered to be a crucial parameter for improved biohydrogen production (Banu et al., 2020). In this aspect, wastewaters generated from industries are reported to be potential organic rich, renewable, and inexpensive substrate for biohydrogen production (Rittmann, 2008; Sivagurunathan et al., 2017). Industrial wastewaters contain highly degradable organic matters that result in net energy production (Rittmann, 2008). The common industrial wastewaters as substrates for biohydrogen production include sugar industry (Jayabalan et al., 2019), food processing industry (Gupta & Pawar, 2018), citric acid (Yang et al., 2006), paper mill, rice mill (Ramprakash & Muthukumar, 2018), distillery (Laurinavichene et al., 2018), chemical (Srikanth et al., 2009), starch processing (Khongkliang et al., 2017), palm oil (Mishra et al., 2016), beverage (Sivagurunathan et al., 2015), cheese whey (Venetsaneas et al., 2009), and pharmaceutical (Shi et al., 2017). The carbohydrate-rich wastewaters have more hydrogen-producing potential than the protein- and fat-rich wastewaters.

Although fermentation to H_2 is straightforward, the main drawback today is that only a small fraction of the electrons in the starting organic material ends up in H_2 (Rittmann, 2008). The problem is that fermenting bacteria channel most of their electron flow to organic products, not to H_2 (Rittmann, 2008). In the H_2-fermentation literature, this is recognized by theoretical "maximum H_2 yield" as 4 mol H_2/mol glucose, which comes from the fermentation of glucose to only H_2 and acetic acid:

$$C_6H_{12}O_6 + 2H_2O \rightarrow 4H_2 + 2CH_3COOH + 2CO_2$$

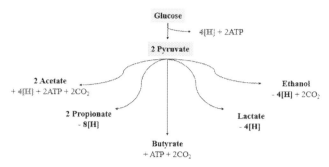

FIG. 5.4 The dominant catabolic pathways of anaerobic glucose fermentation.

This "maximum" reaction results in production of four moles of H_2 per mole of glucose, because two-thirds of the electrons and energy are "misdirected" to acetic acid (Angenent et al., 2004; Rittmann, 2008). Complete oxidation to carbon dioxide and H_2, however, would produce 12 moles of hydrogen per mole glucose (Angenent et al., 2004). The practical H_2 yield is around 2 mol H_2/mol glucose, due to the formation of butyrate, ethanol, propionate, and a few other organic fermentation products (Fig. 5.4) (Rittmann, 2008). Due to these yield problems, traditional H_2 fermentation seems to be "stuck" even before it is applied in a serious way to deal with real residual biomass (Rittmann, 2008). Besides the low H_2 yields resulted from above reason, the dark fermentation commercial feasibility is also limited by the presence of hydrogen-consuming bacteria, such as homoacetogens and methanogens (Puyol et al., 2017). Some pretreatments of the seed sludge are used to stimulate the hydrogen-generating bacteria and suppress the hydrogen-consuming bacteria to improve the hydrogen yield from dark fermentation system (Wang & Wan, 2008; Penteado et al., 2013). The main pretreatment methods reported in the literature are temperature shock, pH shock, chemical pretreatment, physics shock, and combinations of these methods (Penteado et al., 2013). However, extra costs are needed for these pretreatment methods.

A two-phase anaerobic system for high-rate hydrogen production has been demonstrated in numerous studies with great success. In the past decade, various hydrogen-producing two-phase anaerobic systems have emerged to decouple the hydrogen production process from the hydrolysis/acidification step, by employing photofermentation, biophotolysis, or bioelectrolysis as a second stage specifically for hydrogen production (Liu et al., 2006; Shi & Yu, 2006; Tao et al., 2007; Call & Logan, 2008; Chen et al., 2008; Li & Yu, 2011). Through such strategy, it is theoretically possible to completely recover the 12 mol H_2/mol glucose from complex organics (Bartacek et al., 2007; Li & Fang, 2007; Hallenbeck & Ghosh, 2009; Guo et al., 2010). Despite the high performances and great potential of two-phase anaerobic systems; however, there are remaining hurdles (Lee et al., 2010; Li & Yu, 2011). For the hydrolysis/acidification and dark fermentation as a hybrid first stage, it remains a problem in terms of strengthening the competitivity of hydrogen-producing bacteria and improving VFA properties (Brentner et al., 2010; Li & Yu, 2011). For a subsequent photofermentation process, the low light penetration and conversion efficiencies and the high costs of gastight and transparent photobioreactors present some challenges (Tao et al., 2007; Chen et al., 2008; Li & Yu, 2011). Furthermore, many studies have been directed over the years toward photofermentation (Barbosa et al., 2001; Oh, 2004; Rozendal et al., 2006), which use sunlight to overcome this thermodynamical barrier. However, the diffuse nature of solar radiation and the limited conversion efficiencies severely limit the economic feasibility of these processes due to the enormous reactor surface areas that are required (Rozendal et al., 2006).

Microbial electrolysis cell (MEC), as an electrochemical device combined with AD, uses electroactive microorganisms as catalysts to convert the organic matter to hydrogen and provides a novel approach for economically viable H_2 production from a wide range of renewable biomass energy (Wang et al., 2020). In MEC system, H_2 generation is accomplished by the cooperation of a succession of microorganisms under electrochemical conditions. The complex organics are first broken down by anaerobic fermentative bacteria into simple organic matters (such as volatile fatty acids, ethanol, amino acids, etc.). Then, the electroactive microorganisms, which attach on the anode surface and have the ability to facilitate the transfer of electrons from the substrate to the conductive anode, oxidize these simple organic products into CO_2, electrons, and protons with the anode serving as the electron acceptor (Wang et al., 2020). Electroactive microorganisms, such as *Geobacter*, *Shewanella*, and *Pseudomonas* spp., have outer membrane cytochromes and/or nanowires involved in electron transfer, which allow them to have the ability of extracellular electron transfer and transferring electrons to the anode. A small additional voltage is applied in the MEC system to promote the proceeding of the cathodic H^+ ion reduction reaction because it has a lower redox potential than the anode. The electrons then travel through the external circuit, and the protons in the solution pass across the membrane and combine at the cathode to generate hydrogen (Fig. 5.5). The required external potential for an MEC is theoretically 110 mV, much lower than the 1210 mV required for direct electrolysis of water at neutral pH because some

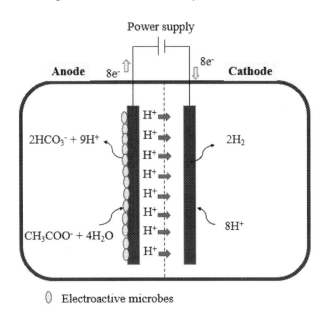

FIG. 5.5   Schematic representation of hydrogen production through biocatalyzed electrolysis of acetate.

energy comes from the biomass oxidation process in the anodic chamber (Du et al., 2007). MECs can potentially produce about 8–9 mol H_2/mol glucose compared to the typical 4 mol H_2/mol glucose achieved in conventional fermentation (Liu et al., 2005; Du et al., 2007). In biohydrogen production using MECs, oxygen is no longer needed in the cathodic chamber (Du et al., 2007). Thus, MEC efficiency improves because oxygen leak to the anodic chamber is no longer an issue (Du et al., 2007). Another advantage is that hydrogen can be accumulated and stored for later usage to overcome the inherent low power feature of the MECs (Du et al., 2007). Therefore, MECs provide a renewable hydrogen source that can contribute to the overall hydrogen demand in a hydrogen economy (Du et al., 2007; Swanson et al., 2016). While, for the electrohydrogenesis, an additional electricity input is required to drive the hydrogen production, and the expensive platinum catalyst for cathode leads to a high reactor cost (Cheng & Logan, 2007; Li & Yu, 2011). Moreover, specific bacteria of photosynthetic bacteria and anode-respiring bacteria are required in such systems (Hallenbeck & Ghosh, 2009; Logan, 2009; Li & Yu, 2011), which are likely to be affected by wastewaters that contain various toxics and competitive microbial consortia if the process is not properly controlled (Li & Yu, 2011).

Renewable hydrogen has many applications, the most prominent ones being for transportation and industry (Logan et al., 2008). Transportation fuels are currently responsible for about 20%–25% of the global fossil fuel consumption (Logan et al., 2008). Because of climate change, and instabilities in the fossil fuel market, there is great interest in hydrogen as a transportation fuel (i.e., the hydrogen economy) (Logan et al., 2008). Moreover, even without a hydrogen economy, there exists a large hydrogen demand (Logan et al., 2008). MECs can contribute significantly to these hydrogen demands by producing large quantities of hydrogen from renewable resources such as biomass and wastewaters. The MEC concept is now well proven, and significant advancements have been made with respect to the performance since its discovery (Logan et al., 2008). To become a mature hydrogen production technology, however, several research questions still need to be addressed (Logan et al., 2008; Rozendal et al., 2008): (1) more experience is required with real organic feedstocks containing complex organic substrates such as polymeric and particulate substances; (2) novel, more cost-effective chemical and/or biological cathodes need to be developed that show low potential losses and are not platinum based; (3) membrane pH gradients need to be eliminated, or membranes should not be used in the reactor; (4) methanogenic consumption of the hydrogen product needs to be prevented (in case of membrane-less MECs and/or MECs with a biocathode). The most critical need is to develop a cost-effective, scalable MEC design (Logan et al., 2008). Furthermore, if MECs are used in combination with new energy sources such as solar energy and wind energy, it is bound to improve the economic performance of MECs, reduce costs, and promote its promotion in industry applications (Wang et al., 2020).

5.3 Biochemicals production

A major limitation to wider application of the bioenergy technologies described in the previous sections is the relative low cost of the current nonrenewable energy sources (Angenent et al., 2004). Government subsidies, or a direct local need to save on energy costs (e.g., biogas that is used directly on-site as a fuel), are necessary to make those processes economically viable (Angenent et al., 2004). In addition, although bioenergy production may reduce the cost of wastewater treatment, it cannot entirely satisfy the energy demand of our society (Angenent et al., 2004). Therefore, the production of high-value chemicals from organic material in wastewater might be more feasible than bioenergy production (Angenent et al., 2004; Bungay, 2004). Industrial wastewaters can become inexpensive raw materials for integrated fermentation processes, among which high-carbohydrate wastewaters that are particularly appropriate for conversion to valuable products in pure-culture or coculture processes. It is the cost efficiency of the bioconversion process that ultimately determines whether a specific waste stream is suitable for production of a specific product (Laufenberg et al., 2003; Angenent et al., 2004). AD is an economical bioconversion process converting organic pollutants in wastewater to biochemicals as intermediates, such as acetate, propionate, ethanol, etc. The poor biochemical yield and limited types of products are perhaps the greatest barrier, and therefore much effort has been devoted to enhancing the amount and types of product formed during anaerobic processes. The first possible approaches to realize this enhancement goal have been discovery of novel microorganisms, manipulation of cell genomes, protein engineering by mutagenesis and molecular evolution, and control of carbon fluxes by the addition of external substrate and inhibitors (Angenent et al., 2004).

Anaerobic bioconversion may also be enhanced by process modification, such as culture immobilization or coupling with other processes (Angenent et al., 2004). For example, anaerobic fermentation combined with aerobic process could convert volatile fatty acids by *Ralstonia eutropha* to enhance the efficiency of PHA production from waste organics (Du & Yu, 2002; Angenent et al., 2004). Currently, as the focus has shifted to microbial reductive processes at the cathode, bioelectrosynthesis combined with anaerobic fermentation, which can couple COD removal and energy recovery from wastewaters with chemical synthesis, is being explored for a number of applications (Wang et al., 2020). For example, microorganisms can catalyze electrochemical reactions of the reduction of carbon dioxide to organics such as acetate, ethanol, etc., even some other higher value compounds. Bioelectrosynthesis is a promising strategy for the microbial conversion of carbon dioxide and other organic feedstocks to transportation fuels and other organic commodities, which holds strong promises for a new concept for biochemicals generation.

5.3.1 Bioethanol production

Bioethanol is produced on a global scale to meet the energy requirements of the modern transportation sector; by using renewable resources for ethanol production, the ecological and environmental impact of drilling, transporting, and processing fossil fuels could, in principle, be reduced (Aditiya et al., 2016; de Azevedo et al., 2017; Nagarajan et al., 2017; Karagoz et al., 2019). Ethanol can also be used as a gasoline substitute to power petrol engines, which leads to increase in its demand (Guo et al., 2015). It is conveyed that instead of burning gasoline, ethanol burning is able to eradicate the release of sulfur dioxide which is the cause of acid rain. Bioethanol can be produced by various raw materials, such as sugars, crops, and lignocelluloses. Sugars (such as molasses) can be directly fermented into ethanol by microorganisms without pretreatment; however, they are only available in limited quantities (Li et al., 2021). Food crops such as barley, wheat, corn stover, cassava, etc. need the essential enzymatic hydrolysis for converting them to fermentable sugars, but the ethanol production of which will bring the potential risk of competing food sources with human

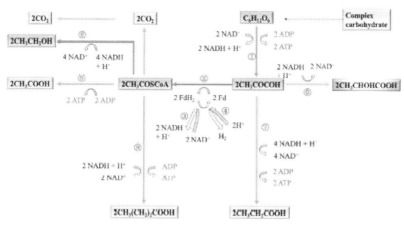

FIG. 5.6 Metabolic pathways involved in NADH/NAD⁺ transformation and ATP synthesis in the anaerobic fermentation of glucose.

beings as well as animals (Li et al., 2021). Lignocellulose-like organics that require a pretreatment prior to the acidic and enzymatic hydrolysis will further increase the operating costs (Li et al., 2021). An alternative to these conventional raw materials for bioethanol production is in demand. High organic load wastewaters can be used efficiently for bioethanol production via AD, and the organic pollutants in wastewaters will be removed at the same time.

As illustrated in Fig. 5.1, AD undergoes three steps: hydrolysis, acidogenesis, and methanogenesis. Commonly, in the mixed cultures of anaerobic fermentation, the organics are initially converted to saccharides, such as glucose, which are then fermented into fatty acids and ethanol via different metabolic pathways (Fig. 5.6). The fermentation products are related directly to the substrate used and the operating conditions, such as hydraulic retention time (HRT), organic loading rate, temperature, and pH (Antoni et al., 2007). The operating pH, in particular, plays a major role on the effluent composition of the acidogenic reactor (Antoni et al., 2007). The best pH condition for ethanol production is 4.0–4.5 during anaerobic fermentation of organic wastewaters, named as ethanol-type fermentation by Ren et al. (2006). Accompanied with the ethanol production, acetic acid, hydrogen, and carbon dioxide are also produced. Therefore, most researchers cared about the ethanol production in anaerobic acidogenic aiming to achieve a high yield of hydrogen.

Recently, some researchers paid their attention especially to ethanol production from some carbohydrate-enriched wastewaters. Apart for the adjustment of the pH condition and HRT to help increase the yield of ethanol, Li et al. (2021) used magnetite to enhance the ethanol-type fermentation process for dairy wastewater via anaerobic fermentation. The primary mechanism was that magnetite stimulated the growth of ethanol-producing Fe(III)-reducing bacteria, such as *Clostridium* species. The extracellular Fe(III) reduction by Fe(III)-reducing bacteria accelerated the intracellular transformation from NADH to NAD⁺ and indirectly preceded with the ATP synthesis for the growth of ethanol-type fermentation bacteria (*Clostridium, Ethanoligenens*). Consequently, a relatively long-term and stable enhancement on bioethanol production was achieved. However, although the yield of ethanol is increased, few researchers focus on the purification of bioethanol from the effluent, which makes a missing step between bioethanol production from AD and the practical application.

5.3.2 Polyhydroxyalkanoates (PHAs) production

Recently, the production of PHAs from wastewater biological treatment systems has recently aroused widespread interests (Wang & Yu, 2006; Kleerebezem & van Loosdrecht, 2007; Bengtsson et al., 2008; Li & Yu, 2011). PHAs are the raw material with interesting characteristics for the production of biodegradable plastics in a significant number of industrial applications (Serafim et al., 2008; Li & Yu, 2011). The most common PHAs are polyhydroxybutyrate (PHB) and polyhydroxyvalerate (PHV) as a class of bioplastics, which can replace conventional petroleum-derived plastics and minimize excess sludge production (Puyol et al., 2017; Zhao et al., 2019). They are stored as granules in the cell cytoplasm by microorganisms under unfavorable growth and nutrient conditions (Anderson & Dawes, 1990; Li & Yu, 2011; Puyol et al., 2017). The biochemical route of substrate (e.g. acetate) conversion to PHA is much shorter and less energy-demanding than the production of biomass. And PHA accumulation is directly associated with good substrate removal and its synthesis could be promoted by the key enzyme and encoded by *phaC* gene, characteristic of microbes with PHA-storing capacity (Puyol et al., 2017; Zhao et al., 2019).

Although PHA production is a well-known process, its production as a bioplastic commodity has some restrictions. First, the use of pure cultures and sterile feedstocks will contribute to high costs production. Second, PHA yield from mixed cultures is low. Third, the extraction and purification methods need to be improved (Puyol et al., 2017). However, due to the considerable interest in the emerging bioeconomy many waste streams and low-value feedstocks are suitable targets for bioplastic production, including municipal wastewater and agro-industrial wastewaters (e.g., molasses, paper mill, oil mill, and dairy), and spent glycerol (Puyol et al., 2017). The use of wastewaters in combination with mixed microbial cultures has increased the feasibility and sustainability of the biopolymers as the costs associated with the feedstocks are decreased, and the operation and maintenance of the process is simplified (Puyol et al., 2017).

Currently, different reactor configurations and microbes are involved in a three-step process to produce PHA. The first step is the prefermentation of the wastewaters, where hydrolytic and fermentative bacteria break down complex organics to readily biodegradable compounds, such as VFA, precursor chemicals that can be used as an easy and readily available carbon source for PHA production, increasing PHA accumulation yields (Puyol et al., 2017). Although the direct use of nonfermented wastewaters is possible, complex substrates may not be completely degraded, therefore decreasing accumulation yields. Moreover, it has been found that complex substrates can promote the growth of non PHA-storing biomass (Albuquerque et al., 2011; Basset et al., 2016; Puyol et al., 2017). Although wastewaters prefermentation can be seen as an extra step that can increase the costs it can also generate opportunities in a circular economy concept.

The second step is the enrichment of PHA-producing microbes, where activated sludge is the most common seed biomass. The previously produced VFA are used under dynamic feeding strategies (e.g., by imposing feast/famine and presence/absence of electron donor) to enrich the seed sludge with microbes that possess a high PHA storing capacity. The final yields are related with several factors, such as the substrate type, nitrogen (ammonia) limitation, the pH, organic loading rate, and the accumulation cycle length (Anderson & Dawes, 1990; Morgan-Sagastume et al., 2014; Puyol et al., 2017). Jiang et al. (2011) obtained up to a 90% of PHA by dry cell content when using activated sludge enriched in sequencing batch reactor (SBR) with lactate as substrate, by *Plasticicumulans acidivorans* a novel *gammaproteobacterium*, which are the nearest relatives found to the genera *Methylocaldum*. However, the monomeric distribution of PHA will directly influence the bioplastic properties, therefore not only it is desirable to have high PHA yields but also a stable and robust biopolymer composition. The biopolymer composition is known to be dependent on the VFA distribution of the feedstock during the accumulation (Laycock et al., 2013; Puyol et al., 2017), where propionate has been correlated with the higher PHV percentages (Serafim et al., 2008; Puyol et al., 2017); but also could be dependent on the metabolic pathways used of the different PHA-storing biomass.

Finally, the third step consists in the PHA accumulation in batch systems, where the biopolymer content of the previously enriched community is maximized (Albuquerque et al., 2013; Moralejo-Gárate et al., 2014; Puyol et al., 2017). The principle objective of the PHA production process researchers are working on is the production of bioplastics (Kleerebezem & van Loosdrecht, 2007). Evidently, the production of chemicals for the chemical industry is more profitable than the production of energy carriers, owing to the low price of fossil fuels (Kleerebezem & van Loosdrecht, 2007). However, the production of PHA as an energy carrier might also be of interest in cases where no complex infrastructure for methane compression or gas engines for electricity generation are available or cannot be established in an economically viable way (Kleerebezem & van Loosdrecht, 2007). In this case nutrient-deficient wastewaters rich in soluble organic material could effectively be treated in a PHA production process where up to 70% of the organic material is concentrated in a solid product with a relatively low nitrogen content (Kleerebezem & van Loosdrecht, 2007).

The PHA production can be combined with other wastewater treatment plant processes such as organic and nutrients removal. For instance, Morgan-Sagastume successfully integrated the PHA-storing enrichment step with the treatment of the readily biodegradable COD from influent wastewater, with average COD removals of 70% in a pilot-scale feast/famine SBR (Morgan-Sagastume et al., 2015). The enriched PHA-storing community was transferred to the PHA accumulation reactor, where it was fed with VFA-rich stream from waste-activated sludge prefermentation obtaining PHA productivities of 38% PHA by dry cell (Puyol et al., 2017). The integration of different operational units within the treatment plant allows to improve the nutrient and energy recovery from the wastewater (Puyol et al., 2017).

5.3.3 Biopesticides production

Biopesticide is another obtainable value-added product from microorganisms during wastewater fermentation. Biopesticides are a broad group of agents that are mass produced and biologically based agents for controlling plant pests (Greaves & Grant, 2010), which are generally highly target specific, leave no toxic residues, and produce a lesser overall impact on the environment than chemical pesticides. One of the most well-known and dominant biopesticides is derived from *Bacillus thuringiensis*, and mainly consists crystal delta endotoxins and other pesticidal substances (Brar et al., 2006; Sanchis & Bourguet, 2008; Li & Yu, 2011). The use of agro-industrial wastewater for *B. thuringiensis* cultivation and biopesticide production has been extensively investigated (Sachdeva et al., 2000; Khuzham-shukurov et al., 2001; Montiel et al., 2001; Vidyarthi et al., 2002; Yezza et al., 2006; Li & Yu, 2011). Wastewaters can provide the necessary nutritional elements to sustain growth, sporulation, and crystal formation for microbes. The most widely used microbial pesticides are subspecies and strains of *B. thuringiensis* (Li & Yu, 2011). Nevertheless, these processes generally suffer from low product yield and complicated composition, attributed to the complex nature of wastewaters and the coexistence of other competitive bacteria. Moreover, most of the nutrients in the wastewaters are unavailable to *B. thuringiensis*. This weakens its ability to produce pores, crystals, and other insecticidal metabolites and leads to low entomotoxicity. Therefore, problems occur in one-stage pesticide fermentation in terms of product harvesting and separation (Brar et al., 2009; Li & Yu, 2011). A two-stage bioconversion process seems to provide a viable solution. Tirado-Montiel found that acid hydrolysis of wastewater sludge can improve the entomotoxicity of *B. thuringiensis* by 24% attributed to the increasing bioavailability of substrate and nutrients (Tirado-Montiel et al., 2001). Meanwhile, a relatively simplified and optimized environment in the subsequent step can favor the predominance of *B. thuringiensis* and lead to higher biopesticide yield and quality. Thus, although no two-step bioconversion process has been experimentally demonstrated for biopesticide production from wastewater so far, it exhibits a high potential to recover biopesticides from wastewaters with enhanced product quality and productivity (Li & Yu, 2011).

5.3.4 Bioflocculant production

Bioflocculant, as another valuable microbe-derived metabolite with high biodegradability and ecological safety, has also recently attracted considerable attention (Salehizadeh & Shojaosadati, 2001; Li & Yu, 2011). Bioflocculant is extracellular biopolymers excreted by microorganisms, including proteins, glycoproteins, polysaccharides, lipids, and glycolipids (Salehizadeh & Shojaosadati, 2003; Li & Yu, 2011). Direct bioflocculants production from organic wastewaters has been performed in several studies with certain success (Wang et al., 2007; Zhang et al., 2007; Zhong et al., 2008; Li & Yu, 2011). For example, *Bacillus* sp. AEMREG7 was shown to produce a bioflocculant with a maximum flocculating activity of 92.6% against kaolin clay suspension, and *Enterobacter cloacae* sp. WD7 and *Pseudomonas alcaligenes* WD22 were shown to produce bioflocculants with flocculating activities of 91% and 55%, respectively. However, limited by a high complexity of wastewater, the low product yield remains a major constriction to its application (Zhang et al., 2007; Li & Yu, 2011). Consequently, the screening of new strains producing highly efficient bioflocculants with reduced production cost became a solution (bioflocculant). In addition, a two-stage bioconversion process may offer another attractive solution (Fujita et al., 2000; Fujita et al., 2001; Li & Yu, 2011). Especially, some low-molecular VFAs such as acetic and propionic acids were found to give a higher bioflocculant yield than other simple organic source (Fujita et al., 2000; Li & Yu, 2011). Therefore, although utilization of VFAs for bioflocculant production has only been demonstrated in the treatment of sludge digestion liquor by far, the application of a two-step bioconversion process for bioflocculant production from various wastewaters is technologically viable and expectable. Most of all, unlike other biochemicals that usually need high product purity for application, bioflocculant can be directly used in biological wastewater treatment systems with not purity requirement. Thus, this process shows great promise for large-scale application. However, like other biochemical recovery processes, the similar barriers of low production yield and high recovery cost need to be overcome to make such processes commercially viable.

5.4 Conclusions and perspectives

The concept of wastewater treatment is evolving from pollutant removal to resource recovery. In this respect, anaerobic bioconversion offers an appealing avenue to recover sustainable bioenergy and valuable biochemicals from wastewaters. Especially, bioconversion of different types of organic wastewaters to methane via AD is a mature process that is being used within full-scale engineering applications around the world. Although methane is a low-value product, AD still represents one of the most economically viable technologies, as biogas can be converted to liquid fuel and high-value products through conventional chemical manufacturing processes. While AD for hydrogen, bioethanol, PHAs, biopesticides, bioflocculants and other biochemicals are yet to be demonstrated at a pilot for even laboratory scale. Several challenges arise from the separation of reaction processes, and many barriers remain for their scaling up and commercialization. Issues concerning the separation of the soluble products from the fermentation broth, and the stability of pure- or coculture fermentation processes remain to be addressed. Furthermore, the effect of reactors design configuration and operating conditions need to be more widely studied and optimized before the processes can be scaled up.

Notwithstanding the many remaining restrictions, the rapid advances in metabolic and biochemical engineering, process optimization, integration strategy as well as cost-effective separation and purification technologies are likely to bring the AD out of the laboratory systems and to substantially promote and extend their application for bioenergy and biochemicals production in the near future. Furthermore, a better understanding of the microorganisms and their respiratory mechanism of AD is expected to benefit the scaling up of the existing systems and the development of new technologies. Therefore, AD represents a promising biotechnology paradigm to meet the dual ends of wastewater treatment and bioenergy/biochemical production, and may finally make "green energy" and "bioproducts" practical and sustainable.

References

Aditiya, H.B., Mahlia, T.M.I., Chong, W.T., Nur, H., Sebayang, A.H., 2016. Second generation bioethanol production: a critical review. Renew. Sust. Energ. Rev. 66, 631–653.

Albuquerque, M.G.E, Carvalho, G., Kragelund, C., Silva, A.F., Barreto-Crespo, M.T., Reis, M.A., Nielsen, P.H., 2013. Link between microbial composition and carbon substrate-uptake preferences in a PHA-storing community. ISME J. 7 (1), 1–12.

Albuquerque, M.G.E, Martino, V., Pollet, E., Averous, L., Reis, M.A.M., 2011. Mixed culture polyhydroxyalkanoate (PHA) production from volatile fatty acid (VFA)-rich streams: effect of substrate composition and feeding regime on PHA productivity, composition and properties. J. Biotechnol. 151 (1), 66–76.

Anderson, A.J., Dawes, E.A., 1990. Occurrence, metabolism, metabolic role, and industrial uses of bacterial polyhydroxyalkanoates.. Microbiol. Rev. 54 (4), 450–472.

Angelidaki, I., Ellegaard, L., Ahring, B.K., 1993. A mathematical model for dynamic simulation of anaerobic digestion of complex substrates: focusing on ammonia inhibition. Biotechnol. Bioeng. 42 (2), 159–166.

Angenent, L.T., Karim, K., Muthanna, H.A., Brian, A.W., Rosa, D., 2004. Production of bioenergy and biochemicals from industrial and agricultural wastewater. Trends Biotechnol. 22 (9), 477–485.

Antoni, D., Zverlov, V.V., Schwarz, W.H., 2007. Biofuels from microbes. Appl. Microbiol. Biotechnol. 77 (1), 23–35.

Appels, L., Baeyens, J., Degreve, J., Dewil, R., 2008. Principles and potential of the anaerobic digestion of waste-activated sludge. Prog. Energy Combust. 34 (6), 755–781.

Appels, L., Lauwers, J., Degreve, J., Helsen, L., Lievens, B., Willems, K., Van Impe, J., Dewil, R., 2011. Anaerobic digestion in global bio-energy production: potential and research challenges. Renew. Sust. Energy Rev. 15 (9), 4295–4301.

Argun, H., Kargi, F., Kapdan, I.K., Oztekin, R., 2008. Biohydrogen production by dark fermentation of wheat powder solution: effects of C/N and C/P ratio on hydrogen yield and formation rate. Int. J. Hydrog. Energy 33 (7), 1813–1819.

Ball, M., Wietschel, M., 2009. The future of hydrogen—opportunities and challenges. Int. J. Hydrog. Energy 34 (2), 615–627.

Banu, J.R., Kavitha, S.K., Kannah, Y., Bhosale, R.R., Kumar, G., 2020. Industrial wastewater to biohydrogen: possibilities towards successful biorefinery route. Bioresour. Technol. 298, 122378.

Barbosa, M.J., Rocha, J.M.S., Tramper, J., Wijffels, R.H., 2001. Acetate as a carbon source for hydrogen production by photosynthetic bacteria. J. Biotechnol. 85 (1), 25–33.

Bartacek, J., Zabranska, J., Lens, P.N.L., 2007. Developments and constraints in fermentative hydrogen production. Biofuel. Bioprod. Bioref. 1 (3), 201–214.

Basset, N., Katsou, E., Frison, N., Malamis, S., Dosta, J., Fatone, F., 2016. Integrating the selection of PHA storing biomass and nitrogen removal via nitrite in the main wastewater treatment line. Bioresour. Technol. 200, 820–829.

Bengtsson, S., Hallquist, J., Werker, A., Welander, T., 2008. Acidogenic fermentation of industrial wastewaters: effects of chemostat retention time and pH on volatile fatty acids production. Biochem. Eng. J. 40 (3), 492–499.

Boone, D.R., Johnson, R.L., Liu, Y., 1989. Diffusion of the interspecies electron carriers H_2 and formate in methanogenic ecosystems and its implications in the measurement of K_m for H_2 or formate uptake. Appl. Environ. Microbiol. 55 (7), 1735–1741.

Brar, S.K., Verma, M., Tyagi, R.D., Valero, J.R., Surampalli, R.Y., 2006. Efficient centrifugal recovery of *Bacillus thuringiensis* biopesticides from fermented wastewater and wastewater sludge. Water Res. 40 (6), 1310–1320.

Brar, S.K., Verma, M., Tyagi, R.D., Surampalli, R.Y., 2009. Value addition of wastewater sludge: future course in sludge reutilization. Pract. Period. Hazard. Toxic Radioact. Waste Manage. 13 (1), 59–74.

Brentner, L.B., Peccia, J., Zimmerman, J.B., 2010. Challenges in developing biohydrogen as a sustainable energy source: Implications for a research agenda. Environ. Sci. Technol. 44 (7), 2243–2254.

Bungay, H.R., 2004. Confessions of a bioenergy advocate. Trends Biotechnol. 22 (2), 67–71.

Call, D., Logan, B.E., 2008. Hydrogen production in a single chamber microbial electrolysis cell lacking a membrane. Environ. Sci. Technol. 42 (9), 3401–3406.

Chan, Y.J., Chong, M.F., Law, C.L., Hassell, D.G., 2009. A review on anaerobic–aerobic treatment of industrial and municipal wastewater. Chem. Eng. J. 155 (1-2), 1–18.

Chen, C.Y., Yang, M.H., Yeh, K.L., Liu, C.H., Chang, J.S., 2008. Biohydrogen production using sequential two-stage dark and photo fermentation processes. Int. J. Hydrog. Energy. 33 (18), 4755–4762.

Chen, S., Rotaru, A.E., Liu, F., Philips, J., lovley, D.R., 2014. Carbon cloth stimulates direct interspecies electron transfer in syntrophic co-cultures. Bioresour. Technol. 173, 82–86.

Chen, S.S., Rotaru, A.E., Shrestha, P.M., Malvankar, N.S., Liu, F.H., Fan, W., Nevin, K.P., Lovley, D.R., 2015. Promoting interspecies electron transfer with biochar. Sci. Rep. 4 (1), 5019.

Cheng, S.A., Logan, B.E., 2007. Sustainable and efficient biohydrogen production via electrohydrogenesis. Proc. Natl Acad. Sci. USA 104 (47), 18871–18873.

Choong, Y.Y., Norli, I., abdullah, A.Z., Yhaya, M.F., 2016. Impacts of trace element supplementation on the performance of anaerobic digestion process: a critical review. Bioresour. Technol. 209, 369–379.

Cord-Ruwisch, R., Lovley, D.R., Schink, B., 1998. Growth of *Geobacter sulfurreducens* with acetate in syntrophic cooperation with hydrogen-oxidizing anaerobic partners. Appl. Environ. Microbiol. 64 (6), 2232–2236.

Cruz Viggi, C., Rossetti, S., Fazi, S., Paiano, P., Majone, M., Aulenta, F., 2014. Magnetite particles triggering a faster and more robust syntrophic pathway of methanogenic propionate degradation. Environ. Sci. Technol. 48 (13), 7536–7543.

Das, D., Veziroglu, T., 2008. Advances in biological hydrogen production processes. Int. J. Hydrog. Energy. 33 (21), 6046–6057.

de Azevedo, A., Fornasier, F., de Silva, S.M., Schneider, R.D., Hoelta, M., de Souza, D., 2017. Life cycle assessment of bioethanol production from cattle manure. J. Clean. Prod. 162, 1021–1030.

Demirel, B., Yenigun, O., Onay, T.T., 2005. Anaerobic treatment of dairy wastewaters: a review. Process Biochem. 40 (8), 2583–2595.

Demirel, B., Yenigün, O., 2002. Two-phase anaerobic digestion processes: a review. J. Chem. Technol. Biotechnol. 77 (7), 743–755.

Di Lorenzo, M., Curtis, T.P., Head, I.M., Scott, K., 2009. A single-chamber microbial fuel cell as a biosensor for wastewaters. Water Res. 43 (13), 3145–3154.

Du, G., Yu, J., 2002. Green technology for conversion of food scraps to biodegradable thermoplastic polyhydroxy-alkanoates. Environ. Sci. Technol. 36 (24), 5511–5516.

Du, Z.W., Li, H.R., Gu, T.Y., 2007. A state of the art review on microbial fuel cells: a promising technology for waste-water treatment and bioenergy. Biotechnol. Adv. 25 (5), 464–482.

Fang, F., Liu, X.W., Xu, J., Yu, H.Q., Li, Y.M., 2009. formation of aerobic granules and their PHB production at various substrate and ammonium concentrations. Bioresour. Technol. 100 (1), 59–63.

Fujita, M., Ike, M., Tachibana, S., Kitada, G., Kim, S.M., Inoue, Z., 2000. Characterization of a bioflocculant produced by *Citrobacter sp.* TKF04 from acetic and propionic acids. J. Biosci. Bioeng. 89 (1), 40–46.

Fujita, M., Ike, M., Jang, J.H., Kim, S.M., Hirao, T., 2001. Bioflocculation production from lower-molecular fatty acids as a novel strategy for utilization of sludge digestion liquor. Water Sci. Technol. 44 (10), 237–243.

Gavrilescu, M., Chisti, Y., 2005. Biotechnology—a sustainable alternative for chemical industry. Biotechnol. Adv. 23 (7-8), 471–499.

Greaves, J., Grant, W., 2010. Underperforming policy networks: the biopesticides network in the United Kingdom. British Polit. 5 (1), 14–40.

Guo, M.X., Song, W.P., Buhain, J., 2015. Bioenergy and biofuels: history, status, and perspective. Renew. Sust. Energy Rev. 42, 712–725.

Guo, X.M., Trably, E., Latrille, E., Carrere, H., Steyer, J.P., 2010. Hydrogen production from agricultural waste by dark fermentation: a review. Int. J. Hydrog. Energy 35 (19), 10660–10673.

Gupta, S., Pawar, S.B., 2018. An integrated approach for microalgae cultivation using raw and anaerobic digested wastewaters from food processing industry. Bioresour. Technol. 269, 571–576.

Hahn-Hägerdal, B., Galbe, M., Gorwa-Grauslund, M.F., Liden, G., Zacchi, G., 2006. Bio-ethanol – the fuel of tomorrow from the residues of today. Trends Biotechnol. 24 (12), 549–556.

Hallenbeck, P.C., Ghosh, D., 2009. Advances in fermentative biohydrogen production: the way forward. Trends Biotechnol. 27 (5), 287–297.

Harper, S.R., Pohland, F.G., 1986. Recent developments in hydrogen management during anaerobic biological wastewater treatment. Biotechnol. Bioeng. 28 (4), 585–602.

Jantsch, T.G., Angelidaki, I., Schmidt, J.E., Brana de Hvidsten, B.E., Ahring, B.K., 2002. Anaerobic biodegradation of spent sulphite liquor in a UASB reactor. Bioresour. Technol. 84 (1), 15–20.

Jayabalan, T., Matheswaran, M., Mohammed, S.N., 2019. Biohydrogen production from sugar industry effluents using nickel based electrode materials in microbial electrolysis cell. Int. J. Hydrog. Energy 44 (32), 17381–17388.

Jiang, Y., Sorokin, D.Y., Kleerebezem, R., Muyzer, G., van Loosdrecht, M., 2011. *Plasticicumulans acidivorans* gen. nov., sp. nov., a polyhydroxyalkanoate-accumulating gammaproteobacterium from a sequencing-batch bioreactor. Int. J. Syst. Evol. Microbiol. 61 (9), 2314–2319.

Kalogo, Y., Verstraete, W., 1999. Development of anaerobic sludge bed (ASB) reactor technologies for domestic wastewater treatment: motives and perspectives. World J. Microb. Biotechnol. 15 (5), 523–534.

Kapdan, I.K., Kargi, F., 2006. Bio-hydrogen production from waste materials. Enzyme Microb. Technol. 38 (5), 569–582.

Karagoz, P., Bill, R.M., Ozkan, M., 2019. Lignocellulosic ethanol production: evaluation of new approaches, cell immobilization and reactor configurations. Renew. Energy. 143, 741–752.

Karim, K., Gupta, S.K., 2003. Continuous biotransformation and removal of nitrophenols under denitrifying conditions. Water Res. 37 (12), 2953–2959.

Kato, S., Hashimoto, K., Watanabe, K., 2012. Methanogenesis facilitated by electric syntrophy via (semi)conductive iron-oxide minerals. Environ. Microbiol. 14 (7), 1646–1654.

Khongkliang, P., Kongjan, P., Utarapichat, B., Reungsang, A., O-Thong, S., 2017. Continuous hydrogen production from cassava starch processing wastewater by two-stage thermophilic dark fermentation and microbial electrolysis. Int. J. Hydrog. Energy 42 (45), 27584–27592.

Khuzhamshukurov, N.A., Yusupov, T.Y., Khalilov, I.M., Guzalova, A.G., Muradov, M.M., Davranov, K.D., 2001. The insecticidal activity of *Bacillus thuringiensis* cells. Appl. Biochem. Microb. 37 (6), 596–598.

Kim, S.H., Han, S.K., Shin, H.S., 2004. Two-phase anaerobic treatment system for fat-containing wastewater. J. Chem. Technol. Biotechnol. 79 (1), 63–71.

Kleerebezem, R., van Loosdrecht, M.C., 2007. Mixed culture biotechnology for bioenergy production. Curr. Opin. Biotechnol. 18 (3), 207–212.

Laufenberg, G., Kunz, B., Nystroem, M., 2003. Transformation of vegetable waste into value added products: (A) the upgrading concept; (B) practical implementations. Bioresour. Technol. 87 (2), 167–198.

Laurinavichene, T., Tekucheva, D., Laurinavichius, K., Tsygankov, A., 2018. Utilization of distillery wastewater for hydrogen production in one-stage and two-stage processes involving photofermentation. Enzyme Microb. Tech. 110, 1–7.

Laycock, B., Halley, P., Pratt, S., Werker, A., Lant, P., 2013. The chemomechanical properties of microbial polyhydroxyalkanoates. Prog. Polym. Sci. 38 (3-4), 536–583.

Lee, H., Vermaas, W.F.J., Rittmann, B.E., 2010. Biological hydrogen production: prospects and challenges. Trends Biotechnol. 28 (5), 262–271.

Lee, S., Kang, H., Lee, Y.H., Lee, T.J., Han, K., Choi, Y., Park, H., 2012. Monitoring bacterial community structure and variability in time scale in full-scale anaerobic digesters. J. Environ. Monitor. 14 (7), 1893–1905.

Lettinga, G., Velsen, V.A.F.M., Hobma, S.W., de Zeeuw, W., Klapwijk, A., 1980. Use of the upflow sludge blanket (USB) reactor concept for biological wastewater treatment, especially for anaerobic treatment. Biotechnol. Bioeng. 22 (4), 699–734.

Li, C., Fang, H.H., 2007. Fermentative hydrogen production from wastewater and solid wastes by mixed cultures. Crit. Rev. Environ. Sci. Technol. 37 (1), 1–39.

Li, Q., Xu, M.J., Wang, G.J., Chen, R., Qiao, W., Wang, X.C., 2018. Biochar assisted thermophilic co-digestion of food waste and waste activated sludge under high feedstock to seed sludge ratio in batch experiment. Bioresource Technol. 249, 1009–1016.

Li, W., Yu, H.Q., 2011. From wastewater to bioenergy and biochemicals via two-stage bioconversion processes: a future paradigm. Biotechnol. Adv. 29 (6), 972–982.

Li, Y., Zhao, Z.Q., Yu, Q.L., Sun, C., Wang, M.W., Zhang, Y.B., 2021. High-efficiency ethanol yield from anaerobic fermentation of organic wastes via stimulating growth of ethanol-producing Fe(III)-reducing bacteria with magnetite. ACS Sustain. Chem. Eng. 9 (3), 1246–1253.

Liu, D.W., Liu, D.P., Zeng, R.J., Angelidaki, I., 2006. Hydrogen and methane production from household solid waste in the two-stage fermentation process. Water Res. 40 (11), 2230–2236.

Liu, F., Rotaru, A.E., Shrestha, P., Malvankar, N., Nevin, K.P., Lovley., D.R., 2015. Magnetite compensates for the lack of a pilin-associated c-type cytochrome in extracellular electron exchange. Environ. Microbiol. 17 (3), 648–655.

Liu, F., Rotaru, A.E., Shrestha, P.M., Malvankar, N.S., Nevin, K.P., Lovley, D.R., 2012. Promoting direct interspecies electron transfer with activated carbon. Energy Environ. Sci. 5 (10), 8982.

Liu, H., Grot, S., Logan, B.E., 2005. Electrochemically assisted microbial production of hydrogen from acetate. Environ. Sci. Technol. 39 (11), 4317–4320.

Logan, B.E., 2009. Exoelectrogenic bacteria that power microbial fuel cells. Nat. Rev. Microbiol. 7 (5), 375–381.

Logan, B.E., Call, D., Cheng, S.A., Hamelers, H.V.M., Sleutels, T.H.J.A., Jeremiasse, A.W., Rozendal, R.A., 2008. Microbial electrolysis cells for high yield hydrogen gas production from organic matter. Environ. Sci. Technol. 42 (23), 8630–8640.

Menezes, V.S., Amorim, N.C.S., Macedo, W.V., Amorim, E.L.C., 2019. Biohydrogen production from soft drink industry wastewater in an anaerobic fluidized bed reactor. Water Pract. Technol. 14 (3), 579–586.

Mishra, P., Thakur, S., Singh, L., Wahid, Z.A., Sakinah, M., 2016. Enhanced hydrogen production from palm oil mill effluent using two stage sequential dark and photo fermentation. Int. J. Hydrog. Energy 41 (41), 18431–18440.

Montiel Corona, V., Razo-Flores, E., 2018. Continuous hydrogen and methane production from *Agave tequilana* bagasse hydrolysate by sequential process to maximize energy recovery efficiency. Bioresour. Technol. 249, 334–341.

Montiel, M.D.L.T., Tyagi, R.D., Tyagi, R.D., Valero, J.R., 2001. Wastewater treatment sludge as a raw material for the production of *Bacillus thuringiensis* based biopesticides. Water Res. 35 (16), 3807–3816.

Moralejo-Gárate, H., Kleerebezem, R., Mosquera-Corral, A., Campos, J.L., Palmeiro-Sanchez, T., van Loosdrecht, M.C.M., 2014. Substrate versatility of polyhydroxyalkanoate producing glycerol grown bacterial enrichment culture. Water Res. 66, 190–198.

Morgan-Sagastume, F., Valentino, F., Hjort, M., Cirne, D., Karabegovic, L., Gerardin, F., Johansson, P., Karlsson, A., Magnusson, P., Alexandersson, T., Bengtsson, S., Majone, M., Werker, A., 2014. Polyhydroxyalkanoate (PHA) production from sludge and municipal wastewater treatment. Water Sci. Technol. 69 (1), 177–184.

Morgan-Sagastume, F.M., Hjort, M., Cirne, D., Ferardin, F., Lacroix, S., Gaval, G., Karabegovic, L., Alexandersson, T., Johansson, P., Karlsson, A., Bengtsson, S., Acros-Hernandez, M.V., Magunsson, P., Werker, A., 2015. Integrated production of polyhydroxyalkanoates (PHAs) with municipal wastewater and sludge treatment at pilot scale. Bioresour. Technol. 181, 78–89.

Nagarajan, S., Skillen, N.C., 2017. Cellulose II as bioethanol feedstock and its advantages over native cellulose. Renew. Sust. Energ. Rev. 77, 182–192.

Oh, Y., 2004. Photoproduction of hydrogen from acetate by a chemoheterotrophic bacterium *Rhodopseudomonas palustris* P4. Int. J. Hydrog. Energy 29 (11), 1115–1121.

Park, J.H., Kang, H.J., Park, K.H., Park, H.D., 2018. Direct interspecies electron transfer via conductive materials: a perspective for anaerobic digestion applications. Bioresour. Technol. 254, 300–311.

Peng, H., Zhang, Y.B., Tan, D.M., Zhao, Z.Q., Zhao, H.M., Quan, X., 2018. Roles of magnetite and granular activated carbon in improvement of anaerobic sludge digestion. Bioresour. Technol. 249, 666–672.

Penteado, E.D., Lazaro, C.Z., Sakamoto, I.K., Zaiat, M., 2013. Influence of seed sludge and pretreatment method on hydrogen production in packed-bed anaerobic reactors. Int. J. Hydrog. Energy 38 (14), 6137–6145.

Puyol, D., Batstone, D.J., Hulsen, T., Astals, S., Peces, M., Kromer, J.O., 2017. Resource recovery from wastewater by biological technologies: opportunities, challenges, and prospects. Front. Microbiol. 7, 2106.

Qiao, W., Takayanagi, K., Shofie, M., Niu, Q.G., Yu, H.Q., Li, Y.Y., 2013. Thermophilic anaerobic digestion of coffee grounds with and without waste activated sludge as co-substrate using a submerged AnMBR: system amendments and membrane performance. Bioresour. Technol. 150, 249–258.

Ramm, P., Terboven, C., Neitmann, E., Sohling, U., Mumme, J., Herrmann, C., 2019. Optimized production of biomethane as an energy vector from low-solids biomass using novel magnetic biofilm carriers. Appl. Energy 251, 113389.

Ramprakash, B., Muthukumar, K., 2018. Influence of sulfuric acid concentration on biohydrogen production from rice mill wastewater using pure and coculture of *Enterobacter aerogenes* and *Citrobacter freundii*. Int. J. Hydrog. Energy 43 (19), 9254–9258.

Rashama, C., Ijoma, G., Matambo, T., 2019. Biogas generation from by-products of edible oil processing: a review of opportunities, challenges and strategies. Biomass Convers. Bioresour. 9 (4), 803–826.

Ren, N.Q., Li, J.Z., Li, B.K., Wang, Y., Liu, S.R., 2006. Biohydrogen production from molasses by anaerobic fermentation with a pilot-scale bioreactor system. Int. J. Hydrog. Energy 31 (15), 2147–2157.

Rittmann, B.E., 2008. Opportunities for renewable bioenergy using microorganisms. Biotechnol. Bioenergy 100 (2), 203–212.

Rotaru, A., Shrestha, P.M., Liu, F.H., Shrestha, M., Shrestha, D., Embree, M., Zengler, K., Wardman, C., Nevin, K.P., Lovley, D.R., 2014. A new model for electron flow during anaerobic digestion: direct interspecies electron transfer to Methanosaeta for the reduction of carbon dioxide to methane. Energy Environ. Sci. 7 (1), 408–415.

Rozendal, R.A., Sleutels, T.H.J.A., Hamelers, H.V.M., Buisman, C.J.N., 2008. Effect of the type of ion exchange membrane on performance, ion transport, and pH in biocatalyzed electrolysis of wastewater. Water Sci. Technol. 57 (11), 1757–1762.

Rozendal, R.A., Hamelers, V.M.H., Euverink, G.J.W., Metz, S.J., Buisman, C.J.N., 2006. Principle and perspectives of hydrogen production through biocatalyzed electrolysis. Int. J. Hydrog. Energy 31 (12), 1632–1640.

Sachdeva, V., Tyagi, R.D., 2000. Production of biopesticides as a novel method of wastewater sludge utilization/disposal. Water Sci. Technol. 42 (9), 211–216.

Salehizadeh, H., Shojaosadati, S.A., 2001. Extracellular biopolymeric flocculants: recent trends and biotechnological importance. Biotechnol. Adv. 19 (5), 371–385.

Salehizadeh, H., Shojaosadati, S.A., 2003. Removal of metal ions from aqueous solution by polysaccharide produced from *Bacillus firmus*. Water Res. 37 (17), 4231–4235.

Sanchis, V., Bourguet, D., 2008. *Bacillus thuringiensis*: applications in agriculture and insect resistance management. A review. Agron. Sustain. Dev. 28 (1), 11–20.

Schnürer, A., Zellner, G., Svensson, B.H., 1999. Mesophilic syntrophic acetate oxidation during methane formation in biogas reactors. FEMS Microbiol. Ecol. 29 (3), 249–261.

Serafim, L.S., Lemos, P.C., Albuquerque, M.G.E., Reis, M.A.M., 2008. Strategies for PHA production by mixed cultures and renewable waste materials. Appl. Microbiol. Biotechnol. 81 (4), 615–628.

Shi, X.Y., Yu, H.Q., 2006. Continuous production of hydrogen from mixed volatile fatty acids with *Rhodopseudomonas capsulata*. Int. J. Hydrog. Energy 31 (12), 1641–1647.

Shi, X., Leong, K.Y., Ng, H.Y., 2017. Anaerobic treatment of pharmaceutical wastewater: a critical review. Bioresour. Technol 245, 1238–1244.

Shin, S.G., Han, G., Lim, J., Lee, C., Hwang, S., 2010. A comprehensive microbial insight into two-stage anaerobic digestion of food waste-recycling wastewater. Water Res. 44 (17), 4838–4849.

Sivagurunathan, P., Sen, B., Lin, C.Y., 2015. High-rate fermentative hydrogen production from beverage wastewater. Appl. Energy 147, 1–9.

Sivagurunathan, P., Kumar, G., Mudhoo, A., Rene, E.R., Saratale, G.D., Kobayashi, T., Xu, K.Q., Kim, S., Kim, D., 2017. Fermentative hydrogen production using lignocellulose biomass: an overview of pre-treatment methods, inhibitor effects and detoxification experiences. Renew. Sust. Energy Rev. 77, 28–42.

Srikanth, S., Venkata Mohan, S., Prathima Devi, M., Lenin Babu, M., Sarma, P.N., 2009. Effluents with soluble metabolites generated from acidogenic and methanogenic processes as substrate for additional hydrogen production through photo-biological process. Int. J. Hydrog. Energy 34 (4), 1771–1779.

Stams, A.J.M., Plugge, C.M., 2009. Electron transfer in syntrophic communities of anaerobic bacteria and archaea. Nat. Rev. Microbiol. 7 (8), 568–577.

Storck, T., Virdis, B., Batstone, D.J., 2016. Modelling extracellular limitations for mediated versus direct interspecies electron transfer. ISME J. 10 (3), 621–631.

Summers, Z.M., Fogarty, H.E., 2010. Direct exchange of electrons within aggregates of an evolved syntrophic coculture of anaerobic bacteria. Science 330 (6009), 1413–1415.

Swanson, M., Reguera, G., et al., 2016. Microbe. ASM Press, Washington, DC.

Tao, Y., Chen, Y., Wu, Y.Q., He, Y.L., Zhou, Z.H., 2007. High hydrogen yield from a two-step process of dark- and photo-fermentation of sucrose. Int. J. Hydrog. Energy 32 (2), 200–206.

Vatsala, T.M., Raj, S.M., Manimaran, A., 2008. A pilot-scale study of biohydrogen production from distillery effluent using defined bacterial co-culture. Int. J. Hydrog. Energy. 33 (20), 5404–5415.

Venetsaneas, N., Antonopoulou, G., Stamatelatou, K., Kornaros, M., Lyberatos, G., 2009. Using cheese whey for hydrogen and methane generation in a two-stage continuous process with alternative pH controlling approaches. Bioresour. Technol. 100 (15), 3713–3717.

Verhoeven, J.T.A., Meuleman, A.F.M., 1999. Wetlands for wastewater treatment: opportunities and limitations. Ecol. Eng. 12 (1), 5–12.

Vidal, G., Carvalho, A., Mendez, R., Lema, J.M., 2000. Influence of the content in fats and proteins on the anaerobic biodegradability of dairy wastewaters. Bioresour. Technol. 74 (3), 231–239.

Vidyarthi, A.S., Tyagi, R.D., Valero, J.R., Surampalli, R.Y., 2002. Studies on the production of *B. thuringiensis* based biopesticides using wastewater sludge as a raw material. Water Res. 36 (19), 4850–4860.

Wang, A.J., Liu, W.Z., Zhang, B., Cai, W.W., 2020. Bioelectrosynthesis: Principles and technologies for value-added products, 2020. Wiley, NJ.

Wang, J., Yu, H.Q., 2006. Cultivation of polyhydroxybutyrate-rich aerobic granular sludge in a sequencing batch reactor. Water Sup. 6 (6), 81–87.

Wang, J., Wan, W., 2008. Comparison of different pretreatment methods for enriching hydrogen-producing bacteria from digested sludge. Int. J. Hydrog. Energy 33 (12), 2934–2941.

Wang, S.G., Gong, W.X., Liu, X.W., Tian, L., Yue, Q.Y., Gao, B.Y., 2007. Production of a novel bioflocculant by culture of *Klebsiella mobilis* using dairy wastewater. Biochem. Eng. J. 36 (2), 81–86.

Weiland, P., 2010. Biogas production: current state and perspectives. Appl. Microbiol. Biotechnol. 85 (4), 849–860.

Yang, H.J., Shao, P., Lu, T.M., Shen, J.Q., Wang, D.F., Xu, Z.N., Yuan, X., 2006. Continuous bio-hydrogen production from citric acid wastewater via facultative anaerobic bacteria. Int. J. Hydrog. Energ. 31 (10), 1306–1313.

Yang, Y.F., Zhang, Y.B., Li, Z.Y., Zhao, Z.Q., Quan, X., Zhao, Z.S., 2017. Adding granular activated carbon into anaerobic sludge digestion to promote methane production and sludge decomposition. J. Clean. Prod. 149, 1101–1108.

Yezza, A., Tyagi, R.D., Valero, J.R., Surampalli, R.Y., 2006. Bioconversion of industrial wastewater and wastewater sludge into *Bacillus thuringiensis* based biopesticides in pilot fermentor. Bioresour. Technol. 97 (15), 1850–1857.

Yu, H.Q., Fang, H.H.P., 2002. Acidogenesis of Dairy Wastewater at Various pH Levels. Pergamon Press, Oxford.

Zhang, Z.Q., Lin, B., Xia, S.Q., Wang, X.J., Yang, A.M., 2007. Production and application of a novel bioflocculant by multiple-microorganism consortia using brewery wastewater as carbon source. J. Environ. Sci. 19 (6), 667–673.

Zhao, L.M., Han, D., Yin, Z.C., 2019. Biohydrogen and polyhydroxyalkanoate production from original hydrolyzed polyacrylamide-containing wastewater. Bioresour. Technol. 287, 121404.

Zhao, Z.Q., Li, Y., Quan, X., Zhang, Y.B., 2017. Towards engineering application: potential mechanism for enhancing anaerobic digestion of complex organic waste with different types of conductive materials. Water Res. 115, 266–277.

Zhao, Z.Q., Li, Y., He, J.Y., Zhang, Y.B., 2018. Establishing direct interspecies electron transfer during laboratory-scale anaerobic digestion of waste activated sludge via biological ethanol-type fermentation pretreatment. ACS Sustain. Chem. Eng. 6 (10), 13066–13077.

Zhao, Z.Q., Li, Y., Zhang, Y.B., Lovley, D.E., 2020. Sparking anaerobic digestion: promoting direct interspecies electron transfer to enhance methane production. Iscience. 23 (10179412).

Zhen, G.Y., Pan, Y., Lu, X.Q., Li, Y., Zhang, Z.Y., Niu, C.X., Kumar, G., Koabayashi, T., Zhao, Y.C., Xu, K.Q., 2019. Anaerobic membrane bioreactor towards biowaste biorefinery and chemical energy harvest: recent progress, membrane fouling and future perspectives. Renew. Sust. Energ. Rev. 115, 109392.

Zhong, L., Zhong, S., Lei, H.Y., Chen, R.W., Bai, T., 2008. Production and application of a bioflocculant by culture of *Bacillus licheniformis* X14 using starch wastewater as carbon source. J. Biotechnol. 136, S313.

Algal cultivation and algal residue conversion to bioenergy and valuable chemicals

Pengfei Cheng[a,b], Chengxu Zhou[a], Yanzhang Feng[a], Ruirui Chu[a], Haixia Wang[a], Yahui Bo[a], Yandu Lu[c], Roger Ruan[b], Xiaojun Yan[d]

[a]College of Food and Pharmaceutical Sciences, Ningbo University, Ningbo, Zhejiang, China, [b]Center for Biorefining and Department of Bioproducts and Biosystems Engineering, University of Minnesota-Twin Cities, Saint Paul, MN, USA, [c]State Key Laboratory of Marine Resource Utilization in South China Sea, Hainan University, Hainan, China, [d]Key Laboratory of Marine Biotechnology of Zhejiang Province, Ningbo University, Ningbo, Zhejiang, China

6.1 Introduction

Energy is the lifeblood of today's world economy, and the driving force for social development. Due to the continuous expansion of the world economy and human civilization, the demand for energy is increasing, and traditional fossil energy is facing the danger of gradual depletion of reserves. As an alternative, biofuel has become one of the fastest growing, and most widely used bioenergy products in recent years. However, due to limitations raw materials, lack of economic incentives, and other problems, large-scale development of biofuels has been severely restricted (Searchinger et al., 2008; Kalita, 2008).

Among many organisms, microalgae have the potential to become an excellent source of biodiesel and bio-oil due to their unique composition and structure. "Microalgae" is a general term for a group of single-cell, or simple multicell, organisms that can effectively use solar energy, and carry out photosynthesis of H_2O, CO_2, and inorganic salts, similar to plant cells (Wang et al., 2008). Some microalgae can accumulate a large amount of lipids during the growth and reproduction process, and some produce hydrocarbons in secondary metabolic processes. In addition, microalgae can also be used to produce animal feed, human food, and high value-added biologically active ingredients, as they contain proteins

Biomass, Biofuels, Biochemicals.
DOI: https://doi.org/10.1016/B978-0-323-90633-3.00002-X

and amino acids, carbohydrates, and high, unsaturated fatty acids (Ali et al., 2021). The biofuels produced directly or indirectly, by microalgae include fatty acids, hydrocarbons, methane, bio-hydrogen, and methanol (Waltz, 2009).

6.2 Advantages and development of energy from microalgae

Microalgae are widely distributed, and the most primitive plants on the earth. They grow photoautotrophically, and can synthesize triglycerides up to 20% to 50% of their biomass, in the process of growth (Hu et al., 2008). Therefore, they are considered to be a promising source of raw materials for biodiesel production. They offer a single source of raw materials for renewable fuel production, and the potential to reduce fuel prices. Microalgae cultivation may offer other unique development advantages, which are as follows:

1. High cultivation efficiency, short growth cycle, and can be obtained in a relatively short period of time. Each ton of microalgae can fix about 1.83 tons of biogenic CO_2, and naturally recycle this greenhouse gas (Berberoglu et al., 2009).
2. A large amount of noncultivated land and nonfreshwater resources can be utilized, and thus does not affect agricultural production; the oil yield per unit area of land use may be higher than that of oil crops (Chisti, 2007; Schenk et al., 2008).
3. Microalgae are rich in high value-added byproducts such as polysaccharides, pigments, and proteins. Comprehensive utilization of these byproducts can help to achieve greater overall economic efficiency during the process of energy development. Therefore, vigorously developing microalgae energy will play a positive role in energy supply, low-carbon economy, energy conservation, reduction of emissions, and national security.

The attention to microalgae fuel began in the late 1960s, when Gelpi et al. discovered that algae, such as *Botryococcus braunii*, contained higher hydrocarbons, and the composition was very similar to that of industrial gasoline (Gelpi et al., 1968). This interesting discovery has attracted widespread attention. From 1976 to 1998, the US Department of Energy (US-DOE) invested 25 million dollars to implement the Aquatic Species Program (ASP), attempting to replace fossil fuels with biofuels produced by microalgae. But this effort failed to solve the problems of large-scale cultivation and high cost, and was therefore forced to terminate. Then, in 2008, due to the rise of international crude oil prices and the global call for greenhouse gas emission reductions, the use of microalgae to produce bioenergy was once again entered to address the field of energy research. The US-DOE formulated a new project for microalgal energy development, with a total investment of 800 million dollars, and more than 80 companies joined the effort, triggering a new round of research and development (Wijffels et al., 2010).

Although the technology and application of microalgae have been studied for decades in China, and large-scale production of *Spirulina* and *Chlorella* ranks first in the world, microalgae fuel development is relatively limited at present. With increasing interest in microalgae fuel around the world, it has attracted additional attention of many Chinese researchers, and China launched its first microalgae fuel project (Project 973) in 2011.

At present, reducing the production cost of microalgae fuel is still a major problem for commercial development around the world. The cost includes five aspects: algae species screening, cultivation, harvesting, dehydration, and oil extraction. To reduce the cost and realize commercial benefits, comprehensive research on these five aspects needs further development efforts (Wijffels et al., 2010).

6.2.1   Species of energy microalgae

Based on the unique advantages of microalgae, some species of energy microalgae may play an important role in replacing fossil energy. The oil content directly determines the production capability for microalgae biodiesel. Clearly, when screening potential microalgae species, those with higher oil content are selected. Different types of microalgae differ in oil content. Numerous studies have shown that the algae with higher oil content are mainly concentrated in the *Chlorophyta, Bacillariophyta,* or *Chrysophyta* algal types, and the oil-rich microalgae species include *B. braunii, Nannochloropsis* sp., *Chlorella, Dunaliella* (as shown in Table 6.1).

Under normal conditions, the average oil content of some green algae can reach 25%, and other species can reach 45% when they are cultivated under certain stress conditions. Therefore, some factors, such as the acquisition of high-oil content algae species, and the optimization of cultivation conditions, are essential for the development of biofuels using microalgae.

6.2.2   Efficient cultivation of microalgae

The cultivation of microalgae is photoautotrophic. The unique advantages that microalgae oil production offers are also dependent on the mode of photoautotrophic cultivation. The device used for photoautotrophic cultivation of microalgae is similar in structure to a general bioreactor, which can control and regulate light, temperature, dissolved oxygen, CO_2, pH, nutrients, and other culture conditions. Using microalgae to economically produce biofuels requires an efficient industrial system for large-scale cultivation. The development of an inexpensive and efficient photobioreactor is a crucial step in the production of biofuels from microalgae. Many developed countries have listed the design of efficient cultivation of

TABLE 6.1   Oil content of different microalgae (Chisti 2007).

Microalgae species	Oil content (%)
Botryococcus braunii	25–75
Chlorella sp.	28–32
Dunaliella primolecta	23
Nannochloris sp.	20–35
Nannochloropsis sp.	31–68
Neochloris oleoabundans	35–54
Nitzschia sp.	45–47
Phaeodactylum tricornutum	20–30
Isochrysis sp.	25–33

microalgae as a key development for marine biotechnology. The relatively mature development of large-scale microalgae cultivation systems includes "open" and "closed" systems. However, only the open-track culture tank system and the closed column photo-bioreactor system have been applied on a large scale, to date (Chisti, 2007).

6.2.2.1 Open pond culture systems

The open culture system refers to the "open track pond" which was designed in the 1960s, and is now widely used in the commercial, large-scale cultivation of microalgae. The race-track tank is usually a ring structure. The depth of the culture solution is generally 15–30 cm, and it has two or more circulating water channels. The open runway pond has the advantages of easy construction, low cost, and simple operation. But it is only suitable for algae species that can withstand harsh process conditions, due to its low cultivation efficiency, large area, and susceptibility to contamination by other, ambient microorganisms. It is difficult to adapt to the development of modern microalgae biotechnology (Zhang et al., 2016; Mohsenpour et al., 2014). Therefore, the development of a closed and efficient photobioreactor to realize large-scale and high-density cultivation of microalgae has become one of the urgent tasks to be solved in the industrial production of microalgae.

6.2.2.2 Closed culture system

A photobioreactor refers to a closed culture system; this design has been applied to many microalgae on a large scale. Closed culture systems include flat-plate reactors, tubular reactors, vertical-column reactors, and stirred tank reactors. Compared with an open culture system, the closed photobioreactor has the advantages of high culture efficiency, less pollution, higher specific surface area, and good gas–liquid mass transfer (Carvalho et al., 2006). However, the closed photobioreactor culture system also has some shortcomings. For example, when the density of microalgae cells is high, light diffusion into the culture system is difficult to achieve; the construction costs and culture operating costs are high. Also, there is a problem with biological adhesion on the inner surface of the reactor, which inhibits light diffusion (Scott et al., 2010).

The plate reactor is made of transparent glass or plexiglass plates. It is divided into the light source, circulation device, plate reactor, temperature control system, and CO_2 supply system. It can be adjusted according to the change of sunlight intensity and incident direction. By adjusting the thickness of different reactors to maintain a short light path, it can ensure that the effective liquid layer is fully exposed to light, and is easier to achieve a high-density culture (Janssen et al., 2010). In 1986, Ramos et al. developed a flat-plate photobioreactor for the first time (Ramos and Roux, 1986). The reactor has a high light utilization rate, is easy to scale up, encourage algae growth, and clean the reactor plates. Its internal algal adherent growth and external salt precipitation are easy to handle, and its structure is relatively simple. It can also be adjusted to improve placement angle to achieve the best light extraction effect. Compared with other types of photobioreactors, flat-plate reactors have the characteristics of large light-specific surface area, short optical path, ease of expansion, and small footprint. Therefore, this design has become a popular option for the development of closed photobioreactors.

Tubular reactors generally consist of a row of horizontal, vertical, or inclined transparent plastic or glass pipes, which are bent into different shapes. The transparent pipes are used to produce algae under the illumination of an external light source. Due to the tightness of the pipeline system, it is easy to be matched with other processing equipment. When the algae are produced to a certain stage, they can be transferred to the next process with a pump, thereby realizing the automation of the production process. This reactor was first designed by Davi et al. in the 1950s. The circulation of microalgae in the tubular reactor is driven by a water pump, or by air-lift circulation. The diameter of the pipe is usually about 10 cm. Increasing the pipe diameter will reduce the ratio of the light area to the reactor volume, which directly affects the growth of microalgae (Hao et al., 2020; Converti et al., 2006). However, a series of problems such as the accumulation of O_2 in the culture solution, poor circulation of algae cells in the pipeline, uneven light exposure, and high temperature in the closed pipeline have restricted the large-scale application of this system (Rubio et al., 2015).

Vertical column reactors have always used tubular glass or translucent materials with a larger diameter (compared to tubular photobioreactors) to capture sunlight. Also, it can be modified into an inner guide tube type, air lift type, or bubbling tower. Vertical column photobioreactors are generally small in size and not easy to scale up. Magnification along the diameter direction will increase the optical path, so that the internal algae cannot get sufficient light; the vertical magnification will have higher requirements on the reactor material and reactor supporting equipment. Airlift reactors are driven by air and lower energy input can obtain the required mass transfer coefficient and liquid circulation flow rate when cultivating microalgae in bubble towers (Loubiere et al., 2009). These bubble column reactors and airlift reactors have a small footprint, low construction cost, and simple operation, and are widely used in biological treatment processes such as wastewater treatment, and the chemical industry (Kumar et al., 2012). The airlift reactor has a diversion tube and the microalgae cells can complete the cycle from darkness (the liquid rise stage inside the diversion tube) to light (the liquid fall stage outside the diversion tube). With more liquid microalgal cells in the reactor, the distribution is more uniform. Therefore, compared with the bubble column reactor, the airlift reactor shows superior performance in the cultivation of microalgae, such as the cultivation of *Porphyridium* sp. at a lower airlift aeration rate (1 L/min). The growth rate in the reactor is 30% higher than that of the bubble column reactor (Merchuk et al., 2015). Compared with the open culture system, the vertical column reactor has a lower construction cost, high culture efficiency, and simple operation. At present, the hanging system is used in China, primarily, for secondary relay culture of diet microalgae, while foreign countries have been using this design for large-scale production of diet algae. For example, the British Seasalter Shellfish company has constructed a continuous culture stand-up bag system using 500 L polyvinyl plastic bags, which are supported on the outside by a metal frame; these reactors have been used for the cultivation of diet microalgae (for human consumption).

A stirred reactor is a design that was first proposed to be used in microalgae cultivation by increasing light; it is commonly used in industry and laboratories to cultivate microorganisms (Mohammed et al., 2014; Marianna et al., 2018). The ratio of the illuminated area of the stirred reactor to the volume of the reactor is relatively small. How to increase this ratio without affecting other performance parameters is the main issue to be considered. Judging from current research progress of closed photobioreactors, capital costs are reduced, and the manufacturing and maintenance costs are minimized by considering the materials of construction of the stirred reactor and external support facilities. Also, it strengthens the mixing of the light direction, adjusts the external structure of the reactor, and increases the internal components, so that the fluid in the reactor forms a vortex, which increases the exchange frequency of algae cells in the light and dark areas, thereby increasing the yield of algae cells. The enclosed photobioreactor is more suitable for the rapid and efficient expansion of energy to microalgae seeds, due to its high cell culture density and yield. However, its high cost, high operating and maintenance costs, and difficulty in scaling up are not yet optimized.

6.2.2.3 Immobilized and biofilm attached culture of microalgae

Traditional photobioreactors have low biological yields when cultivating various kinds of microalgae (Table 6.2). Therefore, it is neither feasible for microalgae projects with high value-added products, nor is it feasible for microalgae projects that only produce biofuels. To become more cost-effective, key design factors need further development, both the photobio-reactors designs and cultivation methods.

Using traditional reactors to cultivate microalgae have a great demand for water, which accounts for more than 99% of the entire system. In a nutshell, water generally plays the following roles in the process of microalgae culture: (1) water supports the entire microalgae culture system and provides space for microalgae growth, (2) water acts as a buffer medium for the entire microalgae culture system for stable cultivation and pH control of the liquid, (3) water dissolves the nutrients in the medium to provide a nutritional environment for microalgae growth.

However, immobilized culture is a new type of culture mode that is different from traditional liquid suspension cultures. It refers to batch culture in which algae cells are directly embedded in seaweed glue, polyurethane foam glue, or adsorbed onto ready-made foam. When algae cells are embedded in seaweed gel, they are affected by space barriers, their hydrocarbon production increases, and their photosynthetic system is protected. But their growth doubling-time is longer compared to free cells. The use of polyurethane as an embedding agent is toxic. A few polymers are less toxic, but the metabolic activity of cells and the amount of hydrocarbon production are lowered. However, by adsorbing the algae onto a small piece of foam, the biomass and hydrocarbon production are equivalent to those of free cells. Johnson and Wen et al. cultivated microalgae on plastic foam, and the biological yield was about 2.5 g/m²d (Johnson et al., 2010). Bailliez et al. fixed *B.braunii* onto a calcium alginate gel for cultivation, but the biological logarithmic yield was somewhat inhibited, and the growth phase was reduced by 21% compared with liquid

TABLE 6.2   Biomass productivity of different species of *B. braunii*.

B. braunii sp. (different strains)	Biomass yield	Reference
B. braunii SAG30.81	0.05 g/Ld	Rao et al. (2007)
B. braunii LB572	0.076 g/Ld	Rao et al. (2007)
B .braunii Yayoi	0.042 g/Ld	Saga et al. (2015)
B. braunii Kutz IPPAS H-252	0.15 g/Ld	Natalia et al. (2005)
B. braunii AP103	0.056 g/Ld	Ashokkumar et al. (2012)
B. braunii sp.	0.71 g/m²d	Ozkan et al. (2012)

suspension culture (Metzger et al., 2005). *B. braunii* can be on the surface of agar and used 10% CO_2 and 0.24 mol/L $NaHCO_3$ as a carbon source for cultivation, but the results were not ideal (Lopez et al., 2009). Ozkana et al. attached *Botryococcus* to a cement plate loaded with biofilm for cultivation research, and its biomass yield was as low as 0.71 g/m^2d (Ozkana et al., 2012). Shi et al. developed a twin-layer adherent culture technique, where water flows through the lower glass fiber layer to provide water and nutrients; algae grew in the upper nylon layer, but the biomass yield was low (Shi et al., 2007). In these few cultivation studies noted above, the reduction of algae in the water, the biomass yield, and oil yield of *B. braunii* had not been significantly improved, and it was still not worthy of commercial large-scale cultivation for microalgae energy production.

In recent years, Liu and Cheng's team has proposed a new type of microalgae adherent culture method (Liu et al., 2013; Cheng et al., 2016). The algae cells are directly inoculated on the filter membrane material to form a biofilm. The filter membrane, soaked by the culture medium, provides nutrients and water for the algae cells, and the system is fed with 1% CO_2 in air to provide a carbon source. It was found that *Scenedesmus*, *Botryococcus*, *Pseudochlorococcum*, *Cylindrotheca* (diatoms), and *Spirulina* can achieve good adherent growth. The growth rate had little effect regarding the species of algae, but it was related to the light intensity and the composition of the medium. The oil content of *Scenedesmus* was about 50%, after it was induced by nitrogen deficiency. The study found that in this type of attached culture, the light saturation point is about 100–150 $\mu mol/m^2s$. Considering that sunlight intensity during outdoor cultivation can generally reach 400–2000 $\mu mol \bullet m^2s$, which is much higher than the above-mentioned light saturation point, if the microalgae cell biofilm is directly placed under strong light, it will produce photoinhibition on the algae cells. At the same time, the sunlight cannot be fully utilized. Therefore, a new design principle of light intensity, diluted microalgae, adherent culture reactor has been proposed. It will expand the culture area per unit of incident light, or expand the culture area through the light–dark cycle during the week to achieve light intensity. The light intensity on the incident surface is diluted. Based on this principle, a variety of attached culture reactor structures have been proposed. For example, a plug-in array reactor has been designed, and the average density of microalgae cultured indoors can reach 200–300 g/m^2. The indoor yield of growth can reach 60–90 $g.m^2d$, and outdoor yield can reach 40–60 g/m^2d, both of which are much higher yields than liquid in open ponds, or in photobioreactors (Cheng et al., 2013; Liu et al., 2013). Using this method, water consumption can be reduced by 90%, and at the same time, it has a huge advantage in energy consumption, harvesting, and pollution prevention.

6.3 Microalgae sewage treatment and resource engineering technology

The process of wastewater treatment with microalgae is the absorption and transformation of various nutrients in the wastewater. The nutritional requirements of most algae are simple, and the nitrogen, phosphorus, and other metal ion can be consumed efficiently to promote their own growth. Meanwhile, microalgae can synthesize their own cell materials, and high value products, through photosynthesis with chlorophyll in their cells.

Typically, ammonia nitrogen, phosphorus, and heavy metals are the main pollutants in most wastewater. Microalgae can use a large number of organic nitrogen compounds, and inorganic nitrogen compounds, as nutrients to synthesize amino acids (Fan et al., 2018). There are two main mechanisms to remove NH_3-N: one is the synthesis of amino acids by using the nitrogen source in ammonia nitrogen; the other is consumption of carbon dioxide in the wastewater to produce oxygen to raise pH, so that ammonia nitrogen is converted into ammonia (gas) and released into the air. The removal of ammonia nitrogen by algae, results in the reduction of total nitrogen. Compared with algae, the assimilation and absorption capacity of heterotrophic microorganisms are weaker under the condition of shorter sludge age, so that most ammonia nitrogen is only converted into nitro- or nitroso- nitrogen. Although ammonia nitrogen in water can be removed, the removal of total nitrogen is still poor.

In addition, the dissolved forms of phosphorus in water can also promote the growth of algae in varying degrees (Shang-Bo et al., 2012). The removal of phosphate includes two aspects (Kuenzler, 2010): First, it is directly absorbed by algae cells under aerobic conditions, and transformed into ATP, phospholipid, and other organic materials through horizontal phosphorylation, oxidative phosphorylation, and photosynthetic phosphorylation. Second, the growth of algae leads to an increase in pH, and the alkaline environment causes calcium ions in the water to be dissolved by the phosphate to form calcium hydroxy phosphate precipitation and then to be absorbed. Therefore, microalgal cells can be used to remove nitrogen, phosphorus, and other nutrients enriched in sewage, and store them in the form of organic matter.

Moreover, microalgae can also absorb heavy metals (Svaldenis and Arnas, 2014). There is a layered cell wall that contains a porous structure with multiple layers of microfibrils, such as cellulose, pectin, alginate, and polygalactose sulfate, and the main component inside is cellulose. Its cell structure and physiological characteristics provide shielding for toxic heavy metal and an environment for the treatment of wastewater. Currently, two mechanisms for the absorption of soluble metals by algae have been interpreted. (1) First, is passive absorption, also known as biosorption, which meets the dynamic balance of adsorption and desorption. This process, which uses metal ions instead of monovalent and divalent ions, combines with functional groups on the cell wall, and is a way of enrichment without any metabolic process or energy supply. But the efficiency and selectivity of the enrichment depends on the structure and ion types. In addition, dead algae cells also have the capacity of adsorption, because the active functional groups still exist on the cell wall. (2) The second mechanism is active absorption, which is slow and irreversible. In this process, metal ions will be accumulated on the cell surface and then combine with certain enzymes on the plasma membrane, such as membrane transferase, hydrolase, to be actively transported into the cell for final accumulation. These two mechanisms can work in algae-based wastewater treatment, simultaneously. Their relative importance depends on the type of algae, cultivation conditions, and the chemical nature of the metal ions. Specifically, the time of cultivation and contact with metal ions, ion concentration, competitive ions, and culture medium also affect the metal purification efficiency.

6.3.1 Pretreatment of biogas slurry wastewater

Biogas slurry, which is the liquid effluent from an anaerobic digester, is not only rich in nutrients, but also contains more suspended particles with higher turbidity, and there are great differences in nutrients and physical and chemical properties. In an open environment, miscellaneous algae and predators such as rotifers result in a mixed symbiosis, which makes cells vulnerable to insect damage. Therefore, it is necessary to pretreat the wastewater before microalgae-based wastewater treatment (Passos et al., 2016). It is necessary to take measures for pretreatment due to the following three aspects: reducing turbidity, adjusting nutrient structure, and preventing pests. The high content of suspended particles in the slurry (about 30-40 kg/m^3) leads to increased turbidity, which limits light and its utilization, and reduces biomass accumulation, which affects sewage treatment efficiency. In general, microalgae-based purification of biogas slurry systems are equipped with primary settling tanks and aeration tanks that can initially settle the particulate matter, and then further reduce the turbidity by adding flocculants, followed by filtration. When microalgae are cultivated in open raceway ponds, their growth is inhibited by carbon sources, which in turn cause the microalgae to become carbon source dependent, and can effect carbonate balance in the water. The change of ions promotes the concentration of hydroxide ions, causing the pH to rise, then inhibits the growth of microalgae and aerobic bacteria. By adding CO_2 and cofermenting with aquaculture-rich substances and breeding wastewater, it is possible to effectively increase the carbon content in the biogas slurry, increase the C/N ratio, and avoid the inhibition of algae growth caused by the limitation of carbon sources. A great impact has also been found with the N/P ratio in sewage on the efficient removal of nitrogen and phosphorus. Aslan and Kapdan (2006) believe that the appropriate N/P ratio may enhance the removal of nutrients by *Chlorella* in high-concentration N/P sewage. For microalgae, the suitable N/P ratio may vary with the algae species, so it is very important to screen suitable algae species for different N/P ratios, and establish a database that matches various typical biogas slurries and algae species.

In general, microalgae-based wastewater treatment is carried out in open raceway systems. Pollution from parasites and predators is the common problem faced by all the processes of microalgae scale cultivation. Therefore, how to avoid the pollution and reduce the reproduction of parasites should be considered when microalgae are to be used for wastewater treatment. In the laboratory, many researchers are inclined to use high-temperature sterilization as a pretreatment before algae processing. However, sterilization is not suitable for large-scale cultivation, and an economical, operational method to avoid the outbreak of pests during microalgae culture is necessary. The life cycle and characteristics of eukaryotic parasites were analyzed in the culture process in open raceway ponds (Peter et al., 2016). A possible way to isolate suspended particles in the air to avoid infection and control the phagocytosis of parasites was considered. It is necessary to pretreat the high-concentration and complex wastewater before processing with microalgae.

6.3.2 Breeding and domestication of microalgae

In the purification process for biogas slurry, the selection and domestication of microalgae is necessary for the whole wastewater treatment system. A suitable microalga for large-scale purification should meet the requirements of rapid growth, strong pollution resistance, tolerance of high ammonia nitrogen and organic matter, predator resistance, and resulting in rich, high value-added, products. The selection of algae species follows two aspects: (1) collection, screening, and separation of microalgae from the natural environment, and optimize the physical and chemical characteristics, such as culture temperature, light, pH, and nutrient conditions for growth and biomass composition of microalgae; (2) and genetic modification of the microalgae species.

A number of research studies have reported that *Chlorella* and *Scenedesmus* are widely studied due to their strong pollution resistance. However, different types of *Scenedesmus* and *Chlorella* have different removal effects on ammonia nitrogen, nitrate-nitrogen, and total nitrogen in wastewater. The biodegradability of two kinds of green algae, one blue alga, one euglenoids, and two mixed algae in the symbiotic system, with activated sludge bacteria, with fourfold and eightfold dilution ratio of pig wastewater has been evaluated (Godos et al., 2010). *Chlorella* and *Euglenaviridis* can grow in fourfold and eightfold dilution, while *Scenedesmusobliquus* and the algae isolated from the septic stabilization pond can only grow in the eightfold dilution. However, the *Spirulinaplatensis,* separated from a high turbidity pond cannot grow in these conditions. Further research has shown that some species of *Chlorella* have strong tolerance to high ammonia nitrogen, and can effectively treat wastewater. Researchers have reported that the *Chlorella*, screened from biogas slurry wastewater, can tolerate the high concentration of pollutants (Zhou et al., 2012). After 10 days of treatment, the removal rates of ammonia nitrogen and total phosphorus can reach 94.76% and 80.03% respectively, which resulted in lower effluent concentrations than the highest discharge standard of water pollutants in the international pig industry. These reports show that some species of algae, isolated from sewage plants or natural water bodies, can adapt well to the actual wastewater environment.

6.4   Application of microalgae culture in wastewater treatment

As we know, the utilization of microalgae biotechnology to treat wastewater may include several steps: culture of microalgae, separation and recovery of microalgae, biological refining of microalgae, and the production of high value products. However, any of these process steps could affect the resource utilization of microalgal biomass for wastewater treatment.

6.4.1   Culture of microalgae

According to the different characteristics of algae growth, the culture system can be roughly classified into two categories: liquid suspension culture and biofilm attached culture. Furthermore, suspension culture systems can be further divided into two types: open pond, and closed bioreactor. (1) Open systems refer to various types of pond systems, typically such as high-efficiency algae ponds and runway-type algae ponds. (2) Closed systems refer to various types of photobioreactors, which are divided into categories such as tubular (vertical, horizontal, spiral), cylindrical, and thin plate (Yan et al., 2016). The traditional liquid suspension culture for the treatment of wastewater faces many problems, such as large area requirements, low efficiency, difficulty to control conditions, and high recovery costs. In addition, the growth of microalgae is easily affected by nutrients, light, pH, temperature, inoculation concentration, and other variable conditions. A 530 L runway pond and a 380 L pipeline photobioreactor were used for the treatment of sewage, and the results showed that the removal rate of total nitrogen reached $65.12 \pm 2.87\%$ and $89.68 \pm 3.12\%$, respectively. The values of total phosphorus removal reached $58.78\% \pm 1.17\%$ and $86.71 \pm 0.61\%$, respectively

(Arbib et al., 2013). Under the same conditions, light and temperature became the main factors that limited the growth of microalgae and the resulting removal efficiency.

According to the principle of light dilution and immobilization, biofilm attached culture is a novel method that separates algal cells from the culture medium and fixes them on certain biofilm materials. Then, the liquid culture medium can drip into the back of or inside of the attached porous material that supports the algal cells in a humid state to support algal cell growth under certain light intensity and nutrient concentrations. Because of the specificity of the attached culture device, the high energy consumption process for algal cell separation by centrifugation can be omitted, and the resulting operating costs can be reduced when treating wastewater with microalgae. Among many species of microalgae, *Scenedesmus* and *Chlorella* can accumulate more oil during the cultivation process and have a strong tolerance to sewage, which make them the ideal algae resource for sewage purification. Researchers (Cheng et al., 2016) have applied this method to study the effects of oil-producing algae, *Scenedesmus* and *Chlorella*, on the treatment of swine wastewater. The results show that the biomass yield and oil accumulation rate of *Scenedesmus* and *Chlorella* are similar compared to the normal culture medium, and both can purify the pollution contaminants: (1) ammonia nitrogen, (2) total phosphorus, and (3) chemical oxygen demand in the wastewater. The removal rates for *Scenedesmus* were 96.59%, 74.52%, and 72.47%, respectively; the removal rates for *Chlorella* were 94.90%, 73.5%, and 71.40%, respectively. However, the open culture system is still worth considering the mainstream reactor configuration for microalgae-based sewage treatment, due to the actual situation of sewage treatment (large water volume, and construction and operating costs).

6.4.2 Separation and recovery of microalgae

The separation and harvest of microalgae are very important in the entire process of microalgae-based wastewater treatment and resource utilization, and its economic cost accounts for more than 30% of the total costs (Harith et al., 2010). In general, the size of microalgae cells is less than 30 μm, negatively charged, and has a density close to water. These characteristics often make algal cells a stable suspension state in water and difficult to achieve natural separation by gravity precipitation such as activated sludge. In the process of wastewater treatment with microalgae, it not only contaminates the water twice, but also makes it difficult to maintain a large amount of biomass in the reactor (generally only 0.2~0.6 g/L). Low culture density of microalgae leads to low removal efficiency, which results in poor treatment stability. In this regard, it is necessary to reduce the treatment load and adopt a long hydraulic retention time (HRT), which in turn leads to an increase of the occupied land area. At present, the HRT of an algae pond system is commonly 2–6 days, and the equivalent population covers an area of more than 10 m^2 "per what scale-up"? Obviously, its floor area is much larger than the main unit of secondary/tertiary sewage treatment, which is difficult to accept in cities with limited land use.

From the perspective of energy production, the optimal biomass of algal cell raw materials meeting the requirements of industrial utilization should be 300–400 g/L (dry mass). Therefore, the algal broth under conventional culture systems needs to be concentrated more than 1000 times before it can be used in the industry. This high energy separation and concentration process is the main operating cost in microalgae-based wastewater treatment (accounting for 20%~50% of the total cost). These excessive production costs make practical algae cultivation, coupled with wastewater treatment, similar to the production of fossil fuels (Huang et al., 2012). It can be seen that algae cell separation and harvesting is a challenging process bottleneck, restricting the large-scale industrial application of microalgae-based wastewater treatment (Cheng et al., 2020).

The common methods used for microalgae separation and collection include centrifugation, filtration (consisted of membrane filtration), air flotation, direct gravity sedimentation, and flocculation.

1. Centrifugation is a fast and reliable method of separation and recovery. However, due to the extremely high energy consumption, and investment and operating costs, it is not cost-effective for large-scale engineering applications under the current technical conditions.
2. Filtration when separating filamentous algae has lower energy consumption and cost; however, membrane pollution is easily formed for nonfilamentous algae, and also the costs are high and cannot meet the requirements of efficient and low-cost harvesting.
3. As for the air flotation, it is suitable for the harvesting of single-cell algae, but it is difficult to be applied under the conditions of mixed culture of sewage. In addition, due to the production of considerable tiny bubbles, the investment and operating costs/energy consumption of this method are also high.
4. Although the direct gravity sedimentation is a cheaper method, it takes a longer time and has the worst separation effect and reliability.
5. Flocculation is widely used to separate dispersed and colloidal substances in water, which have been used to harvest microalgae as early as the 1980s. After flocculation, the suspended algae broth can be separated by high-efficiency gravity sedimentation, and the separated algae cells can be directly trapped in the reactor to achieve the purpose of maintaining high biomass concentration and ensuring the quality of the effluent. In this respect, it is economical and feasible to choose flocculation to deal with large amounts of dilute algae. Although algae cannot directly meet the requirements of industrial applications after the process, it can significantly reduce the energy consumption and cost of the subsequent concentration. Therefore, flocculation has been regarded as the best way to achieve large-scale separation and harvesting of microalgae (Pragya et al., 2013). On the other hand, the "attached culture" method of microalgal cell growth and separation from the culture medium offers an economical alternative for microalgae harvesting.

6.4.3 Development of microalgae rich in high value products

After the treatment of wastewater by microalgae, the use of the algae cells to produce high value-added products, or biodiesel, is one of the main driving forces for microalgae-based wastewater treatment. Many types of algal residue are rich in oils and can be used to produce biodiesel (fatty acid methyl esters); other algae are extremely rich in hydrocarbons, with a chemical structure similar to mineral oil, which can be processed into gasoline and diesel after extraction. Moreover, some species of green algae and cyanobacteria can produce hydrogen during photosynthesis in certain conditions. In general, the transformation of microalgae to downstream products can be classified as chemical, biochemical, or thermochemical transformation, or can be used for direct combustion. Also, bioconversion of algal residue to high value algal products is another option for economic utilization.

Compared with other crops, microalgae have many advantages that are as follows:

1. The high photosynthetic efficiency with fast growth rate and short cycle, and oil production which can be 7–30 times than that of seed oil crops;
2. Microalgal biomass has a higher calorific value, which is 1.6 times that of wood or crop straw;
3. The production and processing cost of biomass is low, especially when cultured with sewage as the substrate (Shahid et al., 2017). To achieve ecological sewage treatment and renewable energy production, the cultivation of microalgae has been put forward by many developed countries and regions, as a promising goal.

6.5 Conclusions and perspectives

In the process of using microalgae to purify sewage, researchers have focused their attention to the removal of nutrients, but the large number of algae produced in the process can also be used to extract and produce oil, phycobiliprotein, polysaccharides, and other high value-added products. Culturing microalgae with biogas slurry (from anaerobic digesters) has the potential to simultaneously reduce nutrients in sewage, produce algae cells, and extract bioenergy. Due to continued depletion, and environmental problems caused by the use of fossil fuels, the development of new and environmentally friendly energy sources is promoted. A new generation of biofuels, with microalgae as the raw material can reduce the consumption of water, improve land use, and realize the harmony between human activities and nature.

However, large-scale, commercial production of microalgae still has many technical and economic challenges to be resolved, including culturing, harvesting, downstream processing technologies, and operating costs. These problems can be solved economically by developing high value-added products, combined with microalgae biofuels, as well as using microalgae to treat wastewater. The feasibility of producing microalgae high value-added products has been verified, and algal products, such as polysaccharides and phycobiliprotein, have also been industrialized, and the use of microalgae for wastewater treatment has also been commercialized to a degree. Future commercial, industrial applications of microalgae remain very promising, as research and development work, worldwide, continue to improve these possibilities.

Acknowledgments

This research was supported in part by grants from the State Key Laboratory of Marine Resource Utilization in the South China Sea (Hainan University) (MRUKF2021003), the Open Fund of Key Laboratory of Experimental Marine Biology, Chinese Academy of Sciences (KF2019NO3), the Natural Science Foundation of Zhejiang Province (LY20D060003), the Ningbo Municipal Science and Technology Project (2019C10071), University of Minnesota MnDrive Environment Program MNE12, and University of Minnesota Center for Biorefining.

References

Ali, Z., Subeshan, B., Alam, M.A., Asmatulu, E., Xu, J., 2021. Recent progress in extraction/transesterification techniques for the recovery of oil from algae biomass. Biomass Convers. Bioresour. 157, 25–42.

Arbib, Z., Ruiz, J., álvarez-Díaz, P., et al., 2013. Long term outdoor operation of a tubular airlift pilot photobioreactor and a high rate algal pond as tertiary treatment of urban wastewater. Ecol. Eng. 52, 143–153.

Ashokkumar, V., Rengasamy, R., 2012. Mass culture of *Botryococcus braunii* Kutz. under open raceway pond for biofuel production. Bioresour. Technol. 104, 394–399.

Aslan, S., Kapdan, I.K., 2006. Batch kinetics of nitrogen and phosphorus removal from synthetic wastewater by algae. Ecol. Eng. 28 (1), 64–70.

Berberoglu, H., Gomez, P.S., Pilon, L., 2009. Radiation characteristics of *Botryococcus braunii*, *Chlorococcum littorale*, and *Chlorella sp.* used for CO_2 fixation and biofuel production. J. Quant. Spectrosc. Radiat. Transf. 110 (17), 1879–1893.

Carvalho, A.P., Meireles, L.A., Malcata, F.X., 2006. Microalgal reactors: a review of enclosed system designs and performances. Biotechnol. Progr. 22 (6), 1490–1506.

Cheng, P.F., Cheng, J.J., Cobb, K., Zhou, C.X., et al., 2020. *Tribonema sp.* and *Chlorella zofingiensis* co-culture to treat swine wastewater diluted with fishery wastewater to facilitate harvest. Bioresour. Technol. 297, 122516.

Cheng, P.F., Ji, B., Gao, L.L., et al., 2013. The growth, lipid and hydrocarbon production of *Botryococcus braunii* with attached cultivation. Bioresour. Technol. 138, 95–100.

Cheng, P.F., Wang, Y., Yang, Q.Y., et al., 2016. Coupling treatment of cobalt-containing industrial wastewater and hydrocarbon production by adherent culture of Botryococcus biofilm. Environ. Sci. 037 (007), 2666–2672.

Chisti, Y., 2007. Biodiesel from microalgae. Biotechnol. Adv. 25 (3), 294–306.

Converti, A., Lodi, A., Del Borghi, A., et al., 2006. Cultivation of Spirulina platensis in a combined airlift-tubular reactor system. Biochem. Eng. J. 32 (1), 8–13.

Fan, L., Brett, M.T., Song, M., 2018. The bioavailability of different dissolved organic nitrogen compounds for the freshwater algae Raphidocelis subcapitata. Sci. Total Environ. 618, 479–486.

Gelpi, E., Oro, J., Schneide, H., et al., 1968. Olefins of high molecular weight in 2 microscopic algae. Science 161 (3842), 700–702.

Godos, I.D., Vargas, V.A., Saúl, B., et al., 2010. A comparative evaluation of microalgae for the degradation of piggery wastewater under photosynthetic oxygenation. Bioresour. Technol. 101 (14), 5150–5158.

Hao, C., Qian, F., Qiang, L., Chao, X., Yun, H., et al., 2020. Modeling for hydrothermal hydrolysis of microalgae slurry in tubular reactor: two phase flow and heat transfer effects. Appl. Therm. Eng. 180, 115784.

Harith, Z.T., Yusoff, F.M., Shariff, M., et al., 2010. Effect of different separation techniques and storage temperatures on the viability of marine microalgae, *Chaetoceros calcitrans*, during storage. Biotechnology. 9 (3), 387–391.

Huang, D., Zhou, H., Lin, L., 2012. Biodiesel: an alternative to conventional fuel. Energy Procedia 16, 1874–1885.

Hu, Q., Sommerfeld, M., Jarvis, E., et al., 2008. Microalgal triacylglycerols as feedstocks for biofuel production: perspectives and advances. Plant J. 54 (4), 621–639.

Janssen, M., Tramper, J., Mur, L.R., et al., 2010. Enclosed outdoor photobioreactors: light regime, photosynthetic efficiency, scale-up, and future prospects. Biotechnol. Bioeng. 81 (2), 193–210.

Johnson, M.B., Wen, Z.Y., 2010. Development of an attached microalgal growth system for biofuel production. Appl. Microbiol. Biotechnol. 85 (3), 525–534.

Kalita, D., 2008. Hydrocarbon plant—new source of energy for future. Renew. Sustain. Energy Rev. 12 (2), 455–471.

Kuenzler, E.J., 2010. Glucose-6-phosphate utilization by marine algae. J. Phycol. 1 (4), 156–164.

Kumar, K., Das, D., 2012. Growth characteristics of *chlorella sorokiniana* in airlift and bubble column photobioreactors. Bioresour. Technol. 116, 307–313.

Liu, T.Z., Wang, J.F., Hu, Q., et al., 2013. Attached cultivation technology of microalgae for efficient biomass feedstock production. Bioresour. Technol. 127, 216–222.

Lopez, C.V.G., Fernandez, F.G.A., Sevilla, J.M.F., et al., 2009. Removal of CO_2 from flue gases coupled to the photosynthetic generation of organic matter by cyanobacteria. New Biotechnol. 25, S265.

Loubiere, K., Olivo, E., Bougaran, G., et al., 2009. A new photobioreactor for continuous microalgal production in hatcheries based on external-loop airlift and swirling flow. Biotechnol. Bioeng. 102 (1), 132–147.

Marianna, D., Tsolcha, O.N., Tekerlekopoulou, A.G., Dimitrios, B., et al., 2018. Fish farm effluents are suitable growth media for *Nannochloropsis gaditana*, a polyunsaturated fatty acid producing microalga.. Eng. Life Sci. 18 (11), 851–860.

Merchuk, J., Gluz, M., Mukmenev, I., 2015. Comparison of photobioreactors for cultivation of the red microalga *Porphyridium* sp. J. Chem. Technol. Biotechnol. 75 (12), 1119–1126.

Metzger, P., Largeau, C., 2005. Botryococcus braunii: a rich source for hydrocarbons and related ether lipids. Appl. Microbiol. Biotechnol. 66 (5), 486–496.

Mohammed, K., Ahammad, S.Z., Sallis, P.J., Mota, C.R., 2014. Energy-efficient stirred-tank photobioreactors for simultaneous carbon capture and municipal wastewater treatment. Water Sci. Technol. 69 (10), 2106–2112.

Mohsenpour, Fatemeh, S., 2014. Development of luminescent photobioreactors for improved cultivation of microalgae. New Biotechnol. 31 S26-S26.

Ozkana, A., Kinney, K., Katz, L., et al., 2012. Reduction of water and energy requirement of algae cultivation using an algae biofilm photobioreactor. Bioresour. Technol. 114, 542–548.

Passos, F., Hom-Diaz, A., Blanquez, P., et al., 2016. Improving biogas production from microalgae by enzymatic pretreatment. Bioresour. Technol. 199, 347–351.

Peter, M., Letcher, Philip, A., et al., 2016. An ultrastructural study of *Paraphysoderma sedebokerense* (Blastocladiomycota), an epibiotic parasite of microalgae. Fungal Biol. 120 (3), 324–337.

Pragya, N., Pandey, K.K., Sahoo, P.K., 2013. A review on harvesting, oil extraction and biofuels production technologies from microalgae. Renew. Sust. Energ. Rev. 24, 159–171.

Rao, A.R., Ravishankar, G.A., 2007. Influence of CO_2 on growth and hydrocarbon production in Botryococcus braunii. J. Microbiol. Biotech. 17 (3), 414–419.

Ramos, O.A., Roux, J.C., 1986. Production of Chlorella biomass in different types of flat bioreactors in temperate zones. Biomass 10, 141–156.

Rubio, F.C., Fernández, F.G.A., Pérez, J.A.S., Camacho, F.G., et al., 2015. Prediction of dissolved oxygen and carbon dioxide concentration profiles in tubular photobioreactors for microalgal culture. Biotechnol. Bioeng. 62 (1), 71–86.

Saga, K., Hasegawa, F., Miyagi, S., Atobe, S., Okada, S., et al., 2015. Comparative evaluation of wet and dry processes for recovering hydrocarbon from *botryococcus braunii*. Appl. Energy. 141, 90–95.

Schenk, P.M., Thomas-Hall, S.R., Stephens, E., et al., 2008. Second generation biofuels: high-efficiency microalgae for biodiesel production. Bioenerg. Res. 1 (1), 20–43.

Scott, S.A., Davey, M.P., Dennis, J.S., et al., 2010. Biodiesel from algae: challenges and prospects. Curr. Opin. Biotechnol. 21 (3), 277–286.

Searchinger, T., Heimlich, R., Houghton, R.A., et al., 2008. Use of US croplands for biofuels increases greenhouse gases through emissions from land-use change. Science 319 (5867), 1238–1240.

Shahid, A., Khan, A.Z., Liu, T., et al., 2017. Production and processing of algal biomass. In: Zia, K.M., Zuber, M., Ali, M. (Eds.), Production and processing of algal biomass. Algae Based Polymers Blends & Composites, 273–299.

Shang-Bo, H., Guang-Ying, W.U., 2012. Variation of three forms of phosphorus with algae growth in the backwater area of Daninghe river of the three gorges reservoir. Environ. Sci. Surv. 31 (06), 115–119.

Shi, J., Podola, B., Melkonian, M., 2007. Removal of nitrogen and phosphorus from wastewater using microalgae immobilized on twin layers: an experimental study. J. Appl. Phycol. 19 (5), 417–423.

Svaldenis, A., 2014. Cultivating Algae in a Photobioreactor: CO_2 fixation, synthetic wastewater nutrient removal and biomass production using the green algae species *Chlorella pyrenoidosa*. Bachelor's Degree. Helsinki Metropolia University of Applied Sciences 19 May 2014.

Waltz, E., 2009. Biotech's green gold? Nat. Biotechnol. 27 (1), 15–18.

Wang, B., Li, Y.Q., Wu, N., et al., 2008. CO_2 bio-mitigation using microalgae. Appl. Microbiol. Biot. 79 (5), 707–718.

Wijffels, R.H., Barbosa, M.J., 2010. An outlook on microalgal biofuels. Science 329 (5993), 796–799.

Yan, C., Zhu, L., Wang, Y., 2016. Photosynthetic CO_2 uptake by microalgae for biogas upgrading and simultaneously biogas slurry decontamination by using of microalgae photobioreactor under various light wavelengths, light intensities, and photoperiods. App. Energy. 178 (15), 9–18.

Zhang, C.D., Li, W., Shi, Y.H., Li, Y.G., Huang, J.K., et al., 2016. A new technology of CO_2 supplementary for microalgae cultivation on large scale—a spraying absorption tower coupled with an outdoor open runway pond. Bioresour. Technol. 209, 351–359.

Zhou, W., Li, Y., Min, M., et al., 2012. Growing wastewater-born microalga *Auxenochlorella protothecoides* UMN280 on concentrated municipal wastewater for simultaneous nutrient removal and energy feedstock production. App. Energy 98, 433–440.

Additive strategies for enhanced anaerobic digestion for bioenergy and biochemicals

Pengshuai Zhang[a,b], Chicaiza-Ortiz Cristhian[a,b,c], Jingxin Zhang[a]

[a]China-UK Low Carbon College, Shanghai Jiao Tong University, Shanghai, China, [b]School of Environmental Science and Engineering, Shanghai Jiao Tong University, Shanghai, China, [c]Faculty of Life Sciences, Amazon State University (UEA), Puyo, Pastaza, Ecuador

7.1   Introduction

Due to the increasing number of the worldwide population, as well as the rapid urbanization, particularly in developing countries, more and more problems are showing up nowadays, such as climate change, environmental pollution, etc. The increasing organic waste produced from daily life has threatened environmental safety severely. It is not only an environmental issue, but also a sustainable development challenge, which has been approached by different technologies in recent years, such as landfilling, incineration, or composting. However, these technologies are generally causing some adverse environmental loads (Xu et al., 2018). Anaerobic digestion (AD) is a promising and cost-effective technology, as it can address organic waste stability while achieving bioenergy and nutrient recovery (Batstone and Virdis, 2014, Zhen et al., 2017). Besides, a wide range of substrates applicable to AD, even those with low solids contents, can be run in large- and small-scale digesters (Xu et al., 2018). After the AD, the biogas can be collected as bioenergy, the digestion slurry can be used as a fertilizer. However, there are some drawbacks for AD due to the complexity of feedstock, poor system stability, and low reactor efficiency, which limit its development.

AD can be divided into four phases: hydrolysis, acidogenesis, acetogenesis, and methanogenesis phase (Fig. 7.1). During the hydrolysis phase, the high molecular mass is broken down into a low molecular mass, including insoluble or soluble materials such as carbohydrates, polysaccharides, proteins, and nucleic acids, which are degraded into amino acids, glycerol, and carboxylic acids. It follows that the end products from the hydrolysis phase are further degraded into volatile fatty acid (VFA) by acid-forming anaerobes including acetic

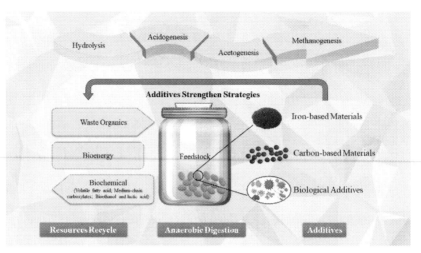

FIG. 7.1   Additive strategies for enhanced anaerobic digestion for bioenergy and biochemical.

acid, butyric acid, propionic acid, ethanol along with others. This step can be used to recover biochemical compounds, such as medium-chain carboxylates, bioethanol, or lactic acid. When the target recovered product is methane, the ethanol and acid generated in the acidogenesis phase are converted into acetate and used as a substrate by methanogens for methane production (Arif et al., 2018). Considering the above-mentioned phases of the AD process, additives can be added to the system for improving the AD's overall performance. The additives can enhance the AD performance by optimizing the system environment, such as pH buffering and oxidation–reduction potential (ORP), or facilitating changes in the microbial community structure. However, the challenge is that there are many types of additives available, therefore suitable additives must be selected for different specific AD conditions.

Therefore, this chapter introduced a brief description of the basic concept of the AD process and the additives used in AD, including iron-based materials, carbon-based materials, biological materials for bioenergy, and biochemicals recovery (Fig. 7.1). Moreover, a detailed discussion of the primary mechanism behind the additive strategies and the comparison of different additives (Table 7.1) are presented to address the AD performance more efficiently.

7.2   Additive strategies for enhanced AD for bioenergy

7.2.1   Iron-based additives

Iron is a multivalent element that is abundant in nature, has good environmental compatibility, and is biologically benign. It is one of the essential elements for microorganisms growth and a key component of several enzymes. Due to its unique characteristics, iron has been widely used in biological treatment to improve contaminant removal over the last decades (Tian and Yu, 2020). Regarding the iron-based AD process, iron can enhance energy recovery

TABLE 7.1 The comparison of different additives.

Additives		Main function/advantages	Disadvantage	Reference
Iron-based additives	Zero-valent iron	Reduce the ORP; act as a cofactor in some critical enzymatic reactions; promote DIET; alleviating ammonia inhibition and the accumulation of VFA	Biomass inhibition by overdosing; associated costs; precipitation	(Ye et al., 2021, Zang et al., 2020, Zhang et al., 2020a)
	Fe(II)	Triggers a higher protein concentration in EPS; promote F_{420}-reducing hydrogenase	Overdose may be toxic to bacteria; difficult to reuse	(Ganzoury and Allam, 2015, Wang et al., 2016)
	Fe(III)	Serve as an electron acceptor for IRB; promote DIET	Overdose may be toxic to bacteria; difficult to reuse	(Honetschlägerová et al., 2018)
	Magnetite	Facilitating IET and DIET; wide substrate applicability for methane production;	-	(Baek et al., 2015, Xu et al., 2019a)
	Nano ZVI	Higher specific surface area and reduction rate; formation of stable 7 Fe-EPS complex; promote DIET	Overdose may be toxic to bacteria; difficult to reuse	(Ajay et al., 2020, Yu et al., 2016)
Carbon-based additives	Activated carbon	High specific surface area, porosity, and adsorption; high electrical conductivity; promote DIET	Low Cost-effectiveness; difficult to reuse	(Capson-Tojo et al., 2018, Wu et al., 2020b, Yang et al., 2020)
	Biochar	Less expensive; wide source; high specific surface area, porosity, and adsorption; high electrical conductivity; promote DIET	Difficult to recycle; complex preparation process	(Pan et al., 2019, Wambugu et al., 2019, Wang et al., 2018, Zielińska and Oleszczuk, 2015)
	Multiwalled carbon nanotubes	Enrich the functional microbe; promote DIET	Low Cost-effectiveness;	(Zhang et al., 2017a)
	Graphene	Promote DIET	Low cost-effectiveness; it might decrease the soluble organic substrates availability due to the strong absorption	(Dong et al., 2019a, Lin et al., 2018, Tian et al., 2017)

(continued)

TABLE 7.1 Cont'd

Additives		Main function/advantages	Disadvantage	Reference
Composite material	ZVI-biochar	Enrich the functional microbe; promote DIET; alleviating ammonia inhibition and the accumulation of VFA	Complex preparation process	(Zhang et al., 2020a)
	Magnetite-contained biochar	High capacitance and excellent conductivity, promote DIET; high cost-effectiveness; environmental friendly	Complex preparation process	(Peng et al., 2018, Wang et al., 2021)
Biological additives	Microbial inoculum	Provide a balanced microbial population; promote hydrolysis process	Scalability is hard work	(Pakarinen et al., 2008, Paritosh et al., 2020)
	Enzymes	Better response to stressful and transitory states; efficiently; promote hydrolysis process		(Guilford et al., 2019, Jiang et al., 2021)
Other additives	Trace metals	Improve stability		(Cai et al., 2018, Takashima, 2018)
	Zeolite	Enrich microbial consortium; reduction of dissolved ammonium from the digestate	Ineffective for energy recovery	(Wijesinghe et al., 2018, Zhang et al., 2016)

efficiency in multiple ways. For instance, iron may act as electron donors for microbes such as methanogens and is directly involved in the catabolic and anabolic stages of microbial metabolism. In addition, the characteristic of multivalent iron could serve as the oxidant for the degradation of toxic substances such as hydrogen sulfide, humic acid (HA), and certain metabolic byproducts. Generally, iron can enhance AD performance in a variety of forms, including zero-valent iron (ZVI), Fe(II), and Fe(III).

7.2.1.1 ZVI addition–enhanced AD

The mechanism of ZVI affects the AD

Among all iron forms utilized to enhance methane production by AD, ZVI is the most widely studied and applied (Ye et al., 2021, Zang et al., 2020, Zhang et al., 2020a). ZVI has been added to the system to carry out various morphological changes and to aid in the degradation of organic matter and the removal of contaminants. In one type, benefiting from the anaerobic corrosion of iron, the ZVI addition could reduce the ORP, thereby regulating the type of fermentation and influencing the structural and enzymatic activities of the AD microbial community. Generally, complex carbon-containing compounds found in organic waste are broken down until they are reduced to simple methane molecules (Zhang et al., 2020a). This result is achieved through a series of complex microbial processes, including the following stages: hydrolysis, acidogenesis, and methanogenesis. The addition of ZVI could effectively improve CH_4 production by promoting synergies between functional microorganisms, which has been shown to promote methanogenesis. For instance, in ZVI-based AD system, ZVI could promote the conversion of CO_2 to methane, thus capturing carbon by methanogen. Carbon dioxide to methane conversion is significant for regulating the global carbon cycle and achieving sustainable energy recovery, particularly in petroleum reservoirs. Carbon neutrality of the ZVI-AD system might be reached (Ma et al., 2018) because ZVI can act as an electron donor in the conversion of carbon dioxide to methane (Wei et al., 2018c, Wu et al., 2015). The principal methanogenic pathways are hydrogenotrophic and acetotrophic methanogenesis. Hydrogenotrophic methanogens use the $H_2/[H^+]$ induced by ZVI addition (through the corrosion reaction (Eq. 7.1)) for methane generation (Ma et al., 2018, Zhang et al., 2019c). *Methanosarcina* spp. is one of the genders that plays a significant role. This microorganism utilizes carbon dioxide and hydrogen predominantly under anaerobic conditions, can resist a wide range of pressures, and propagate rapidly (Qin et al., 2019). In terms of acetotrophic methanogenesis, *Clostridiaceae* converts CO_2 and H_2 into acetic acid via the acetyl-CoA pathway. Simultaneously, CO_2 can form $FeCO_3$ (Eq. 7.2) in the presence of ZVI, indicating a potential pathway for extended immobilization (Ma et al., 2019). In the ZVI-AD system, to promote the methanogenesis process via carbon dioxide capture, small ZVI particles with a larger specific surface area can be chosen to promote hydrogen production and methanogenesis.

$$Fe^0 + 2H_2O \rightarrow Fe^{2+} + 2OH^- + H_2 \tag{7.1}$$

$$Fe^0 + CO_2 + H_2O \rightleftarrows FeCO_3 + H_2 \tag{7.2}$$

At the same time, ZVI could act as a cofactor in a number of critical enzymatic reactions. The methanogenesis process is one of the most metal-rich enzymatic pathways, with iron being the most abundant metal in the synthesis of cytochromes and oxidases during the methanogens stage (Glass and Orphan, 2012). Concerning the hydrolysis-acidification stage, the ZVI dosage facilitates biogas production by enhancing the activity of protease and cellulase, which decompose proteins into amino acids and catalyze the hydrolysis of polysaccharides to monosaccharides, respectively. Likewise, the activity of other key enzymes (acetate kinase (AK), phosphotransacetylase (PTA), butyrate kinase (BK), and phosphotransbutyrylase (PTB)) can also be increased during the hydrolysis and acidification of sludge with ZVI addition (Feng et al., 2014). Moreover, in the methanogenic stage of ZVI-based AD, ZVI could remarkably enhance pyruvate-ferredoxin oxidoreductase activity, which might facilitate the formation of Fe-S clusters (Ganzoury and Allam, 2015, Liu et al., 2012b). The Fe-S cluster is an active group in F_{420}-reducing hydrogenase (Frh) $[H_2 + F_{420} + NADPH = F_{420}H_2 + NADP]$, which uses coenzyme F_{420} to convert CO_2 and H_2 into CH_4.

Moreover, ZVI addition may help stabilize the AD system by alleviating ammonia inhibition and the accumulation of VFAs, especially propionate. Ammonia inhibition often occurs in AD systems with a high solid substrate and an unbalanced C/N of a substrate, especially for the nitrogen-rich animal wastes, for example, swine manure (Yang et al., 2019b). Besides, the accumulation of propionate, which is thermodynamically unfavorable to convert into acetate, is a significant concern in AD, as excess propionate is harmful to the microbe (Ferrer et al., 2010, Qiao et al., 2013). When a proper dose of ZVI is added to the AD system, ZVI may act as the conductive mediator, promoting the direct interspecies electron transfer (DIET). As a result, the percentage of methyl coenzyme-M reductase (mcrA) gene copies to 16S also improved, which might reflect the methanogen abundance. Methanogens could resist ammonia inhibition due to their abundance. Furthermore, the improved DIET process results in an increase of *Syntrophomonas,* which plays a critical role in propionate degradation during AD (Zhang et al., 2019b). Hence, the addition of ZVI to organic waste could improve the energy recovery by alleviating ammonia inhibition and enhancing the utilization of thermodynamically unfavorable propionate (Meng et al., 2020). As a result, ZVI has demonstrated its potential to improve AD performance through several mechanisms.

Application of ZVI in AD and its operating parameters

ZVI is frequently used in AD systems with organic wastewater and organic solid waste as substrates. Manure is a common organic solid waste, and its quantity is significantly increasing simultaneously with the increase of population and large-scale development of livestock, which has caused a considerable load to the environment. AD is an effective way to treat the manure and recovery the resource from manure. However, the ammonia and sulfur released during manure AD can lead a treatment failure. The addition of ZVI could alleviate this inhibition by stimulating the acidification process and decreasing the proportion of propionic acid, which is less easily converted to methane compared to acetic and butyric acid (Meng et al., 2020, Yang et al., 2019a, Yang et al., 2018).

Besides, municipal wastewater sludge is another type of solid waste and entails a potential risk to human health. However, the low hydrolytic ability of sludge, which is caused by complex structural components in sludge, limits its application in AD. Therefore, in recent years, research on AD of sludge by enhanced ZVI has been focused on methane production (Ye et al., 2021, Zhao et al., 2018b). However, it should be noted that the primary mechanism by which ZVI effects sludge is through its ability to promote hydrogenotrophic methanogenesis, while causing only a slight effect on the solubilization, hydrolysis, and acidification processes (Zhao et al., 2018b). FW (food waste) is also often used as a substrate for AD. However, due to its high carbohydrate content and rapid hydrolysis rate, VFA accumulation occurs during the AD of FW. Thereby, the pH value of the system is lowered, and the system is unable to maintain AD stability; this results in a reduction in methane yield. Furthermore, the ZVI-based AD system could facilitate the hydrolytic–acidogenic process and shortened the lag phase of methanogenic stages in this circumstance. Additionally, ZVI can be used to recover energy from wastewater. For instance, blackwater, which contains concentrated organic materials, can be anaerobically digested to recover bioenergy. However, the high free ammonia in blackwater is the defect during AD. The ZVI addition may improve methane production due to a reduced ORP and improved hydrolysis-acidification (Xu et al., 2019b). Moreover, as for the preconcentrated domestic wastewater rich in organic matter, ZVI can also be applied. And the suggested ZVI dosage at fixed inoculum ratio (substrate/inoculum = 0.5) is 6 g/L (Zang et al., 2020).

Accordingly, ZVI has demonstrated its potential in the AD treatment of several types of organic wastes. Numerous factors can affect the performance of ZVI-based AD, including the properties of iron, the dosage of ZVI addition, and the inclusion in different stages. Particle size is an essential factor that affects the ZVI-based AD performance. Generally, a smaller particle size ZVI has a large specific surface area corresponding to a faster corrosion reaction rate (Dong et al., 2019b). A powder ZVI showed superior AD performance compared to the scrap ZVI (Charalambous and Vyrides, 2021). However, when considering the economic value, a proper dosage of scrap ZVI seems more cost-effective (Charalambous and Vyrides, 2021). Moreover, the nano-ZVI presents a higher specific surface area, reduction rate, and excellent activity to promote H_2 production during the corrosion process in comparison with ZVI, thus nano-ZVI could increase the abundance of hydrolysis and acidogenesis bacteria. Moreover, the H_2 can be used as an electron donor for methane production by methanogens (Zhou et al., 2020). Besides, the small size of nano-ZVI clusters and the formation of stable Fe-EPS complex promote the dissolution of nano-ZVI and thereby improve the bioavailability of iron. The extracellular polymeric substances (EPS) could interact with nano-ZVI and work as a sink to accept electrons from nano-ZVI, favoring the direct interspecific electron transfer (He et al., 2020a). Among the considerations of nano-ZVI, it is easy to aggregate into larger particles during the reaction, weakening the catalyst performance; sometimes nano-ZVI is more toxic to bacteria, causes oxidative stress to them, and influences methane production (Summer et al., 2020). In contrast, a proper dosage is critical to maximizing the effect of ZVI. The methane production is significantly correlated with the dose of ZVI. Due to the fact that ZVI can improve the pH of the system, an excess of ZVI may cause the pH to continue rising in a suitable pH range. A considerable number of studies indicated that a higher dosage of ZVI could inhibit the growth activity of methanogens (Ye et al., 2021). Likewise, the different addition stage has an effect on the ZVI performance. For instance, as for the swine manure, the ZVI addition in the methanogenic stage is more beneficial for CH_4 production than in the acidogenic stage (Yang et al., 2019a).

Furthermore, the environment may affect the ZVI performance in which ORP is another factor that defines the removal of pollutants in the bio-ZVI system. The reviewed literature in this chapter shows that, among several of the characteristics that impact AD performance, ORP is one of the most decisive characteristics. For anaerobic microorganisms, a lower oxygen and ORP concentration is required, as a higher potential could ultimately reduce the cytochrome, impacting extracellular electron transport and microorganisms growth (Li et al., 2019b). Besides, the ORP could be used to determine the fermentation type. In general, as the species of both butyric acid–type and ethanol-type fermentation are strict anaerobes, they are able to increase methane generation. While facultative anaerobes, that is, species of propionic acid fermentation, are inappropriate with methane production. Hence, some researchers have been controlling the fermentation type by adjusting the ORP (Chen et al., 2015, Liu et al., 2012b). Moreover, evaluating the ORP of sludge helps to determine the operability of AD. ORP has the ability to regulate the AD process by modifying the $NADH/NAD^+$ ratio, which is critical for increasing methane generation (Wei et al., 2018a, Xi et al., 2014). As a result, ORP should be adjusted continuously during the AD process. ZVI demonstrates a great ability to regulate the system ORP. Overall, many factors influence the ZVI performance, which also varies depending on the substrate type. Further research needs to be conducted to design the better ZVI-based AD, as well as measuring its environmental loads in the different stages utilizing tools such as life cycle assessment (LCA) to determine which configuration results in the best outcome for the environment. Besides, it may be considered to conduct a life cycle cost for economic viability and a social life cycle assessment (S-LCA) to evaluate the social and sociological aspects of the outcomes.

7.2.1.2 *The addition of Fe(II) and Fe(III) for the enhancement of AD*

Fe(II) is required for microorganism proliferation and participates in the removal of various pollutants and the recovery of energy from organic waste. Moreover, it has a synthetic effect on key enzyme activity involved in methanogenesis, including AK, PTA, BK, and PTB (Wei et al., 2018b). The addition of dissolved Fe(II) promotes the protein content in the EPS, enriching methanogens and preventing the accumulation of excessive propionic and butyric acid accumulation in the system, which increases the microorganisms ability to deal with extreme environments (Ganzoury and Allam, 2015, Wang et al., 2016). The Fe-S cluster is an active group in F_{420}-reducing hydrogenase (Frh), which uses the coenzyme F_{420} to transfer CO_2 and H_2 into CH_4, thus it is necessary for the catalysis of Frh. Additionally, iron (II) acts as a cofactor, enhancing the action of coenzyme F_{420} in indirectly converting carbon dioxide and hydrogen to methane. And the Fe-S cluster is significantly stimulated by Fe (II) as a cofactor (Wang et al., 2016, Zhao et al., 2018a). When sulfur is reduced to sulfate, iron(II) acts as a food for sulfate-reducing bacteria and is combined with hydrogen sulfide (H_2S) to form iron sulfate (FeS), thus removing sulfur (Ruan et al., 2017). However, it should be noted that when a specific concentration of sulfur is present in the system, Fe(II) can precipitate to form FeS, reducing the Fe(II) bioavailability. This circumstance can be resolved by adding chelating agents such as ethylenediamine-N,N′-disuccinic acid ethylenediaminetetraacetic acid (EDDS), ethylenediaminetetraacetic acid, and nitrilotriacetic acid (Bartacek et al., 2012, Thanh et al., 2017, Vintiloiu et al., 2013), which can protect Fe^{2+} from sulfide precipitation. Finally, Thanh et al. (2017) demonstrated that the EDDS outperforms the other two agents.

Through an electrochemical mechanism, iron-reducing bacteria (IRB) may decompose organic compounds and create iron(II) oxide. While iron oxide is insoluble in water under normal conditions, when mixed with an electron acceptor (often Fe(III) oxide), IRB reduced iron to Fe(II) (Honetschlägerová et al., 2018). This approach begins with direct interaction between toxins and pollutants and continues with the use of electron shuttles and bacteria nanowires (Zhang et al., 2014b). When Fe(III) oxide and the reductase located in the extracellular membrane region come into direct contact, direct electron transfer occurs. The enzymes mainly comprise multiheme c-type cytochromes and are involved in metal reduction, which oxidizes organic substances to short-chain or long-chain fatty acids, while simultaneously reducing Fe(III) to Fe(II) (Shi et al., 2009). The second alternative is dominated by IRB and is devoid of multiheme c-type cytochromes. IRB and iron oxide do not need to come in direct contact with the bacteria to commence electron transfer, as the bacteria generate conductive pili (Liu et al., 2016). Likewise, in the third pathway, electrons are transported by exogenous and/or endogenous electron shuttles, such as metal chelators and flavins, S^0, biochar, and Fe(II) (Dong et al., 2019c, Feng et al., 2012, Qiao et al., 2018). CO_2 and water are also generated when IRB oxidized organic materials. Thus, when oxidized HA is reduced by accepting displaced electrons from the process, the Fe(III) is reduced by taking electrons reduction simultaneously (Yang et al., 2017b).

7.2.1.3 Iron oxide addition–enhanced AD

In AD conditions, ZVI is frequently oxidized to Fe(II). In addition, if ZVI is introduced into the reaction system under strictly anaerobic conditions, it may react with water and oxidant, forming Fe(III) oxide attached to the ZVI surface (Zhang et al., 2020b). Moreover another possible combination is the result of $Fe(OH)_3$ and citrate, which results in an increase in phenol degradation rates when compared to $Fe(OH)_3$ alone. Therefore, the Fe (III)-reducing genera involves *Petrimonas* and *Shewanella*, that might be involved in the syntrophic decomposition of phenol by methanogens via (DIET) (Li et al., 2019c). On the other hand, the application of nanoparticles leads to a higher production of biogas. Apart from the financial implications, large-scale plants might involve environmental and health-related complications. These variables have not been thoroughly considered in-depth yet (Dehhaghi et al., 2019).

Lu et al. (2019) investigated the effects of ferric oxide (Fe_2O_3) addition on swine manure. Using a 75 mmol of Fe_2O_3, they demonstrated a maximum 11.06% increase in accumulative CH_4. The formation of Fe-S precipitates and DIET may be associated with higher CH_4 production. Another result was that Fe_2O_3 had a negligible effect on the microbial community impact and increased polysaccharide and soluble protein concentrations by mid-week 6 as the concentration of soluble iron is reduced rapidly for the adsorption of EPS in the different stages (Lu et al., 2019).

In addition, semiconductive iron oxides are crucial and complex components of all iron oxide addition in AD. Semiconductive iron oxides (e.g., magnetite, hematite, maghemite, goethite) can serve as electrodes for a donor or accept extracellular electrons in microbial energy metabolism (Xu et al., 2019a). Moreover, recent research has found that those electrically semiconductive iron oxides can accelerate the DIET process by mean of the formation of an electric field, which is linked to DIET of microbial cultures and indirect electron transfer; however certain conductive materials (CMs), such as Ag, Mg, ferrihydrite, and carbon black can act as an inhibitor of methane production (Martins et al., 2018). Furthermore, the different surface electron densities of Fe(II) and c-type cytochromes (c-Cyts) adsorbed on the surface of hematite accelerate the electron transfer from Fe(II) to c-Cyts (Liu et al., 2020). Accelerated DIET can help waste/wastewater treatment to maintain a more efficient and stable AD system.

7.2.2   Carbon-based functional materials

Carbon-based functional materials are widely used in AD for strengthening AD performance because they provide a space for microbial immobilization and facilitates the DIET between different species (Lovley, 2017). In general, the carbon-based materials used in AD were diverse, ranging from highly defined graphite to less defined carbonaceous materials, with various morphologies such as AC (activated carbon), biochar, nanographene, among others (Lu et al., 2020).

7.2.2.1   AC additives

AC productions and its effect on AD

AC is a porous material enclosed by carbon atoms (Marsh and Rodríguez-Reinoso, 2006), and it is generally derived from carbonaceous sources such as coal, bamboo, and coconut husk; these products are activated later by mean of physical or chemical pathways (Bagheri et al., 2021). Over the last 25–30 years, AC has been widely used in biological sewage treatment to increase microorganism activity and efficiently remove organic matter. AC is used in the AD as an additive to improve the energy recovery and the removal of toxic substances. Many studies have ratified their numerous advantages. Yang et al. (2017a) explored the effect of AC on AD of waste-activated sludge and found that by adding AC from 0 to 5.0 g (based 150 mL substrate), the methane production increased by 17.4%, and the sludge reduction rate increased as well from 39.1% to 45.2%. Similarly, when municipal solid waste (MSW) is used as a substrate, methane production can be significantly enhanced when AC is amended (Ayodele et al., 2021, Dang et al., 2017). Furthermore, AC can be used to mitigate the toxic shock in AD, such as pentachlorophenol (PCP), one of the most toxic compounds for methanogenesis (Xiao et al., 2015). Its high adsorptive capacity for phenolic compounds has been demonstrated. Besides, according to the LCA approach, the addition of AC is proved to have environmental benefits in both midpoint (which has a stronger relation to the environmental flows and comes in general with lower parameter uncertainty) and endpoint levels (which is easier to interpret in terms of the relevance of the environmental flows) (Hauschild and Huijbregts, 2015, Zhang et al., 2020b).

The mechanism behind the addition of AC in AD

Several mechanisms can facilitate AD performance, that is, the surface features of AC, including physical support, specific surface area, porosity, adsorption, among others. The porosity of AC can provide a physical support for microbial attachment, thereby enhancing microbial activity (Wu et al., 2020b). During the AD process, metabolic disruption such as HA from microbes may inhibit methane production, whereas AC's high specific surface area and porosity enable the adsorption of these inhibitors (Capson-Tojo et al., 2018, Yang et al., 2020). Moreover, a better microorganism's attachment to AC can also improve the syntrophic interactions between functional microorganisms (Capson-Tojo et al., 2018). Previously published results proved that AC as an immobilization matrix improves consortia growth for biogas production (Zhang et al., 2017b).

   On the other hand, AC presents high electrical conductivity, which refers to its ability to act as an electron acceptor or donor. It can promote the DIET between bacteria and methanogens (Liu et al., 2012a, Martins et al., 2018). In the AD process, the complex organics are oxidized to simple organic molecules sequentially, during the hydrolysis and acidogenic stages, and then to methane during the methanogenic stage. The interspecies electron transfer (IET) rate and route among the functional microorganisms determine the overall process efficiency. Moreover, the IET has been carried out via three ways: (1) electron shuttles such as hydrogen and formate, (2) DIET by electrically conductive pili or direct contact, and (3) direct electron transfer mediated by AC (Martins et al., 2018). The presence of AC can promote the DIET via a conduction-based mechanism. It serves as a mediator of electron transfer between bacteria and methanogens in an efficient route, thereby accelerating the IET rate. Martins et al. (2018) found that AC amended may be a more effective electron transfer medium than electrically conductive pili and associated c-type cytochrome involved in biological interspecies electrical connections regarding the electron transfer process. An intuitive phenomenon of the AC is that it can reduce the start-up days to initiate methanogenesis in the digester (Capson-Tojo et al., 2018).

Parameters that affect the efficiency of the addition AC in AD

The AD performance with amended AC varies with different parameters such as particular size, microporous structure, and dose (Martins et al., 2018). Ma et al. (2020b) found that powdered AC was superior in enhancing the methane production process than granular AC when ethanol and glucose were treated because powdered AC provides more abundant micropore–mesopore structures for microorganisms to colonize. Also, powdered AC exhibits a better performance in the PCP removal during AD (Xiao et al., 2015). Moreover, an adequate dose of AC is crucial for AD performance. An overdosed AC may have negative effects on AD performance, destroying the diversity of microorganisms drastically (Zhang et al., 2018).

Although the AC has a positive effect on methane production performance, its major shortcoming is its cost effectiveness. Therefore, it is necessary to develop an effective and cheaper alternative to AC for energy recovery from AD.

7.2.2.2 Biochar additives

Biochar is another additive material that can improve AD performance. A large number of studies have reported that biochar has a reinforcing effect on energy recovery. Biochar is a carbon-rich solid material derived from the thermal conversion of biomass with an oxygen-free environment. In comparison to some current commercial ACs and other fine materials, biochar is less expensive and has a greater adsorption capability for pollutants, which is likely attributable to the abundance of functional groups on the biochar surface (Huang et al., 2021). Generally, biochar can be produced from biomass waste through pyrolysis, gasification, hydrothermal carbonization, or torrefaction. Biomass is derived from various sources such as FW, sludge, MSW, agricultural waste, among others. Hence, its composition varies with biomass types, but it is predominantly composed of cellulose, hemicellulose, and lignin, with smaller amounts of pectin, protein, extractives, and ashes (Fabbri and Torri, 2016). The complex structure of biomass is decomposition via a series of reactions such as dehydration, cross-linking, depolymerization, fragmentation, rearrangement, repolymerization, condensation, and carbonization at different temperatures (Masebinu et al., 2019), resulting in the formation of biochar with different properties.

Biochar production and features

The biochar features include elemental composition, surface functional groups, porosity, and specific surface area determined by the biomass used and the devolatilization temperature. The biomass can range from lignocellulosic biomass and biogenic materials to organic wastes such as animal manure and sewage sludge (Sun et al., 2020). Furthermore, pyrolysis is a widely accepted process for biochar production, which decomposes organic materials thermally under oxygen-free conditions at a temperature ranging from 300 to 900 °C. The parameters that influence the biochar characteristic include reaction temperature, heating rate, and residence time. Generally, the biochar heating value and pH increased with the increased pyrolysis temperature, and the syngas yield increases simultaneously (CO, CO_2, H_2, and C1-C2 hydrocarbons). Besides, pyrolysis can be divided into slow pyrolysis and fast pyrolysis, which depends on the rate of temperature increase. For example, Inguanzo et al. (2001) found that a higher temperature rising rate resulted in a lower volatile matter content and a higher biochar yield, indicating that a high temperature rising rate is effective in terms of the biochar's quality. In addition, the residence time can affect the characteristic of biochar. Hanif et al. (2020) evaluated the effect of different residence times on a specific surface area and pore area of biochar, their results indicated that increasing the residence time resulted in a significant decrease in both the specific surface area and pore volume. Another derived method is gasification, a thermochemical partial oxidation process that produces gaseous elements; thereby, biochar yields are only approximately 5%–10% of the raw biomass material mass. Drying, pyrolysis, oxidation, and gasification are all steps in the process. Generally, carbon conversion rates, as well as the composition and heating value of the product gas are dependent on the gasification agent used. Hydrothermal carbonization is the other commonly used biochar-derived technology; it is more suitable for biomass with high moisture contents compared to pyrolysis and gasification. As for the process of hydrothermal carbonization, the biomass is mixed with water, and heat up to a specific temperature (generally above 100 °C) in a confined space. This process produces biochar with high carbon content. Thus, the reaction temperature, pressure, residence time, and water-biomass ratio, all influence the biochar characteristics.

Biochar as a stabilizing agent for AD

The biochar derived from these different technologies can affect the biochar characteristics, which act as a powerful additive in the energy recovery field. These functional characteristics include pH buffer capability, minerals and trace metals, surface functional groups, porosity and specific surface area, electron donor and acceptor, among others. Biochar could serve as a stabilizing agent to balance the physical and chemical properties of an AD system. Biochar has superior pH buffer capability due to the ash-inorganic alkalis and organic alkalis functional groups. It can alleviate the abrupt drop in pH that occurs during the hydrolysis stages of AD, particularly when the easily hydrolyzed organic FW are used as substrates. In addition, it could accelerate the conversion of macromolecular substances to dissolved substrates, reducing the soluble salts, total ammonia nitrogen, and free ammonia levels in the AD system (Ma et al., 2020a). Hence, it is possible to protect AD from VFA accumulation while improving the energy recovery efficiency (Pan et al., 2019, Wang et al., 2018). Besides, the alkalinity of biochar facilitates the carbonization of CO_2 to HCO_3^- and CO_3^{2-} which could also be used for methane production by hydrogenotrophic methanogens (Jang et al., 2018). Moreover, as an additive with high porosity and large specific surface area, biochar addition could mitigate the decrease of porosity during the high-solid waste AD, which is required for liquid recirculation and mass transfer rate enhancement (André et al., 2015, Indren et al., 2020).

Effect of biochar addition on microorganism metabolism in AD

Biochar can create a retention trap for microbial colonies and boost the growth of immobilized bacteria in AD due to its derived materials, porosity, and conductivity, thereby enhancing the removal of organic contaminants and energy recovery. Biochar is generally derived from waste biomass, and it contains a high concentration of minerals and trace metals such as Fe, Pb, Cd, Zn, Cu, Ni, and Cr, which may supplement or replace nutrients required by microorganisms (Wambugu et al., 2019, Zielińska and Oleszczuk, 2015). Recent research concluded that biochar could significantly multiply the bioavailability of TE (trace elements) and improve the relative gene abundance and enzyme activity, all of which is beneficial for methane production (Qi et al., 2021). These biochar-enriched microorganisms were found to contribute efficiently to methanogenesis archaea, such as hydrogenotrophic methanogens, which can consume H_2 and CO_2 from the syngas obtaining CH_4 as the main product (He et al., 2020b, Zhang and Wang, 2020). For example, Lü et al. (2016) published that the archaea were found tightly bound on biochar, and its proportion was higher than in the control without biochar. On the other hand, the high porosity and specific surface area of biochar allow the remotion of inhibitors from substrate degradation such as NH_3, phenols, limonene, and other metabolites (Pan et al., 2019). Also, the surface area and porosity of biochar can cause a random collision with microorganisms and accelerate the secretion of cell surface protein or extracellular polysaccharides in AD, which can form a tight biofilm to resist inhibition in the external environment such as low pH and high content of ammonia (Liu et al., 2017, Mumme et al., 2014).

Besides, biochar's electrical conductivity could serve as an efficient electron donor and acceptor between syntrophic microorganisms, thereby strengthening the DIET of biochar (Chen et al., 2014, Qi et al., 2021). During the AD system, IET between syntrophic microorganisms plays a crucial role in organic matter oxidation and CO_2 to CH_4 reduction, and this is a vital process to break through the thermodynamic barrier to maintain breeding between syntrophic communities of anaerobic bacteria and archaea (Stams and Plugge, 2009). Besides, microbial cells were capable of exchanging electrons over long distances through electrical connections (Wegener et al., 2015). Moreover, thereby, the electrical conductivity of the system is linked to the IET rate. According to previous research, the electrical conductivities of AD with amended biochar are 9.97–28.20 mS/cm, which is significantly higher than that of the electrical conductivities of AD without biochar addition (Franke-Whittle et al., 2014), implying that biochar addition could enhance the interspecies electron transferability. Additionally, recent research found that methane production in the AD of sludge was positively correlated with biochar's electron-donating capacity rather than its bulk electrical conductivity (Shen et al., 2021), which means that the electron transfer mediated by the redox-active functional groups play an important role in IET.

Furthermore, signaling compounds such as the resuscitation-promoting factors and quorum sensing are another way of microbial communication that can be facilitated by biochar. The resuscitating viable but nonculturable bacteria via signaling compounds could be immobilized on the biochar and promote AD efficiency (Wang et al., 2020c).

Overall, biochar could enhance the energy recovery from AD due to its unique physico-chemical properties. Additional and novel biochar applications require a better understanding of different biochar physiochemical properties that are formed as a result of biochar sources and production methods. Experiments with biochar characterizations are essential for a better understanding of the relationship between biochar and its potential applications. Moreover, it is essential to recognize the degree of natural variations in the chemical and structural properties of biochar to develop strategies for mitigating and/or minimizing the potential impacts of natural variations on the features of biochar materials.

7.2.2.3 *Other carbon-based functional materials*

Besides the AC and biochar used in AD, other carbon-based functional materials could further improve the AD by modifying the structure and area functional group. Multiwalled carbon nanotubes are allotropes of carbon with cylindrical nanostructures and can be formed both naturally and artificially. It was verified that in the presence of multiwalled carbon nanotubes, the methanogenic activity of pure cultures of methanogens increased significantly and that it was correlated with variations in the redox potential (ORP) (Hao et al., 2019, Martins et al., 2018). It accelerates the methane production and yield in an anaerobic system by positively modifying the composition of the residential microbial community, particularly the abundance of *Bacteroidetes, Methanobacteria*, and *Methanobacterium* at the phylum, class, and genus levels, respectively (Hao et al., 2019). Moreover, ZVI biochar is a composite material made by modifying the biochar with ZVI on its surface. The ZVI-biochar combines the advantage of ZVI and biochar. Zhang et al. (2019c) investigated the effects ZVI and biochar on microbial nitrate reduction and found that the coexistence of ZVI and biochar resulted in greater nitrate removal; after 14 days, nitrate levels in the mixed-waste culture decreased 37% in the biochar-ZVI configuration, but only 29% and 18% with ZVI alone and control reactor, respectively. Future research would help in determining whether ZVI-biochar aids in other AD parameters; particularly, performing materials and energy balance around ZVI, biochar, and control configurations for methanogenic AD, which would be useful in determining the most favorable configuration. Li et al. (2019a) reported that the bioreduction of nano-zero valent iron (NZVI) can be enhanced effectively by biochar with *Shewanella putrefaciens* CN32 through the electron transport in the degradation of PCP.

Further study should focus on bioreduction processes mediated by biochar for the purpose of mitigating the negative effects of ZVI. It is worth noting, however, that coenhancement with ZVI-biochar is a possibility, but caution must be practiced to ensure that only required material and energy inputs are realized. Certain coenhancement, such as increasing the number of ZVI application points, lowering the total ZVI requirement, or improving waste conversion to high-grade methane, are preferable to optimizing waste treatment and energy output alone at the expense of excessive materials input. Critically, more time would be saved in discovering the processes underlying the potential benefits of coenhancement if unsuccessful configurations of coenhancement were avoided.

7.2.3   Biological additives

7.2.3.1   *Microbial inoculum*

During the AD process, microbes dominate energy recovery performance. However, due to the lack of diversified microorganisms, the majority of AD substrates, such as FW and MSW, frequently failed to recover energy (Li et al., 2018, Martin-Ryals et al., 2015). Moreover, even if there are functional flora in the anaerobic system, another factor that may contribute to the failure of the AD process is the microbial community shift occurring during transitional phases or in response to stress conditions. Therefore, a suitable inoculum is crucial for an efficient AD system because it provides microbe or enzymes, and a specific nutrient and buffer capacity (Pakarinen et al., 2008, Paritosh et al., 2020). In this regard, the appropriate inoculum selection is relevant, as the substrate/inoculum, to obtain a balanced microbial population. In general, a consortium of microorganisms such as hydrolytic and methanogens microorganisms is needed in the inoculum. These microorganisms are capable of converting organic waste to smaller monomers by hydrolysis process and eventually convert to methane via methanogenesis process (Chandra et al., 2012, Gujer and Zehnder, 1982). Microorganisms, for instance, are commonly inoculated with liquid AD effluent, waste-activated sludge, and anaerobic granular sludge (Ge et al., 2016, Lin and Li, 2017, Zhang et al., 2014a). Additionally, some research confirmed that inoculating with communities enriched in members of the robust *Methanosarcina*, during the start-up phase may improve overall performance in acetate-rich systems (Lins et al., 2014). Besides, the substrate/inoculum ratio is the main factor that affects the overall AD performance. In this sense, a higher substrate/inoculum ratio will prolong the lag time during the start-up phase of AD, in terms of the same substrate. Conversely, a lower substrate/inoculum ratio will reduce the reactor utilization efficiency (Shi et al., 2014).

To boost the rate of AD hydrolysis, some specific hydrolysis microorganisms can be added to the system to achieve the target of specific stubborn organic waste such as cellulosic biomass (Martin-Ryals et al., 2015). This bioaugmentation process generally uses the microorganism including fungi, that is, *Orpinomyces* sp. (Akyol et al., 2019) and *Piromyces rhizinflata YM600* (Nkemka et al., 2015); bacteria, that is, *Methanoculleus thermophilus* (Tian et al., 2019) and *Acetobacteroides hydrogenigenes* (Zhang et al., 2015). A consortium of anaerobic ruminal fungi and fermentative bacteria can be added as a bioaugmentation microorganism (Ferraro et al., 2018). Besides, the bioaugmentation process can be used to alleviate ammonia inhibition in the AD system. Ammonia has been reported to have a strong effect on the microorganisms in AD, especially for those involved in the syntrophic acetate oxidation pathway (Yenigün and Demirel, 2013). Therefore, a potential solution to mitigate the inhibitory effect of ammonia in AD processes is to supply the ammonia-tolerant syntrophic acetate oxidation methanogenic consortia (Fotidis et al., 2013). However, it should be undertaken that in the majority of continuous operation cases, the enhanced methane production was not sustained over time, as the bioaugmented microorganisms competed with the indigenous microbial community. As a consequence, the ability to maintain a stable bioaugmented microbial community has been a significant challenge (Martin-Ryals et al., 2015).

7.2.3.2   *Enzymes*

The increase in methane is related to the addition of a synergistic amount of fibers to the FW. Several facts as the hydrolytic enzymes from fresh FW and constant leaching have contributed to the reduction of fiber (Guilford et al., 2019). Also, the complementary effect of CMs is related to a balanced C/N ratio, a buffering ability pH, and the presence of TE, which stimulate enzyme and coenzyme synthesis. Moreover, TE can facilitate the synthesis of critical coenzymes or cofactors in methanogenic pathways (Bong et al., 2018). In this sense, extracellular enzymes regulate the hydrolysis of organic waste in a digester, that is, esterase, protease, and cellulase activities are strongly correlated with the chemical oxygen demand (COD) degradation performance (Guilford et al., 2019).

Jiang et al. (2021) found that the addition of garbage enzyme (GE) decreases NH_3 emissions by 40.9%. GE is a combination of different enzymes derived from fermented waste, brown sugar, and water; it has been used in the bioremediation process due to its decomposition and catalytic properties, GE operates over a comprehensive temperature range. On the other hand, the digestion of hemicelluloses and cellulose is catalyzed via distinct enzyme systems; hemicellulose hydrolyzes fastly than cellulose and requires additional energy; cellulase enzymes might be classified as exoglucanases, endoglucanases, and glucosidases (Zhuang et al., 2021).

Recent publications deal with genetically engineered enzymes to enhance biogas and ensure stabilization; Thus, enzymatic approaches apply only to single cells and complex microbial populations. Moreover, some researchers show that the C/N ratio and particle size can significantly influence the rates of the AD enzyme-catalyzed waste treatments; also, the temperature is an essential operating parameter for enzymes and coenzymes as high temperatures might denature enzymes, failing the process (Zhang et al., 2019a).

Methanogenesis is a highly metalloenzyme-dependent process; there are three types of methanogenic reactions: acetoclastic, hydrogenotrophic, and methylotrophic. Metal requirements for several pathways vary in function to the enzymes needed. Moreover, iron is a needed cofactor for enzymes (Hassaneen et al., 2020). Several enzymes, together with metal ions, coenzymes, and the required cofactors during AD, are crucial; that is, some cofactors of different enzymes are formyl-MF-dehydrogenase, hydrogenases, CO dehydrogenase (Li et al., 2021). Protein degradation plays a decisive role in biogas production. Thus, the enzymatic hydrolysis barrier and molecular toxicity effects limit the use of sewage sludge. Traditional pretreatments have been proposed as potential strategies for targeted protein degradation enhancement in sewage sludge; they could improve hydrolysis, overcome challenges, regulate microorganisms, secretion and activity of key enzymes, and IET (Chen et al., 2021). The presence of eggshell waste in FW fermentation has an influence on the concentration and distribution of VFAs. The microbial distribution of FW fermentation is altered by eggshell conditioning, thereby influencing the microbial contributions of metabolic enzymes found in VFA producers (Luo et al., 2021).

In addition, iron is required for enzymatic biosynthesis and hydrogenase synthesis. For example, pyruvate ferredoxin oxidoreductase is a critical hydrogen production enzyme in which the active center contains three (4Fe-4S) clusters. Additionally, critical trace metals (Na, Mg, K, Ca, and Fe) have been supplied via hybrid-Fe bioreactors, enhancing enzymatic activity and microbial growth (Zhang et al., 2020c). Hydrogenases, which are composed of [Fe-Fe]-hydrogenases and [Ni-Fe] hydrogenase, can promote proton-to-hydrogen reduction during AD (Zhao et al., 2021). The increasing recognition of additives such as Fe-based nanomaterials, carbon-based materials, and composites demonstrates the promising scale potential for application (Li et al., 2021).

7.3 Additive strategies for enhanced AD for biochemicals

7.3.1 VFA production from AD

AD for the production of VFA is also a potential strategy for recovering energy from organic wastes (Yuan et al., 2019). The characteristic of a short production cycle, a wide range usage area, and high market demand for VFA make it a critical bioproduct. In general, VFA production occurred during the second phase of the AD process is called acidogenesis (Leng et al., 2018). Acid-forming bacteria, including acetogenic and homoacetogenic bacteria, store hydrolyzed organics such as monosaccharides and amino acids intracellularly as significant intermediate products during this stage. The pyruvates is converted to acetyl-CoA for the generation of VFA (Leng et al., 2018). More research has focused on VFA production. There are many additives used to improve the efficiency of VFA production via AD.

As previously stated, iron is a critical aspect of microbial growth. Therefore, iron-related additives such as ZVI, ferric oxide, and magnetite are generally used (Wang et al., 2020a). ZVI can alleviate the drop of pH, stabilize the ORP at an anaerobic level, and promote the dissolution of insoluble organics. In this regard, the ZVI could stimulate functional enzyme activity and the abundance of hydrolysis bacteria. In addition, ferric oxide addition could increase the ORP, stimulating the enrichment of hydrolysis bacteria such as *Pseudomonas* (Wang et al., 2020a). However, iron additives should be used with caution, as some other researchers found that iron has an inhibitory effect on protein release. The iron coagulant species in sludge can "fix" or "lock-up" substrates, which results in a retarding the conversion of organic matter to VFA (Ping et al., 2021). Fortunately, recent research revealed that CaO_2 beads addition could achieve the VFA recovery from iron-rich sludge. The CaO_2 beads can effectively enrich dissimilatory IRB and promote iron reduction–related genes (Ping et al., 2021). Besides, among all VFAs, butyric acid has the highest market value (Bhatia and Yang, 2017), and is frequently used in chemical, pharmaceutical, food, beverage, and textile, industries as a solvent, conservative, diluent, drug, plasticizer, perfume, fiber, additive, and raw material (Atasoy and Cetecioglu, 2020). However, during the butyric fermentation process, the other metabolic pathway produces acetic acid, propionic acid, and ethanol, thereby reducing the generation of butyric acid. Therefore, some strategies can be used to optimize butyric production while also lowering the overall production cost. Adding *Clostridium* as a bioaugmentation process is an ideal measure (Atasoy and Cetecioglu, 2020) as it is one of the most valuable butyric acid producers because of its high productivity and relative stability (Wenzel et al., 2018).

7.3.2 Medium-chain carboxylates production from AD

Medium-chain carboxylates are a type of highly valuable commercial chemicals that include hexanoate, caprylate, and others. They can be generated as value-added products from AD (Liu et al., 2017). Compared with short-chain fatty acids, the medium-chain carboxylates have potential advantages in many aspects (Ge et al., 2015). On one hand, the lower oxygen/carbon ratios than short-chain fatty acids (873.7–2793.9 KJ/mol) make it have a higher energy density (3492.4–4798.7 KJ/mol), making them useful in a variety of daily life applications such as fragrances (Kenealy et al., 1995), food additives (Xu et al., 2015), animal feed additives

(Cavalcante et al., 2017), and antimicrobial agents (Desbois, 2012). Moreover, it can use the waste fatty as the substrate in the AD process, which can avoid the inhibitory effect of oil or glycerol in the substrate on microorganisms (Dams et al., 2018, Liu and Jiang, 2020). Besides, due to the fact that it contains long hydrophobic carbon chains and medium-chain carboxylates, it is easily separated from AD liquor at lower costs (Agler et al., 2012).

During this process, the short chain of fatty acids can be used as an electron acceptor to interact with acetyl-CoA under the dominance of anaerobic microbe, extending its carbon chain length with two carbons at each cycle and producing a medium-chain of carboxylates or medium-chain of fatty acids (Wu et al., 2020a). Therefore, when considering the process of production, many additives can improve its products such as biochar, iron, and ethanol. Biochar addition could create a more restrictive system with a more stable microorganism community structure for chain elongation, which alleviates the inhibition caused by products or substrates. Besides, the functional microbe can be adsorbed on the rough surfaces or pores of biochar, facilitating electron transfer (Liu et al., 2017). Moreover, the ethanol can also be an additive in the form of a cosubstrate to provide the electron donor for chain elongation in this process because it not only could be produced from a large amount of industrial and agricultural waste (e.g., brewing wastewater and lignocellulose), but also could lead to relatively high medium-chain carboxylates productivity (Wu et al., 2020a). Some biological additives can be used as bioaugmentation factors to strengthen the production efficiency of medium-chain carboxylates. The genus *Clostridium* is generally considered to be a producer of medium-chain carboxylates. The detailed mechanism behind this genus is that *Clostridium* could improve the production of n-butyric, which is an essential intermediate in the chain elongation process when ethanol is used as an electron donor (Dams et al., 2018).

7.3.3   Bioethanol and lactic acid production from AD

Except for the VFA and medium-chain carboxylate, bioethanol and lactic acid production from the AD process are hotspots. Bioethanol is a clean alternative for transportation fuel, which can reduce negative environmental impacts, including greenhouse gas (GHGs) and carbon dioxide (CO_2) emission (Hafid et al., 2017, Mussatto et al., 2010). Lactic acid is an important value-added production widely used in the food, pharmaceutical, and chemical industries (Mazzoli et al., 2014). Generally, the different fermentation products can be obtained by adjusting the AD inoculum and environmental conditions (Wang et al., 2020a). For example, controlling pH and ORP could shift the fermentation types (Chen et al., 2015, Feng et al., 2018). Ethanol-type fermentation generally occurs at pH 4.0–4.5, whereas lactic acid takes part mainly when pH is below 4.0 (Wang et al., 2020a). As for bioethanol fermentation, the biochar can be additive in the fermentation broth to avoid the inhibition of byproducts from lignocellulosic biomass hydrolysis, improving bioethanol production (Wang et al., 2020b). Moreover, it should be denoting that it is not adsorptive detoxification to alleviate the inhibition effect but the cell immobilization on biochar. When the fermentation target product is lactic acid, the ZVI and ferric oxide could be effective additives (Li et al., 2017). In the AD system, the behavior of iron→Fe(II)→Fe(III) could reduce the AD ORP, which can reduce the amount of energy required for lactic acid fermentation, thereby promoting the conversion of pyruvic acid to lactic acid (Jin et al., 2019). And the Fe^{3+} released from ZVI could improve lactate dehydrogenase activity, which is beneficial for the conversion of pyruvate to lactic acid (Zou et al., 2012). In addition, ferric oxide has been shown to improve lactic acid production as it can improve the abundance of *Lactobacillus panis*, which belong to homofermentative species, and can enhance lactate production (Wang et al., 2020a).

7.4 Conclusions and perspectives

Global population growth and rapid urbanization, particularly in developing countries, have resulted in a continuously growing total waste generation. Additionally, AD is a promising waste treatment and resource recovery technology. An additive strategy is a valuable measure to improve the performance of AD. As for bioenergy recovery, the most frequently used common additives include CM such as iron-based materials, carbon-based materials, and biological additives, that is, enzymes and inoculum. The iron-based materials can enhance the AD performance by changing the system environment or microbial activity, which benefits the DIET process and enhances methane production. The Fe^{2+}/Fe^{3+} cycle is critical to the process of AD enhancement. In addition, carbon-based materials such as active carbon and biochar can act as a carrier for the growth of functional microorganisms; the conductivity and surface functional groups of carbon-based materials could enhance the DIET process by removing inhibitors from the AD feedstock. The biological additives, such as functional microorganisms and enzymes, are generally used to treat substrates that are difficult to handle, such as cellulose and hemicellulose. It has the potential to improve the hydrolysis efficiency of AD. Moreover, when the AD target production is biochemical such as bioethanol or medium-chain carboxylates, these additives still work.

Nevertheless, even if utilizing additives can bring about high benefits, knowledge gaps still exist hereon, and further research should focus on its in-depth mechanism and practical application. As for the carbon-based materials, the current study mostly investigated the relationship between additives and microbial by exclusive strategies. The quantification of the relationship is difficult to clarify. For example, most researchers have found that the biochar could enrich the functional microorganisms related to AD. However, the *in-situ* determination of DNA composition on biochar is difficult. Moreover, the direct evidence of DIET behavior is difficult to observe, which hinders the exploration of the internal mechanism by carbon-based additives. Hence, future research should focus on how to directly observe the impact of biochar on interspecies electron transfer from various angles. As for the iron-based materials, because the microorganisms are sensitive to the form of iron oxide, the iron oxides with different forms may have different effect mechanisms on AD. Therefore, when the iron-based materials were added to the AD system, how the various forms of iron oxides are formed and the interactions between iron oxides and functional microorganisms or their excretive appendages need to be explored.

In addition, as for the practical application, it is difficult to use the additives in large-scale plants as the dosage is greater. In terms of economic benefits, it is inappropriate to add too many additives, especially for iron-based materials. And the iron-based material may take minor effect on the operation of the reactor due to the corrosion of iron. Therefore, future applications of additives may consider compounding different materials to maximize the value of each one and avoid overdosage of additives into the digester. At the same time, more research should also be focused on the recycling and reuse of additives. For example, whether the magnetic field can also be used to promote the recovery of ZVI residues during AD requires further research. Moreover, considering the AD residue matter is generally used as a fertilizer; therefore, the concerns of environmental health risks, practical utilization of anaerobic digestate as a biofertilizer is still needed to be considered. And the effect of additives on soil and plants may generate great potential threats for environmental health risks. More attention should be also focused on how to alleviate the effect of additives on environmental health risks. Finally, due to the rising concerns of global climate change, it is worth noting that the LCA approach or GHG emission evaluation should be considered to avoid additive waste and secondary environmental pollution.

References

Agler, M., Spirito, C., Usack, J., Werner, J., Angenent, L., 2012. Chain elongation with reactor microbiomes: upgrading dilute ethanol to medium-chain carboxylates. Energy Environ. Sci. 5, 8189–8192.

Ajay, C.M., Mohan, S., Dinesha, P., Rosen, M.A., 2020. Review of impact of nanoparticle additives on anaerobic digestion and methane generation. Fuel 277, 118234.

Akyol, Ç., Ince, O., Bozan, M., Ozbayram, E.G., Ince, B., 2019. Fungal bioaugmentation of anaerobic digesters fed with lignocellulosic biomass: what to expect from anaerobic fungus *Orpinomyces* sp. Bioresour. Technol. 277, 1–10.

André, L., Durante, M., Pauss, A., Lespinard, O., Ribeiro, T., Lamy, E., 2015. Quantifying physical structure changes and non-uniform water flow in cattle manure during dry anaerobic digestion process at lab scale: implication for biogas production. Bioresour. Technol. 192, 660–669.

Arif, S., Liaquat, R., Adil, M., 2018. Applications of materials as additives in anaerobic digestion technology. Renew. Sustain. Energy Rev. 97, 354–366.

Atasoy, M., Cetecioglu, Z., 2020. Butyric acid dominant volatile fatty acids production: Bio-augmentation of mixed culture fermentation by *Clostridium butyricum*. J. Environ. Chem. Eng. 8 (6), 104496.

Ayodele, O.O., Adekunle, A.E., Adesina, A.O., Pourianejad, S., Zentner, A., Dornack, C., 2021. Stabilization of anaerobic co-digestion of biowaste using activated carbon of coffee ground biomass. Bioresour. Technol. 319, 10.

Baek, G., Kim, J., Cho, K., Bae, H., Lee, C., 2015. The biostimulation of anaerobic digestion with (semi) conductive ferric oxides: their potential for enhanced biomethanation. Appl. Microbiol. Biotechnol. 99 (23), 10355–10366.

Bagheri, M., Jafari, S.M., Eikani, M.H., 2021. Ultrasonic-assisted production of zero-valent iron-decorated graphene oxide/activated carbon nanocomposites: chemical transformation and structural evolution. Mater. Sci. Eng. C 118, 111362.

Bartacek, J., Fermoso, F.G., Vergeldt, F., Gerkema, E., Maca, J., van As, H., Lens, P.N.L., 2012. The impact of metal transport processes on bioavailability of free and complex metal ions in methanogenic granular sludge. Water Sci. Technol. 65 (10), 1875–1881.

Batstone, D.J., Virdis, B., 2014. The role of anaerobic digestion in the emerging energy economy. Curr. Opin. Biotechnol. 27, 142–149.

Bhatia, S., Yang, Y.-h., 2017. Microbial production of volatile fatty acids: current status and future perspectives. Rev. Environ. Sci. Bio/Technol. 16, 327–345.

Bong, C.P.C., Lim, L.Y., Lee, C.T., Klemeš, J.J., Ho, C.S., Ho, W.S., 2018. The characterisation and treatment of food waste for improvement of biogas production during anaerobic digestion—a review. J. Clean. Prod. 172, 1545–1558.

Cai, Y.F., Wang, J.G., Zhao, Y.B., Zhao, X.L., Zheng, Z.H., Wen, B.T., Cui, Z.J., Wang, X.F., 2018. A new perspective of using sequential extraction: to predict the deficiency of trace elements during anaerobic digestion. Water Res. 140, 335–343.

Capson-Tojo, G., Moscoviz, R., Ruiz, D., Santa-Catalina, G., Trably, E., Rouez, M., Crest, M., Steyer, J.P., Bernet, N., Delgenes, J.P., Escudie, R., 2018. Addition of granular activated carbon and trace elements to favor volatile fatty acid consumption during anaerobic digestion of food waste. Bioresour. Technol. 260, 157–168.

Cavalcante, W.d.A., Leitão, R.C., Gehring, T.A., Angenent, L.T., Santaella, S.T., 2017. Anaerobic fermentation for N-caproic acid production: a review. Process Biochem. 54, 106–119.

Chandra, R., Takeuchi, H., Hasegawa, T., 2012. Methane production from lignocellulosic agricultural crop wastes: a review in context to second generation of biofuel production. Renew. Sustain. Energy Rev. 16 (3), 1462–1476.

Charalambous, P., Vyrides, I., 2021. In situ biogas upgrading and enhancement of anaerobic digestion of cheese whey by addition of scrap or powder zero-valent iron (ZVI). J. Environ. Manage. 280, 111651.

Chen, S., Gao, J., Dong, B., 2021. Bottlenecks of anaerobic degradation of proteins in sewage sludge and the potential targeted enhancing strategies. Sci. Total Environ. 759, 143573.

Chen, S., Rotaru, A.E., Shrestha, P.M., Malvankar, N.S., Liu, F., Fan, W., Nevin, K.P., Lovley, D.R., 2014. Promoting interspecies electron transfer with biochar. Sci. Rep. 4, 5019.

Chen, X., Yuan, H., Zou, D., Liu, Y., Zhu, B., Chufo, A., Jaffar, M., Li, X., 2015. Improving biomethane yield by controlling fermentation type of acidogenic phase in two-phase anaerobic co-digestion of food waste and rice straw. Chem. Eng. J. 273, 254–260.

Dams, R.I., Viana, M.B., Guilherme, A.A., Silva, C.M., dos Santos, A.B., Angenent, L.T., Santaella, S.T., Leitão, R.C., 2018. Production of medium-chain carboxylic acids by anaerobic fermentation of glycerol using a bioaugmented open culture. Biomass Bioenergy 118, 1–7.

Dang, Y., Sun, D., Woodard, T.L., Wang, L.Y., Nevin, K.P., Holmes, D.E., 2017. Stimulation of the anaerobic digestion of the dry organic fraction of municipal solid waste (OFMSW) with carbon-based conductive materials. Bioresour. Technol. 238, 30–38.

Dehhaghi, M., Tabatabaei, M., Aghbashlo, M., Kazemi Shariat Panahi, H., Nizami, A.S., 2019. A state-of-the-art review on the application of nanomaterials for enhancing biogas production. J. Environ. Manage. 251, 109597.

Desbois, A., 2012. Potential applications of antimicrobial fatty acids in medicine, agriculture and other industries. Recent Pat. Anti-Infect. Drug Discov. 7, 111–122.

Dong, B., Xia, Z.H., Sun, J., Dai, X.H., Chen, X.M., Ni, B.J., 2019a. The inhibitory impacts of nano-graphene oxide on methane production from waste activated sludge in anaerobic digestion. Sci. Total Environ. 646, 1376–1384.

Dong, D., Aleta, P., Zhao, X., Choi, O.K., Kim, S., Lee, J.W., 2019b. Effects of nanoscale zero valent iron (nZVI) concentration on the biochemical conversion of gaseous carbon dioxide (CO_2) into methane (CH_4). Bioresour. Technol. 275, 314–320.

Fabbri, D., Torri, C., 2016. Linking pyrolysis and anaerobic digestion (Py-AD) for the conversion of lignocellulosic biomass. Curr. Opin. Biotechnol. 38, 167–173.

Feng, C., Yue, X., Li, F., Wei, C.-H., 2012. Bio-current as an indicator for biogenic Fe(II) generation driven by dissimilatory iron reducing bacteria. Biosens. Bioelectron. 39, 51–56.

Feng, K., Li, H., Zheng, C., 2018. Shifting product spectrum by pH adjustment during long-term continuous anaerobic fermentation of food waste. Bioresour. Technol. 270, 180–188.

Feng, Y., Zhang, Y., Quan, X., Chen, S., 2014. Enhanced anaerobic digestion of waste activated sludge digestion by the addition of zero valent iron. Water Res. 52, 242–250.

Ferraro, A., Dottorini, G., Massini, G., Mazzurco Miritana, V., Signorini, A., Lembo, G., Fabbricino, M., 2018. Combined bioaugmentation with anaerobic ruminal fungi and fermentative bacteria to enhance biogas production from wheat straw and mushroom spent straw. Bioresour. Technol. 260, 364–373.

Ferrer, I., Vázquez, F., Font, X., 2010. Long term operation of a thermophilic anaerobic reactor: process stability and efficiency at decreasing sludge retention time. Bioresour. Technol. 101 (9), 2972–2980.

Fotidis, I.A., Karakashev, D., Angelidaki, I., 2013. Bioaugmentation with an acetate-oxidising consortium as a tool to tackle ammonia inhibition of anaerobic digestion. Bioresour. Technol. 146, 57–62.

Franke-Whittle, I.H., Walter, A., Ebner, C., Insam, H., 2014. Investigation into the effect of high concentrations of volatile fatty acids in anaerobic digestion on methanogenic communities. Waste Manage. (Oxford) 34 (11), 2080–2089.

Ganzoury, M.A., Allam, N.K., 2015. Impact of nanotechnology on biogas production: a mini-review. Renew. Sustain. Energy Rev. 50, 1392–1404.

Ge, S., Usack, J., Spirito, C., Angenent, L., 2015. Long-term N-caproic acid production from yeast-fermentation beer in an anaerobic bioreactor with continuous product extraction. Environ. Sci. Technol. 49 (13), 8012–8021.

Ge, X., Xu, F., Li, Y., 2016. Solid-state anaerobic digestion of lignocellulosic biomass: recent progress and perspectives. Bioresour. Technol. 205, 239–249.

Glass, J.B., Orphan, V.J., 2012. Trace metal requirements for microbial enzymes involved in the production and consumption of methane and nitrous oxide. Front. Microbiol. 3, 61.

Guilford, N.G.H., Lee, H.P., Kanger, K., Meyer, T., Edwards, E.A., 2019. Solid-state anaerobic digestion of mixed organic waste: the synergistic effect of food waste addition on the destruction of paper and cardboard. Environ. Sci. Technol. 53 (21), 12677–12687.

Gujer, W., Zehnder, A.J.B., 1982. Conversion processes in anaerobic digestion. Water Sci. Technol 15 (8-9), 127–167.

Hafid, H.S., Rahman, N.A.A., Shah, U.K.M., Baharuddin, A.S., Ariff, A.B., 2017. Feasibility of using kitchen waste as future substrate for bioethanol production: a review. Renew. Sustain. Energy Rev. 74, 671–686.

Hanif, M.U., Zwawi, M., Capareda, S.C., Iqbal, H., Algarni, M., Felemban, B.F., Bahadar, A., Waqas, A., 2020. Influence of pyrolysis temperature on product distribution and characteristics of anaerobic sludge. Energies 13 (1), 79.

Hao, Y., Wang, Y., Ma, C., White, J.C., Zhao, Z., Duan, C., Zhang, Y., Adeel, M., Rui, Y., Li, G., Xing, B., 2019. Carbon nanomaterials induce residue degradation and increase methane production from livestock manure in an anaerobic digestion system. J. Clean. Prod. 240, 118257.

Hassaneen, F.Y., Abdallah, M.S., Ahmed, N., Taha, M.M., Abd ElAziz, S.M.M., El-Mokhtar, M.A., Badary, M.S., Allam, N.K., 2020. Innovative nanocomposite formulations for enhancing biogas and biofertilizers production from anaerobic digestion of organic waste. Bioresour. Technol. 309, 123350.

Hauschild, M.Z., Huijbregts, M.A.J., 2015. Introducing life cycle impact assessment. In Hauschild, M.Z., Huijbregts, M.A.J. (Eds.), Life Cycle Impact Assessment, Springer Netherlands, Dordrecht, pp. 1–16.

He, C.S., Ding, R.R., Chen, J.Q., Li, W.Q., Li, Q., Mu, Y., 2020a. Interactions between nanoscale zero valent iron and extracellular polymeric substances of anaerobic sludge. Water Res. 178, 115817.

He, P., Zhang, H., Duan, H., Shao, L., Lü, F., 2020b. Continuity of biochar-associated biofilm in anaerobic digestion. Chem. Eng. J. 390, 124605.

Honetschlägerová, L., Škarohlíd, R., Martinec, M., Šír, M., Luciano, V., 2018. Interactions of nanoscale zero valent iron and iron reducing bacteria in remediation of trichloroethene. Int. Biodeterior. Biodegrad. 127, 241–246.

Huang, W.-H., Lee, D.-J., Huang, C., 2021. Modification on biochars for applications: a research update. Bioresour. Technol. 319, 124100.

Indren, M., Birzer, C.H., Kidd, S.P., Medwell, P.R., 2020. Effect of total solids content on anaerobic digestion of poultry litter with biochar. J. Environ. Manage. 255, 109744.

Inguanzo, M., Menéndez, J.A., Fuente, E., Pis, J.J., 2001. Reactivity of pyrolyzed sewage sludge in air and CO_2. J. Anal. Appl. Pyrolysis 58-59, 943–954.

Jang, H.M., Choi, Y.-K., Kan, E., 2018. Effects of dairy manure-derived biochar on psychrophilic, mesophilic and thermophilic anaerobic digestions of dairy manure. Bioresour. Technol. 250, 927–931.

Jiang, J., Wang, Y., Yu, D., Zhu, G., Cao, Z., Yan, G., Li, Y., 2021. Comparative evaluation of biochar, pelelith, and garbage enzyme on nitrogenase and nitrogen-fixing bacteria during the composting of sewage sludge. Bioresour. Technol. 333, 125165.

Jin, Y., Gao, M., Li, H., Lin, Y., Wang, Q., Tu, M., Ma, H., 2019. Impact of nanoscale zerovalent iron on volatile fatty acid production from food waste: key enzymes and microbial community. J. Chem. Technol. Biotechnol. 94 (10), 3201–3207.

Kenealy, W.R., Cao, Y., Weimer, P.J., 1995. Production of caproic acid by cocultures of ruminal cellulolytic bacteria and *Clostridium kluyveri* grown on cellulose and ethanol. Appl. Microbiol. Biotechnol. 44 (3), 507–513.

Leng, L., Yang, P., Singh, S., Zhuang, H., Xu, L., Chen, W.-H., Dolfing, J., Li, D., Zhang, Y., Zeng, H., Chu, W., Lee, P.-H., 2018. A review on the bioenergetics of anaerobic microbial metabolism close to the thermodynamic limits and its implications for digestion applications. Bioresour. Technol. 247, 1095–1106.

Li, H., Chen, S., Ren, L.Y., Zhou, L.Y., Tan, X.J., Zhu, Y., Belver, C., Bedia, J., Yang, J., 2019a. Biochar mediates activation of aged nanoscale ZVI by *Shewanella putrefaciens* CN32 to enhance the degradation of pentachlorophenol. Chem. Eng. J. 368, 148–156.

Li, X., Zhang, W., Xue, S., Lai, S., Li, J., Chen, H., Liu, Z., Xue, G., 2017. Enrichment of D-lactic acid from organic wastes catalyzed by zero-valent iron: an approach for sustainable lactate isomerization. Green Chem. 19 (4), 928–936.

Li, Y., Chen, Y., Wu, J., 2019b. Enhancement of methane production in anaerobic digestion process: a review. Appl. Energy 240, 120–137.

Li, Y., Ren, C., Zhao, Z., Yu, Q., Zhao, Z., Liu, L., Zhang, Y., Feng, Y., 2019c. Enhancing anaerobic degradation of phenol to methane via solubilizing Fe(III) oxides for dissimilatory iron reduction with organic chelates. Bioresour. Technol. 291, 121858.

Li, Y., Wang, Y., Yu, Z., Lu, J., Li, D., Wang, G., Li, Y., Wu, Y., Li, S., Xu, F., Li, G., Gong, X., 2018. Effect of inoculum and substrate/inoculum ratio on the performance and methanogenic archaeal community structure in solid state anaerobic co-digestion of tomato residues with dairy manure and corn stover. Waste Manage. (Oxford) 81, 117–127.

Li, Y., Zhao, J., Krooneman, J., Euverink, G.J.W., 2021. Strategies to boost anaerobic digestion performance of cow manure: laboratory achievements and their full-scale application potential. Sci. Total Environ. 755 (Pt 1), 142940.

Lin, L., Li, Y., 2017. Sequential batch thermophilic solid-state anaerobic digestion of lignocellulosic biomass via recirculating digestate as inoculum—Part I: reactor performance. Bioresour. Technol. 236, 186–193.

Lin, R.C., Cheng, J., Ding, L.K., Murphy, J.D., 2018. Improved efficiency of anaerobic digestion through direct inter-species electron transfer at mesophilic and thermophilic temperature ranges. Chem. Eng. J. 350, 681–691.

Lins, P., Reitschuler, C., Illmer, P., 2014. *Methanosarcina* spp., the key to relieve the start-up of a thermophilic anaerobic digestion suffering from high acetic acid loads. Bioresour. Technol. 152, 347–354.

Liu, D., Zhang, Q., Wu, L., Zeng, Q., Dong, H., Bishop, M.E., Wang, H., 2016. Humic acid-enhanced illite and talc formation associated with microbial reduction of Fe(III) in nontronite. Chem. Geol. 447, 199–207.

Liu, F., Rotaru, A.-E., Shrestha, P.M., Malvankar, N.S., Nevin, K.P., Lovley, D.R., 2012a. Promoting direct interspecies electron transfer with activated carbon. Energy Environ. Sci. 5 (10), 8982.

Liu, N., Jiang, J., 2020. Valorisation of food waste using salt to alleviate inhibition by animal fats and Vegetable oils during anaerobic digestion. Biomass Bioenergy 143, 105826.

Liu, T., Wang, Y., Liu, C., Li, X., Cheng, K., Wu, Y., Fang, L., Li, F., Liu, C., 2020. Conduction band of hematite can mediate cytochrome reduction by Fe(II) under dark and anoxic conditions. Environ. Sci. Technol. 54 (8), 4810–4819.

Liu, Y., He, P., Shao, L., Zhang, H., Lü, F., 2017. Significant enhancement by biochar of caproate production via chain elongation. Water Res. 119, 150–159.

Liu, Y., Zhang, Y., Quan, X., Li, Y., Zhao, Z., Meng, X., Chen, S., 2012b. Optimization of anaerobic acidogenesis by adding Fe0 powder to enhance anaerobic wastewater treatment. Chem. Eng. J. 192, 179–185.

Lovley, D., 2017. Syntrophy goes electric: direct interspecies electron transfer. Annu. Rev. Microbiol. 71, 643–664.

Lü, F., Luo, C., Shao, L., He, P., 2016. Biochar alleviates combined stress of ammonium and acids by firstly enriching Methanosaeta and then Methanosarcina. Water Res. 90, 34–43.

Lu, J.S., Chang, J.S., Lee, D.J., 2020. Adding carbon-based materials on anaerobic digestion performance: a mini-review. Bioresour. Technol. 300, 122696.

Lu, T., Zhang, J., Wei, Y., Shen, P., 2019. Effects of ferric oxide on the microbial community and functioning during anaerobic digestion of swine manure. Bioresour. Technol. 287, 121393.

Luo, J., Huang, W., Zhang, Q., Guo, W., Xu, R., Fang, F., Cao, J., Wu, Y., 2021. A preliminary metatranscriptomic insight of eggshells conditioning on substrates metabolism during food wastes anaerobic fermentation. Sci. Total Environ. 761, 143214.

Ma, H., Hu, Y., Kobayashi, T., Xu, K.-Q., 2020a. The role of rice husk biochar addition in anaerobic digestion for sweet sorghum under high loading condition. Biotechnol. Rep. 27, e00515.

Ma, J., Wei, H., Su, Y., Gu, W., Wang, B., Xie, B., 2020b. Powdered activated carbon facilitates methane productivity of anaerobic co-digestion via acidification alleviating: microbial and metabolic insights. Bioresour. Technol. 313, 123706.

Ma, L., Zhou, L., Mbadinga, S.M., Gu, J.-D., Mu, B.-Z., 2018. Accelerated CO_2 reduction to methane for energy by zero valent iron in oil reservoir production waters. Energy 147, 663–671.

Ma, L., Zhou, L., Ruan, M.-Y., Gu, J.-D., Mu, B.-Z., 2019. Simultaneous methanogenesis and acetogenesis from the greenhouse carbon dioxide by an enrichment culture supplemented with zero-valent iron. Renew. Energy 132, 861–870.

Marsh, H., Rodríguez-Reinoso, F., 2006. Activated Carbon (Origins). In Marsh, H., Rodríguez-Reinoso, F. (Eds), Activated Carbon, Elsevier Science Ltd, Oxford, pp. 13–86.

Martin-Ryals, A., Schideman, L., Li, P., Wilkinson, H., Wagner, R., 2015. Improving anaerobic digestion of a cellulosic waste via routine bioaugmentation with cellulolytic microorganisms. Bioresour. Technol. 189, 62–70.

Martins, G., Salvador, A.F., Pereira, L., Alves, M.M., 2018. Methane production and conductive materials: a critical review. Environ. Sci. Technol. 52 (18), 10241–10253.

Masebinu, S.O., Akinlabi, E.T., Muzenda, E., Aboyade, A.O., 2019. A review of biochar properties and their roles in mitigating challenges with anaerobic digestion. Renew. Sustain. Energy Rev. 103, 291–307.

Mazzoli, R., Bosco, F., Mizrahi, I., Bayer, E.A., Pessione, E., 2014. Towards lactic acid bacteria-based biorefineries. Biotechnol. Adv. 32 (7), 1216–1236.

Meng, X., Sui, Q., Liu, J., Yu, D., Wang, Y., Wei, Y., 2020. Relieving ammonia inhibition by zero-valent iron (ZVI) dosing to enhance methanogenesis in the high solid anaerobic digestion of swine manure. Waste Manage. (Oxford) 118, 452–462.

Mumme, J., Srocke, F., Heeg, K., Werner, M., 2014. Use of biochars in anaerobic digestion. Bioresour. Technol. 164, 189–197.

Mussatto, S.I., Dragone, G., Guimarães, P.M.R., Silva, J.P.A., Carneiro, L.M., Roberto, I.C., Vicente, A., Domingues, L., Teixeira, J.A., 2010. Technological trends, global market, and challenges of bio-ethanol production. Biotechnol. Adv. 28 (6), 817–830.

Nkemka, V.N., Gilroyed, B., Yanke, J., Gruninger, R., Vedres, D., McAllister, T., Hao, X., 2015. Bioaugmentation with an anaerobic fungus in a two-stage process for biohydrogen and biogas production using corn silage and cattail. Bioresour. Technol. 185, 79–88.

Pakarinen, O., Lehtomäki, A., Rissanen, S., Rintala, J., 2008. Storing energy crops for methane production: effects of solids content and biological additive. Bioresour. Technol. 99 (15), 7074–7082.

Pan, J., Ma, J., Liu, X., Zhai, L., Ouyang, X., Liu, H., 2019. Effects of different types of biochar on the anaerobic digestion of chicken manure. Bioresour. Technol. 275, 258–265.

Paritosh, K., Yadav, M., Chawade, A., Sahoo, D., Kesharwani, N., Pareek, N., Vivekanand, V., 2020. Additives as a support structure for specific biochemical activity boosts in anaerobic digestion: a review. Front. Energy Res. 8, 88.

Peng, H., Zhang, Y., Tan, D., Zhao, Z., Zhao, H., Quan, X., 2018. Roles of magnetite and granular activated carbon in improvement of anaerobic sludge digestion. Bioresour. Technol. 249, 666–672.

Ping, Q., Zhang, Z., Dai, X., Li, Y., 2021. Novel CaO$_2$ beads used in the anaerobic fermentation of iron-rich sludge for simultaneous short-chain fatty acids and phosphorus recovery under ambient conditions. Bioresour. Technol. 322, 124553.

Qi, Q., Sun, C., Zhang, J., He, Y., Tong, Y, W., 2021. Internal enhancement mechanism of biochar with graphene structure in anaerobic digestion: the bioavailability of trace elements and potential direct interspecies electron transfer. Chem. Eng. J. 406, 126833.

Qiao, J.-t., Li, X.-m., Li, F.-b., 2018. Roles of different active metal-reducing bacteria in arsenic release from arsenic-contaminated paddy soil amended with biochar. J. Hazard. Mater. 344, 958–967.

Qiao, W., Takayanagi, K., Niu, Q., Shofie, M., Li, Y.Y., 2013. Long-term stability of thermophilic co-digestion sub-merged anaerobic membrane reactor encountering high organic loading rate, persistent propionate and detectable hydrogen in biogas. Bioresour. Technol. 149, 92–102.

Qin, Y., Chen, L., Wang, T., Ren, J., Cao, Y., Zhou, S., 2019. Impacts of ferric chloride, ferrous chloride and solid retention time on the methane-producing and physicochemical characterization in high-solids sludge anaerobic digestion. Renew. Energy 139, 1290–1298.

Ruan, R., Cao, J., Li, C., Zheng, D., Luo, J., 2017. The influence of micro-oxygen addition on desulfurization performance and microbial communities during waste-activated sludge digestion in a rusty scrap iron-loaded anaerobic digester. Energies 10 (2), 258.

Shen, Y., Yu, Y., Zhang, Y., Urgun-Demirtas, M., Yuan, H., Zhu, N., Dai, X., 2021. Role of redox-active biochar with distinctive electrochemical properties to promote methane production in anaerobic digestion of waste activated sludge. J. Cleaner Prod. 278, 123212.

Shi, J., Xu, F., Wang, Z., Stiverson, J.A., Yu, Z., Li, Y., 2014. Effects of microbial and non-microbial factors of liquid anaerobic digestion effluent as inoculum on solid-state anaerobic digestion of corn stover. Bioresour. Technol. 157, 188–196.

Shi, L., Richardson, D.J., Wang, Z., Kerisit, S.N., Rosso, K.M., Zachara, J.M., Fredrickson, J.K., 2009. The roles of outer membrane cytochromes of Shewanella and Geobacter in extracellular electron transfer. Environ. Microbiol. Rep. 1 (4), 220–227.

Stams, A.J.M., Plugge, C.M., 2009. Electron transfer in syntrophic communities of anaerobic bacteria and archaea. Nat. Rev. Microbiol. 7 (8), 568–577.

Summer, D., Schöftner, P., Watzinger, A., Reichenauer, T.G., 2020. Inhibition and stimulation of two perchloroethene degrading bacterial cultures by nano- and micro-scaled zero-valent iron particles. Sci. Total Environ. 722, 137802.

Sun, X., Atiyeh, H.K., Li, M., Chen, Y., 2020. Biochar facilitated bioprocessing and biorefinery for productions of biofuel and chemicals: a review. Bioresour. Technol. 295, 122252.

Takashima, M., 2018. Effects of thermal pretreatment and trace metal supplementation on high-rate thermophilic anaerobic digestion of municipal sludge. J. Environ. Eng. 144 (3), 8.

Thanh, P.M., Ketheesan, B., Yan, Z., Stuckey, D., 2017. Effect of Ethylenediamine-N,N'-disuccinic acid (EDDS) on the speciation and bioavailability of Fe(2+) in the presence of sulfide in anaerobic digestion. Bioresour. Technol. 229, 169–179.

Tian, H., Yan, M., Treu, L., Angelidaki, I., Fotidis, I.A., 2019. Hydrogenotrophic methanogens are the key for a successful bioaugmentation to alleviate ammonia inhibition in thermophilic anaerobic digesters. Bioresour. Technol. 293, 122070.

Tian, T., Qiao, S., Li, X., Zhang, M.J., Zhou, J.T., 2017. Nano-graphene induced positive effects on methanogenesis in anaerobic digestion. Bioresour. Technol. 224, 41–47.

Tian, T., Yu, H.-Q., 2020. Iron-assisted biological wastewater treatment: synergistic effect between iron and microbes. Biotechnol. Adv. 44, 107610.

Vintiloiu, A., Boxriker, M., Lemmer, A., Oechsner, H., Jungbluth, T., Mathies, E., Ramhold, D., 2013. Effect of ethylenediaminetetraacetic acid (EDTA) on the bioavailability of trace elements during anaerobic digestion. Chem. Eng. J. 223, 436–441.

Wambugu, C., Rene, E., van de Vossenberg, J., Dupont, C., van Hullebusch, E., 2019. Role of biochar in anaerobic digestion based biorefinery for food waste. Front. Energy Res. 7, 14.

Wang, D., Ma, W., Han, H., Li, K., Xu, H., Fang, F., Hou, B., Jia, S., 2016. Enhanced anaerobic degradation of Fischer–Tropsch wastewater by integrated UASB system with Fe-C micro-electrolysis assisted. Chemosphere 164, 14–24.

Wang, G., Li, Q., Gao, X., Wang, X.C., 2018. Synergetic promotion of syntrophic methane production from anaerobic digestion of complex organic wastes by biochar: performance and associated mechanisms. Bioresour. Technol. 250, 812–820.

Wang, M., Zhao, Z., Zhang, Y., 2021. Magnetite-contained biochar derived from fenton sludge modulated electron transfer of microorganisms in anaerobic digestion. J. Hazard. Mater. 403, 123972.

Wang, Q., Feng, K., Li, H., 2020a. Nano iron materials enhance food waste fermentation. Bioresour. Technol. 315, 123804.

Wang, W.-t., Dai, L.-c., Wu, B., Qi, B.-f., Huang, T.-f., Hu, G.-q., He, M.-x., 2020b. Biochar-mediated enhanced ethanol fermentation (BMEEF) in *Zymomonas mobilis* under furfural and acetic acid stress. Biotechnol. Biofuels 13 (1), 28.

Wang, Y., Wang, H., Wang, X., Xiao, Y., Zhou, Y., Su, X., Cai, J., Sun, F., 2020c. Resuscitation, isolation and immobilization of bacterial species for efficient textile wastewater treatment: a critical review and update. Sci. Total Environ. 730, 139034.

Wegener, G., Krukenberg, V., Riedel, D., Tegetmeyer, H.E., Boetius, A., 2015. Intercellular wiring enables electron transfer between methanotrophic archaea and bacteria. Nature 526 (7574), 587–590.

Wei, J., Hao, X., van Loosdrecht, M.C.M., Li, J., 2018a. Feasibility analysis of anaerobic digestion of excess sludge enhanced by iron: a review. Renew. Sustain. Energy Rev. 89, 16–26.

Wei, W., Cai, Z., Fu, J., Xie, G.-J., Li, A., Zhou, X., Ni, B.-J., Wang, D., Wang, Q., 2018b. Zero valent iron enhances methane production from primary sludge in anaerobic digestion. Chem. Eng. J. 351, 1159–1165.

Wei, Y., Zhao, Y., Lu, Q., Cao, Z., Wei, Z., 2018c. Organophosphorus-degrading bacterial community during composting from different sources and their roles in phosphorus transformation. Bioresour. Technol. 264, 277–284.

Wenzel, J., Fiset, E., Batlle-Vilanova, P., Cabezas, A., Etchebehere, C., Balaguer, M., Colprim, J., Puig, S., 2018. Microbial community pathways for the production of volatile fatty acids from CO_2 and electricity. Front. Energy Res. 6, 15.

Wijesinghe, D.T.N., Dassanayake, K.B., Scales, P.J., Sommer, S.G., Chen, D.L., 2018. Effect of Australian zeolite on methane production and ammonium removal during anaerobic digestion of swine manure. J. Environ. Chem. Eng. 6 (1), 1233–1241.

Wu, D., Zheng, S., Ding, A., Sun, G., Yang, M., 2015. Performance of a zero valent iron-based anaerobic system in swine wastewater treatment. J. Hazard. Mater. 286, 1–6.

Wu, S.-L., Wei, W., Sun, J., Xu, Q., Dai, X., Ni, B.-J., 2020a. Medium-chain fatty acids and long-chain alcohols production from waste activated sludge via two-stage anaerobic fermentation. Water Res. 186, 116381.

Wu, Y., Wang, S., Liang, D., Li, N., 2020b. Conductive materials in anaerobic digestion: from mechanism to application. Bioresour. Technol. 298, 122403.

Xi, Y., Chang, Z., Ye, X., Xu, R., Du, J., Chen, G., 2014. Methane production from wheat straw with anaerobic sludge by heme supplementation. Bioresour. Technol. 172, 91–96.

Xiao, Y., De Araujo, C., Sze, C.C., Stuckey, D.C., 2015. Controlling a toxic shock of pentachlorophenol (PCP) to anaerobic digestion using activated carbon addition. Bioresour. Technol. 181, 303–311.

Xu, F., Li, Y., Ge, X., Yang, L., Li, Y., 2018. Anaerobic digestion of food waste—challenges and opportunities. Bioresour. Technol. 247, 1047–1058.

Xu, H., Chang, J., Wang, H., Liu, Y., Zhang, X., Liang, P., Huang, X., 2019a. Enhancing direct interspecies electron transfer in syntrophic-methanogenic associations with (semi)conductive iron oxides: effects and mechanisms. Sci. Total Environ. 695, 133876.

Xu, J., Guzman, J., Andersen, S., Rabaey, K., Angenent, L., 2015. In-line and selective phase separation of medium-chain carboxylic acids using membrane electrolysis. Chem. Commun. 51, 6847–6850.

Xu, R., Xu, S., Zhang, L., Florentino, A.P., Yang, Z., Liu, Y., 2019b. Impact of zero valent iron on blackwater anaerobic digestion. Bioresour. Technol. 285, 121351.

Yang, B., Xu, H., Liu, Y., Li, F., Song, X., Wang, Z., Sand, W., 2020. Role of GAC-MnO$_2$ catalyst for triggering the extracellular electron transfer and boosting CH$_4$ production in syntrophic methanogenesis. Chem. Eng. J. 383, 123211.

Yang, Y., Wang, J., Zhou, Y., 2019a. Enhanced anaerobic digestion of swine manure by the addition of zero-valent iron. Energy Fuels 33 (12), 12441–12449.

Yang, Y., Yang, F., Huang, W., Huang, W., Li, F., Lei, Z., Zhang, Z., 2018. Enhanced anaerobic digestion of ammonia-rich swine manure by zero-valent iron: with special focus on the enhancement effect on hydrogenotrophic methanogenesis activity. Bioresour. Technol. 270, 172–179.

Yang, Y., Zhang, Y., Li, Z., Zhao, Z., Quan, X., Zhao, Z., 2017a. Adding granular activated carbon into anaerobic sludge digestion to promote methane production and sludge decomposition. J. Cleaner Prod. 149, 1101–1108.

Yang, Z., Wang, W., Liu, C., Zhang, R., Liu, G., 2019b. Mitigation of ammonia inhibition through bioaugmentation with different microorganisms during anaerobic digestion: selection of strains and reactor performance evaluation. Water Res. 155, 214–224.

Yang, Z., Wang, X.-l., Li, H., Yang, J., Zhou, L.-Y., Liu, Y.-d, 2017b. Re-activation of aged-ZVI by iron-reducing bacterium *Shewanella putrefaciens* for enhanced reductive dechlorination of trichloroethylene. J. Chem. Technol. Biotechnol. 92 (10), 2642–2649.

Ye, W., Lu, J., Ye, J., Zhou, Y., 2021. The effects and mechanisms of zero-valent iron on anaerobic digestion of solid waste: a mini-review. J. Cleaner Prod. 278, 123567.

Yenigün, O., Demirel, B., 2013. Ammonia inhibition in anaerobic digestion: a review. Process Biochem. 48 (5), 901–911.

Yu, B., Huang, X.T., Zhang, D.L., Lou, Z.Y., Yuan, H.P., Zhu, N.W., 2016. Response of sludge fermentation liquid and microbial community to nano zero-valent iron exposure in a mesophilic anaerobic digestion system. RSC Adv. 6 (29), 24236–24244.

Yuan, Y., Hu, X., Chen, H., Zhou, Y., Zhou, Y., Wang, D., 2019. Advances in enhanced volatile fatty acid production from anaerobic fermentation of waste activated sludge. Sci. Total Environ. 694, 133741.

Zang, Y., Yang, Y., Hu, Y., Ngo, H.H., Wang, X.C., Li, Y.-Y., 2020. Zero-valent iron enhanced anaerobic digestion of pre-concentrated domestic wastewater for bioenergy recovery: characteristics and mechanisms. Bioresour. Technol. 310, 123441.

Zhang, C., Liu, S., Xie, T., Hai, R., Wang, X., Li, Y., 2017a. Effect of multiwalled carbon nanotube exposure in short term on anaerobic granular sludge. Chin. J. Environ. Eng. 11 (10), 5375–5380.

Zhang, F.-F., Wang, W., Yuan, S.-J., Hu, Z.-H., 2014a. Biodegradation and speciation of roxarsone in an anaerobic granular sludge system and its impacts. J. Hazard. Mater. 279, 562–568.

Zhang, J., Guo, R.-B., Qiu, Y.-L., Qiao, J.-T., Yuan, X.-Z., Shi, X.-S., Wang, C.-S., 2015. Bioaugmentation with an acetate-type fermentation bacterium *Acetobacteroides hydrogenigenes* improves methane production from corn straw. Bioresour. Technol. 179, 306–313.

Zhang, J., Loh, K.-C., Lee, J., Wang, C.-H., Dai, Y., Wah Tong, Y., 2017b. Three-stage anaerobic co-digestion of food waste and horse manure. Sci. Rep. 7 (1), 1269.

Zhang, J., Qu, Y., Qi, Q., Zhang, P., Zhang, Y., Tong, Y.W., He, Y., 2020a. The bio-chemical cycle of iron and the function induced by ZVI addition in anaerobic digestion: a review. Water Res. 186, 116405.

Zhang, J., Tian, H., Wang, X., Tong, Y.W., 2020b. Effects of activated carbon on mesophilic and thermophilic anaerobic digestion of food waste: process performance and life cycle assessment. Chem. Eng. J. 399, 125757.

Zhang, L., Loh, K.-C., Zhang, J., 2019a. Enhanced biogas production from anaerobic digestion of solid organic wastes: current status and prospects. Bioresour. Technol. Rep. 5, 280–296.

Zhang, L., Zhang, J., Loh, K.C., 2018. Activated carbon enhanced anaerobic digestion of food waste—laboratory-scale and pilot-scale operation. Waste Manage. (Oxford) 75, 270–279.

Zhang, M., Ma, Y., Ji, D., Li, X., Zhang, J., Zang, L., 2019b. Synergetic promotion of direct interspecies electron transfer for syntrophic metabolism of propionate and butyrate with graphite felt in anaerobic digestion. Bioresour. Technol. 287, 121373.

Zhang, M., Wang, Y., 2020. Effects of Fe-Mn-modified biochar addition on anaerobic digestion of sewage sludge: biomethane production, heavy metal speciation and performance stability. Bioresour. Technol. 313, 123695.

Zhang, N., Stanislaus, M.S., Hu, X.H., Zhao, C.Y., Zhu, Q., Li, D.W., Yang, Y.N., 2016. Strategy of mitigating ammonium-rich waste inhibition on anaerobic digestion by using illuminated bio-zeolite fixed-bed process. Bioresour. Technol. 222, 59–65.

Zhang, S., Zhao, Y., Yang, K., Liu, W., Xu, Y., Liang, P., Zhang, X., Huang, X., 2020c. Versatile zero valent iron applied in anaerobic membrane reactor for treating municipal wastewater: performances and mechanisms. Chem. Eng. J. 382, 123000.

Zhang, Y., Douglas, G.B., Kaksonen, A.H., Cui, L., Ye, Z., 2019c. Microbial reduction of nitrate in the presence of zero-valent iron. Sci. Total Environ. 646, 1195–1203.

Zhang, Y., Feng, Y., Yu, Q., Xu, Z., Quan, X., 2014b. Enhanced high-solids anaerobic digestion of waste activated sludge by the addition of scrap iron. Bioresour. Technol. 159, 297–304.

Zhao, C., Sharma, A., Ma, Q., Zhu, Y., Li, D., Liu, Z., Yang, Y., 2021. A developed hybrid fixed-bed bioreactor with Fe-modified zeolite to enhance and sustain biohydrogen production. Sci. Total Environ. 758, 143658.

Zhao, Z., Zhang, Y., Li, Y., Quan, X., Zhao, Z., 2018. Comparing the mechanisms of ZVI and Fe_3O_4 for promoting waste- activated sludge digestion. Water Res. 144, 126–133.

Zhen, G., Lu, X., Kato, H., Zhao, Y., Li, Y.-Y., 2017. Overview of pretreatment strategies for enhancing sewage sludge disintegration and subsequent anaerobic digestion: current advances, full-scale application and future perspectives. Renew. Sustain. Energy Rev. 69, 559–577.

Zhou, J., You, X., Niu, B., Yang, X., Gong, L., Zhou, Y., Wang, J., Zhang, H., 2020. Enhancement of methanogenic activity in anaerobic digestion of high solids sludge by nano zero-valent iron. Sci. Total Environ. 703, 135532.

Zhuang, H., Lee, P.H., Wu, Z., Jing, H., Guan, J., Tang, X., Tan, G.A., Leu, S.Y., 2021. Genomic driven factors enhance biocatalyst-related cellulolysis potential in anaerobic digestion. Bioresour. Technol. 333, 125148.

Zielińska, A., Oleszczuk, P., 2015. The conversion of sewage sludge into biochar reduces polycyclic aromatic hydrocarbon content and ecotoxicity but increases trace metal content. Biomass Bioenergy 75, 235–244.

Zou, H., Wang, Q.H., Liu, J.G., Wang, S., 2012. The influence of Zn^{2+} and Fe^{3+} on lactic acid production by *Lactobacillus amylophilus* GV6 under open fermentation. Zhongguo Huanjing Kexue/China Environ. Sci. 32, 499–503.

Bioreactors for enhanced anaerobic digestion for bioenergy and biochemicals

Tengyu Zhang[a,b], Endashaw Workie[a], Jingxin Zhang[a]

[a]China-UK Low Carbon College, Shanghai Jiao Tong University, Shanghai, China,
[b]Department of Chemical and Biological Engineering, University of Sheffield, Sheffield, UK

8.1   Introduction

Energy shortages and environmental pollution are common challenges facing all mankind, as set out to be minimized by the United Nations Sustainable Development Goals (Losch, 2020). With the rapid population growth and economic development, it is estimated that 2.2 billion tons of municipal solid waste (MSW) will be generated worldwide by 2025 (Rogoff & Screve, 2019). As high as 70% of the organic solid waste (OSW) produces emissions of greenhouse gas (GHG) that contributes to global climate change (Luis Fernando Marmolejo Rebellón, 2017). The authorities are being urged to adopt techneconomic and political solutions to manage the increasing amount of MSW due to this environmental threat. To solve the problem, converting organic waste to energy (WTE) is regarded as one of the reasonable management methods to reduce consumption of virgin fossil-based resources by making up for energy shortage. More than 70% of Asian countries have established waste management systems to reduce the load of air and water environment caused by emission and relieve the pressure of climate change caused by the burning of fossil fuels. For example, in China, biogas is supplied to 96,000 rural households. In addition to resource generation, this project saves 58,780 tons of CO_2-equivalent emissions annually (Luis Fernando Marmolejo Rebellón, 2017). In the European Union (EU), 27% of OSW is used for energy, while Sweden, Norway, and Denmark have a waste energization rate of more than 50%. Up to now, there are 2179 WTE factories in the EU (Rogoff & Screve, 2019). In Africa, WTE plants can supply 30% of household energy.

Bioenergy and biofuel are the main products of energy conversion. Organic waste can be a valuable source of renewable energy (Tyagi et al., 2018), including organic components of MSW, food waste, animal excrement, and waste sludge of sewage treatment plants, etc. As of now, anaerobic digestion (AD) technology for biogas production constitutes the most sustainable way to use the energy present in organic solid wastes. AD is a biological resource utilization technology, in which organic matter is decomposed by microorganisms under anaerobic conditions into clean energy—biogas (methane, hydrogen, and carbon dioxide). Biogas production through AD has become a commercial method for organic waste reuse and energy production. As an alternative energy, biogas has been with legislation supports in Germany, the United States, the United Kingdom, and other countries. The EU expects biogas production to account for 3% of natural gas by 2030, and Poland has a large potential capacity for biogas generation, which could cover around 47% of the domestic demand for natural gas. Europe has more than 14,500 biogas plants installed (Lora Grando et al., 2017), and the United States with 2200 plants. AD is also widely applied in China (40 million plants), India (5 million plants), and Nepal (300,000 plants) (Tyagi et al., 2018). And the numbers are still on the rise. In general, countries with developed AD factories prefer to build small factories. Compared with small factories, large factories have advantages in economic scale but more complex operation conditions. In Denmark, biogas is used in combined heat and power plants, while in Sweden it is normally converted into vehicle fuel. Since the EU has set the target of having 50% of all the country manure being processed to generate renewable energy, 90% of the plant capacity built is in the range of 15,000–80,000 t/y and the average values of capacity are 38,000 t/y. It is worth noting that some plants with a capacity exceeding 200,000 t/y are present in addition to injecting natural gas distribution network, also for fuel cell or vehicle fuel (Lora Grando et al., 2017).

In addition to the productivity of AD plants, the great potential of AD treatment of organic waste is also reflected in the production of biological byproducts with different substrate characteristics. Some organic wastes such as high nitrogenous waste are more suitable for producing biochemicals (Vanwonterghem et al., 2015). AD application to produce organic acids, alcohols, and other chemicals can be seen as another way. The world is now within the scope of using special organic wastes as the substrate of the AD produces organic acids, ethanol, and biodiesel. This production process is performed in a sealed vessel known as a reactor, which is designed and developed in different forms. To date, several reactors have been developed and applied for AD process particularly to produce biogas, ethanol, biodiesel, and alcohols. This article mainly aims to enhance the target product by introducing a new type of AD reactor. Generally, the types of AD reactors are mainly divided into wet and dry digestion according to the water content of the substrate. Continuous mode according to the feeding style: single-stage, two-stage, and batch mode. The reactor is also categorized according to the temperature of the digestion process (mesophilic or thermophilic) and the shape of the reactor (vertical or horizontal) (Nayono, 2010).

The technical innovation of an AD reactor is the key to improve the energy utilization of waste. AD is divided into four stages: hydrolyzation, acid production, acetic acid production, and methane production. The main functional microbiota of each stage has different growth preferences and metabolic mechanisms. Novel AD bioreactor provides a better living environment for microorganisms by changing the structure of the reactor on the basis of the existing theories. It improves the output efficiency of the target product by solving the acidification inhibition problem and stimulating the electron transfer between microorganisms. Novel reactor provides a more economical and effective way for WTE conversion (Lora Grando et al., 2017).

In the following sections, we will introduce the new AD reactors for the enhanced production of different targets on the basis of the common AD reactors. Novel AD reactors with different target products can be classified into bioenergy (methane and hydrogen) and biochemicals (organic acids, alcohols, and biodiesel). Methane and hydrogen as the general products of AD reactor enhance the biomass (methane and hydrogen production of a new type of AD reactor than multistage AD and zero-valent iron (ZVI) AD, bioelectrochemical AD, microbubble AD, microbial electrolysis cell AD, and anaerobic down-flow structured-bed reactor, the reactor structure, function, and principle and application scope are introduced in Section 8.2. As biochemical products, organic acids also show their conversion value, such as butyric acid and polyhydroxyalkanoate (PHA) production. At the same time, alcoholization reaction and esterification reaction also occur in the anaerobic reactor to produce ethanol and biodiesel, respectively. Therefore, the production of intermediate products of AD has been emphasized and efforts are being made to enhancing production of it. The principle and application of new reactors for enhancing organic acids, alcohols, and biodiesel are described in Section 8.3.

The main purposes of this chapter are (1) to introduce the structural characteristics of new AD reactors in the last 5 years to enhance the production of bioenergy and biochemicals, and (2) to analyze the development direction and application prospect of AD reactor in the future.

8.2   Bioreactors for enhanced AD for bioenergy

8.2.1   A brief history of the anaerobic digesters

Traditional AD reactor structures can be divided into rectangular, cylindrical, and egg shapes. Further, the reactor can be mainly divided into the single-stage continuous process from processing unit, the multistage continuous processing and batch processing. Each processing pattern from the feed water content is divided into a high solid AD reactor and wet AD reactor, which is defined as if the total solid substrate concentration less than 15% of the wet process, if the concentration of up to 20%–40% for the dry process. In the wet digestion process, the liquid must be recycled or digested with more liquid waste, which is more suitable for the treatment of organic solid waste, such as food waste, animal waste, municipal sludge and other waste streams combined waste streams (Nayono, 2010).

In 2008, more than 90% of factories in Europe used single-stage AD reactors. The main disadvantage of a single-stage AD system is that it cannot meet the different growth conditions of microorganisms in different reaction stages of AD, which can easily cause the accumulation and inhibition of organic acids in the process and prevent the reaction progress. Among them, the multistage AD reactor separates the reaction of different stages and solves the problems of acidification inhibition. Stimulated by the increasing demand for OSW from AD plants, several designs for commercial AD plants have been developed over the past two decades. Methods are added to enhance mixing such as gas injection or mechanical stirrers. Currently, the reactor types commonly used in AD plants around the world including Biocel (batch system), Dranco, Valorga, Kompogas (one-stage dry system), Waasa, BTA (one-stage wet system), Schwarting Uhde (two-stage wet system), and Linde BRV (two-stage dry system) are briefly presented in this subchapter (Nayono, 2010).

8.2.2   Novel anaerobic digesters for enhanced methane yield

8.2.2.1   *Multistage AD reactors*

The process of AD of methane production is mainly restricted by the complex nature of substrates, unfavorable environmental conditions and the lack of effective microbial community utilization in the reaction process. In this case, the one-stage AD has been criticized for its low performance due to the accumulation of volatile fatty acids (VFAs), unbalanced nutrient profiles, and byproduct inhibition (Zhang et al., 2019) and most importantly brings the absence of controlling the optimum thriving condition in each of four digestion stages.

Multistage AD separates the integrated sludge treatment process from a new perspective, aiming to improve the above restrictions on gas production and optimize the reaction rate and gas production. Multistage AD is an acid–gas system where the hydrolysis and volatile acid fermentation process and methane production are separately performed. Currently, multistage AD consists of two-stage and three-stage digestion systems (Fig. 8.1). Separation of AD reaction was realized in a two-stage digestion system, conducting hydrolysis and methanogenesis in two reactors. The three-stage digestion system is, however, more complicated as all steps including hydrolysis, acidogenesis/acetogenesis, and methanogenesis are carried out in different chambers.

This separating AD operation offers many advantages over the commonly used monodigestion system. It can improve the efficiency of different steps in the process by separating them in space or time and optimizing the operating conditions for each trophic group of microorganisms in which they occur thereby improving the overall reaction rate and methane yield. Multistage systems can also prevent the growth balance disruption between the physiological properties and nutrient needs of acetogens and methanogens which are otherwise severely inhibited by environmental changes (Van et al., 2020). The unstable AD process corresponds to lower methane output (Hegde & Trabold, 2019; Rabii et al., 2019). Hence, multistage digestion can stabilize the process by reducing the fluctuations in waste heterogeneity, organic loading, and substrate degradation rate. Nevertheless, due to high operation

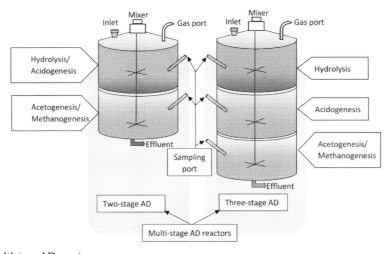

FIG. 8.1   **Multistage AD reactors.**

and maintenance requirements as compared to the monodigestion system, it has not yet been widely applied for pilot scales (Van et al., 2020)

In previous studies (Zhang et al., 2019), the hydrolysis part was separated from the part producing acid and methane, forming new leachate and flowing anaerobic sludge reactor system. In the experiment of codigestion of food waste and garden waste with a high-solid AD two-phase system, the reaction time was shortened by 18 days, and the biogas output was increased by 83.25% (Zhang et al., 2017a). In a three-stage AD reactor, it was conducive to the selective enrichment of methane-producing bacteria that can process a wider range of biomass constituents (Rabii et al., 2019) and significantly improved the bioenergy conversion rate (Zhang et al., 2017b). They used a three-stage AD reactor for codigestion of food waste and horse manure, which improved the dissolution efficiency of solid organic matter and the generation of VFAs, increased the rate of methane production by 11%~23%. Moreover, the study showed that compared with conventional first-stage reactors, the methane production rate could increase by 24%~54% under high organic load conditions (Zhang et al., 2017b).

Anaerobic baffle reactor (ABR) is one of the reactor types to be applied with a multistage digestion mechanism. It is a more traditional vertical segmented multistage reactor composed of alternating baffles that promote the upward or downward flow of substrates and enable long-term separation of various stages of AD and microbial communities. In the latest experimental progress of adding a three-phase separator (Sayedin et al., 2018), the high recovery rate of the new ABR reduced the VFAs concentration and significantly improved the reactor efficiency.

In addition to the separation of reactors, the combination of two or more reactors can also achieve multiphase reaction conditions. In this case, the leach bed reactor (LBR) is simple, flexible, and suitable for dry or high solid-phase AD applications. Compared with the wet reactor, the LBR dry reactor can reduce the reactor volume and the heating mixing energy consumption (Sayedin et al., 2018). It can be coupled with an external bioreactor such as an up-flow anaerobic sludge bed (UASB) or anaerobic filter to obtain a higher microbial retention rate and thus improve the performance of AD in methane production. High microbial density increases methane production by forming biofilms on the medium surface, trapping the suspended microorganisms in the filter bed, and helping the reactor to maintain stable performance under high organic load conditions (Chakraborty & Venkata Mohan, 2019).

8.2.2.2 ZVI AD reactors

AD efficiency in methane production is subject to various biochemical reactions. If these reactions go uncontrolled, the digestion process can slow down and cause reactor failure due to the entrance of inhibitors mainly including VFAs, ammonia, sulfur, and heavy metals (Paritosh et al., 2020). Additive assisted reactors are therefore applied to get rid of the inhibitor effect and enhance the methane yield. Recently, ZVI has been proved to promote AD by increasing enzyme activity and interspecific electrons. It becomes a popular additive to enhance methane production due to its viability and ability to lower redox reactions. ZVI can act as an electron donor, causing the microbes to disintegrate by changing the pH in the reactor (Geng et al., 2020) meanwhile producing H_2 that can be used by hydrogenotrophic methanogens and sulfate-reducing bacteria. This process also buffers the acid released by acidogens to maintain a stable and favorable environment for methanogenesis (Puyol et al., 2018). Thus, reasonable iron ion concentration accelerates the electron transfer and sludge granulation rate (Zhang et al., 2020), thereby the methane-producing rate significantly increased.

More research shows that adding the ZVI to accelerate the hydrolysis and fermentation stage can not only speed up the electron transfer between microorganisms but also increase the accumulation of iron ions and delay the toxicity of H₂S excessive acidification degradation. Besides, the inhibitory effect of sulfate on AD was also reduced (Zhang et al., 2020). Wang et al. (2019) proposed that adding ZVI stimulated the activity and diversity of the microbial community, and the number of microorganisms was higher than that of an AD reactor without adding it. In general, heat treatment and alkali pretreatment have inhibitory effects on AD. The study of Zhang et al. (2015) found that AD of ZVI can be used to improve the methanogenic activity of anaerobic sludge digesters under the above two methane-inhibiting conditions.

Thus, the concept of using ZVI to transform the reactor to improve the bioenergy yield came into being and was realized in several experiments. Liu et al. (2015) used an acrylic plastic UASB AD reactor with a working volume of about 10 L as the basis and installed a cylindrical ZVI bed two-thirds of the depth of the reactor (Fig. 8.2). The ZVI bed was filled with cylindrical stainless-steel metal mesh and 300 g of waste iron. The contact between the packed tower and the suspended sludge layer formed the ZVI–UASB reactor. The reactor has achieved high performance in degrading propionic acid and improving the activity of methanogens. Moreover, the ZVI in the reactor is suitable for replacement within 0.5–1 year (Liu et al., 2015). Likewise, as early as 2011 (Zhang et al., 2011), the ZVI–AD reactor pilot model was put forward in the ZVI bacteria abundance of the packed bed is much higher than the other part. In previous pilot tests (Li et al., 2017), a ZVI bed (Φ 1200 mm × 300 mm) was inserted at two-thirds the depth in A UASB (Φ 1200 mm × 4500 mm) to form a ZVI-AD reactor. The ZVI bed is a construct of cylindrical stainless-steel mesh wrapped in 200 kg scrap iron. In the experiment of this reactor, the advantages of chemical oxygen demand (COD) removal rate are also discovered while improving methane production rate.

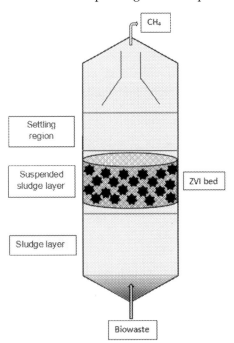

FIG. 8.2   Zero-valent iron-anaerobic digestion reactor.

8.2.2.3 Bioelectrochemical AD reactors

A combination of bioelectrochemical with AD reactor (BEAD) has been proposed as one of the novel reactors for methane production by exchanging electrons directly with the electrodes. The anode oxidizes organic matter to produce e^-, H^+, and CO_2 while the cathode uses these products as a substrate to produce methane through direct bioelectrochemical, hydrogen-mediated, and carbon compounds–mediated methanation (Park et al., 2020). In the experiment of Song et al. (2016), it was proved that methane recovery was as high as 69.1%–98.7% when BEAD was used to treat sewage sludge. Under the high load production of methane, the electrochemical removal rate of organic matter increases the number of electroactive microbial AD and COD removal rate (Park et al., 2019b) and improves the acidification efficiency (Zhang et al., 2015). The electrochemical strengthening of exogenous active bacteria and bacterial VFA oxidation prevents the accumulation of VFAs (Park et al., 2019a) and the negative impact of the methanogenesis process. In addition, in the case of high organic load, rapid electrochemical reduction of hydrogen ions and adjustment of pH value (Park et al., 2019b) are important conditions for methane generation.

In general, anodic oxidation of methane production is more efficient, cathode current production is the bigger, the methane production rate 304.5 mL CH_4/L reactor/day, and a maximum current of 10 A. The maximum COD, total oxygen carbon (TOC), and carbohydrate removal efficiencies were 92.1%, 64.2%, and 98.9% (Sangeetha et al., 2017), respectively. Relatively speaking, studies have shown that carbon dioxide can be directly reduced by using cathode to produce methane (Rodríguez-Alegre et al., 2019). Cai et al. (2016) introduced a new combination of two AD reactors isolated by anion exchange. The inner tube is cathode AD, and the outer tube is anode AD (Fig. 8.3). The AD effect of 0.070 mL

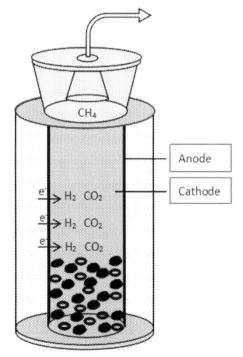

FIG. 8.3 Bioelectrochemical anaerobic digestion reactor.

CH_4/mL reactor/day was 2.4 times that of the anode. The degradation effect of cathode AD on glucose and fermentation broth was also improved. In general, electrochemical AD is of great significance for effluent control, biogas upgrading, and nutrient recovery (De Vrieze et al., 2018). It is worth noting that in Park's experiment, the electrochemical effect of AD on granular activated carbon electrode at 10 °C was evaluated, and the improvement direction of AD reactor in cold areas (Park et al., 2019c) was pointed out. The reactor simulation was implemented in a plexiglass box in the laboratory. These frames are cylindrical and are fitted with 8.0 mL collecting ducts. The solution (20 mL) in each reactor was continuously stirred with magnets. For the BEAD reactor, an anode and a cathode chamber were fabricated on both sides of the reactor in one reactor. The electrodes, made of granular activated carbon, are housed in a cylindrical cage and separated from the chamber using a spacer. A graphite rod is inserted as a collector in both the anode and cathode chambers. Methane production under this process can be increased by 5.3–6.6 times. The study also discussed the difference between auxiliary bioelectrochemical reactor (ABER) and electrochemical AD by a rotating impeller with cathode anode, circular wall of the reactor, and separate anode. AD reactor set the reaction device. The results showed that ABER indirect voltage could increase the hydrogen-producing bacteria and produce hydrogen, methane bacteria in the reactor are similar to the effects of supply voltage directly. This provides an idea for the commercial and flexible application of electrochemical AD reactors (Park et al., 2019d). However, because the surface area of the electrode and the working volume of the reactor are relatively lower than those of other types of reactors, the proportion of electrochemical methane production in the total methane production rate is also lower. Furthermore, the optimization of electrode materials placed in AD is the key point of microbial electrolysis cells to promote the future development of AD.

8.2.2.4 *Microbubble AD reactors*

Microbubbles have been demonstrated to enhance biogas production in AD. In the case of AD, microbubbles are created by supplying pure CO_2 into the reactor and remain to facilitate an efficient conversion. As methane has low water solubility and is thus easy to adhere to the microbial membranes of the organic phase, forming a gas layer in the organic phase until a large bubble helps to extract methane from the reactor (Al-mashhadani et al., 2016). Briefly, the injection of microbubbles can prevent thermal stratification while maintaining pH balance, increase the contact between feed and microorganisms, and use the fluid shock to generate microbubbles to strip methane generated in the bioreactor, which is more conducive to the formation and collection of methane from the perspective of thermodynamics (Gilmour & Zimmerman, 2020).

In AD, carbon dioxide has been shown to promote the production of methane by acetyl chips. Previous studies have shown that daily AD methane production can be increased 2.4 times by injecting CO_2-enriched microbubbles (Al-mashhadani et al., 2016; Mulakhudair et al., 2017). A new type of carbon dioxide-AD reactor was designed by Mulakhudair to increase production of methane: Bajon semicontinuous AD reactor installation based on 1-m-high tower of about 10 cm in diameter of carbon dioxide. Carbon dioxide on the downstream side of the reactor through the porous plate continuous injection of digestive juices. Final, wire mesh is used to enhance the gas–liquid mass transfer, speed up carbon dioxide dissolved in the digestive juice (Fig. 8.4) (Mulakhudair et al., 2017).

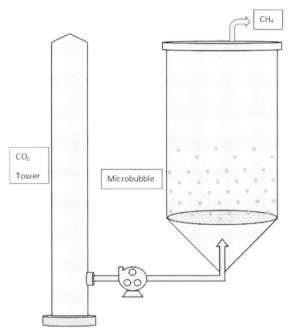

FIG. 8.4 Microbubble anaerobic digestion reactor.

8.2.3 Novel anaerobic digesters for enhanced hydrogen yield

There are different types of bioreactors used in biological hydrogen production, including anaerobic downward structure bed reactor, up-flow anaerobic sludge blanket reactor and continuously stirred tank reactor. Besides, a continuous reactor model, such as an external loop bioreactor consists of a single hydrogen fermentation tank using the traditional bioreactor design has shown promising results (Khan et al., 2016).

8.2.3.1 Microbial electrolysis cell-anaerobic digesters

As a bioelectrochemical method, a microbial electrolytic cell (MEC) is a new technology that consumes a small amount of electric energy and promotes the decomposition of organic matter to generate fuels and chemicals. Compared with the input energy, the overall efficiency of MEC AD exceeds 400% (Hassanein et al., 2017). Microbial catalytic electrolysis was used to produce hydrogen, mainly through hydrogen hydrolysis to enhance the coupled AD process and produce methane (Gajaraj et al., 2017). However, in previous studies, an MEC was proved not only to improve methane, when the electrode voltage was 0.3 V, the methane production rate increased by 22.4%. However, when the voltage rose to 0.6 V, the methane production rate decreased and the hydrogen production increased, which may be due to the inhibition of methanogens by increase cathode hydrogen ions (Gajaraj et al., 2017).

Similarly, electrochemical technology combined with an AD reactor can be used in a continuous stirred tank reactor to study the effect of continuous removal of VFAs on hydrogen production. The VFAs in the reactor were continuously removed by electrodialysis (ED). The amount of food waste produced by ED increases, and the hydrogen rate increases by 3.5 times (Hassan et al., 2019).

The apparatus comprises a gas circulation loop returning from the top of the fermenter to the liquid phase. The circuit includes a sensor to measure the composition of the gas, a thermostatic heating pad to maintain the temperature, and automatic addition of 1 M NaOH to maintain the pH of the fermenter and stirring the liquid phase continuously with a magnetic agitator. In addition, a liquid circulation loop was added into the bioreactor design to realize the removal of VFA by ED. The membrane layer in the cell consists of a series of alternating anion exchange membranes (AEMs) and cation exchange membranes. The active area of each membrane is 64 cm², and the membrane reactor contains 20 pairs of anion and cationic membranes. The ED battery is powered by a direct current and maintains a constant voltage of 18V. The liquid is periodically withdrawn from the loop through a cross-flow microfiltration membrane with an aperture of 0.1 mm. The filtered liquid is dialyzed and returned to a recirculation loop connected to the reactor; VFAs removed from the reactor content flow are collected in a concentrated flow containing phosphoric acid (Hassan et al., 2019). Moreover, studies have shown that electrochemical AD has a good performance for sanitation with the high removal rate of carbon and nitrogen from waste streams. A microbial fuel cell (MFC)-MEC coupling with an ABR has been applied to the treatment of fecal sewage at the same time recycling energy (Fig. 8.5) (Hongbo Liu et al., 2020). However, the cost of this method is relatively high, and how to improve the economic benefit is still the focus of the future development of MEC AD reactor. Moreover, it is essential to maintain high mass/electron transfer efficiency in MEC-AD (Kadier et al., 2020).

8.2.3.2 *Anaerobic down-flow structured-bed reactors*

Upflow fixed-bed reactors are considered suitable for hydrogen production from wastewater because they both provide new biomass and avoid scouring by hydrogen-producing microorganisms compared to fully mixed reactors (Fig. 8.6). Anzola-Rojas et al., (2016) showed that hydrogen could be continuously and steadily produced in the new reactor due to the constant renewal of biomass (Anzola-Rojas et al., 2016). Anaerobic downflow structural bed reactor was used to improve the production capacity of hydrogen, which was mainly produced by ethanol fermentation. The maximum value of the total energy conversion rate is 23.40 KJ/hL. The devices mentioned above had a volume of 4.35 L and are constructed from an acrylic

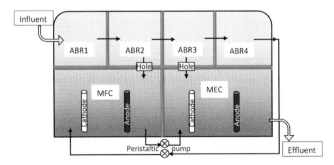

FIG. 8.5   Microbial electrolysis cell-AD.

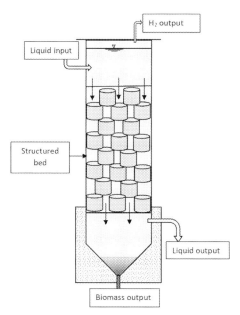

FIG. 8.6   **Anaerobic down-flow structured-bed reactor.**

pipe with a diameter of 80 mm and a length of 800 mm. The packing consists of small cylinders, 20 mm in diameter and 25 mm in length, arranged in alternating columns and cantilever with stainless steel rods. The material used for biomass adhesion is recycled low-density polyethylene, and hot water is recycled over the reactor body to keep the temperature at around 25 °C. Although compared with other anaerobic fixed-bed systems, a new type of anaerobic downlink structural bed reactor has shown its feasibility and stability in terms of long-term sustainable hydrogen production, so far, both the volume rate and hydrogen production are too low (Anzola-Rojas et al., 2016).

8.3   Bioreactors for enhanced AD for biochemicals

8.3.1   A brief history of the anaerobic digesters for biochemicals

Another method of anaerobic treatment of organic waste has gained attention, which is the production of biochemicals such as organic acids, alcohols, and biodiesel. It was found that, in addition to the recycling of bioenergy (biogas), direct recycling of OSW for fermentation or posttreatment can yield other organic acid molecules (polyhydroxylalkanes, or medium-chain long fatty acids) that may yield more valuable end-products.

VFAs and other short-chain organic acids, such as lactic acid and pyruvate, are intermediate products of anaerobic organic degradation. The process for volatile acid production was invented as early as the 1980s. The main VFAs are formic acid (C1), acetic acid (C2), propionic acid (C3), pyruvate (C3), lactic acid (C3), butyric acid (C4), and valerate (C5). In recent years, there has been an increase in the production of organic acids as the main research recovery route, especially the realization of the value of end products, such as PHA, carbon-chain VFAs, and other organic molecules, under the condition of biotransformation or membrane separation to achieve the collection process (Kleerebezem et al., 2015).

The key step in obtaining organic acids and ethanol as final products is to inhibit the generation of methane. The accumulation of organic acids leads to a decrease in pH, which is an inhibition of the generation of methane. The multistage digestion method was proposed in the early stage, and the main design idea is based on the separation and treatment unit and the culture of different functional bacteria to separate target production respectively.

In addition, substrate preacidification, control of VFAs composition, and effluent organic concentration are also the main means to optimize the production of AD chemicals. The two most commonly used techniques for VFAs production are attachment growth and suspension growth (Khan et al., 2016). Many bioreactors are used in VFAs production, such as packed bed biofilm column reactors, anaerobic leach-bed reactors, two-stage thermophilic anaerobic membrane bioreactors, and continuous flow fermentation reactors (Khan et al., 2016).

At the same time of acidification, alcoholization and esterification reactions also take place in AD. In the acidification stage, the hydrolyzed monomer will be further decomposed into alcohols by microorganisms, and the production of bioethanol requires the codigestion of organic waste and cellulosic biomass waste. In addition, the waste oil in a solid organic waste can replace rapeseed oil to produce biodiesel through ester conversion reaction, which can reduce the food hidden danger caused by biodiesel production. In this section, a novel AD reactor for the acquisition and utilization of organic acids, alcohols, biodiesel, and other substances is studied.

8.3.2   Novel anaerobic digesters for enhanced organic acid yield

8.3.2.1   Multistage AD reactors

A two-stage AD reactor separates the hydro-acid and methane-producing processes and produces and accumulates organic acids during the first stage of acidification. The results showed that the VFAs, mainly acetic acid, butyric acid, and propionic acid, increased by 49.3% when the cysteine dosage was 50 mg/L (Liu & Chen, 2018). A sequencing batch reactor (SBR) can also be used as an anaerobic reactor to produce VFAs and realize polyphase reaction. VFAs were enriched and produced in a microbial SBR with a working volume of 8 L and an exchange ratio of 20%. The treatment system inhibited the growth of methanogens with VFAs as substrates and increased the yield of VFA. Under the condition of no methane gas production, the enhanced reactor achieved stable VFA accumulation, and the average VFA yield increased by 43.8% compared with the original VFA content (Lim & Vadivelu, 2019). Multiple anaerobic sequence batch reactors (ASBR) are also suitable for organic acid production under acidic conditions. This method can increase the concentration of VFAs at low pH (4.5–5.5) by shortening the retention time of solid and liquid sludge, increasing the specific surface area of sludge, and reducing the mass transfer limit. About 100 mL of sludge was inoculated with laboratory-scale ASBR with a height of 150 cm and an inner column diameter of 6.5 cm. After the start-up phase, the system operation sequence intermittent reactor as an operating cycle of 2 h, consisting of a feed phase of 17 min, a reaction phase of 95 min, a sedimentation phase of 2 min, and the last 3 min of wastewater phase, of which half of the reactor liquid provided a retention time of 4 h. Solid retention time is 1–2.5 days, set up by manual removal of biomass and solid concentration of wastewater (Tamis et al., 2015).

8.3.2.2   *Microbial electrosynthesis AD reactors*

It was found that the external electrochemical stimulation affected the rate kinetics by improving the bioelectric catalytic capacity and regulating the electron flow direction of the product biosynthesis. Anaerobic electrofermentation reactor was used for the production of VFAs. The reactor was continuously fed and continuously stirred by magnetic force, operated with a graphite electrode at room temperature. Comparing with the experimental group, the yield of organic acid was increased to nearly two times by electric fermentation. The yield of acetic acid was the highest, followed by butyric acid and propionic acid (Shanthi Sravan et al., 2018). Microbial electrosynthesis (MES) can be seen as an emerging technology that breaks new ground in the field of waste biorefinery by integrating bioprocesses for platform chemicals production. Sensors were added to the electrolytic cell to monitor the acetic acid yield concentration representing VFAs to reflect the effect of different operating conditions such as temperature, conductive medium, and current on the production of organic acids. A novel biosensor based on a three-compartment microbiological electrochemical system was placed in a microbiological cell to monitor acetate in the AD process. In this system, acetate is first transferred from the sample chamber to the anode via AEM under the action of concentration difference and then oxidized by the anode biofilm as the substrate to generate current. The design avoids the influence of the fluctuation of waste properties in the cathode reaction and studies the response of current density to different acetate concentrations. The effects of selectivity, sample temperature, and applied resistance were also evaluated. The value between current density and acetic acid concentration (160 mM) is 0.99 at a specific reaction time (2 ~ 5 h). The current density increased by about 20% after 5 h at higher sample temperatures (e.g., 37 and 55 °C). The detection range increases with the decrease of external resistance. Made physically separated by an AEM, a three-chamber ABER is constructed by adding a sample, anode, and anode with the same dimensions (8 cm × 8 cm × 4 cm) to a traditional two-chamber MFC reactor (Fig. 8.7). These rectangular compartments are made of nonconductive polycarbonate plates with three-electrode ports, and rubber tubes are inserted into the medium to fill, inflate, and sample. Assemble the reactor with stainless steel screws and rubber gaskets to prevent leakage.

Carbon fiber brushes were treated at 450 °C for 30 min as an anode electrode. The cathode electrode was made of graphite plate (5.2 cm × 5.2 cm) and coated with 0.5 mg/cm² platinum. Both electrodes are connected to an external resistor (Sun et al., 2019).

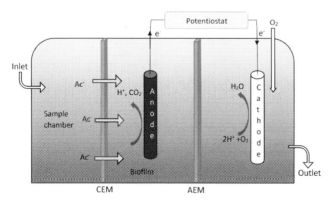

FIG. 8.7   **Microbial electrosynthesis anaerobic digestion reactor.**

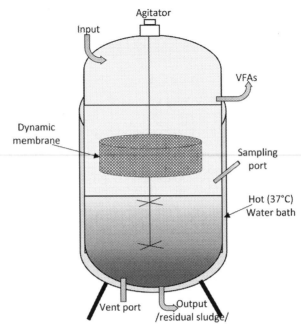

FIG. 8.8   Anaerobic dynamic membrane reactor.

8.3.2.3   Anaerobic dynamic membrane reactors

The retention of slow-growing anaerobic microbes is still challenging organic acid production. The combination of AD and membrane separation could solve the problems regarding the potential anaerobic microorganisms. Briefly, liquid fermentation in an anaerobic dynamic membrane reactor (AnDMBR) (Fig. 8.8) was considered as a new strategy to improve VFAs collection in sludge. The working principle is by replacing the external separation system through immersed membrane directly in the bioreactor, the energy costs and footprint will be greatly reduced. The submerged membranes use coarse bubble airing to limit fouling and transferring oxygen to the feedstock and keeping the solids suspended (Fan et al., 2017). Common organic acids such as itaconic, propionic, and lactic acids can be produced by a fermentation process using AnDMBR.

The experimental results showed that the concentration and substrate conversion rate reached 7.8 kg VFA–COD/m^3 d, 60 g/L, and 0.38 kg VFA–COD/kg VS, respectively. The dynamic membrane runs stably for about 70 days. During operation, membrane flux increased from 6.25 L/m^{2}d to 25 L/m^{2}d. A dynamic biofilm anaerobic digester with a biofilm rotary wheel at the top of the tank driven by screws, sludge feeding on the side, and sludge discharge was set at the bottom. The external circulating water temperature of the tank was kept at 37 °C and the VFA was eventually collected on the biofilm using a backpipe. Microbial analysis results showed that with the increase of membrane separation and organic loading rate, the abundance and uniformity of bacteria in the membrane bioreactor increased. Also, residual solids were not conducive to improving the yield of VFAs. Energy consumption will be reduced by avoiding "useless" residue entering the fermenter (Liu et al., 2019).

8.3.3 Novel anaerobic digesters for enhanced alcohol yield

Conventional reactors and fermentation systems have limitations due to substrate sterilization before fermentation, and commercial enzyme requirements made it relatively costly. Roukas & Kotzekido (2020) used a rotary biofilm reactor (RBR) to produce bioethanol from the nonsterilized beet molasses in repeated-batch fermentation. The reactor comprises a glass cylinder with two stainless-steel plates and a series of circular discs in which microorganisms make a biofilm on the spinning discs, which are half-soaked in the medium during fermentation. Through this reactor, 52.3 g/L of bioethanol has been produced for 60 days in repeated-batch culture and a positive energy balance of 1.63 MJ/MJ was obtained. The authors stated that this reactor could be easily extended and maintained and relatively operated with low energy (Roukas & Kotzekidou, 2020).

The fermentation of a mixed anaerobic medium can also promote the production of bioethanol through the accumulation of VFAs and the reduction of pH value. Under these conditions, acidifying microorganisms can be converted from the acid production stage to the alcohol production stage due to their ability of biphasic fermentation. MES can overcome the thermodynamic limitations of traditional fermentation and improve the production of bioethanol by adding a small amount of external energy to the anode and cathode in the fermentation medium. MES bioalcohols rely on microbial degradation of VFAs attached to solid cathodes by electroactive biofilms, which act as an electron donor. Thus, controlling microbial redox reactions and increasing available reduced nicotinamide adenine dinucleotide (NADH) have been reported to reduce metabolite production for more favorable conditions. In a two-compartment MES reactor, water is oxidized at the anode side without any living strain and can be decomposed into protons, oxygen, and electrons. The generated electrons and protons move to the cathode via an external circuit and a proton exchange membrane, respectively. On the cathode, electrons are absorbed by microorganisms, and the final electron acceptors (acetic and butyric acids) are bioreduced using available electrons and protons (VFAs + NE^- + $nH^+ \rightarrow$ alcohols) (Gavilanes et al., 2019).

8.3.4 Novel anaerobic digesters for enhanced biodiesel yield

Biodiesel is an alternative fuel source that can be produced from a wide range of biomaterials, food, and waste oils. It is manufactured through materials pretreatment, transesterification, and posttreatment process. The purpose of this process is to transform corn straw, livestock manure, and food waste into the esters that form the biodiesel fuel and develop an integrated system of AD and aerobic fungal fermentation. Conventional chemical reactors such as rotating, microwave, plug-flow reactor, and separation reactors are commonly used to produce biodiesel, and their efficiency is limited mainly due to high energy consumption and long reaction time. Consequently, other novel reactors are being developed to enhance the biodiesel yield by facilitating favorable transesterification reactions (Fig. 8.9). Each reactor operating conditions vary depending on the feedstock properties (Raheem et al., 2020). An intensified ultrasonic cavitation reactor is by far effective in completing the reaction within a short time and ambient temperature and also requires a low cost of production (Gholami et al., 2021). The working principle of this reactor is a high density of energy transmitted by the formation and subsequent collapse of cavities over a small area led to the chemical

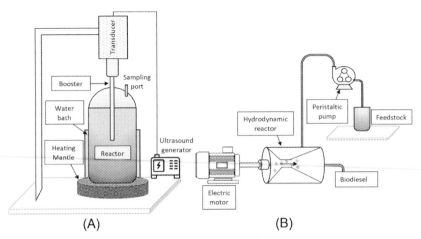

FIG. 8.9   **Novel anaerobic digesters for enhanced biodiesel yield.** (A) Ultrasonic cavitation reactor; (B) ultrasonic-hydrodynamic combined cavitation reactor.

reaction. The cavitation caused the mixing process by the shock waves, acoustic streaming, and microturbulent eddies. Farvardin et al. (2019) and Mohiddin et al. (2021) also reported the application of an ultrasonic-hydrodynamic combined cavitation reactor was able to enhance biodiesel production. The study used waste oil as a feedstock and the result indicated that increasing the power of ultrasound leads to a constantly increased yield of biodiesel.

The high temperatures required in the reaction process often lead to catalyst disabling due to active sintering. Here, the microchannel reactor (MCR), by reducing the blocking and overpressure risk from an uncontrolled output, has the potential to increase the productivity of the process. Furthermore, MCR provides an effective mass and heat transfer rate through a bunch of fine channels, allowing a short diffusion distance and a high surface-to-volume ratio. Due to an increased mixing rate within the methanol slug phase, a continuing process using an MCR was shown to achieve 98.6% oil conversion efficiency in 40 s (Mohd Laziz et al., 2020). In ambient room environments, the reaction was successfully performed without heating element in the reactor that reduces design complexity and makes the operation safer and more energy-efficient.

8.4   Conclusions and perspectives

The study of the new AD reactor for the energy utilization of organic waste is of great significance to the realization of sustainable development. It not only mitigates the climate change caused by the GHG emission but also makes a great contribution to the energy shortage as alternative clean energy. The purpose of this chapter is to collect and summarize the novel AD reactor in the past 5 years, sort out the latest technical direction, and explore the development prospect of improving the reactor to enhance the energy utilization of AD.

Based on the existing theory, the novel AD bioreactor changes the three stages of AD: hydrolysis, producing acetic acid and hydrogen, producing methane, each stage the main function of microbiota environment parameters (temperature, pH, redox potential, etc.) and reaction time, to provide a better living environment, solve the problem of acidification and inhibition, stimulate microbial species of electron transfer and other methods to enhance the target product output efficiency and achieve more economic and efficient waste energy. In this chapter, the multistage AD reactor to separate different microbial functional communities, the ZVI-AD reactor to promote electron transfer, and the microbubble AD reactor to balance pH and achieve purification of target products are mentioned in the above improved methods. In addition to producing physical energy sources (methane and hydrogen), biochemicals (organic acids, bioalcohols, and biodiesel), as an alternative source of energy in AD, are valued for their nontrivial value as biological resources. As an intermediate product, organic acids often inhibit the AD reaction process due to excessive accumulation. However, short-chain acids (acetic acid, butyric acid) and medium-chain acids in organic acids are valuable biochemicals, which can also be processed into environmentally friendly plastics (PHAs). In the process of AD with organic wastes as substrates, alcohols and biodiesel can also be obtained as AD products except of acid. The types of new AD reactors targeted at the above biochemical products include multistage AD reactor, MEC reactor, AnDMBR, RBR, and intensified ultrasonic cavitation reactor.

The central goal of this chapter is to research novel enhancement AD reactor for bioenergy and biochemical production, first introduced the concept of AD as a means of dealing with organic waste biological resources recycling, AD is the first choice of the organic part of urban living garbage processing method, not only for safe disposal of pollution, to protect vulnerable to pollution area, prevention of infectious diseases, also to increase the value and market of waste to energy and products. Second, the structural characteristics and enhancement principle of the novel AD reactor for the generation of biomass energy and biochemicals are introduced to classify and summarize the improvements made in the recent 5 years of bioreactors and show the clear technical development direction. Finally, to recent research, there are a lot of research focused on establishing a new type of AD reactor (products mainly methane of hydrogen) to reinforce the waste energy, biological chemicals as a byproduct also gradually discovered its value; however, the target mainly for chemical research is currently more stay in theory research and pilot phase, perhaps because of unclear economic evaluation and immature industrial processes, plants that actually use AD reactors to produce chemicals in large quantities is still lack. The use of safe and reliable technology to maximize resource recovery remains a major objective of our future work.

References

Al-mashhadani, M.K.H., Wilkinson, S.J., Zimmerman, W.B., 2016. Carbon dioxide rich microbubble acceleration of biogas production in anaerobic digestion. Chem. Eng. Sci. 156, 24–35.

Anzola-Rojas, M., del, P., Zaiat, M., 2016. A novel anaerobic down-flow structured-bed reactor for long-term stable H_2 energy production from wastewater. J. Chem. Technol. Biotechnol. 91 (5), 1551–1561.

Cai, W., Han, T., Guo, Z., Varrone, C., Wang, A., Liu, W., 2016. Methane production enhancement by an independent cathode in integrated anaerobic reactor with microbial electrolysis. Bioresour. Technol. 208, 13–18.

Chakraborty, D., Venkata Mohan, S., 2019. Efficient resource valorization by co-digestion of food and vegetable waste using three stage integrated bioprocess. Bioresour. Technol. 284, 373–380.

De Vrieze, J., Arends, J.B.A., Verbeeck, K., Gildemyn, S., Rabaey, K., 2018. Interfacing anaerobic digestion with (bio) electrochemical systems: potentials and challenges. Water Res. 146, 244–255.

Fan, R., Ebrahimi, M., Czermak, P., 2017. Anaerobic membrane bioreactor for continuous lactic acid fermentation. Membranes 7 (2), 1–14.

Farvardin, M., Hosseinzadeh Samani, B., Rostami, S., Abbaszadeh-Mayvan, A., Najafi, G., Fayyazi, E., 2019. Enhancement of biodiesel production from waste cooking oil: ultrasonic- hydrodynamic combined cavitation system. Energy SourcesA Recovery Util. Environ. Eff 28 (1), 1–15. doi: 10.1080/15567036.2019.1657524.

Gajaraj, S., Huang, Y., Zheng, P., Hu, Z., 2017. Methane production improvement and associated methanogenic assemblages in bioelectrochemically assisted anaerobic digestion. Biochem. Eng. J. 117, 105–112.

Gavilanes, J., Reddy, C.N., Min, B., 2019. Microbial Electrosynthesis of bioalcohols through reduction of high concentrations of volatile fatty acids. Energy Fuels 33 (5), 4264–4271.

Geng, S., Song, K., Li, L., Xie, F., 2020. Improved Algal sludge methane production and dewaterability by zerovalent iron-assisted fermentation. ACS Omega 5 (11), 6146–6152.

Gholami, A., Pourfayaz, F., Maleki, A., 2021. Techno-economic assessment of biodiesel production from canola oil through ultrasonic cavitation. Energy Rep. 7, 266–277.

Gilmour, D.J., Zimmerman, W.B., 2020. Microbubble intensification of bioprocessing. In: Poole, R. (Ed.), Advances in Microbial Physiology 1st ed., 77. Elsevier Ltd, Amsterdam, pp. 1–35.

Hassan, G.K., Massanet-Nicolau, J., Dinsdale, R., Jones, R.J., Abo-Aly, M.M., El-Gohary, F.A., Guwy, A., 2019. A novel method for increasing biohydrogen production from food waste using electrodialysis. Int. J. Hydrog. Energy 44 (29), 14715–14720.

Hassanein, A., Witarsa, F., Guo, X., Yong, L., Lansing, S., Qiu, L., 2017. Next generation digestion: complementing anaerobic digestion (AD) with a novel microbial electrolysis cell (MEC) design. Int. J. Hydrog. Energy 42 (48), 28681–28689.

Hegde, S., Trabold, T.A., 2019. Anaerobic digestion of food waste with unconventional co-substrates for stable biogas production at high organic loading rates. Sustainability (Switzerland); 11 (14), 3875.

Kadier, A., Jain, P., Lai, B., Kalil, M.S., Kondaveeti, S., Alabbosh, K.F.S., Abu-Reesh, I.M., Mohanakrishna, G., 2020. Biorefinery perspectives of microbial electrolysis cells (MECs) for hydrogen and valuable chemicals production through wastewater treatment. Biofuel Res. J. 7 (1), 1128–1142.

Khan, M.A., Ngo, H.H., Guo, W.S., Liu, Y., Nghiem, L.D., Hai, F.I., Deng, L.J., Wang, J., Wu, Y., 2016. Optimization of process parameters for production of volatile fatty acid, biohydrogen and methane from anaerobic digestion. Bioresour. Technol. 219, 738–748.

Kleerebezem, R., Joosse, B., Rozendal, R., Loosdrecht, M.C.M.Van, 2015. Anaerobic digestion without biogas? Rev. Environ. Sci. Bio/Technol. 14 (4), 787–801.

Li, Y., Zhang, J., Zhang, Y., Quan, X., 2017. Scaling-up of a zero valent iron packed anaerobic reactor for textile dye wastewater treatment: a potential technology for on-site upgrading and rebuilding of traditional anaerobic wastewater treatment plant. Water Sci. Technol. 76 (4), 823–831.

Lim, J.X., Vadivelu, V.M., 2019. Enhanced volatile fatty acid production in sequencing batch reactor: microbial population and growth kinetics evaluation, Proc. AIP Conference, 2124 (July).

Liu, Hongbo, Lv, Y., Xu, S., Chen, Z., Lichtfouse, E., 2020. Configuration and rapid start-up of a novel combined microbial electrolytic process treating fecal sewage. Sci. Total Environ. 705, 135986.

Liu, Hongbo, Wang, L., Zhang, X., Fu, B., Liu, H., Li, Y., Lu, X., 2019. A viable approach for commercial VFAs production from sludge: liquid fermentation in anaerobic dynamic membrane reactor. J. Hazard. Mater. 365, 912–920.

Liu, Hui, Chen, Y., 2018. Enhanced methane production from food waste using cysteine to increase biotransformation of l -monosaccharide, volatile fatty acids, and biohydrogen. Environ. Sci. Technol. 52 (6), 3777–3785.

Liu, Y., Zhang, Y., Ni, B.J., 2015. Zero valent iron simultaneously enhances methane production and sulfate reduction in anaerobic granular sludge reactors. Water Res. 75, 292–300.

Lora Grando, R., de Souza Antune, A.M., da Fonseca, F.V., Sánchez, A., Barrena, R., Font, X., 2017. Technology overview of biogas production in anaerobic digestion plants: a European evaluation of research and development. Renew. Sustain. Energy Rev. 80 (May), 44–53.

Losch, A., 2020. Developing our planetary plan with an 18th United Nations Sustainable Development Goal: space environment. HTS Theol. Stud. 76 (1), 1–7.

Mohd Laziz, A., KuShaari, K.Z., Azeem, B., Yusup, S., Chin, J., Denecke, J., 2020. Rapid production of biodiesel in a microchannel reactor at room temperature by enhancement of mixing behaviour in methanol phase using volume of fluid model. Chem. Eng. Sci. 219, 115532.

Mohiddin, M.N., Bin, Tan, Y.H., Seow, Y.X., Kansedo, J., Mubarak, N.M., Abdullah, M.O., Chan, Y.S., Khalid, M., 2021. Evaluation on feedstock, technologies, catalyst and reactor for sustainable biodiesel production: a review. J. Ind. Eng. Chem. 98, 60–81.

Mulakhudair, A.R., Al-Mashhadani, M., Hanotu, J., Zimmerman, W., 2017. Inactivation combined with cell lysis of *Pseudomonas putida* using a low pressure carbon dioxide microbubble technology. J. Chem. Technol. Biotechnol. 92 (8), 1961–1969.

Nayono, S., 2010. Anaerobic digestion of organic solid waste for energy production. Karlsruhe: KIT Scientific Publishing. doi: 10.5445/KSP/1000015038.

Paritosh, K., Yadav, M., Chawade, A., Sahoo, D., Kesharwani, N., Pareek, N., Vivekanand, V., 2020. Additives as a support structure for specific biochemical activity boosts in anaerobic digestion: a review. Front. Energy Res. 8, 1–17.

Park, J.G., Jiang, D., Lee, B., Jun, H.B., 2020. Towards the practical application of bioelectrochemical anaerobic digestion (BEAD): Insights into electrode materials, reactor configurations, and process designs. Water Res. 184, 116214.

Park, J.G., Lee, B., Kwon, H.J., Jun, H.B., 2019a. Contribution analysis of methane production from food waste in bulk solution and on bio-electrode in a bio-electrochemical anaerobic digestion reactor. Sci. Total Environ. 670, 741–751.

Park, J.G., Lee, B., Kwon, H.J., Park, H.R., Jun, H.B., 2019b. Effects of a novel auxiliary bio-electrochemical reactor on methane production from highly concentrated food waste in an anaerobic digestion reactor. Chemosphere 220, 403–411.

Park, J.G., Lee, B., Park, H.R., Jun, H.B., 2019c. Long-term evaluation of methane production in a bio-electrochemical anaerobic digestion reactor according to the organic loading rate. Bioresour. Technol. 273, 478–486.

Park, J., Shin, W., Shi, W., Jun, H., 2019d. Changes of bacterial communities in an anaerobic digestion and a bio-electrochemical anaerobic digestion reactors according to organic load. Energies 12, 2958.

Puyol, D., Flores-Alsina, X., Segura, Y., Molina, R., Padrino, B., Fierro, J.L.G., Gernaey, K.V., Melero, J.A., Martinez, F., 2018. Exploring the effects of ZVI addition on resource recovery in the anaerobic digestion process. Chem. Eng. J. 335, 703–711.

Rabii, A., Aldin, S., Dahman, Y., Elbeshbishy, E., 2019. A review on anaerobic co-digestion with a focus on the microbial populations and the effect of multi-stage digester configuration. Energies 12 (6).

Raheem, I., Mohiddin, M.N., Bin, Tan, Y.H., Kansedo, J., Mubarak, N.M., Abdullah, M.O., Ibrahim, M.L., 2020. A review on influence of reactor technologies and kinetic studies for biodiesel application. J. Ind. Eng. Chem. 91, 54–68.

Rebellón, L.F.F.M., 2017. Waste Management, Intechopen, London, pp. 195–200. https://doi.org/10.5772/3150.

Rodríguez-Alegre, R., Ceballos-Escalera, A., Molognoni, D., Bosch-Jimenez, P., Galí, D., Licon, E., Pirriera, M.Della, Garcia-Montaño, J., Borràs, E., 2019. Integration of membrane contactors and bioelectrochemical systems for CO_2 conversion to CH_4. Energies 12 (3), 1–19.

Rogoff, M.J., Screve, F. (Eds.), 2019. Introduction and overview. In: Waste-To-energy, 3rd edn. Elsevier, Amsterdam, pp. 1–12.

Roukas, T., Kotzekidou, P., 2020. Rotary biofilm reactor: a new tool for long-term bioethanol production from non-sterilized beet molasses by *Saccharomyces cerevisiae* in repeated-batch fermentation. J. Cleaner Prod. 257, 120519.

Sangeetha, T., Guo, Z., Liu, W., Gao, L., Wang, L., Cui, M., Chen, C., Wang, A., 2017. Energy recovery evaluation in an up flow microbial electrolysis coupled anaerobic digestion (ME-AD) reactor: role of electrode positions and hydraulic retention times. Appl. Energy 206, 1214–1224.

Sayedin, F., Kermanshahi-pour, A., He, S., 2018. Anaerobic digestion of thin stillage of corn ethanol plant in a novel anaerobic baffled reactor. Waste Manage. (Oxford) 78, 541–552.

Shanthi Sravan, J., Butti, S.K., Sarkar, O., Vamshi Krishna, K., Venkata Mohan, S., 2018. Electrofermentation of food waste—regulating acidogenesis towards enhanced volatile fatty acids production. Chem. Eng. J. 334, 1709–1718.

Song, Y.C., Feng, Q., Ahn, Y., 2016. Performance of the bio-electrochemical anaerobic digestion of sewage sludge at different hydraulic retention times. Energy Fuels 30 (1), 352–359.

Sun, H., Angelidaki, I., Wu, S., Dong, R., Zhang, Y., 2019. The potential of bioelectrochemical sensor for monitoring of acetate during anaerobic digestion: focusing on novel reactor design. Front. Microbiol. 10, 1–10.

Tamis, J., Joosse, B.M., van Loosdrecht, M.C.M., Kleerebezem, R., 2015. High-rate volatile fatty acid (VFA) production by a granular sludge process at low pH. Biotechnol. Bioeng. 112 (11), 2248–2255.

Tyagi, V.K., Fdez-Güelfo, L.A., Zhou, Y., Álvarez-Gallego, C.J., Garcia, L.I.R., Ng, W.J., 2018. Anaerobic co-digestion of organic fraction of municipal solid waste (OFMSW): progress and challenges. Renew. Sustain. Energy Rev. 93, 380–399.

Van, D.P., Fujiwara, T., Tho, B.L., Toan, P.P.S., Minh, G.H., 2020. A review of anaerobic digestion systems for biodegradable waste: configurations, operating parameters, and current trends. Environ. Eng. Res. 25 (1), 1–17.

Vanwonterghem, I., Jensen, P.D., Rabaey, K., Tyson, G.W., 2015. Temperature and solids retention time control microbial population dynamics and volatile fatty acid production in replicated anaerobic digesters. Sci. Rep. 5, 1–8.

Wang, S., Zhou, A., Zhang, J., Liu, Z., Zheng, J., Zhao, X., Yue, X., 2019. Enhanced quinoline removal by zero-valent iron-coupled novel anaerobic processes: performance and underlying function analysis. RSC Adv. 9 (3), 1176–1186.

Zhang, J., Loh, K.C., Lee, J., Wang, C.H., Dai, Y., Wah Tong, Y., 2017a. Three-stage anaerobic co-digestion of food waste and horse manure. Sci. Rep. 7 (1), 1–10.

Zhang, J., Loh, K.C., Li, W., Lim, J.W., Dai, Y., Tong, Y.W., 2017b. Three-stage anaerobic digester for food waste. Appl. Energy 194, 287–295.

Zhang, J., Qu, Y., Qi, Q., Zhang, P., Zhang, Y., Tong, Y.W., He, Y., 2020. The bio-chemical cycle of iron and the function induced by ZVI addition in anaerobic digestion: a review. Water Res. 186, 116405.

Zhang, J., Zhang, Y., Quan, X., Chen, S., 2015. Enhancement of anaerobic acidogenesis by integrating an electrochemical system into an acidogenic reactor: effect of hydraulic retention times (HRT) and role of bacteria and acidophilic methanogenic Archaea. Bioresour. Technol. 179, 43–49.

Zhang, L., Loh, K.C., Zhang, J., Mao, L., Tong, Y.W., Wang, C.H., Dai, Y., 2019. Three-stage anaerobic co-digestion of food waste and waste activated sludge: identifying bacterial and methanogenic archaeal communities and their correlations with performance parameters. Bioresour. Technol. 285, 121333.

Zhang, Y., Feng, Y., Quan, X., 2015. Zero-valent iron enhanced methanogenic activity in anaerobic digestion of waste activated sludge after heat and alkali pretreatment. Waste Manage. (Oxford) 38 (1), 297–302.

Zhang, Y., Jing, Y., Zhang, J., Sun, L., Quan, X., 2011. Performance of a ZVI-UASB reactor for azo dye wastewater treatment. J. Chem. Technol. Biotechnol. 86 (2), 199–204.

Bioaugmentation strategies *via* acclimatized microbial consortia for bioenergy production

Le Zhang[a,b], Hailin Tian[a,b], Jonathan T.E. Lee[a,b], Jun Wei Lim[a,b], Kai-Chee Loh[b,c], Yanjun Dai[b,d], Yen Wah Tong[a,b,c]

[a]NUS Environmental Research Institute, National University of Singapore, Singapore
[b]Energy and Environmental Sustainability for Megacities (E2S2) Phase II, Campus for Research Excellence and Technological Enterprise (CREATE), Singapore
[c]Department of Chemical and Biomolecular Engineering, National University of Singapore, Singapore
[d]School of Mechanical Engineering, Shanghai Jiao Tong University, Shanghai, China

9.1 Introduction

Anaerobic digestion (AD) is an efficient technology for simultaneous waste management and bioenergy generation from organic wastes, resulting in more environmentally sustainable practices (Kougias & Angelidaki, 2018; Sawatdeenarunat et al., 2015; Xu et al., 2018). AD technologies have attracted global attention within the last several decades, particularly in the context of sustainable development. However, substantial research activities associated with AD indicated that the digestion of various organic wastes frequently encountered process deficiency due to the accumulation of various inhibitory substances (Chen et al., 2008), including high ammonia concentration (e.g., > 3 g/L NH_4^+-N or > 0.15 g/L NH_3-N) (Amha et al., 2018; Capson-Tojo et al., 2020; Jiang et al., 2019; Rajagopal et al., 2013), accumulated volatile fatty acids (VFA) in excess (Zhao et al., 2020), recalcitrant lignocellulosic components (Akyol et al., 2019), and emerging pollutants (e.g. plastics) that hitherto remain undegradable (Rodríguez-Rodríguez et al., 2014). More specifically, in an AD system fed with nitrogen-rich substrates (e.g. food waste, manure, algal biomass, and sludge), the accumulation of high concentration of ammonia inhibited the normal growth of methanogenic archaea and syntrophic acetate oxidizing bacteria (SAOB), which led to an inhibition of the methanogenesis step and a decrease in methane production, ultimately presenting process failure. In an AD system fed with lignocellulosic substrates, the rigid 3-D matrix comprising hemicellulose, cellulose, and lignin resisted microbial and enzymatic attack, which led to low hydrolysis rate and ineffectual decomposition of the feedstock. Furthermore, small quantities of pollutants (e.g., plastics) in the organic wastes and accumulation of VFA caused by high organic loadings also inhibited AD processes by decreasing the efficacy of microbial decomposition.

To overcome the aforementioned inhibitory effects on AD processes as well as to improve the bioenergy production efficiency, many strategies such as parameter optimization (Khan et al., 2016), additive amendment (Ajay et al., 2020; Romero-Güiza et al., 2016), multistage bioreactor design (Srisowmeya et al., 2020; Zhang et al., 2019c), and bioaugmentation methods employing anaerobic microbes (Nzila, 2017; Vinzelj et al., 2020) have been investigated. Among these strategies, bioaugmentation employing microbial consortium is attracting increasing attention, due to the fact that AD is essentially a microbe-mediated multistep biological process (i.e., hydrolysis, acidogenesis, acetogenesis and methanogenesis) governed by microbiota (Tao et al., 2020). During the bioaugmentation processes, exogenous prea-dapted consortium, or pure culture, or genetically modified microorganisms harboring specific metabolic activities are added into the AD systems (Atasoy et al., 2018; Herrero & Stuckey, 2015).

Over the past decade, bioaugmentation strategies employing microbial consortia have been used to mitigate inhibition caused by ammonia (Yang et al., 2020) and VFA (Zhao et al., 2020), to enhance biodegradation of lignocellulosic components (Valdez-Vazquez et al., 2019) and emerging pollutants (Rodríguez-Rodríguez et al., 2014), and to accelerate methanogenesis for biogas production (Li et al., 2018b; Zhang et al., 2018a), as well as to improve production of hydrogen and VFA (Eder et al., 2020). For instance, it has been shown that bioagumentation with *Methanoculleus* sp. (DTU887) enhanced methane yield by 21% in anaerobic digesters under extremely high ammonia level of 13.5 g NH_4^+-N/L (Yan et al., 2020b). Regarding bioaugmentation in AD of lignocellulosic substrates, Lee et al. (2017; 2020a) had acclimated a diverse starting inoculum to the AD of *Axonopus compressus* cow grass, compared communities to identify putative microorganisms for augmentation, and optimized the bioaugmentation with an increase of 21% in methane production. Moreover, bioaugmentation of photosynthetic bacteria (e.g., *Ectothiorhodospira*) effectively recovered an overloaded AD system from VFA inhibition (pH 6.0) (Zhao et al., 2020). Bioaugmentation with *Thermoanaerobacterium thermosaccharolyticum* W16 had enhanced hydrogen yields by varying degrees for different types of seed sludge (Zhang et al., 2019a). Going forward, bio-augmentation using microbes with a specific biodegradation function could potentially be used to degrade emerging pollutants, such as plastics (Roager & Sonnenschein, 2019), skatole (Ma et al., 2020), and pharmaceuticals (Rodríguez-Rodríguez et al., 2014) in AD systems. In addition, recent studies have greatly contributed knowledge on long-term operations with bioaugmentation, microbiome functional reconstruction, long-term storage of acclimated consortium, and integrated bioaugmentation methods with pretreatments.

Fig. 9.1 illustrates the increasing number of reports that dealt with bioaugmentation studies on AD over the past 12 years. Notwithstanding, many of these studies are diverse and not con-solidated, and had not been particularly useful for identifying the state-of-the-art key findings and to integrate them within a specific AD system to achieve higher bioenergy yields. To date, only bioaugmentation strategies for enhancing the biological wastewater treatment (Raper et al., 2018; Stephenson & Stephenson, 1992), remediation of polluted soil (Cycoń et al., 2017;

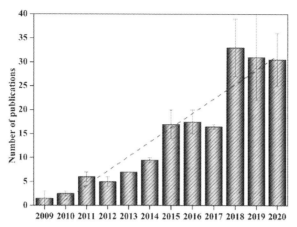

FIG. 9.1 **Number of annual publications from 2009 to 2020 on "bioaugmentation" and "anaerobic digestion" (based on search results in Scopus database and Web of Science database).** Linear regression line was added to display the growth trend.

El Fantroussi & Agathos, 2005; Lebeau et al., 2008), and treatment of biorefractory organic compounds (Semrany et al., 2012; Zhan et al., 2020) have been well documented.

This chapter aimed to provide a more comprehensive overview of the applications, advancements, and challenges of bioaugmentation technologies in AD systems in the recent years, with the hope of achieving higher process efficiencies by mitigating various inhibitory factors. The theoretical basis of employing bioaugmentation in various AD processes and the popular operational procedures of the strategies are first analyzed in Section 9.2. This is followed by a comprehensive summary of various bioaugmentation practices for enhancing AD: including mitigating ammonia inhibition during AD processes (Section 9.3.1), enhancing biodegradability of lignocellulosic components (Section 9.3.2), ameliorating pressure from high VFA or heavy duty AD (Section 9.3.3), enhancing AD for biohydrogen production (Section 9.3.4), and enhancing biodegradation of emerging pollutants (e.g. plastics) in AD bioreactors (Section 9.3.5). The limitations and prospects of bioaugmentation strategies to enhance AD processes for biofuel production are finally proposed in Section 9.4.

9.2 Theoretical basis and operational procedures of bioaugmentation strategies

9.2.1 Theoretical basis

AD of organic wastes relies on multistep biochemical reactions of hydrolytic and fermentative bacteria and methanogenic archaea, and can therefore be improved by the augmentation with microbes. Fig. 9.2 depicts the different forms of inhibition in AD processes of heterogeneous organic wastes and corresponding putative target sites of bioaugmentation strategies.

The degradation of nitrogen-rich organic wastes (e.g., manure and protein-rich algal residue) generates NH_4^+ and/or NH_3 in the AD systems, leading to potential inhibition to normal function of microbial communities, when ammonia concentration exceeds a certain

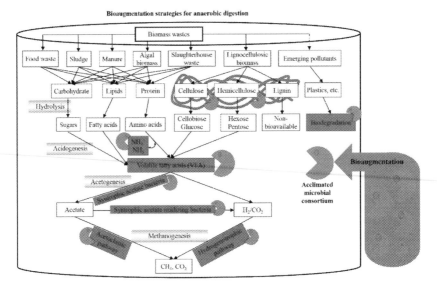

FIG. 9.2 Different forms of inhibition in AD processes of heterogeneous organic wastes and potential putative target sites of bioagumentation strategies.

threshold (i.e., > 3 g NH_4^+-N/L or 0.15 g NH_3-N/L) (Fotidis et al., 2013). More specifically, excess ammonia causes inhibition to microorganisms in several aspects, including affecting intracellular pH via NH_4^+, increasing energy consumption in proton balancing, and inhibiting key enzymatic reactions (Yang et al., 2019). It has been reported that methanogenic archaea and SAOB are more susceptible to ammonia inhibition (Ruiz-Sánchez et al., 2018). Hence, during bioaugmentation operations for mitigating ammonia inhibition (Fig. 9.2), the ammonia-tolerant microbial strains can be cultivated and added into the ammonia-inhibited AD systems to perform methanogenesis for methane production.

Regarding bioaugmentation for enhancing degradation of lignocellulosic biomass, the main aim is to increase the bottleneck hydrolysis rate of cellulose and hemicellulose. Previous studies have demonstrated that lignocellulosic biomass is composed of three different organic polymers (i.e. cellulose, hemicellulose, and lignin) entangled together into a lignin-encrusted superstructure, which is resistant to microbial degradation. Lignin, a complex phenyl propane polymer, is a practically nondegradable component in AD systems as its degradation requires an aerobic environment (Khan & Ahring, 2019). Hence, pretreatments are frequently used to remove the lignin prior to subsequent AD processes (Hu et al., 2016). Cellulose polymer is composed of β-D-glucose monomers while hemicellulose polymer consists of a mix of hexose and pentose monomers (Fig. 9.2). The complete decomposition of cellulose and hemicellulose needs efficient biochemical reactions catalyzed by a variety of hydrolytic enzymes (e.g. endoglucanases, exoglucanases, and β-glucosidases) secreted by cellulolytic microbes (Strang et al., 2017). As hydrolysis is considered to be the rate-limiting step during AD of lignocellulosic material, the cellulose degrading bacterial consortia can first be enriched on α-cellulose and then introduced to AD systems fed with lignocellulosic feedstock to facilitate the breakdown of lignocellulosic components. For instance, anaerobic fungi can produce cellulolytic, hemicellulolytic, carbohydrate hydrolyzing, and proteolytic enzymes for enhancing hydrolysis and anaerobic fermentation of lignocellulosic compounds (Yıldırım et al., 2017).

Regarding bioaugmentation for recovering overloaded AD digesters, the idea is to reduce VFA inhibition that is caused by digester overloading (Li et al., 2018c), unbalanced C/N ratios, and drastic temperature changes (e.g. from mesophilic to thermophilic) (Zhu et al., 2018). Indeed, overloading is a common issue encountered by AD systems, which leads to an accumulation of VFA that cannot be efficiently utilized by methanogens due to the toxicity of excessive VFA (De Vrieze et al., 2012). Therefore, beneficial microorganisms such as propionate-degrading enrichment cultures, VFA-degrading cultures, SAOB, and hydrogenotrophic methanogens can be added to the overloaded anaerobic digesters to restore function, strengthen the digester performance, and prevent imminent process failure.

In addition to the above bioaugmentation strategies, methane yields can also be enhanced by promoting methanogenesis through bioaugmentation with methanogens. Indeed, two types of methanogens, namely hydrogenotrophic methanogens and acetoclastic methanogens, can be utilized for bioaugmentation (Fig. 9.2). Nevertheless, thanks to the sensitivity of methanogens to oxygen, the long-term storage and effective dosing methods of the bioaugmented methanogens require more studies. For a successful bioaugmentation operation, the bioaugmented methanogens have to be able to coexist with the indigenous microorganisms resulting in a new steady state and thereby enhance methanogenesis (Venkiteshwaran et al., 2016). Alternatively, hydrogen-producing strains can be added to AD digesters fed with various organic wastes to strengthen metabolic pathways leading to hydrogen production (Sharma & Melkania, 2018). The theoretical foundation of bioaugmentation for enhancing hydrogen production is that hydrogen fermentation relies on hydrogen fermenting bacteria and that the bioaugmented facultative anaerobic microorganisms could remarkably improve hydrogen production capacities of augmented AD systems.

Finally, various emerging pollutants (e.g., plastics, pyridine, etc.) existing in the heterogeneous organic wastes are considered potentially harmful to the AD systems (Lim et al., 2018; Wen et al., 2013). Furthermore, the AD digesters are not specifically designed for the removal of these pollutants. Therefore, bioaugmentation of bacteria capable of degrading such pollutants can be performed in the AD digesters, thereby reducing the hazardous effects to anaerobic communities and enhancing the digester performance. Hitherto, the understanding of microbial biodegradation of plastics and other pollutants is still nascent, but ever-increasing knowledge on interactions between microbes and pollutants presents huge application potential of bioaugmentation strategies in the AD digesters for enhancing biodegradation of emerging pollutants.

9.2.2 Common operational procedures

Based on the overview of substantial studies on bioaugmentation strategies, a frequently used procedure of bioaugmentation in anaerobic digesters is proposed (Fig. 9.3). Obtaining microbial cultures is a prerequisite for performing bioaugmentation. In this regard, mixed microbial consortium can be obtained from the industrial-scale biogas digesters (Strang et al., 2017) while the pure strains can be purchased from the commercial companies such as DSMZ (Deutsche Sammlung von Mikroorganismen und Zellkulturen GmbH Company, Germany)

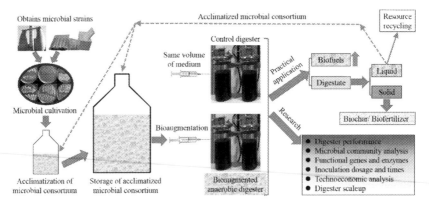

FIG. 9.3   **The flow chart of frequently used operation procedures of bioaugmentation strategies in anaerobic digesters.**

(Yang et al., 2019), ATCC (American type culture collection) (Ecem Öner et al., 2018), and National Collection of Microbial Strains and Cell Cultures (Mexico City) (Valdez-Vazquez et al., 2019), etc. In addition, acclimated consortia can also be obtained from the discharged digestate. All these cultures can be stored at −80 °C until the next step, namely, cultivation of microbial cultures.

Cultivation of microbial cultures is one of the most key steps during bioaugmentation, as cultivation of different microbial species usually requires different conditions. The sequential cultivation can be performed to attain subgeneration from seed generation during the exponential growth period. Prior to bioaugmentation operations, the prepared cultures can be acclimated to specific conditions (e.g., high ammonia levels) to induce microorganisms into acquiring/presenting the corresponding ability (e.g., ammonia tolerance). Optical density at 600 nm can be used as an indicator of dynamic microbial concentration during microbial cultivation. Subsequently, the acclimated microbial consortium can be added in the bioaugmented AD digesters, while the same volume of medium should be added in the control digesters (Zhu et al., 2018). Notably, control digesters may not be needed for application purpose. On the one hand, several aspects are frequently investigated in the bioaugmentation studies, including digester performance, microbial community analysis, economic analysis, and digester scale-up. On the other hand, during the practical application of bioaugmentation strategies, the biofuel yields could be enhanced due to the fact that the acclimated microbial consortium enhanced the AD process through several aspects (Fig. 9.2). Furthermore, the anaerobic digestate discharged from the AD digesters can be separated into solid phase and liquid phase (Wang et al., 2020b). The solid phase of digestate can be converted into biochar through a gasification or pyrolysis process (Chang et al., 2020; Kim et al., 2020; Liu et al., 2020), or converted into biofertilizer via a composting process (Slepetiene et al., 2020). Meanwhile, the liquid phase of the digestate can be utilized as a source of acclimated consortium (Hosseini Koupaie et al., 2019). By this means, the anaerobic digester can enter an increasingly cyclic bioaugmentation operation that can maximize biofuel production.

9.3   Key findings in bioaugmentation strategies for enhancing AD

9.3.1   Bioaugmentation for mitigating ammonia inhibition

Ammonia inhibition, one of the major issues associated with AD processes of N-rich biomass wastes, has been well documented (Capson-Tojo et al., 2020; Yenigün & Demirel, 2013). The bioaugmentation studies for mitigating ammonia inhibition during AD processes are summarized chronologically in Table 9.2. The frequently used microbes for bioaugmentation in ammonia-inhibited digesters included several types, such as pure strains (i.e., SAOB, hydrogenotrophic methanogens, obligate aceticlastic methanogens, and facultative aceticlastic methanogens) (Fotidis et al., 2014; Yang et al., 2019) as well as enriched ammonia-tolerant methanogenic cultures (Fotidis et al., 2017; Mahdy et al., 2017). More specifically, the frequently used SAOB included *Clostridium ultunense* spp. nov. and *Methanoculleus* spp. strain MAB1 (Fotidis et al., 2013), *Syntrophaceticu schinkii* (Yang et al., 2019), and *Tepidanaerobacter acetatoxydans* (Yang et al., 2020). Bioaugmentation with single SAOB had no obvious effects on methane production, which was probably due to the fact that single SAOB cannot accelerate the methanogenesis process that was performed by the methanogenic partner of SAOB (Fotidis et al., 2013). Instead, the bioaugmentation of enriched hydrogenotrophic methanogens (*Methanoculleus thermophilus)* successfully stimulated the growth of SAOB (*Thermacetogenium phaeum*) in thermophilic continuous digesters fed with cattle manure (Tian et al., 2019d). In another study (Tian et al., 2019b), another hydrogenotrophic methanogenic culture (*Methanoculleus bourgensis*) was bioaugmented in an ammonia inhibited digester fed with microalgae at an extreme ammonia level (e.g., 11 NH_4^+-N/L). With the help of high throughput 16S rRNA gene sequencing, the bioaugmented *M. bourgensis* established a newly efficient community (Tian et al., 2019b), which can be regarded as a key indicator of the successful bioaugmentation. In addition to hydrogenotrophic methanogens, the positive effects of aceticlastic methanogens have also been confirmed in ammonia inhibited digesters (Yang et al., 2020; Yang et al., 2019). Indeed, bioaugmentation with facultative aceticlastic methanogen *Methanosarcina barkeri* alone was effective in mitigating ammonia inhibition and enhanced the methane production by 59.7% (Yang et al., 2019). Furthermore, the obligate aceticlastic methanogen (*Methanosaeta harundinacea*) and hydrogenotrophic methanogen (*Methanobrevibacter smithii*) and SAOB (*S. schinkii*) were jointly bioaugmented into the ammonia inhibited digesters to balance the aceticlastic and hydrogenotrophic methanogenesis pathways for higher methane production compared to bioaugmentation using individual strains (Yang et al., 2020). Such combinations of two or more strains might be worth investigating due to its potential ability to offer synergistic effects and improve process efficiency through selection of appropriate strains. In addition, it is notable that the rapid development of gene sequencing technologies and bioinformatic analysis tools (Lim et al., 2020b; Zhang et al., 2019b) offered a great opportunity to study the AD-related metabolic pathways (e.g. syntrophic acetate oxidization (SAO), aceticlastic and hydrogenotrophic methanogenesis) at the levels of functional genes and enzymes (Campanaro et al., 2020; Tian et al., 2019c; Zhu et al., 2019). For instance, on the basis of 16S rRNA gene sequencing results and COG (cluster of orthologous groups) and KEGG (Kyoto Encyclopedia of Genes and Genomes) databases, the metabolic potential of functional genes involved in the methane metabolism and the relative abundances of functional enzymes of microbial communities in bioaugmented and

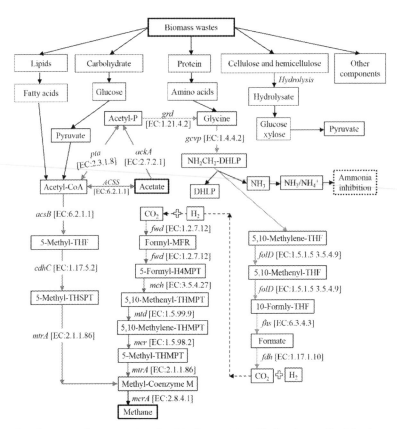

FIG. 9.4 **Functional genes and enzymes involved in the acetate oxidation (green line), hydrogenotrophic methanogenesis pathway (blue line), and aceticlastic methanogenesis pathway (red line).** Adapted from (Bedoya et al., 2019; Yan et al., 2020b; Yang et al., 2020) with permission/license granted by the publisher. Increase of functional species via bioaugmentation using acclimatized ammonia-tolerant microbial consortium in AD digesters could generate positive effects on the enzyme activity to mitigate ammonia inhibition.

nonbioaugmented digesters had been analyzed (Yang et al., 2020). The results demonstrated a stronger total enzyme activity in bioaugmented digesters with *M. barkeri* and digesters with *M. harundinacea + S. schinkii + M. smithii*. Moreover, the key metabolic pathways related to acetate degradation and methanogenesis occurring in the microbial consortium at high ammonia levels have been successfully reconstructed (Yan et al., 2020c). To go a step further, functional enzymes and genes involved in the acetate oxidation, aceticlastic, and hydrogenotrophic methanogenesis pathways were summarized (Fig. 9.4 and Table 9.1) based on reported research. It is foreseeable that more deep analyses on microbial communities and methanogenic pathways in ammonia-inhibited and ammonia -tolerant digesters will be reported in the near future.

Furthermore, information presented in Table 9.2 can also shed light on some general trends and limitations in the current studies on bioaugmentation strategies for mitigating ammonia inhibition. Initially, the most common bioreactors in previous bioaugmentation studies

TABLE 9.1 Functional enzymes and functional genes involved in the acetate oxidation, aceticlastic, and hydrogenotrophic methanogenesis pathways (Bedoya et al., 2019; Yan et al., 2020b; Yang et al., 2020; Zhu et al., 2019).

Gene	Enzyme EC number	Enzyme full name	Related metabolism
grd	[EC:1.21.4.2 1.21.4.3 1.21.4.4]	Glycine/sarcosine/betaine reductase complex component C subunit alpha	Acetate oxidation
gcvp	[EC:1.4.4.2]	Glycine dehydrogenase	Acetate oxidation
folD	[EC:1.5.1.5 3.5.4.9]	Methylenetetrahydrofolate dehydrogenase (NADP+) / methenyltetrahydrofolate cyclohydrolase	Acetate oxidation
fhs	[EC:6.3.4.3]	Formate–tetrahydrofolate ligase	Acetate oxidation
fdh	[EC:1.17.1.10]	Formate dehydrogenase (NADP+) alpha subunit	Acetate oxidation
fwdE, fmdE	[EC:1.2.7.12]	Formylmethanofuran dehydrogenase subunit E	Hydrogenotrophic methanogenesis
mch	[EC:3.5.4.27]	Methenyltetrahydromethanopterin cyclohydrolase	hydrogenotrophic methanogenesis
mtd	[EC:1.5.99.9]	Methylenetetrahydromethanopterin dehydrogenase	Hydrogenotrophic methanogenesis
mer	[EC:1.5.98.2]	5,10-methylenetetrahydromethanopterin reductase	Hydrogenotrophic methanogenesis
mtrA	[EC:2.1.1.86]	Tetrahydromethanopterin S-methyltransferase subunit A	Hydrogenotrophic methanogenesis
mcrA	[EC:2.8.4.1]	Methyl-coenzyme M reductase alpha subunit	Hydrogenotrophic methanogenesis
ackA	[EC:2.7.2.1]	Acetate kinase	Hydrogenotrophic methanogenesis
pta	[EC:2.3.1.8]	Phosphate acetyltransferase	Aceticlastic methanogenesis
ACSS	[EC:6.2.1.1]	Acetyl-CoA synthetase	Aceticlastic methanogenesis
acsB	[EC:6.2.1.1]	Acetyl-CoA synthetase	Aceticlastic methanogenesis
cdhC	[EC:1.17.5.2]	Caffeine dehydrogenase subunit gamma	Aceticlastic methanogenesis
mtrA	[EC:2.1.1.86]	Tetrahydromethanopterin S-methyltransferase subunit A	Aceticlastic methanogenesis

related to ammonia inhibition alleviation were lab-scale (0.25–4.5 L) continuously stirred tank reactor (CSTR) and batch reactors (Table 9.2); however, the practical engineering experience obtained from tests using larger scale (100–1000 L) reactors is an essential prerequisite for the reactor scale-up and potential industrial application of the bioaugmentation strategies (Kumar et al., 2016; Tabatabaei et al., 2020). Furthermore, the reported bioaugmentation studies for mitigating ammonia inhibition focused mainly on reactor performance and microbial community analysis. Nevertheless, a relatively high cost of pure microbial strains and chemicals for medium preparation might be a big challenge for large-scale applications of bioaugmented microbial cultures. To tackle the high cost issue, ammonia-tolerant inocula

TABLE 9.2 Bioaugmentation studies for mitigating ammonia inhibition during AD process.

Bioaugmented cultures	Substrate	Reactor	Experimental conditions	Performance	Key findings	References
Syntrophic acetate-oxidizing consortium (*Clostridium ultunense* spp. nov. and *Methanoculleus* spp. strain MAB1), hydrogenotrophic methanogen (*Methanoculleus bourgensis* MS2^T)	Basal anaerobic medium used as feedstock	250 mL UASB reactors	Fed-batch mode, 37 °C, 3–5 g NH$_4^+$-N/L	SAO coculture had no effect in a UASB reactor; SAO coculture and hydrogenotrophic methanogen led to a 42% higher growth rate	Methanogen was a key partner during bioaugmentation of SAO cultures	(Fotidis et al., 2013)
Hydrogenotrophic methanogen (*Methanoculleus bourgensis* MS2^T)	Cattle manure	2.3 L CSTR	37 °C, HRT 24 days, OLR 1.74 gVS/L/d, 3–5 g NH$_4^+$-N/L	A 31.3% increase in methane yield in bioaugmented reactors	Confirmed that bioaugmentation can be performed without interrupting the continuous AD operations	(Fotidis et al., 2014)
Enriched methanogenic propionate degrading microbial consortia	Sodium propionate	1.5 L CSTR	36 ± 1 °C, 100 rpm, HRT 15 d	Methane recovery rate was enhanced by 21%	A higher dosage of bioaugmentation can shorten the recovery time	(Li et al., 2017)
An enriched ammonia-tolerant methanogenic culture	Pig and cattle manure	2.3 L CSTR	1.74 gVS/L/d, HRT 24 days, 1.65–5 g NH$_4^+$-N/L	Methane production was improved by 36%	Enriched culture was 25% more efficiently than pure culture in alleviating ammonia inhibition	(Fotidis et al., 2017)
An ammonia acclimatized (>4.5 g NH$_4^+$-N/L) inoculum	Microalgae (*Chlorella vulgaris*) + cattle manure	320 mL batch reactors; 2.3 L CSTR	55 °C; 37 °C, HRT 23 days, OLR 2.33–4.66 gVS/L, 3.7%–4.2 NH$_4^+$-N/L	Coupling codigestion and bioaugmentation with ammonia-tolerant inoculum showed the highest methane yield	Ammonia-tolerant inocula could be a promising approach to enhance biogas production from microalgae	(Mahdy et al., 2017)
Hydrogenotrophic *Methanoculleus bourgensis*	Cattle manure, microalgae *C. vulgaris*	2.3 L CSTR	37 °C, HRT 23 d, OLR 2 gVS/L/d, 11 NH$_4^+$-N/L	28% increase in methane production	Bioaugmented *M. bourgensis* established a newly efficient community	(Tian et al., 2019b)

Bioaugmented cultures	Substrate	Reactor	Experimental conditions	Performance	Key findings	References
Hydrogenotrophic methanogens (*Methanoculleus thermophilus*)	Cattle manure	2.3 L CSTR	53 ± 1 °C, HRT 15 days, OLR 2.5, 1.42–5.0 NH_4^+-N/L	Methane yield increased by 11%–13%; VFA decreased by 45%–52%	Bioaugmented *Methanoculleus thermophilus* stimulated growth of SAOB (*Thermacetogenium phaeum*)	(Tian et al., 2019d)
Seven pure strains (obligate aceticlastic methanogen, facultative aceticlastic methanogen, SAOB, hydrogenotrophic methanogen)	Basal anaerobic medium as feedstock; glucose as a carbon source	500 mL batch reactors	OLR 1 g glucose/L/d, HRT 10 days, 4 NH_4^+-N/L, Initial pH 7.8	Hydrogenotrophic methanogen *Methanobrevibacter smithii* and SAOB *Syntrophaceticu schinkii* enhanced methane yield by 71.1%	Balancing hydrogenotrophic and aceticlastic methanogenic pathways is very critical	(Yang et al., 2019)
Methanoculleus bourgensis MS2 culture (DSM 3045)	Organic fraction of municipal solid waste	4.5 L CSTR	OLR 3.4 gVS/L/d, HRT 20 days, 9.5–13.5 NH_4^+-N/L	Bioaugmentation enhanced methane production by 21% and decreased VFA by 10%	Key microorganisms formed a syntrophy-supported food web after bioaugmentation	(Yan et al., 2020b)
SAOB (*Syntrophaceticu schinkii*, *Tepidanaerobacter acetatoxydans*), Obligate aceticlastic methanogen (*Methanosaeta harundinacea*), Facultative aceticlastic methanogen (*Methanosarcina barkeri*), Hydrogenotrophic methanogen (*Methanobrevibacter smithii*)	Glucose	1.2 L reactors	37 °C, OLR 1 g-glucose/L/d, HRT 10 days, 130 rpm, 4 g NH_4^+-N/L	*Methanosaeta harundinacea*+ *Syntrophaceticu schinkii*+ *Methanobrevibacter smithii* showed a 49% increase in methane yield; *Methanosarcina barkeri* or *syntrophaceticu schinkii*+ *Methanobrevibacter smithii* enhanced methane production by 35%	Enhancement of both hydrogenotrophic and aceticlastic methanogenic pathways should be considered; bioaugmented microbial species should be properly selected to generate synergies	(Yang et al., 2020)

was instead obtained from full-scale reactors for use in ammonia inhibited digesters (Mahdy et al., 2017), resulting in a relatively higher methane yield of 431 mL/gVS than the nonbioaugmented digester at a substrate to inoculum ratio of 0.25. The results demonstrate that existing ammonia-tolerant inocula, as a low-cost alternative of pure microbial strains, could be a promising option in bioaugmentation operations (Mahdy et al., 2017). Similarly, an enriched methanogenic propionate degrading microbial consortia was successfully bioaugmented in digesters under ammonia stress (3 g N/L), with the finding that a double dosage (0.6 g dry cell weight/L/d) of bioaugmentation for 45 days efficiently prevented the unstable digestion process against further deterioration and shortened the recovery time, especially when the inhibition was severe (Li et al., 2017). Therefore, repeated bioaugmentation of low-cost ammonia-acclimatized microbial consortia obtained from operational digesters could be an easily applicable approach to simultaneously alleviate ammonia inhibition, enhance biofuel production, and reduce operation costs. Indeed, repeated bioaugmentation regimes was found to be more beneficial to shift indigenous bacterial community composition than one-time bioaugmentation pattern, leading to a successful introduction of the added microbes into the digesters (Yang et al., 2016). However, the main limitation of the repeated bioaugmentation technique that needs to be overcome is the long-term (6–12 months) storage of ammonia-acclimatized microbial consortia. On the one hand, a low temperature (4 °C) and an oxygen-free environment are recommended for further studies as the former could inhibit microbial metabolic activity and the latter could improve the survival of anaerobic microbes (e.g. methanogens) (Yan et al., 2020a). On the other hand, it might be feasible to integrate the acclimated microbial consortium with some supporting media such as biochar, zeolites, activated carbon, or chitosan (He et al., 2020; Poirier et al., 2017) and manufacture them into granular microbial agents for long-term storage and commercial use.

9.3.2 Bioaugmentation for enhancing biodegradation of lignocellulosic biomass

Bioaugmentation techniques using various acclimatized microbial consortia have been used extensively in AD digesters fed with lignocellulosic biomass to enhance hydrolysis of cellulose and hemicellulose polymers so as to provide more hydrolytic intermediates (e.g., cellobiose, glucose, hexose and pentose, etc.) for subsequent acidogenesis/acetogenesis and methanogenesis. Table 9.3 presents in a chronological manner bioaugmentation studies for enhancing biodegradation of lignocellulosic biomass wastes. The lignocellulosic substrates for biofuel production through bioaugmented AD include cowgrass (Lee et al., 2020a), wheat straw (Ferraro et al., 2020; Ferraro et al., 2019), rice straw (Shetty et al., 2020), oil palm empty fruit bunches (Suksong et al., 2019), cereal crops residues (Akyol et al., 2019), mushroom spent straw (Ferraro et al., 2018), corn stover (Strang et al., 2017), manure (Tsapekos et al., 2017), excess sludge (Hu et al., 2016), microalgae biomass (Aydin et al., 2017), and brewery spent grain (Čater et al., 2015). The commonly used microbes in bioaugmentation in the last 5 years include two types, namely pure microbial strains and mixed microbial consortia. The former include cellulose-hydrolyzing strains *Clostridium cellulolyticum* and *Clostridium cellulovorans* (Lee et al., 2020a), fiber digesting anaerobic fungi (e.g. species *Orpinomyces joyonii* (Shetty et al., 2020), phylum *Neocallimastigomycota* (Ferraro et al., 2018), and genera *Anaeromyces* and *Piromyces* (Nkemka et al., 2015), fermentative strain such as *Clostridium acetobutylicum* (Valdez-Vazquez et al., 2019), thermophilic cellulolytic species such as

TABLE 9.3   Bioaugmentation for enhancing biodegradation of lignocellulosic biomass.

Bioaugmented microbial cultures	Substrate	Reactor	Experimental conditions	Performance	Key findings	References
Ruminococcus flavefaciens 007C, Pseudobutyrivibrio xylanivorans Mz5T, Fibrobacter succinogenes S85, and *Clostridium cellulovorans*	Brewery spent grain	1 L reactors	37 °C, 120 rpm, OLR 0.413 gVS/L, bioaugmented with 5% (v/v) enriched cultures	Optimal bioaugmentation with *Pseudobutyrivibrio xylanivorans* Mz5T enhanced methane yields by 18%	Bioaugmentation operations caused greater changes in bacterial community than in archaeal community	(Čater et al., 2015)
Highly active lignocellulolytic microorganisms (*Clostridium stercorarium* and *Bacteroides cellulosolvens*)	Excess sludge	0.5 L reactors	35 °C, 100 rpm, sludge retention time 20 days, pretreatment to remove lignin incrustation	Thermal pretreatment was the most effective method; pretreatment and bioaugmentation enhanced degradation of cellulose, hemicellulose and lignin, leading to a twofold higher methane yield	Thermal pretreatment was recommended before bioaugmentation in AD digesters with lignocellulosic biomass waste	(Hu et al., 2016)
Cellulolytic bacterial cultures enriched from sheep rumen fluid	Wheat straw	0.5 L glass reactors	37 °C, bioaugmented enriched culture at 2%–4%, fermentation for 30 days	Addition of 2% enriched culture had no effect on methane yield, while addition of 4% enriched culture enhanced methane yield by 27%	Acclimatization process increased abundance of *Ruminococcaceae* and *Lachnospiraceae*	(Ozbayram et al., 2017)
Clostridium thermocellum and *Melioribacter roseus*	Wheat straw	0.3 L glass reactors, 5 L CSTR reactors	53 °C, 150 rpm (2 min ON/OFF), HRT 15 days, daily bioaugmented pure culture at 3.3% (v/v) of working volume	Bioaugmentation with *C. thermocellum* enhanced methane yield by 34%; *M. roseus* only increased methane yield by 10%	Bioaugmentation strategies did not change significantly indigenous microbial communities	(Tsapekos et al., 2017)
Thermophilic cellulolytic consortia	Cellulose and corn stover	0.5 L glass reactors	55 °C, bioaugmented with 10% (v/v)	Bioaugmentation enhanced the methane yield by 22%–24%; dominant strains were orders *Thermoanaerobacterales* and *Clostridiales*	The members of the acclimatized consortium in low abundance had significant role in enhancing methane yield	(Strang et al., 2017)

(continued)

TABLE 9.3 (Cont'd)

Bioaugmented microbial cultures	Substrate	Reactor	Experimental conditions	Performance	Key findings	References
Clostridium thermocellum	Wheat straw and cow manure (1:1 on VS basis)	0.1 L glass reactors	55 °C, inoculum to substrate ratio 1:1, bioaugmented volume at 5%–20% of working volume	Bioaugmented strategy enhanced methane yield by 39%; bioaugmentation at the ratio of 20% increased family *Ruminococcaceae* and decreased families *Bacteroidaceae* and *Synergistaceae*	Metabolic products of bioaugmented strains greatly affected the diversity of the archaeal community and increased abundance of *Methanomicrobiales*	(Ecem Öner et al., 2018)
Anaerobic ruminal fungi (phylum *Neocallimastigomycota*, *Neocallimastix* sp.) and fermentative bacteria (*Orpynomyces* sp.)	Wheat straw and mushroom spent straw	0.12 L glass reactors	37 °C, 120 rpm, OLR 6.5 gVS/L, bioaugmented volume at 8%–20% of working volume	Bioaugmentation with fungi and fermentative bacteria enhanced the methane yields	Microbial analysis in bioaugmented digestered confirmed a central core of methanogen *Methanosaeta*	(Ferraro et al., 2018)
Hydrolytic strain *Clostridium cellulovorans*; fermentative strain *Clostridium acetobutylicum*	Wheat straw	0.25 L reactors	37 °C, OLR 15 gTS/kg/d, pH 6.5, bioaugmented with 26 mL of single and cocultures	Both strains established a synergistic relationship and showed two- to threefold higher hydrogen yield than single culture	Bioaugmentation with *Clostridium cellulovoran* enhanced xylanolytic activity for the microbial consortium	(Valdez-Vazquez et al., 2019)
Anaerobic fungus *Orpinomyces* sp.	Cow manure, cereal crops	1 L glass reactors	37 °C, substrate to inoculum ratio 1:1 on VS basis, 100 rpm, bioaugmented consortium 10% v/v	Bioaugmentation enhanced the methane yield by 15%–33% (0.43 L/gVS)	Fungal bioaugmentation can be a promising method; repetitive bioaugmentation in continuous AD deserves further studies	(Akyol et al., 2019)
Thermotolerant cellulolytic *Clostridiaceae* and *Lachnospiraceae* rich consortium	Oil empty fruit bunches	0.5 L reactors	40 °C, substrate to inoculum ratio 2:1, initial pH 7.2–7.5, substrate to bioaugmented consortium 500/1 to 5000/1 on VS basis	Bioaugmented reactor with *Clostridiaceae*-rich consortium had a methane yield of 0.217 L/gVS; prehydrolysis with *Lachnospiraceae* rich consortium showed a methane yield of 0.349 L/gVS	*Lachnospiraceae*-rich consortium is suitable for two-stage AD for process enhancement; *Clostridiaceae*-rich consortium is more suitable for single-stage AD for process enhancement	(Suksong et al., 2019)

Bioaugmented microbial cultures	Substrate	Reactor	Experimental conditions	Performance	Key findings	References
Fiber-digesting anaerobic fungus *Orpitomyces joyonii*	Rice straw	2 L glass reactors	37 °C, HRT 15 days, bioaugmented fungus volume 25% v/v in startup inoculum	Bioaugmentation enhanced average methane yield by 38%; reduction in cellulose and hemicellulose were 82% and 97%, respectively	Bioaugmentation using fungus *Orpinomyces joyonii* can be an environment friendly method for enhancing biomethanation	(Shetty et al., 2020)
Cellulose-hydrolyzing *Clostridium cellulolyticum* and *Clostridium cellulovorans*; acetogens *Clostridium aceticum* and *Mesotoga infera*; methanogens *Methanosarcina barkeri* and *Methanosaeta concilii*	*Axonopus compressus* cowgrass	0.5–1 L glass reactors	37 °C, 200 rpm, single; double and triple bioaugmentation	Bioaugmentation with mix of three kinds of microbes (hydrolyzing bacteria, acetogens, and methanogens) enhanced 21% more methane yield	Optimal bioaugmentation regime is one that includes microbes involved simultaneously hydrolysis, acetogenesis, and methanogenesis	(Lee et al., 2020a)
Acclimatized inocula from full-scale AD reactor treating manure	Wheat straw	1.2 L glass reactors	36 °C, OLR 6.5 gVS/L, sequential inoculation for three times (24 h, 48 h, and 96 h, 10 mL each time)	Inoculation for 24 h and 48 h improved hydrogen production while inoculation for 96 h enhanced methane production	Habitat-based selection approach could enhance substrate degradability for biofuel production	(Ferraro et al., 2020)

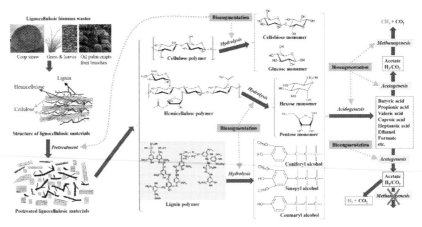

FIG. 9.5   Conversion of lignocellulosic biomass wastes to biofuels via bioaugmented anaerobic digestion using acclimatized microbial consortia.

Clostridium thermocellum (Ecem Öner et al., 2018), and lignocellulolytic microorganisms such as *Clostridium stercorarium* and *Bacteroides cellulosolvens* (Hu et al., 2016). The mixed microbial consortia include acclimatized mesophilic inocula (Ferraro et al., 2020) and thermophilic cellulolytic consortia (Strang et al., 2017) enriched from different full-scale AD reactors. In contrast, the mixed microbial consortia have higher economic feasibility than the pure strains, especially in larger scale application. Furthermore, bioaugmentation can be applied in several steps during AD of lignocellulosic substrates, including pretreatment/hydrolysis, acidogenesis, acetogenesis, and methanogenesis (Fig. 9.5).

During the AD process of lignocellulosic substrates, hydrolysis is the slowest, and is therefore regarded as the rate-determining step. To overcome the limitation in hydrolysis of lignocellulosic biomass, Poszytek et al. (2016) used cellulolytic mixed cultures with high cellulase activity in maize silage AD digester, which enhanced the biogas production by 38%. Similarly, the influence of cellulolytic *Clostridiaceae* and *Lachnospiraceae* rich consortium on anaerobic fermentation of oil palm empty fruit bunches to produce methane was investigated by Suksong et al. (2019), who demonstrated that *Lachnospiraceae*-rich consortium was suitable for enhancing two-stage AD (i.e. prehydrolysis and fermentation) with a methane yield of 0.349 L/gVS while *Clostridiaceae*-rich consortium was more suitable for enhancing single-stage AD with a methane yield of 0.217 L/gVS at a temperature of 40 °C. Indeed, both high temperature (53 °C) and bioaugmentation of thermotolerant hydrolytic microbes (e.g. *C. thermocellum*) in the acclimatized consortium favored the generation of VFA by increasing the degradation efficiency of lignocellulosic components, thereby leading to an increase in methane yield (Tsapekos et al., 2017). Additionally, it was found that dominant species were not the only key factors in thermophilic bioaugmented AD digesters; the members of the acclimatized consortium in low abundance also had a significant role in enhancing methane yields (Strang et al., 2017). The diversity of the archaeal communities in the bioaugmented digesters can be greatly affected by the metabolic products of bioaugmented strains (Ecem Öner et al., 2018). Nevertheless, results in some other studies showed that bioaugmentation strategies did not change significantly indigenous microbial communities (Tsapekos et al., 2017), which could be attributed to environmental stresses stemming from different operational conditions (e.g., pH) and the competition for nutrients between bioaugmented species and the indigenous microbes.

In addition, although thermophilic AD was reported to be advantageous in the degradation of lignocellulosic substrates, most of the related bioaugmented AD were operated at mesophilic conditions (35–37 °C) (Table 9.3), which could be ascribed to the fact that mesophilic AD consumed less energy to maintain constant temperature of digesters than thermophilic AD. To maximize the efficient decomposition of the recalcitrant 3-D structure (Fig. 9.5), especially at mesophilic conditions, some pretreatments can be performed prior to fermentation and methanogenesis process. Based on surveyed studies (Costa et al., 2012; Lavrič et al., 2017; Ortigueira et al., 2019; Villa Montoya et al., 2020), a roadmap for AD of lignocellulosic biomass integrated with pretreatments and bioaugmentation is proposed here. Initially, the raw lignocellulosic biomass such as crop straw can be cut into smaller pieces through simple physical pretreatment (e.g., milling). Subsequently, hydrothermal pretreatment is recommended for pretreatment of lignocellulosic biomass prior to fermentation and methanogenesis process due to the fact that hydrothermal pretreatment has been shown to be the most effective among acid, alkali, hydrothermal, and ultrasonic pretreatments (Hu et al., 2016). Indeed, hydrothermal pretreatment can decrease the degree of polymerization of cellulose, facilitate the formation of lignocellulosic biomass-degrading byproducts, and accelerate the conversion of insoluble fractions to soluble fractions (Lee & Park, 2020; Wang et al., 2018). Compared to chemical pretreatments, hydrothermal pretreatment is more environmentally friendly as it does not involve the use of chemicals. A suitable temperature for the hydrothermal pretreatment of lignocellulosic materials was determined as 150–180 °C (Lee & Park, 2020; Phuttaro et al., 2019). The inclusion of oxygen in hydrothermal techniques has also been shown to result in increased methane yield (Lee et al., 2020b). Notably, wet explosion pretreatment makes lignin more accessible for AD (Usman Khan & Kiaer Ahring, 2020). Furthermore, current bioaugmentation studies focused mainly on the technical aspects at lab-scale; the corresponding economic evaluation of each bioaugmentation approach and larger scale testing should be conducted to identify the commercial potential in full-scale application.

9.3.3 Bioaugmentation for relieving pressure from high concentration of VFA or overloaded AD

In AD systems with the aim to achieve higher biofuel production, organics overloading is one of the most typical process imbalance situations encountered, especially when the C/N ratio is unbalanced, resulting in the accumulation of a high concentration of VFA, pH drop, and an inhibition or even cessation of methanogenesis (Regueiro et al., 2015). In addition, the high levels of salt, in particular sodium (Chen et al., 2020; Liu et al., 2019), poor inoculum selection (Town & Dumonceaux, 2016), and significant variation in temperature (rapid change from mesophilic to thermophilic condition) (Zhu et al., 2018) could also induce the heavy duty AD and inhibit or stop methane production. Hitherto, bioaugmentation methods have been successfully applied to relieve pressure from high VFA or heavy duty AD using acclimatized microbial consortia (Table 9.4).

The bioaugmented consortia utilized were an acid-degrading microbial consortium (Li et al., 2017) and acid-tolerant methanogenic consortium (Li et al., 2018c) that can decrease acid concentrations and convert the VFA to methane, respectively, thereby relieving the high

TABLE 9.4 Bioaugmentation for relieving pressure from high VFA or heavy duty AD.

Bioaugmented cultures	Substrate	Reactor	Experimental conditions	Performance	Key findings	References
Genus *Methanosarcina*	CH_3COOH	0.1 L reactors	52 °C, inoculation ratio 1:10 (v/v)	Gradually adapted inocula for 4 weeks led to a significant increase of *Methanosarcina* spp. during the subsequent fermentation	*Methanosarcina* spp. was an essential driving factor for acetate degradation	(Lins et al., 2014)
Anaerobic and aerated, methanogenic propionate cultures	Synthetic wastewater	0.16 L serum bottles	35 °C, HRT 10 days, overloaded digester OLR 32 g COD/L/d, bioaugmentation dosage 1.5% on VS basis	Digesters bioaugmented with moderately aerated and anaerobic cultures recovered 100 and 25 days before control digesters	Moderately aerated, methanogenic propionate enrichment cultures could be more beneficial for bioaugmentation in overloaded digesters	(Tale et al., 2015)
Acetate-catabolizing methanogenic consortium	Thin stillage	0.1 L glass reactors	55 °C, pH 7.1, 1 mL of bioaugmentation consortium	Bioaugmentation had significantly reduced acetate accumulation and the methane content increased from 0.2% to 74.4%	Acetoclastic methanogen phylogenetically similar to *Methanosarcina sp.* increased more than 100-fold, which contributed to reactor recovery	(Town & Dumonceaux, 2016)
Enriched methanogenic propionate degrading microbial consortia	Sodium propionate	1.5 L reactors	36 °C, 100 rpm, HRT 15 days, dosage of 0.3–0.6 g dry cell weight/L/d	Propionic acid degradation rate was enhanced by 51%	Bioaugmentation promoted significantly the abundance of *Methanosaetaceae*, which contributed to methane production	(Li et al., 2017)
Enriched methanogenic culture	Artificial feedstock containing rice flour, whole egg powder, and milk powder	2 L reactors	OLR 0.5–2.0 g VS/L/d, bioaugmented volume 10 mL/d	Routine bioaugmentation accelerated the degradation of acetate and methane production	Consecutive microbial consortium addition could help reconstruct the methanogens community	(Li et al., 2018b)

Bioaugmented cultures	Substrate	Reactor	Experimental conditions	Performance	Key findings	References
Acid-tolerant methanogenic enrichment	Artificial food waste containing rice flour, whole egg powder, and milk powder	1–2 L CSTR reactors	OLR 0.5–2.0 gVS/L/d, bioaugmented volume of 20 mL/d	Routine bioaugmentation recovered effectively the failing digester by degrading accumulated VFA	Bioaugmentation could enhance the function of genes related to cell motility, signal transduction, energy production, and conversion	(Li et al., 2018c)
A specific axenic thermophilic methanogenic culture	Potato juice	1.2–1.4 L reactors	37 °C, 55 °C, OLR 3–4 gVS/L/d, HRT 7–8 days, bioaugmentation volume of 40 ml for 6 days	Bioaugmentation of axenic methanogenic cultures enhanced the methane production rate by 40%	Bioaugmentation could be an efficient method to accelerate transition of AD digesters from mesophilic to thermophilic conditions	(Zhu et al., 2018)
Photosynthetic bacteria (*Ectothiorhodospira*)	Glucose	0.5 L reactors	37 °C, OLR 0.33–4.76 gCOD/L/d, bioaugmentation volume 30 mL per time	Bioaugmentation enhanced methane content by 82%–85%	Bioaugmentation effectively relieved VFA accumulation and stimulate methane production	(Zhao et al., 2020)

concentration of VFA and recovering the overloaded digesters. Reportedly, acetate-catabolizing and propionate-degrading methanogenic consortia were the most representative acid-degrading microbes (Table 9.4), which could be ascribed to the fact that acetic and propionic acid are two of the most dominant VFA components in AD digesters (Zhang et al., 2020e). For instance, bioaugmentation with an acetate-catabolizing methanogenic consortium in overloaded thermophilic digesters fed with thin stillage had significantly reduced acetate accumulation and enhanced the methane content from 0.2 to 74.4% (Town & Dumonceaux, 2016). The results also showed that an acetoclastic methanogen phylogenetically similar to *Methanosarcina* sp. increased more than 100-fold, which contributed to reactor recovery in the bioaugmented digesters. Similarly, Li et al. (2018a) obtained an improved AD of grass at an organic loading rate (OLR) of 4.0 gVS/L/d by bioaugmentation with methanogenic propionate–utilizing enrichment, and confirmed the dominance of both protein- and amino acid–utilizing bacteria such as *Proteiniphilum*, *Thermovirga*, and *Lutaonella*, syntrophic propionate and butyrate oxidization bacteria such as *Syntrophomonas*, *Syntrophobacter*, and *Syntrophorhabdus*, and methanogens such as *Methanosarcina* and *Methanocella* in the bioaugmented digester. Indeed, *Methanosarcina* spp. was more robust toward various impairments than other methanogens due to the fact that it can convert both acetate and hydrogen/carbon dioxide into methane through both the acetoclastic and hydrogenotrophic pathways, respectively (Lin et al., 2018; Lin et al., 2017; Zhang et al., 2020d). Thus, *Methanosarcina* species can still grow, reproduce, and generate methane at relatively high organic loading rates, low retention times, and high ammonia inhibition (De Vrieze et al., 2012). Hence, *Methanosarcina* was regarded as a key microorganism in recovering reactor overloading, and was frequently identified in bioaugmented digesters. Lins et al. (2014) demonstrated that the acclimatized inoculum after gradual adaptation to 0.15 mol/L CH₃COONa for 4–6 weeks had great potential in relieving high acetic acid loadings during the vulnerable start-up phase of a thermophilic AD process and that the enriched *Methanosarcina* was an essential driving factor for acetate degradation. In addition, it was reported that *Methanosarcina* and *Methanosaeta* were also enriched after acclimation to stepwise increasing NaCl concentration from 1 g/L to 16 g/L for 70 days, which indicated that the acclimatized microbial consortia enriched by both *Methanosarcina* and *Methanosaeta* could be potentially applied to overloaded AD digesters under high-salinity stress for improved methane production (Zhang et al., 2020a). However, given that the available digestate in a full-scale biogas plant is acclimatized to one specific temperature (mesophilic or thermophilic), it could be a challenging but important issue to quickly adapt the inocula to a disparate desired temperature due to the possible drop in biogas yield and significant VFA accumulation (Lim et al., 2020a). To mitigate this limitation, Zhu et al. (2018) evaluated three different bioaugmentation strategies (i.e. thermophilic adapted granules, 40 mL AD digestate daily, and 40 mL pure culture daily) in converting upflow sludge blanket reactors operating temperature from mesophilic to thermophilic, and demonstrated that bioaugmentation of pure methanogenic culture could be an efficient method to accelerate transition of AD digesters from mesophilic to thermophilic conditions. Meanwhile, the results of microbial analysis in bioaugmented digesters revealed that exogenous hydrogenotrophic methanogens could be protected by the encapsulated granular structures and concomitantly promote microbial growth (Zhu et al., 2018).

Aforementioned studies confirmed the potential of bioaugmentation of acclimatized consortia in relieving pressure in overloaded AD digesters; however, the interactions between the bioaugmented microbes and indigenous species remains unclear, and more comparative studies via bioinformatic analyses of microbial communities in successful and failed bioaugmented digesters are warranted. For instance, understanding the partnership between *Methanosarcina* sp. and the SAOB could be beneficial to the enhancement of systems integrated traditional AD and bio-electrochemical systems (e.g., microbial electrolysis cells) (Beegle & Borole, 2018; De Vrieze et al., 2018; Yu et al., 2018). Furthermore, the aforementioned bioaugmentation studies mainly focused on lab-scale AD operations in a short period (< 1 year), thereby longer term system performance in the larger scale bioaugmented AD systems should be investigated to provide reliable supporting information for potential implementation in full-scale anaerobic digesters. A higher dosage of bioaugmentation could reduce the recovery time of severely overloaded AD (Li et al., 2017). Nevertheless, it might be difficult to control and quantitatively augment the optimal numbers of functional microbes, due to limitations in estimating the population of active microbes in a complex microbial consortium. Hence, efficient and economically viable approaches to "quantify" the microbial consortia warrants in depth exploration.

9.3.4 Bioaugmentation for enhancing AD for biohydrogen production

Biohydrogen can be produced through a microbial process entitled AD dark fermentation using many available biomass substrates such as lignocellulosic biomass that do not compete with food supply (Lazaro & Hallenbeck, 2019). As hydrogen-producing microorganisms play an essential role in the formation of hydrogen, bioaugmentation with the appropriate bacteria could be a promising method to enhance the hydrogen yield (Kumar et al., 2016). Thus far, the popular hydrogen-producing bacteria include cocultures of *C. acetobutylicum* X_9 and *Ethanoigenens harbinense* B_{49} (Wang et al., 2008), pure cultures of *Enterobacter cloacae* (Acs et al., 2015; Kotay & Das, 2009), *T. thermosaccharolyticum* (Zhang et al., 2019a), *Escherichia coli* (Kumar et al., 2015), *Cupriavidus necator* (Poirier et al., 2020), *Bacillus subtilis* and *Enterobacter aerogenes* (Sharma & Melkania, 2018), *Bacillus cereus* and *Enterococcus faecalis* (Eder et al., 2020), as well as *Bacillus coagulans* IIT-BT S1 and *Citrobacter freundii* IIT-BT L139 (Kotay & Das, 2009). An excellent review on enhancing biohydrogen production through microbial augmentation in dark fermentation had been published in 2016 (Kumar et al., 2016). Therefore, this book chapter mainly focuses on the novel proceedings on bioaugmented biohydrogen production reported in the ensuing years. Table 9.5 presents the bioaugmentation studies for enhancing AD for hydrogen production.

Recent studies on bioaugmentation for hydrogen production have focused on optimization of dosage, mixed cultures augmentation, and efficient utilization of lignocellulosic substrates. Augmentation with too low a dosage could fail due to competition for nutrients between the bioaugmented species and the indigenous microbial population, while too high a dosage would increase the operational cost and impede the practical application of the technology (Jiang et al., 2020). To identify an optimal bioaugmentation dosage, Sharma & Melkania (2018) investigated the effects of bioaugmentation with hydrogen-producing bacteria at different dosages ranging from 5% to 40% (volume basis) on hydrogen production, and demonstrated that 20% could be an optimal dosage for the bioaugmented AD (32.9-37.1 mL H_2/g-Carbo). Although the obtained "optimal dosage" is probably not applicable to all the other bioaugmented studies, it is useful as a reference to begin optimizing

TABLE 9.5 Bioaugmentation studies for enhancing AD for hydrogen production.

Bioaugmented cultures	Substrate	Reactor	Experimental conditions	Performance	Key findings	References
Escherichia coli, Bacillus subtilis, and *Enterobacter aerogenes*	Organic fraction of municipal solid waste	0.5 L reactors	37 °C, pH 5.5, bioaugmented dosage of 5%–40%	Bioaugmentation with *Bacillus subtilis* showed the highest hydrogen yield (i.e., 43.7 mL H_2/ gCarbon)	Selecting appropriate bacteria is a promising method to enhance biohydrogen production	(Sharma & Melkania, 2018)
Thermoanaerobacterium thermosaccharolyticum W16	Corn stover hydrolysate	0.1 L reactors	55 °C, 120 rpm, pH 7.0, bioaugmented dosage 5%	Bioaugmentation enhanced production of hydrogen (from 8.50 to 9.16 mmol/g-sugar) and acetate	Bioaugmentation was an efficient method to enhance thermophilic hydrogen production from lignocellulosic hydrolysate	(Zhang et al., 2019a)
Thermotoga neapolitana	Glucose	2 L reactors	70 °C, pH 6.5, 150 rpm,	Hydrogen production rate of the bioaugmented digester increased with glucose concentration from 5.6 to 55.5 mmol/L	Preadaptation of bioaugmented species via three successive batch incubations aided their coexistence with indigenous species	(Okonkwo et al., 2020)
Acclimatized microbial consortium in batch reactors	Coffee-processing waste (lignocellulosic material)	0.1 L reactors	35 °C, pH 6.0, bioaugmentation with 10% (v/v) of the microbial consortium	Bioaugmentation microorganisms fermented lignocellulose into hydrogen, ethanol, acetic, and butyric acids	Bioaugmentation can be integrated with other enhancing strategies (e.g., codigestion and hydrothermal pretreatment) for better performance	(Villa Montoya et al., 2020)
Bacillus cereus, Enterococcus faecalis, and *Enterobacter aerogenes*	Sugarcane vinasse	0.06 L glass reactors	37 °C, pH 6.0, bioaugmented volume of 10% (v/v)	Bioaugmentation with *Bacillus cereus* showed an excellent performance	Validated the potential of the isolated strains *B. cereus* and *E. faecalis* in hydrogen production	(Eder et al., 2020)
Clostridium acetobutylicum, Clostridium pasteurianum, E. coli; Lactobacillus bulgaris, Enterococcus casseliflavus, Desulfovibrio vulgaris, and *Cupriavidus necator*	Growth medium (e.g., yeast extract, glucose, etc.)	0.6 L reactors	37 °C, batch mode, bioaugmented dosage at 0.1% of the total biomass of the mixed culture	Bioaugmentation with *C. acetobutylicum, C. pasteurianum,* and *L. bulgaris* induced trophic competition with indigenous species	The species (e.g., *E. coli* and *C. necator*) with minor abundance act as keystone species on biohydrogen production	(Poirier et al., 2020)

the dosage. Furthermore, bioaugmentation studies using mixed cultures or cocultures indicated that it could induce a stronger trophic competition, leading to an unstable hydrogen production. Reportedly, the main liquid metabolite in bioaugmented digesters with cocultures was acetic acid, while the bioaugmented digesters with single species augmentation generated butyric acid (Eder et al., 2020). As such, bioaugmentation with different strains could induce different metabolic pathways such as pyruvate formate lyase pathway or pyruvate:ferredoxin oxidoreductase pathway (Lazaro & Hallenbeck, 2019). However, the reported studies on the bioaugmentation mechanisms at the level of metabolic pathways are very limited due to the complex hydrogenases-catalyzed hydrogen production routes in dark fermentation. Therefore, more microbial community analysis at the levels of genes and functional enzymes can be performed in the bioaugmented digesters to provide deeper insights into enhancing mechanisms of the dominant hydrogen-producing bacteria. Notably, in addition to the predominant bacterial components, some species (e.g., *E. coli* or *C. necator*) with minor abundance also played a vital role in maintaining a stable and high hydrogen production while reducing the metabolic variability through more diverse metabolic pathways (Poirier et al., 2020). Regarding hydrogen production through AD of lignocellulosic substrates, the combined technology that integrates microbial bioaugmentation and hydrothermal pretreatment holds great potential due to the fact that the substrate pretreatment could produce more biodegradable intermediates (e.g. monosaccharides and polysaccharides) while bioaugmentation could facilitate the subsequent fermentation of these intermediates for the production of hydrogen, acetic, and butyric acids (Villa Montoya et al., 2020). Currently, bioaugmented digesters for biohydrogen production are at lab scale (Table 9.5), therefore further experiments utilizing larger scale bioreactor systems are recommended to assess the economic feasibility of the bioaugmented hydrogen fermentation processes.

Hybrid two-stage or three-stage systems are likely to be at the forefront of the development of dark fermentation biofuel systems (Baeyens et al., 2020; Fu et al., 2020; Tian et al., 2019a; Yue et al., 2020; Zhang et al., 2020f), which could be promising technologies for integrating biohydrogen production from the biomass substrates after simple pretreatments (e.g. milling, enzymatic, or hydrothermal). Fig. 9.6 shows the schematic diagrams of different hybrid fermentation systems employing bioaugmentation strategies for biohydrogen production. For the two-stage systems, the dark fermenter in the first stage can be coupled with a biogas digester (Fig. 9.6A) for coproduction of hydrogen and methane, or a microbial electrolysis cell (Fig. 9.6B) for higher hydrogen production. In three-stage systems (Fig. 9.7), the VFA-rich effluent discharged from the first-stage dark fermenter can be centrifuged for solid-liquid separation. Following that, the supernatant can be utilized in a photo fermenter for hydrogen production while the obtained solid phase can be further anaerobically digested in an AD digester for methane production. The discharged digestate from the photo fermenter can also be injected into the AD digester for more energy extraction from the biomass feedstock. All the aforementioned stages are microbial catalyzed processes, thereby allowing enhancement by bioaugmentation strategies using hydrogen-producing bacteria (Lertsriwong & Glinwong, 2020), acclimatized methanogenic consortium (Tian et al., 2019c), photo fermentative bacteria (Cai et al., 2019), or enriched mixed electroactive culture (Chiranjeevi & Patil, 2020), to increase the overall energy production in terms of hydrogen or hydrogen/methane mixture.

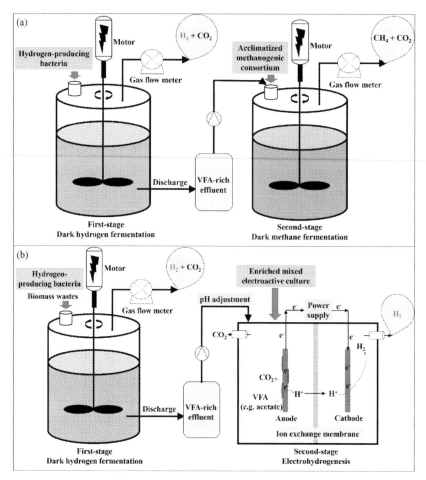

FIG. 9.6   Schematic diagrams of different hybrid two-stage fermentation systems employing bioaugmentation strategies for biohydrogen production: (A) two-stage system combining dark hydrogen fermentation and dark methane fermentation, (B) two-stage system combining dark hydrogen fermentation and electrohydrogenesis. Modified based on (Ding et al., 2020; Varanasi & Das, 2020) with permission/license granted by the publisher.

Among the aforementioned three types of hybrid system, the two-stage system that integrated the dark fermenter with an anaerobic digester is at the highest technology readiness level with several pilot-scale demonstration plants (Balachandar et al., 2020; Seengenyoung et al., 2019). Regarding other two scenarios (Figs. 9.6B and 9.7), some barriers such as low conversion efficiency and high operating costs should be overcome first before they can be employed at a practical level. The economic feasibility, as one of the most key issues, could be increased through bioaugmentation strategies using acclimatized microbial consortia. Nevertheless, the hybrid systems with bioaugmentation strategies are currently under development and require more research to investigate mechanisms of enhancement, optimize operating parameters, cultivate appropriate bioaugmented consortia, improve overall efficiency, and scale-up the systems.

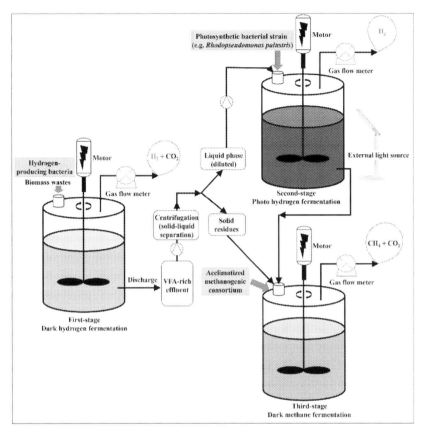

FIG. 9.7   **Schematic diagram of hybrid three-stage fermentation system employing bioaugmentation strategies for biohydrogen production.** The process covers dark hydrogen fermentation, photo hydrogen fermentation, and dark methane fermentation. Modified based on (Ding et al., 2017) with permission/license granted by the publisher.

9.3.5   Bioaugmentation for enhancing biodegradation of pollutants (e.g. plastics) in AD digesters

Among the organic fraction of municipal solid wastes, there is a small fraction of contaminants such as plastics (Zhang et al., 2019d), pharmaceuticals (Gonzalez-Salgado et al., 2020), and skatole (Ma et al., 2020) that can potentially arise in operational issues in anaerobic digesters. For instance, plastics are widely used in food packaging and likely to be included inadvertently as a contaminant in the source-segregated food waste streams, which are usually targeted to be treated in AD plants (Zhang et al., 2018b). The commonly presented microplastics (polyethylene terephthalate, polyethylene, polypropylene, and polystyrene) in wastewater could enter sludge and act as the vector for toxic substances such as antibiotics and persistent organic pollutants, affecting the microbial communities in the AD digester (Zhang et al., 2020g). Yet, the traditional microbial-driven AD plants are not specifically designed for the biodegradation of these pollutants; hence the need for bioaugmentation strategies to enhance biodegradation of pollutants in AD digesters.

Currently, bioaugmentation technology for biodegradation of these pollutants is still in its infancy. Even so, the AD performance of biomass wastes containing the above-mentioned pollutants and the potential pollutant-degradation strains reported in recent years may have laid a foundation for the related bioaugmentation studies. For instance, Li et al. (2020) investigated the effect of microplastic on AD of wasted activated sludge and found that microplastic at a dosage of 1000–200,000 polyester particle per kilogram activated sludge reduced the methane production by 4.9%–11.5% compared to the control. Notably, microbial community structures showed no significant difference with and without microplastics (Li et al., 2020). Similarly, the impact of polystyrene micro- and nanoplastics on the methane generation during AD of organic wastewater was studied by Zhang et al. (2020b), who demonstrated that the 80 nm and 5 μm polystyrene micro-plastics at the concentrations of 0.2 g/L or lower had no significant effect on the cumulative methane production. However, the same microplastics at a higher concentration (i.e., 0.25 g/L) led to a decrease in methane production by 17.9%–19.3%, which might be ascribed to the severe inhibition on the methanogenesis process caused by the polystyrene micro/nanoplastics (Zhang et al., 2020b). Reportedly, 200 and 2000 μg/L polystyrene nanoplastics can cause growth inhibition, morphological damage, and physiological disturbance in marine microalgae cells (Wang et al., 2020a). Taken together, micro/nanoplastics at a sufficiently high level can lead to negative effects on the anaerobic microorganisms and biofuel production. Regarding the potential pollutant-degradation strains for bioaugmentation, Ma et al. (2020) recently confirmed the feasibility of *Burkholderia* sp. IDO3 as a bioaugmentation agent in the sludge system to efficiently degrade skatole, a malodorous odorous pollutant. Roager & Sonnenschein (2019) suggested several bacterial candidates such as *Kocuria palustris* M16, *Rhodococcus* sp. 36, *B. cereus* BF20, *B. cereus*, *Bacillus gottheilii*, *Bacillus pumilus* M27, *Bacillus* sp. 27, *Bacillus sphaericus* Alt, and *B. subtilis* H1584 for biodegradation of plastic debris. In a biodegradation study of plastics by Syranidou et al. (2019), the molecular weight of polystyrene pieces decreased by 27%–33% by enriched genera *Bacillus* and *Pseudonocardia*, which indicated their ability to degrade polystyrene. Yet, the observed biodegradation capacity is very limited so far and needs to be enhanced. More recently, how biogeochemical processes impact the microbe-plastic interactions was discussed by Rogers et al. (2020), who provided more comprehensive summaries of identified microorganisms on the plastic particles in different environments (e.g., marine, freshwater and experimental incubator) and the known plastic degraders. However, more reproducible data should be collected and analyzed to further confirm the capability of aforementioned bacterial candidates for biodegradation of plastic. Hence, considering that it might not be easy to remove the micro/nanoplastics pollution in a short term due to their wide distribution and small particle size, the concentration of micro/nanoplastics in the AD systems should be carefully controlled. Meanwhile, more research efforts should be made to search highly efficient plastic-degrading bacteria or consortia, elucidate biofilm development, investigate the process and mechanism of microbial degradation of plastics, study pretreatments-aided AD of plastics-containing biomass substrates, as well as optimize and implement plastic-degrading bacteria and their enzymes in the larger scale digesters.

9.4   Challenges and opportunities of bioaugmentation to enhance AD for biofuel production

Bioaugmentation strategies for enhancing AD for the sustainable production of biofuels have risen in popularity in the past few years. The aforementioned studies have validated the great potential of bioaugmentation strategies in mitigating ammonia inhibition, enhancing degradation of lignocellulosic biomass, recovering heavy loaded AD digesters, enhancing biohydrogen production, and enhancing biodegradation of pollutants such as plastics. However, there are some technical challenges that need to be addressed before bioaugmentation using microbial consortia is widely accepted and applied.

First, regarding bioaugmentation for mitigating ammonia inhibition, understanding the functional enzymes and genes involved in the acetate oxidation, aceticlastic, and hydrogenotrophic methanogenesis pathways (Fig. 9.4) would be helpful to elucidate the mechanism of mitigation of ammonia inhibition. A deeper analysis of microbial communities and methanogenic pathways in terms of the related genes and enzymes in ammonia-inhibited and ammonia-tolerant digesters can be conducted with the plethora of widely available sequencing techniques. It has been noted that not all bioaugmentations produce positive outcomes due to differences in the added microbes. Hence, careful selection and acclimatization of the microbial strains or consortia are essential prior to the bioaugmentation. Aceticlastic and hydrogenotrophic methanogenic pathways should both be taken into consideration during bioaugmentation as the balance of these two pathways can lead to better reactor performance. In terms of the hydrogenotrophic pathway, more SAOB species as significant partners of the methanogens should be isolated and evaluated in the future.

Second, to overcome the limitations in hydrolysis rate of lignocellulosic biomass wastes, prior to the bioaugmented AD process, hydrothermal pretreatment of the substrate is recommended due to its advantages in increasing efficiency and environmental protection. In the bioaugmentation operations for recovering heavy loaded AD digesters, *Methanosarcina* as the key microorganism for recovering overloading should be further investigated due to its high resistance to the otherwise detrimental concentrations of nutrients (high VFA) via genetic and metabolic engineering approaches. It is recommended, for instance, to dig up genes and enzymes that work against the environmental stress in *Methanosarcina*. Also, the mode of interactions between the bioaugmented microbes and indigenous flora remains unclear, and deserves more studies through bioinformatic analysis of the microbial communities.

Regarding dark fermentation, several pilot-scale demonstrations of the hybrid systems (Fig. 9.6) have demonstrated their technical feasibility to increase the overall energy conversion efficiency in terms of hydrogen or hydrogen/methane mixture. However, the relatively high operating costs reduce the economic viability of these hybrid systems, thus impeding their potential to be commercialized. Therefore, bioaugmentation strategies using acclimatized microbial consortia should be investigated. To date, hybrid systems with bioaugmentation strategies are seldom reported. Hence, more effort should be made to cultivate appropriate bioaugmentation consortia, investigate mechanisms of enhancement, and optimize the process parameters, which would be beneficial to improve the overall efficiencies and scale up the promising hybrid systems for hydrogen production.

Additionally, the current understanding of biodegradation of organic pollutants such as plastics or microplastics is still nascent, but the importance to carefully control the pollutant concentration to support a normal biofuel production process has been confirmed by the ever-increasing reports. Although some microbial candidates (see Section 9.3.5) have been reported as plastic degraders, more reproducible data should be collected and utilized to confirm their capabilities of degrading plastics or other pollutants. To go a step further, more research should be undertaken to elucidate biofilm development, analyze microbial communities in the AD systems fed with plastics-containing biomass substrates, as well as evaluate key plastic-degrading microorganisms and their metabolites in larger scale systems.

There are several limitations and opportunities common to the technologies surveyed. First, the reported bioaugmentation studies for enhancing AD processes are frequently limited to lab scale, which is not sufficient for industrial application. Hence, more pilot-scale and larger scale tests should be conducted to confirm the technical feasibility. Second, the related economic analyses of bioaugmentation strategies are currently very limited. The corresponding economic feasibility of the bioaugmentation approaches should be evaluated to justify their commercial potential in full-scale application. For instance, an economic analysis of an up-scaled (1528 m^3) biogas plant fed with food waste was recently conducted by Jiang et al. (2020), who reported that an appropriate bioaugmentation process increased the profit to about 537 US$ per day when treating capacity was 21 ton/d. Reutilization of filtered digestate containing the active microbial consortium for bioaugmentation in the AD system could simultaneously provide a comprehensive microbes for the targeted AD process and decrease the operating cost (Zhu et al., 2018), thus this multiple inoculation technique for bioaugmentation might hold great potential for industrial application. Additionally, coupling bioaugmentation strategies with other strategies such as pretreatment, multistage AD (Zhang et al., 2020c), and addition of conductive materials (Xiao et al., 2019) are also recommended. Furthermore, the long-term storage of the acclimated consortium is currently an understudied issue. To confront this issue, the development of encapsulated microbial consortium additive as a potentially commercial product for the relatively long-term storage of functional microorganisms prior to various bioaugmentation applications might be a potential opportunity. Another issue of the bioaugmentation strategies might be how to control and add the optimal numbers of functional microbes. A dosage of 5%–20% is suggested as a starting point for optimization. From the practical application perspective, the development of acclimatized microbial consortia (e.g. multifunctional cultures) that can cope with multiple inhibition issues could also be useful because a specific AD system may simultaneously encounter several inhibition issues (e.g. ammonia inhibition, VFA inhibition, and plastic pollution). Alternatively, the synergism of different functional species might be a research and development trend for future bioaugmentation technology. Penultimately, the modeling of bioaugmented AD processes and the automation of bioaugmentation operations should also be investigated, which would be beneficial to promote potential commercial application.

Lastly, to maximize the generalization abilities of various acclimatized microbes as "biocatalysts" in the bioaugmented AD systems, the genetic and metabolic engineering approaches can be utilized to investigate the related genes and enzymes related to various metabolic pathways. More specifically, with the help of the tools of molecular microbiology, synthetic biology, and predictive biology (Lopatkin & Collins, 2020; Zhao et al., 2019), more effort can be made to develop various engineered strains, with higher tolerance to elevated concentrations of ammonia or acids, or with higher fermentation efficiency of biomethane or biohydrogen, or with higher degradation ability of lignocellulosic substrates or emerging pollutants. In this case, the engineered strains would revolutionize significantly the bioaugmentation research field for biofuel production.

9.5 Conclusions and perspectives

This chapter has demonstrated the great potential of bioaugmentation technologies in mitigating ammonia inhibition, enhancing degradation of lignocellulosic biomass, recovering heavy loaded AD digesters, enhancing biohydrogen production, and enhancing biodegradation of pollutants (e.g., plastics). In order for bioaugmentation technologies to be highly reliable and cost-effective, technical limitations of the aforementioned specific processes have to be satisfactorily addressed. A number of promising technical highlights have been identified, including repeated bioaugmentation, functional genes and enzymes, balancing aceticlastic and hydrogenotrophic methanogenic pathways, pretreatments-coupled bioaugmentation, hybrid hydrogen fermentation systems, acid-resistant *Methanosarcina*, and plastic-degrading microbial candidates. To broaden their application in AD of biomass wastes to biofuels, more experimental data, especially of the pilot- and larger scale digesters, should be collected to evaluate their technical and economic feasibilities and environmental sustainability. Moreover, genetic engineering and metabolic approaches, high-throughput sequencing and bioinformatic technologies, as well as bioreactor and process engineering should be taken into consideration to systematically optimize the bioaugmentation strategies for enhancing biofuel production from biomass wastes.

Acknowledgments

This book chapter was funded by the National Research Foundation, Prime Minister's Office, Singapore under its Campus for Research Excellence and Technological Enterprise (CREATE) Program. For all copyrighted figures including Fig. 9.4, Figs. 9.6 and 9.7, the permission attained from the publishers for adaptation and reuse in the present work has been acknowledged.

References

Acs, N., Bagi, Z., Rakhely, G., Minarovics, J., Nagy, K., Kovacs, K.L., 2015. Bioaugmentation of biogas production by a hydrogen-producing bacterium. Bioresour. Technol. 186, 286–293.

Ajay, C., Mohan, S., Dinesha, P., Rosen, M.A., 2020. Review of impact of nanoparticle additives on anaerobic digestion and methane generation. Fuel 277, 118234.

Akyol, Ç., Ince, O., Bozan, M., Ozbayram, E.G., Ince, B., 2019. Fungal bioaugmentation of anaerobic digesters fed with lignocellulosic biomass: what to expect from anaerobic fungus *Orpinomyces sp*. Bioresour. Technol. 277, 1–10.

Amha, Y.M., Anwar, M.Z., Brower, A., Jacobsen, C.S., Stadler, L.B., Webster, T.M., Smith, A.L., 2018. Inhibition of anaerobic digestion processes: applications of molecular tools. Bioresour. Technol. 247, 999–1014.

Atasoy, M., Owusu-Agyeman, I., Plaza, E., Cetecioglu, Z., 2018. Bio-based volatile fatty acid production and recovery from waste streams: current status and future challenges. Bioresour. Technol. 268, 773–786.

Aydin, S., Yıldırım, E., Ince, O., Ince, B., 2017. Rumen anaerobic fungi create new opportunities for enhanced methane production from microalgae biomass. Algal Res 23, 150–160.

Baeyens, J., Zhang, H., Nie, J., Appels, L., Dewil, R., Ansart, R., Deng, Y., 2020. Reviewing the potential of bio-hydrogen production by fermentation. Renew. Sustain. Energy Rev. 131, 110023.

Balachandar, G., Varanasi, J.L., Singh, V., Singh, H., Das, D., 2020. Biological hydrogen production via dark fermentation: a holistic approach from lab-scale to pilot-scale. Int. J. Hydrog. Energy 45 (8), 5202–5215.

Bedoya, K., Coltell, O., Cabarcas, F., Alzate, J.F., 2019. Metagenomic assessment of the microbial community and methanogenic pathways in biosolids from a municipal wastewater treatment plant in Medellín, Colombia. Sci. Total Environ. 648, 572–581.

Beegle, J.R., Borole, A.P., 2018. Energy production from waste: Evaluation of anaerobic digestion and bioelectrochemical systems based on energy efficiency and economic factors. Renew. Sustain. Energy Rev. 96, 343–351.

Cai, J., Zhao, Y., Fan, J., Li, F., Feng, C., Guan, Y., Wang, R., Tang, N., 2019. Photosynthetic bacteria improved hydrogen yield of combined dark- and photo-fermentation. J. Biotechnol. 302, 18–25.

Campanaro, S., Treu, L., Rodriguez-R, L.M., Kovalovszki, A., Ziels, R.M., Maus, I., Zhu, X.Y., Kougias, P.G., Basile, A., Luo, G., Schluter, A., Konstantinidis, K.T., Angelidaki, I., 2020. New insights from the biogas microbiome by comprehensive genome-resolved metagenomics of nearly 1600 species originating from multiple anaerobic digesters. Biotechnol. Biofuels 13 (1), 25.

Capson-Tojo, G., Moscoviz, R., Astals, S., Robles, Á., Steyer, J.-P., 2020. Unraveling the literature chaos around free ammonia inhibition in anaerobic digestion. Renew. Sustain. Energy Rev. 117, 109487.

Čater, M., Fanedl, L., Malovrh, Š., Marinšek Logar, R., 2015. Biogas production from brewery spent grain enhanced by bioaugmentation with hydrolytic anaerobic bacteria. Bioresour. Technol. 186, 261–269.

Chang, S., Zhang, Z., Cao, L., Ma, L., You, S., Li, W., 2020. Co-gasification of digestate and lignite in a downdraft fixed bed gasifier: effect of temperature. Energy Convers. Manage. 213, 112798.

Chen, Q., Liu, C., Liu, X., Sun, D., Li, P., Qiu, B., Dang, Y., Karpinski, N.A., Smith, J.A., Holmes, D.E., 2020. Magnetite enhances anaerobic digestion of high salinity organic wastewater. Environ. Res. 189, 109884.

Chen, Y., Cheng, J.J., Creamer, K.S., 2008. Inhibition of anaerobic digestion process: a review. Bioresour. Technol. 99 (10), 4044–4064.

Chiranjeevi, P., Patil, S.A., 2020. Strategies for improving the electroactivity and specific metabolic functionality of microorganisms for various microbial electrochemical technologies. Biotechnol. Adv. 39, 107468.

Costa, J.C., Barbosa, S.G., Sousa, D.Z., 2012. Effects of pre-treatment and bioaugmentation strategies on the anaerobic digestion of chicken feathers. Bioresour. Technol. 120, 114–119.

Cycoń, M., Mrozik, A., Piotrowska-Seget, Z., 2017. Bioaugmentation as a strategy for the remediation of pesticide-polluted soil: a review. Chemosphere 172, 52–71.

De Vrieze, J., Arends, J.B.A., Verbeeck, K., Gildemyn, S., Rabaey, K., 2018. Interfacing anaerobic digestion with (bio) electrochemical systems: potentials and challenges. Water Res. 146, 244–255.

De Vrieze, J., Hennebel, T., Boon, N., Verstraete, W., 2012. *Methanosarcina*: the rediscovered methanogen for heavy duty biomethanation. Bioresour. Technol. 112, 1–9.

Ding, L., Cheng, J., Lin, R., Deng, C., Zhou, J., Murphy, J.D., 2020. Improving biohydrogen and biomethane co-production via two-stage dark fermentation and anaerobic digestion of the pretreated seaweed *Laminaria digitata*. J. Clean. Prod. 251, 119666.

Ding, L., Cheng, J., Lu, H., Yue, L., Zhou, J., Cen, K., 2017. Three-stage gaseous biofuel production combining dark hydrogen, photo hydrogen, and methane fermentation using wet *Arthrospira platensis* cultivated under high CO_2 and sodium stress. Energy Convers. Manage. 148, 394–404.

Ecem Öner, B., Akyol, Ç., Bozan, M., Ince, O., Aydin, S., Ince, B., 2018. Bioaugmentation with *Clostridium thermocellum* to enhance the anaerobic biodegradation of lignocellulosic agricultural residues. Bioresour. Technol. 249, 620–625.

Eder, A.S., Magrini, F.E., Spengler, A., da Silva, J.T., Beal, L.L., Paesi, S., 2020. Comparison of hydrogen and volatile fatty acid production by *Bacillus cereus*, *Enterococcus faecalis* and *Enterobacter aerogenes* singly, in co-cultures or in the bioaugmentation of microbial consortium from sugarcane vinasse. Environ. Technol. Inno. 18, 100638.

El Fantroussi, S., Agathos, S.N., 2005. Is bioaugmentation a feasible strategy for pollutant removal and site remediation? Curr. Opin. Microbiol. 8 (3), 268–275.

Ferraro, A., Dottorini, G., Massini, G., Mazzurco Miritana, V., Signorini, A., Lembo, G., Fabbricino, M., 2018. Combined bioaugmentation with anaerobic ruminal fungi and fermentative bacteria to enhance biogas production from wheat straw and mushroom spent straw. Bioresour. Technol. 260, 364–373.

Ferraro, A., Massini, G., Mazzurco Miritana, V., Rosa, S., Signorini, A., Fabbricino, M., 2020. A novel enrichment approach for anaerobic digestion of lignocellulosic biomass: process performance enhancement through an inoculum habitat selection. Bioresour. Technol. 313, 123703.

Ferraro, A., Massini, G., Mazzurco Miritana, V., Signorini, A., Race, M., Fabbricino, M., 2019. A simplified model to simulate bioaugmented anaerobic digestion of lignocellulosic biomass: biogas production efficiency related to microbiological data. Sci. Total Environ. 691, 885–895.

Fotidis, I.A., Karakashev, D., Angelidaki, I., 2013. Bioaugmentation with an acetate-oxidising consortium as a tool to tackle ammonia inhibition of anaerobic digestion. Bioresour. Technol. 146, 57–62.

Fotidis, I.A., Treu, L., Angelidaki, I., 2017. Enriched ammonia-tolerant methanogenic cultures as bioaugmentation inocula in continuous biomethanation processes. J. Clean. Prod. 166, 1305–1313.

Fotidis, I.A., Wang, H., Fiedel, N.R., Luo, G., Karakashev, D.B., Angelidaki, I., 2014. Bioaugmentation as a solution to increase methane production from an ammonia-rich substrate. Environ. Sci. Technol. 48 (13), 7669–7676.

Fu, S.-F., Liu, R., Sun, W.-X., Zhu, R., Zou, H., Zheng, Y., Wang, Z.-Y., 2020. Enhancing energy recovery from corn straw via two-stage anaerobic digestion with stepwise microaerobic hydrogen fermentation and methanogenesis. J. Clean. Prod. 247, 119651.

Gonzalez-Salgado, I., Cavaillé, L., Dubos, S., Mengelle, E., Kim, C., Bounouba, M., Paul, E., Pommier, S., Bessiere, Y., 2020. Combining thermophilic aerobic reactor (TAR) with mesophilic anaerobic digestion (MAD) improves the degradation of pharmaceutical compounds. Water Res. 182, 116033.

He, P., Zhang, H., Duan, H., Shao, L., Lü, F., 2020. Continuity of biochar-associated biofilm in anaerobic digestion. Chem. Eng. J. 390, 124605.

Herrero, M., Stuckey, D., 2015. Bioaugmentation and its application in wastewater treatment: a review. Chemosphere 140, 119–128.

Hosseini Koupaie, E., Azizi, A., Bazyar Lakeh, A.A., Hafez, H., Elbeshbishy, E., 2019. Comparison of liquid and dewatered digestate as inoculum for anaerobic digestion of organic solid wastes. Waste Manage. 87, 228–236.

Hu, Y., Hao, X., Wang, J., Cao, Y., 2016. Enhancing anaerobic digestion of lignocellulosic materials in excess sludge by bioaugmentation and pre-treatment. Waste Manage. 49, 55–63.

Jiang, J., Li, L., Li, Y., He, Y., Wang, C., Sun, Y., 2020. Bioaugmentation to enhance anaerobic digestion of food waste: dosage, frequency and economic analysis. Bioresour. Technol. 307, 123256.

Jiang, Y., McAdam, E., Zhang, Y., Heaven, S., Banks, C., Longhurst, P., 2019. Ammonia inhibition and toxicity in anaerobic digestion: a critical review. J. Water Process Eng. 32, 100899.

Khan, M., Ngo, H.H., Guo, W., Liu, Y., Nghiem, L.D., Hai, F.I., Deng, L., Wang, J., Wu, Y., 2016. Optimization of process parameters for production of volatile fatty acid, biohydrogen and methane from anaerobic digestion. Bioresour. Technol. 219, 738–748.

Khan, M.U., Ahring, B.K., 2019. Lignin degradation under anaerobic digestion: influence of lignin modifications—a review. Biomass Bioenergy 128, 105325.

Kim, J.-H., Oh, J.-I., Tsang, Y.F., Park, Y.-K., Lee, J., Kwon, E.E., 2020. CO_2-assisted catalytic pyrolysis of digestate with steel slag. Energy 191, 116529.

Kotay, S.M., Das, D., 2009. Novel dark fermentation involving bioaugmentation with constructed bacterial consortium for enhanced biohydrogen production from pretreated sewage sludge. Int. J. Hydrog. Energy 34 (17), 7489–7496.

Kougias, P.G., Angelidaki, I., 2018. Biogas and its opportunities—a review. Front. Environ. Sci. Eng. 12 (3), 14.

Kumar, G., Bakonyi, P., Kobayashi, T., Xu, K.-Q., Sivagurunathan, P., Kim, S.-H., Buitrón, G., Nemestóthy, N., Bélafi-Bakó, K., 2016. Enhancement of biofuel production via microbial augmentation: the case of dark fermentative hydrogen. Renew. Sustain. Energy Rev. 57, 879–891.

Kumar, G., Bakonyi, P., Sivagurunathan, P., Kim, S.-H., Nemestóthy, N., Bélafi-Bakó, K., Lin, C.-Y., 2015. Enhanced biohydrogen production from beverage industrial wastewater using external nitrogen sources and bioaugmentation with facultative anaerobic strains. J. Biosci. Bioeng. 120 (2), 155–160.

Lavrič, L., Cerar, A., Fanedl, L., Lazar, B., Žitnik, M., Logar, R.M., 2017. Thermal pretreatment and bioaugmentation improve methane yield of microalgal mix produced in thermophilic anaerobic digestate. Anaerobe 46, 162–169.

Lazaro, C.Z., Hallenbeck, P.C., 2019. Fundamentals of biohydrogen production. In: Pandey, A., Mohan, S.V., Chang, J.S., Hallenbeck, P.C., Larroche, C. (Eds.), Biohydrogen 2nd Edn. Elsevier, Cham, pp. 25–48.

Lebeau, T., Braud, A., Jézéquel, K., 2008. Performance of bioaugmentation-assisted phytoextraction applied to metal contaminated soils: a review. Environ. Pollut. 153 (3), 497–522.

Lee, J., Park, K.Y., 2020. Impact of hydrothermal pretreatment on anaerobic digestion efficiency for lignocellulosic biomass: influence of pretreatment temperature on the formation of biomass-degrading byproducts. Chemosphere 256, 127116.

Lee, J.T., He, J., Tong, Y.W., 2017. Acclimatization of a mixed-animal manure inoculum to the anaerobic digestion of *Axonopus compressus* reveals the putative importance of *Mesotoga infera* and *Methanosaeta concilii* as elucidated by DGGE and Illumina MiSeq. Bioresour. Technol. 245, 1148–1154.

Lee, J.T., Wang, Q., Lim, E.Y., Liu, Z., He, J., Tong, Y.W., 2020a. Optimization of bioaugmentation of the anaerobic digestion of *Axonopus compressus* cowgrass for the production of biomethane. J. Clean. Prod. 258, 120932.

Lee, J.T.E., Khan, M.U., Tian, H., Ee, A.W.L., Lim, E.Y., Dai, Y., Tong, Y.W., Ahring, B.K., 2020b. Improving methane yield of oil palm empty fruit bunches by wet oxidation pretreatment: mesophilic and thermophilic anaerobic digestion conditions and the associated global warming potential effects. Energy Convers. Manage. 225, 113438.

Lertsriwong, S., Glinwong, C., 2020. Newly-isolated hydrogen-producing bacteria and biohydrogen production by *Bacillus coagulans* MO11 and *Clostridium beijerinckii* CN on molasses and agricultural wastewater. Int. J. Hydrog. Energy 45, 26812–26821.

Li, L., Geng, S., Li, Z., Song, K., 2020. Effect of microplastic on anaerobic digestion of wasted activated sludge. Chemosphere 247, 125874.

Li, L., Ying, L., Yongming, S., Zhenhong, Y., Xihui, K., Yi, Z., Gaixiu, Y., 2018a. Effect of bioaugmentation on the microbial community and mono-digestion performance of *Pennisetum hybrid*. Waste Manage. 78, 741–749.

Li, Y., Li, L., Sun, Y., Yuan, Z., 2018b. Bioaugmentation strategy for enhancing anaerobic digestion of high C/N ratio feedstock with methanogenic enrichment culture. Bioresour. Technol. 261, 188–195.

Li, Y., Yang, G., Li, L., Sun, Y., 2018c. Bioaugmentation for overloaded anaerobic digestion recovery with acid-tolerant methanogenic enrichment. Waste Manage. 79, 744–751.

Li, Y., Zhang, Y., Sun, Y., Wu, S., Kong, X., Yuan, Z., Dong, R., 2017. The performance efficiency of bioaugmentation to prevent anaerobic digestion failure from ammonia and propionate inhibition. Bioresour. Technol. 231, 94–100.

Lim, J.W., Park, T., Tong, Y.W., Yu, Z., 2020. The microbiome driving anaerobic digestion and microbial analysis. In: Li, Y., Khanal, S.K. (Eds.). Advances in Bioenergy, 5. Elsevier, Cham, pp. 1–61.

Lim, J.W., Ting, D.W.Q., Loh, K.-C., Ge, T., Tong, Y.W., 2018. Effects of disposable plastics and wooden chopsticks on the anaerobic digestion of food waste. Waste Manage. 79, 607–614.

Lim, J.W., Wong, K., S.W., D., Y., Tong, Y.W., 2020a. Effect of seed sludge source and start-up strategy on the performance and microbial communities of thermophilic anaerobic digestion of food waste. Energy 203, 117922.

Lin, R., Cheng, J., Ding, L., Murphy, J.D., 2018. Improved efficiency of anaerobic digestion through direct interspecies electron transfer at mesophilic and thermophilic temperature ranges. Chem. Eng. J. 350, 681–691.

Lin, R., Cheng, J., Zhang, J., Zhou, J., Cen, K., Murphy, J.D., 2017. Boosting biomethane yield and production rate with graphene: the potential of direct interspecies electron transfer in anaerobic digestion. Bioresour. Technol. 239, 345–352.

Lins, P., Reitschuler, C., Illmer, P., 2014. *Methanosarcina spp.*, the key to relieve the start-up of a thermophilic anaerobic digestion suffering from high acetic acid loads. Bioresour. Technol. 152, 347–354.

Liu, J., Huang, S., Chen, K., Wang, T., Mei, M., Li, J., 2020. Preparation of biochar from food waste digestate: pyrolysis behavior and product properties. Bioresour. Technol. 302, 122841.

Liu, Y., Yuan, Y., Wang, W., Wachemo, A.C., Zou, D., 2019. Effects of adding osmoprotectant on anaerobic digestion of kitchen waste with high level of salinity. J. Biosci. Bioeng. 128 (6), 723–732.

Lopatkin, A.J., Collins, J.J., 2020. Predictive biology: modelling, understanding and harnessing microbial complexity. Nat. Rev. Microbiol. 18 (9), 507–520.

Ma, Q., Qu, H., Meng, N., Li, S., Wang, J., Liu, S., Qu, Y., Sun, Y., 2020. Biodegradation of skatole by *Burkholderia sp. IDO3* and its successful bioaugmentation in activated sludge systems. Environ. Res. 182, 109123.

Mahdy, A., Fotidis, I.A., Mancini, E., Ballesteros, M., González-Fernández, C., Angelidaki, I., 2017. Ammonia tolerant inocula provide a good base for anaerobic digestion of microalgae in third generation biogas process. Bioresour. Technol. 225, 272–278.

Nkemka, V.N., Gilroyed, B., Yanke, J., Gruninger, R., Vedres, D., McAllister, T., Hao, X., 2015. Bioaugmentation with an anaerobic fungus in a two-stage process for biohydrogen and biogas production using corn silage and cattail. Bioresour. Technol. 185, 79–88.

Nzila, A., 2017. Mini review: update on bioaugmentation in anaerobic processes for biogas production. Anaerobe 46, 3–12.

Okonkwo, O., Papirio, S., Trably, E., Escudie, R., Lakaniemi, A.-M., Esposito, G., 2020. Enhancing thermophilic dark fermentative hydrogen production at high glucose concentrations via bioaugmentation with *Thermotoga neapolitana*. Int. J. Hydrog. Energy 45 (35), 17241–17249.

Ortigueira, J., Martins, L., Pacheco, M., Silva, C., Moura, P., 2019. Improving the non-sterile food waste bioconversion to hydrogen by microwave pretreatment and bioaugmentation with *Clostridium butyricum*. Waste Manage. 88, 226–235.

Ozbayram, E.G., Kleinsteuber, S., Nikolausz, M., Ince, B., Ince, O., 2017. Effect of bioaugmentation by cellulolytic bacteria enriched from sheep rumen on methane production from wheat straw. Anaerobe 46, 122–130.

Phuttaro, C., Sawatdeenarunat, C., Surendra, K.C., Boonsawang, P., Chaiprapat, S., Khanal, S.K., 2019. Anaerobic digestion of hydrothermally-pretreated lignocellulosic biomass: influence of pretreatment temperatures, inhibitors and soluble organics on methane yield. Bioresour. Technol. 284, 128–138.

Poirier, S., Madigou, C., Bouchez, T., Chapleur, O., 2017. Improving anaerobic digestion with support media: mitigation of ammonia inhibition and effect on microbial communities. Bioresour. Technol. 235, 229–239.

Poirier, S., Steyer, J.-P., Bernet, N., Trably, E., 2020. Mitigating the variability of hydrogen production in mixed culture through bioaugmentation with exogenous pure strains. Int. J. Hydrog. Energy 45 (4), 2617–2626.

Poszytek, K., Ciezkowska, M., Sklodowska, A., Drewniak, L., 2016. Microbial consortium with high cellulolytic activity (MCHCA) for enhanced biogas production. Front. Microbiol. 7, 324.

Rajagopal, R., Massé, D.I., Singh, G., 2013. A critical review on inhibition of anaerobic digestion process by excess ammonia. Bioresour. Technol. 143, 632–641.

Raper, E., Stephenson, T., Anderson, D.R., Fisher, R., Soares, A., 2018. Industrial wastewater treatment through bioaugmentation. Process Saf. Environ. Protection 118, 178–187.

Regueiro, L., Lema, J.M., Carballa, M., 2015. Key microbial communities steering the functioning of anaerobic digesters during hydraulic and organic overloading shocks. Bioresour. Technol. 197, 208–216.

Roager, L., Sonnenschein, E.C., 2019. Bacterial candidates for colonization and degradation of marine plastic debris. Environ. Sci. Technol. 53 (20), 11636–11643.

Rodríguez-Rodríguez, C.E., Lucas, D., Barón, E., Gago-Ferrero, P., Molins-Delgado, D., Rodríguez-Mozaz, S., Eljarrat, E., Díaz-Cruz, M.S., Barceló, D., Caminal, G., 2014. Re-inoculation strategies enhance the degradation of emerging pollutants in fungal bioaugmentation of sewage sludge. Bioresour. Technol. 168, 180–189.

Rogers, K.L., Carreres-Calabuig, J.A., Gorokh, E., Posth, N.R., 2020. Micro-by-micro interactions: how microorganisms influence the fate of marine microplastics. Limnol. Oceanogr. Lett. 5 (1), 18–36.

Romero-Güiza, M., Vila, J., Mata-Alvarez, J., Chimenos, J., Astals, S., 2016. The role of additives on anaerobic digestion: a review. Renew. Sustain. Energy Rev. 58, 1486–1499.

Ruiz-Sánchez, J., Campanaro, S., Guivernau, M., Fernández, B., Prenafeta-Boldú, F.X., 2018. Effect of ammonia on the active microbiome and metagenome from stable full-scale digesters. Bioresour. Technol. 250, 513–522.

Sawatdeenarunat, C., Surendra, K., Takara, D., Oechsner, H., Khanal, S.K., 2015. Anaerobic digestion of lignocellulosic biomass: challenges and opportunities. Bioresour. Technol. 178, 178–186.

Seengenyoung, J., Mamimin, C., Prasertsan, P., O-Thong, S., 2019. Pilot-scale of biohythane production from palm oil mill effluent by two-stage thermophilic anaerobic fermentation. Int. J. Hydrog. Energy 44 (6), 3347–3355.

Semrany, S., Favier, L., Djelal, H., Taha, S., Amrane, A., 2012. Bioaugmentation: possible solution in the treatment of Bio-refractory organic compounds (Bio-ROCs). Biochem. Eng. J. 69, 75–86.

Sharma, P., Melkania, U., 2018. Effect of bioaugmentation on hydrogen production from organic fraction of municipal solid waste. Int. J. Hydrog. Energy 43 (15), 7290–7298.

Shetty, D., Joshi, A., Dagar, S.S., Ks hirsagar, P., Dhakephalkar, P.K., 2020. Bioaugmentation of anaerobic fungus *Orpinomyces joyonii* boosts sustainable biomethanation of rice straw without pretreatment. Biomass Bioenergy 138, 105546.

Slepetiene, A., Volungevicius, J., Jurgutis, L., Liaudanskiene, I., Amaleviciute-Volunge, K., Slepetys, J., Ceseviciene, J., 2020. The potential of digestate as a biofertilizer in eroded soils of *Lithuania*. Waste Manage. 102, 441–451.

Srisowmeya, G., Chakravarthy, M., Devi, G.N., 2020. Critical considerations in two-stage anaerobic digestion of food waste—a review. Renew. Sustain. Energy Rev. 119, 109587.

Stephenson, D., Stephenson, T., 1992. Bioaugmentation for enhancing biological wastewater treatment. Biotechnol. Adv. 10 (4), 549–559.

Strang, O., Ács, N., Wirth, R., Maróti, G., Bagi, Z., Rákhely, G., Kovács, K.L., 2017. Bioaugmentation of the thermophilic anaerobic biodegradation of cellulose and corn stover. Anaerobe 46, 104–113.

Suksong, W., Kongjan, P., Prasertsan, P., O-Thong, S., 2019. Thermotolerant cellulolytic *Clostridiaceae* and *Lachnospiraceae* rich consortium enhanced biogas production from oil palm empty fruit bunches by solid-state anaerobic digestion. Bioresour. Technol. 291, 121851.

Syranidou, E., Karkanorachaki, K., Amorotti, F., Avgeropoulos, A., Kolvenbach, B., Zhou, N.-Y., Fava, F., Corvini, P.F.X., Kalogerakis, N., 2019. Biodegradation of mixture of plastic films by tailored marine consortia. J. Hazard. Mater. 375, 33–42.

Tabatabaei, M., Aghbashlo, M., Valijanian, E., Kazemi Shariat Panahi, H., Nizami, A.-S., Ghanavati, H., Sulaiman, A., Mirmohamadsadeghi, S., Karimi, K., 2020. A comprehensive review on recent biological innovations to improve biogas production, Part 2: Mainstream and downstream strategies. Renew. Energy 146, 1392–1407.

Tale, V.P., Maki, J.S., Zitomer, D.H., 2015. Bioaugmentation of overloaded anaerobic digesters restores function and archaeal community. Water Res. 70, 138–147.

Tao, Y., Ersahin, M.E., Ghasimi, D.S., Ozgun, H., Wang, H., Zhang, X., Guo, M., Yang, Y., Stuckey, D.C., van Lier, J.B., 2020. Biogas productivity of anaerobic digestion process is governed by a core bacterial microbiota. Chem. Eng. J. 380, 122425.

Tian, H., Li, J., Yan, M., Tong, Y.W., Wang, C.-H., Wang, X., 2019a. Organic waste to biohydrogen: a critical review from technological development and environmental impact analysis perspective. Appl. Energy 256, 113961.

Tian, H., Mancini, E., Treu, L., Angelidaki, I., Fotidis, I.A., 2019b. Bioaugmentation strategy for overcoming ammonia inhibition during biomethanation of a protein-rich substrate. Chemosphere 231, 415–422.

Tian, H., Treu, L., Konstantopoulos, K., Fotidis, I.A., Angelidaki, I., 2019c. 16s rRNA gene sequencing and radioisotopic analysis reveal the composition of ammonia acclimatized methanogenic consortia. Bioresour. Technol. 272, 54–62.

Tian, H., Yan, M., Treu, L., Angelidaki, I., Fotidis, I.A., 2019d. Hydrogenotrophic methanogens are the key for a successful bioaugmentation to alleviate ammonia inhibition in thermophilic anaerobic digesters. Bioresour. Technol. 293, 122070.

Town, J.R., Dumonceaux, T.J., 2016. Laboratory-scale bioaugmentation relieves acetate accumulation and stimulates methane production in stalled anaerobic digesters. Appl. Microbiol. Biotechol. 100 (2), 1009–1017.

Tsapekos, P., Kougias, P., Vasileiou, S., Treu, L., Campanaro, S., Lyberatos, G., Angelidaki, I., 2017. Bioaugmentation with hydrolytic microbes to improve the anaerobic biodegradability of lignocellulosic agricultural residues. Bioresour. Technol. 234, 350–359.

Usman Khan, M., Kiaer Ahring, B., 2020. Anaerobic digestion of biorefinery lignin: effect of different wet explosion pretreatment conditions. Bioresour. Technol. 298, 122537.

Valdez-Vazquez, I., Castillo-Rubio, L.G., Pérez-Rangel, M., Sepúlveda-Gálvez, A., Vargas, A., 2019. Enhanced hydrogen production from lignocellulosic substrates via bioaugmentation with *Clostridium strains*. Ind. Crops Prod. 137, 105–111.

Varanasi, J.L., Das, D., 2020. Maximizing biohydrogen production from water hyacinth by coupling dark fermentation and electrohydrogenesis. Int. J. Hydrog. Energy 45 (8), 5227–5238.

Venkiteshwaran, K., Milferstedt, K., Hamelin, J., Zitomer, D.H., 2016. Anaerobic digester bioaugmentation influences quasi steady state performance and microbial community. Water Res. 104, 128–136.

Villa Montoya, A.C., da Silva, C., Mazareli, R., Silva, E.L., Varesche, M.B.A., 2020. Improving the hydrogen production from coffee waste through hydrothermal pretreatment, co-digestion and microbial consortium bioaugmentation. Biomass Bioenergy 137, 105551.

Vinzelj, J., Joshi, A., Insam, H., Podmirseg, S.M., 2020. Employing anaerobic fungi in biogas production: challenges & opportunities. Bioresour. Technol. 300, 122687.

Wang, A., Ren, N., Shi, Y., Lee, D.-J., 2008. Bioaugmented hydrogen production from microcrystalline cellulose using co-culture—*Clostridium acetobutylicum* X9 and *Ethanoigenens harbinense* B49. Int. J. Hydrog. Energy 33 (2), 912–917.

Wang, D., Shen, F., Yang, G., Zhang, Y., Deng, S., Zhang, J., Zeng, Y., Luo, T., Mei, Z., 2018. Can hydrothermal pretreatment improve anaerobic digestion for biogas from lignocellulosic biomass? Bioresour. Technol. 249, 117–124.

Wang, S., Liu, M., Wang, J., Huang, J., Wang, J., 2020a. Polystyrene nanoplastics cause growth inhibition, morphological damage and physiological disturbance in the marine microalga *Platymonas helgolandica*. Mar. Pollut. Bull. 158, 111403.

Wang, X., Wang, W., Zhou, B., Xu, M., Wu, Z., Liang, J., Zhou, L., 2020b. Improving solid-liquid separation performance of anaerobic digestate from food waste by thermally activated persulfate oxidation. J. Hazard. Mater. 398, 122989.

Wen, D., Zhang, J., Xiong, R., Liu, R., Chen, L., 2013. Bioaugmentation with a pyridine-degrading bacterium in a membrane bioreactor treating pharmaceutical wastewater. J. Environ. Sci. 25 (11), 2265–2271.

Xiao, L., Sun, R., Zhang, P., Zheng, S., Tan, Y., Li, J., Zhang, Y., Liu, F., 2019. Simultaneous intensification of direct acetate cleavage and CO_2 reduction to generate methane by bioaugmentation and increased electron transfer. Chem. Eng. J. 378, 122229.

Xu, F., Li, Y., Ge, X., Yang, L., Li, Y., 2018. Anaerobic digestion of food waste - challenges and opportunities. Bioresour. Technol. 247, 1047–1058.

Yan, M., Fotidis, I.A., Jéglot, A., Treu, L., Tian, H., Palomo, A., Zhu, X., Angelidaki, I., 2020a. Long-term preserved and rapidly revived methanogenic cultures: microbial dynamics and preservation mechanisms. J. Clean. Prod. 263, 121577.

Yan, M., Treu, L., Campanaro, S., Tian, H., Zhu, X., Khoshnevisan, B., Tsapekos, P., Angelidaki, I., Fotidis, I.A., 2020b. Effect of ammonia on anaerobic digestion of municipal solid waste: inhibitory performance, bioaugmentation and microbiome functional reconstruction. Chem. Eng. J. 401, 126159.

Yan, M., Treu, L., Zhu, X., Tian, H., Basile, A., Fotidis, I.A., Campanaro, S., Angelidaki, I., 2020c. Insights into ammonia adaptation and methanogenic precursor oxidation by genome-centric analysis. Environ. Sci. Technol. 54 (19), 12568–12582.

Yang, Z., Guo, R., Xu, X., Wang, L., Dai, M., 2016. Enhanced methane production via repeated batch bioaugmentation pattern of enriched microbial consortia. Bioresour. Technol. 216, 471–477.

Yang, Z., Sun, H., Zhao, Q., Kurbonova, M., Zhang, R., Liu, G., Wang, W., 2020. Long-term evaluation of bioaugmentation to alleviate ammonia inhibition during anaerobic digestion: process monitoring, microbial community response, and methanogenic pathway modeling. Chem. Eng. J. 399, 125765.

Yang, Z., Wang, W., Liu, C., Zhang, R., Liu, G., 2019. Mitigation of ammonia inhibition through bioaugmentation with different microorganisms during anaerobic digestion: selection of strains and reactor performance evaluation. Water Res. 155, 214–224.

Yenigün, O., Demirel, B., 2013. Ammonia inhibition in anaerobic digestion: a review. Process Biochem. 48 (5-6), 901–911.

Yıldırım, E., Ince, O., Aydin, S., Ince, B., 2017. Improvement of biogas potential of anaerobic digesters using rumen fungi. Renew. Energy 109, 346–353.

Yu, Z., Leng, X., Zhao, S., Ji, J., Zhou, T., Khan, A., Kakde, A., Liu, P., Li, X., 2018. A review on the applications of microbial electrolysis cells in anaerobic digestion. Bioresour. Technol. 255, 340–348.

Yue, L., Cheng, J., Hua, J., Dong, H., Zhou, J., Li, Y.-Y., 2020. Improving fermentative methane production of glycerol trioleate and food waste pretreated with ozone through two-stage dark hydrogen fermentation and anaerobic digestion. Energy Convers. Manage. 203, 112225.

Zhan, H., Huang, Y., Lin, Z., Bhatt, P., Chen, S., 2020. New insights into the microbial degradation and catalytic mechanism of synthetic pyrethroids. Environ. Res. 182, 109138.

Zhang, J., Zhang, R., He, Q., Ji, B., Wang, H., Yang, K., 2020a. Adaptation to salinity: response of biogas production and microbial communities in anaerobic digestion of kitchen waste to salinity stress. J. Biosci. Bioeng. 130 (2), 173–178.

Zhang, J., Zhao, M., Li, C., Miao, H., Huang, Z., Dai, X., Ruan, W., 2020b. Evaluation the impact of polystyrene micro and nanoplastics on the methane generation by anaerobic digestion. Ecotox. Environ. Safe. 205, 111095.

Zhang, K., Cao, G.-L., Ren, N.-Q., 2019a. Bioaugmentation with *Thermoanaerobacterium thermosaccharolyticum* W16 to enhance thermophilic hydrogen production using corn stover hydrolysate. Int. J. Hydrog. Energy 44 (12), 5821–5829.

Zhang, L., Kuroki, A., Loh, K.-C., Seok, J.K., Dai, Y., Tong, Y.W., 2020c. Highly efficient anaerobic co-digestion of food waste and horticultural waste using a three-stage thermophilic bioreactor: performance evaluation, microbial community analysis, and energy balance assessment. Energy Convers. Manage. 223, 113290.

Zhang, L., Lim, E.Y., Loh, K.-C., Ok, Y.S., Lee, J.T., Shen, Y., Wang, C.-H., Dai, Y., Tong, Y.W., 2020d. Biochar enhanced thermophilic anaerobic digestion of food waste: focusing on biochar particle size, microbial community analysis and pilot-scale application. Energy Convers. Manage. 209, 112654.

Zhang, L., Loh, K.-C., Dai, Y., Tong, Y.W., 2020e. Acidogenic fermentation of food waste for production of volatile fatty acids: bacterial community analysis and semi-continuous operation. Waste Manage. 109, 75–84.

Zhang, L., Loh, K.-C., Lim, J.W., Zhang, J., 2019b. Bioinformatics analysis of metagenomics data of biogas-producing microbial communities in anaerobic digesters: a review. Renew. Sustain. Energy Rev. 100, 110–126.

Zhang, L., Loh, K.-C., Zhang, J., Mao, L., Tong, Y.W., Wang, C.-H., Dai, Y., 2019c. Three-stage anaerobic co-digestion of food waste and waste activated sludge: identifying bacterial and methanogenic archaeal communities and their correlations with performance parameters. Bioresour. Technol. 285, 121333.

Zhang, S., Chang, J., Liu, W., Pan, Y., Cui, K., Chen, X., Liang, P., Zhang, X., Wu, Q., Qiu, Y., 2018a. A novel bioaugmentation strategy to accelerate methanogenesis via adding *Geobacter sulfurreducens* PCA in anaerobic digestion system. Sci. Total Environ. 642, 322–326.

Zhang, T., Jiang, D., Zhang, H., Jing, Y., Tahir, N., Zhang, Y., Zhang, Q., 2020f. Comparative study on bio-hydrogen production from corn stover: photo-fermentation, dark-fermentation and dark-photo co-fermentation. Int. J. Hydrog. Energy 45 (6), 3807–3814.

Zhang, W., Heaven, S., Banks, C.J., 2018b. Degradation of some EN13432 compliant plastics in simulated mesophilic anaerobic digestion of food waste. Polym. Degrad. Stabil. 147, 76–88.

Zhang, W., Torrella, F., Banks, C.J., Heaven, S., 2019d. Data related to anaerobic digestion of bioplastics: images and properties of digested bioplastics and digestate, synthetic food waste recipe and packaging information. Data Brief 25, 103990.

Zhang, X., Chen, J., Li, J., 2020g. The removal of microplastics in the wastewater treatment process and their potential impact on anaerobic digestion due to pollutants association. Chemosphere 251, 126360.

Zhao, C., Zhang, Y., Li, Y., 2019. Production of fuels and chemicals from renewable resources using engineered *Escherichia coli*. Biotechnol. Adv. 37 (7), 107402.

Zhao, W., Huang, J.J., Hua, B., Huang, Z., Droste, R.L., Chen, L., Wang, B., Yang, C., Yang, S., 2020. A new strategy to recover from volatile fatty acid inhibition in anaerobic digestion by photosynthetic bacteria. Bioresour. Technol. 311, 123501.

Zhu, X., Campanaro, S., Treu, L., Kougias, P.G., Angelidaki, I., 2019. Novel ecological insights and functional roles during anaerobic digestion of saccharides unveiled by genome-centric metagenomics. Water Res. 151, 271–279.

Zhu, X., Treu, L., Kougias, P.G., Campanaro, S., Angelidaki, I., 2018. Converting mesophilic upflow sludge blanket (UASB) reactors to thermophilic by applying axenic methanogenic culture bioaugmentation. Chem. Eng. J. 332, 508–516.

Microbial fermentation *via* genetically engineered microorganisms for production of bioenergy and biochemicals

Kang Zhou, Jie Fu J. Zhou

Department of Chemical and Biomolecular Engineering, National University of Singapore, Singapore

10.1   Introduction

Genetic engineering allows humans to manipulate the genetic instructions in microbes in a programmable manner, enabling rational modification of cellular metabolism for utilizing new feedstocks and/or making novel products (Zhou et al., 2020). It is particularly useful if the desired phenotype cannot be easily obtained by screening wild-type or randomly mutagenized microbes. With genetic engineering, it is possible to rely on well-studied and widely used microbial species (model microbes, such as *Escherichia coli* and *Saccharomyces cerevisiae*), instead of exotic ones that need extensive characterizations before large-scale deployment (Zhang et al., 2021). Without genetic engineering, model microbes cannot utilize many important waste streams, such as carbon dioxide (Gleizer et al., 2019) and plastics (Panda et al., 2021). Spectrum of the products derived from wastes via fermentation also needs to be expanded to sustainably supply more commodities our society needs (Wang et al., 2020) and to improve the economic viability of existing waste recycling processes (Ma et al., 2020). For example, modifying *S. cerevisiae* to produce succinic acid may be more profitable than making ethanol (Liu et al., 2021).

To make a product that is new to a given host strain, it should be determined if the product can be synthesized by any other organisms, including plants and animals. If so, a further question is if the set of enzymatic reactions transforming a starting molecule into the product have been elucidated. The starting molecule should be a metabolite the host strain can naturally produce. The definition of a pathway being elucidated is that sequence of all the involved genes has been determined. Each of the genes may encode an enzyme that catalyzes one transformation in the pathway. The product may be produced when all the needed enzymes are functionally expressed.

When a step in a biosynthetic pathway cannot be catalyzed by known enzymes, bioinformatics would usually be used to identify gene candidates in a repository (Chavali and Rhee, 2018). The search algorithm could be based on sequence homology with enzymes known to catalyze similar reactions. A candidate is often expressed in an engineered microbial host that can synthesize its substrate to examine the transformation of the substrate. *In-vitro* assay may also be set up to characterize the activity of the recombinant enzyme encoded by the candidate. When there is no activity, directed evolution (DE) may be used to tune an enzyme's binding pocket and/or alter its catalytic mechanism. DE usually needs an extensive screening effort to isolate the desired mutants (Grove and Magliery, 2020).

When the objective is to produce a molecule that is not produced by any known organism, a set of transformations must be designed, considering thermodynamic feasibility, the number of transformations, the number of transformations that cannot be catalyzed by known enzymes, the maximal product yield, and the availability of cofactors in the host microbe of choice (Ni et al., 2021). Algorithms based on first principles and/or machine learning have been developed to aid the pathway design (Dale et al., 2010). The designed pathways need to be tested in a wet lab, which may generate new experimental data that can be used to guide improvement of the designs (Zhou et al., 2020).

The principles for enabling a host strain to utilize a new feedstock share many similar components with those for making a new product. For example, one needs to design and construct a reaction pathway connecting the feedstock to a metabolite the host organism can utilize.

The rest of this chapter aims to elaborate the principles articulated above for engineering model microbes to make new products (Section 10.2) and to utilize new feedstocks (Section 10.3). In each subsection, one or a few examples were selected solely for the purpose of discussing the underlying basic principles. We did not aim to provide a comprehensive overview of all the progresses related to the topics. The chapter is concluded by a brief summary and a discussion on the trends of the related technological developments in the near future (Section 10.4).

10.2   Engineering *E. coli* to produce 1,4-butanediol

10.2.1   Pathway to produce 4-hydroxybutyrate from central metabolism

1,4-butanediol (1,4-BDO) is a useful specialty chemical that can be used to synthesize many types of polymers (Liu and Lu, 2015). There is no known wild-type organism that can synthesize it. Yim et al. (from Genomatica) have engineered *E. coli* to produce 1,4-BDO from glucose at a commercial scale (Barton et al., 2015; Burgard et al., 2016; Yim et al., 2011).

4-hydroxybutyrate (4-HB) has high structural similarity with 1,4-BDO and is naturally produced from major cellular building blocks in many species. For example, the model plant species *Arabidopsis thaliana* could derive it from glutamate with 4-aminobutyrate as an intermediate (Shelp and Zarei, 2017). The genes in a few 4-HB biosynthetic pathways have been elucidated. To design a pathway to produce 1,4-BDO from glucose, a reasonable strategy was to combine a 4-HB-producing pathway with a set of reactions that convert 4-HB into 1,4-BDO.

Yim et al. first engineered an *E. coli* strain to produce 4-HB from glucose (Yim et al., 2011). 4-HB can be obtained by reducing 4-oxobutyrate (4-OB) by using reduced nicotinamide adenine dinucleotide (NADH) in the presence of a proper alcohol dehydrogenase (ADH, Fig. 10.1). Although wild-type *E. coli* strains do not catalyze this reaction efficiently, the reaction could be reconstituted when a gene encoding an ADH from *Porphyromonas gingivalis* ([Pg]4Hbd) was expressed. The gene could be obtained by using a gene synthesis service offered by many companies. The gene could be expressed after being introduced into an *E. coli* strain on an expression vector. Details of these standard molecular biology techniques can be found in the literature (Primrose and Twyman, 2013).

The authors considered two routes to derive 4-OB from central metabolism. The first one was reduction of succinyl-CoA (SuCoA) by the action of an aldehyde dehydrogenase from *P. gingivalis* ([Pg]SucD, Fig. 10.1). Succinyl-CoA could be derived from succinate with SucCD as the catalyst. Yim et al. improved the expression of SucCD by increasing their copy number through the use of a replicative plasmid (Yim et al., 2011). Succinate could be derived from the native reductive tricarboxylic acid (TCA) pathway under anaerobic or microaerobic conditions (Fig. 10.1). The second route is decarboxylation of 2-ketoglutarate (2-KG, Fig. 10.1), which can be enabled in *E. coli* by expressing a 2-keto-acid decarboxylase from *Mycobacterium bovis* ([Mb]SucA). 2-KG is a very important building block involved in both carbon and nitrogen metabolism, so growing *E. coli* cells would constantly produce it via the oxidative TCA pathway. Both routes were found to be active after the relevant genes were introduced into a wild-type *E. coli* strain (**W3110** [CGSC 4474] or **MG1655lacI**^{Q} [ATCC 47076]). The engineered *E. coli* strain expressing [Pg]4Hbd and having one of the two routes produced 1–10 mM 4-HB.

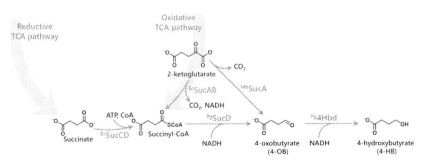

FIG. 10.1 Producing 4-hydroxybutyrate (4-HB) from central metabolism in engineered *Escherichia coli*. 4-HB could be obtained by reduction of 4-oxobutyrate (4-OB) by using NADH. 4-OB could be derived from succinyl-CoA through another reductive reaction. The two reactions described above could be catalyzed by two alcohol dehydrogenases from *Porphyromonas gingivalis* ([Pg]SucD and [Pg]4Hbd). 4-OB could also be obtained through decarboxylation of 2-ketoglutarate (catalyzed by [Mb]SucA, a decarboxylase from *Mycobacterium bovis*). Both succinyl-CoA and 2-ketoglutarate are intermediates of tricarboxylic acid (TCA) cycle. Blue lines with arrowhead indicate the reactions that were new to the host cell (*E. coli*). The thin green lines with arrowhead indicate endogenous reactions. The thick transparent green lines with arrowhead indicate two major pathways in *E. coli*.

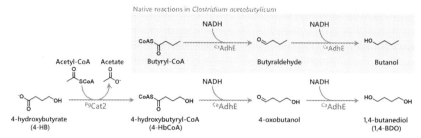

FIG. 10.2   Establishing a pathway for transforming 4-hydroxybutyrate (4-HB) into 1,4-butanediol (1,4-BDO) in engineered *Escherichia coli*. A CoA transferase (^{Pg}Cat2 from *Porphyromonas gingivalis*) was known for activating 4-HB into 4-hydroxybutyryl-CoA (4-HbCoA) by using acetyl-CoA as the CoA donor. An alcohol/aldehyde dehydrogenase (^{Ca}AdhE) from *Clostridium acetobutylicum* was tested for reducing 4-HbCoA into 1,4-BDO, because the native substrate of ^{Ca}AdhE was butyryl-CoA, which was structurally similar to 4-HbCoA. ^{Ca}AdhE was found to catalyze the transformations despite of low efficiency.

10.2.2   Pathway to produce 1,4-BDO from central metabolism

Yim et al. then established a pathway to convert 4-HB into 1,4-BDO (Yim et al., 2011). Chemically, this pathway reduces a carboxylic acid into its corresponding alcohol. To make the reactions thermodynamically feasible, the carboxylic acid needs to be activated by converting it into its CoA thioester. The thioester could be reduced into the alcohol by using NADH in two steps in the presence of proper ADHs (Fig. 10.2). The activation of 4-HB into 4-hydroxybutyryl-CoA (4-HbCoA) was well studied and could be achieved by using acetyl-CoA as the CoA donor in the presence of a CoA transferase (Cat2 from *P. gingivalis*, encoded by ^{Pg}*cat2*). The new enzymes that the researchers needed to develop were the ADHs.

The search for the ADHs was based on prior knowledge. It had been known that *Clostridium acetobutylicum* was able to transform exogenous 4-HB into 1,4-BDO (Jewell et al., 1986). *Clostridium acetobutylicum* is well known for its ability of producing butanol via butyryl-CoA as an intermediate. Due to the structural similarity between 4-HbCoA and butyryl-CoA (Fig. 10.2), the gene responsible for the butyryl-CoA-to-butanol reaction (^{Ca}*adhE*) was tested for converting 4-HbCoA into 1,4-BDO. An *E. coli* strain coexpressing ^{Pg}*cat2* and ^{Ca}*adhE* indeed produced 0.014 mM 1,4-BDO when 10 mM 4-HB was fed (Yim et al., 2011). Despite of the low yield, the critical ADH (^{Ca}AdhE) was identified (it can catalyze both reductive steps to transform 4-HbCoA into 1,4-BDO).

When the succinyl-CoA-based 4-HB-producing pathway was combined with the downstream pathway (converting 4-HB into 1,4-BDO), a much higher titer of 1,4-BDO was achieved (0.6 mM). This was >40-fold higher than what had been achieved by feeding 4-HB. Although the authors attributed the improvement to the changes they made to the gene arrangement on the expression vectors and to the host strain, it could also be related to the transport of 4-HB. Exogenous 4-HB may not be taken up by the cells effectively, while using the 4-HB producing pathway avoided the issue by producing 4-HB intracellularly. When Yim et al. further introduced the 2-KG decarboxylation route, the 1,4-BDO titer was improved to 1.3 mM, suggesting the importance of this decarboxylation reaction, which is not commonly used by organisms that naturally produce 4-HB (Yim et al., 2011). This example highlights the power and the flexibility of genetic engineering in introducing new function to host organisms.

10.2.3 Improving 1,4-BDO production from glucose

After the pathway was established to convert glucose into 1,4-BDO, Yim et al. modified the central metabolism to (1) eliminate/reduce formation of some byproducts, (2) create a stronger driving force for producing 1,4-BDO, and (3) enhance the expression/activity of key genes in the desired metabolic pathways (Yim et al., 2011).

It was introduced above that both oxidative and reductive TCA pathways can be used to supply precursor of 1,4-BDO. The reductive TCA pathway was not desired because it competed with the formation of 1,4-BDO for NADH. This pathway was inactivated by deleting *mdh* (Fig. 10.3), which encodes malate dehydrogenase. Other key NADH-consuming reactions under anaerobic or microaerobic conditions were also repressed or inactivated by gene deletion (Fig. 10.3): lactate formation was eliminated by deleting *ldhA*, which encodes lactate dehydrogenase; ethanol formation was repressed by deleting *adhE*, which encodes an ADH (ethanol was still produced because of the undesired activity of ^{Ca}AdhE; further discussed below); the nonoxidative reaction of converting pyruvate into acetyl-CoA was inactivated by deleting *pflB*, which encodes pyruvate formate lyase (this reaction does not consume NADH but it reduces NADH generation).

After the above modifications to central metabolism were implemented, 1,4-BDO was primarily derived from 2-KG (Fig. 10.3). Then the full pathway from glucose to 1,4-BDO can be easily described as follows: one glucose was converted into two phosphoenolpyruvate (PEP) through glycolysis; one PEP was condensed with one hydrogen carbonate to form one oxaloacetate (OAA); another PEP was converted into one pyruvate, which was oxidatively decarboxylated into one acetyl-CoA; the OAA was condensed with the acetyl-CoA to form one citrate, which was converted into one 2-KG via the oxidative TCA pathway; one 2-KG was converted into one 1,4-BDO by the established pathways (Fig. 10.2). The overall reaction equation for converting glucose into 1,4-BDO is

$$\text{Glucose} \rightarrow \rightarrow \rightarrow 1,4-\text{BDO} + 2\text{CO}_2 + \text{NADH} + \text{ATP} \tag{10.1}$$

Eq. (10.1) indicates that the full pathway from glucose to 1,4-BDO regenerates one NADH and one adenosine triphosphate (ATP) per 1,4-BDO. The produced ATP can be used for cell growth and maintenance, giving incentive to the host cell to use this pathway. The accumulated NADH must be oxidatively recycled back into NAD^+ to maintain the redox balance. The engineered cells could not grow under strict anaerobic condition due to NADH accumulation—other anaerobic NADH-consuming reactions had been inactivated. Yim et al. had to provide oxygen at a limited rate to create a microaerobic condition, in which the accumulated NADH could be oxidized through respiration, which could also regenerate more ATP through the well-known proton-pumping ATPase. Under the microaerobic condition, the pathway is thus self-sufficient (it does not require additional consumption of glucose to provide electrons and energy). Eq. (10.1) thus sets the theoretical yield to be 0.5 g/g (gram 1,4-BDO per gram glucose). Supplying oxygen at a higher rate (aerobic condition) is not desired because then respiration would compete with the 1,4-BDO pathway for NADH and

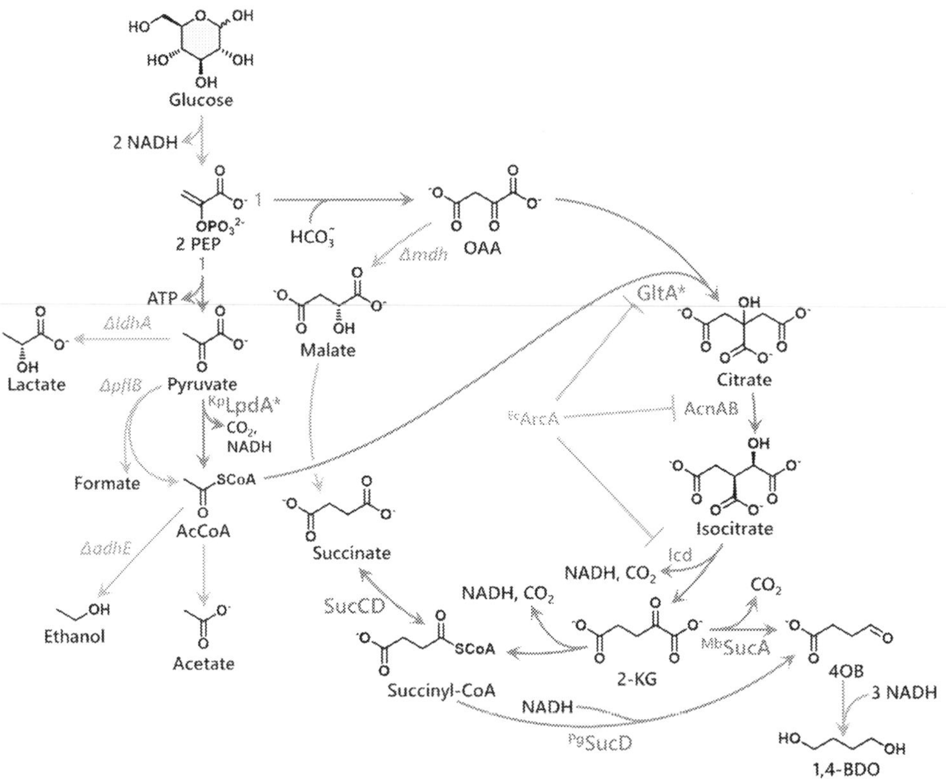

FIG. 10.3   Engineering *Escherichia coli* central metabolism to improve production of 1,4-butanediol (1,4-BDO). The green line with arrowhead indicates a single-step endogenous reaction. A blue line with arrowhead indicates a single-step reaction new to the host. The purple line with arrowhead indicates a multistep pathway. The gray line with arrowhead indicates an inactive or repressed native reaction. ArcA represses transcription of many genes involved in the tricarboxylic acid (TCA) cycle and the transcription inhibition was removed by deleting *arcA* in the genome. The number in green font indicates the desired flux distribution at branch point. PEP, phosphoenolpyruvate; AcCoA, acetyl-CoA; OAA, oxaloacetate; 2-KG, 2-ketoglutarate; 4-OB, 4-oxobutyrate; *ldhA* encodes lactate dehydrogenase. *pflB* encodes pyruvate formate lyase. *adhE* encodes an alcohol dehydrogenase. ᴷᴾLpdA is lipoamide dehydrogenase from *Klebsiella pneumonia*.
*An enzyme being mutated to overwrite undesired regulation. GltA, citrate synthase; AcnAB, aconitate hydratase; Icd, isocitrate dehydrogenase; ᴹᵇSucA, 2-keto-acid decarboxylase from *Mycobacterium bovis*; ᴾᵍSucD, an aldehyde dehydrogenase from *P. gingivalis*.

reduce the incentive for the cells to use the 1,4-BDO pathway. Under the microaerobic condition the engineered cells relied on the 1,4-BDO pathway to maintain redox balance, because the supplied oxygen was insufficient to oxidize all the NADH generated by the glucose-to-2-KG segment and the cells could not produce other terminal products to sink the electrons stored in the NADH (Fig. 10.4).

In the glucose-to-1,4-BDO pathway described above, a number of reactions may be inhibited under the microaerobic condition. Lipoamide dehydrogenase (involved in converting pyruvate into acetyl-CoA, encoded by *lpdA*), and citrate synthase (encoded by *gltA*) were

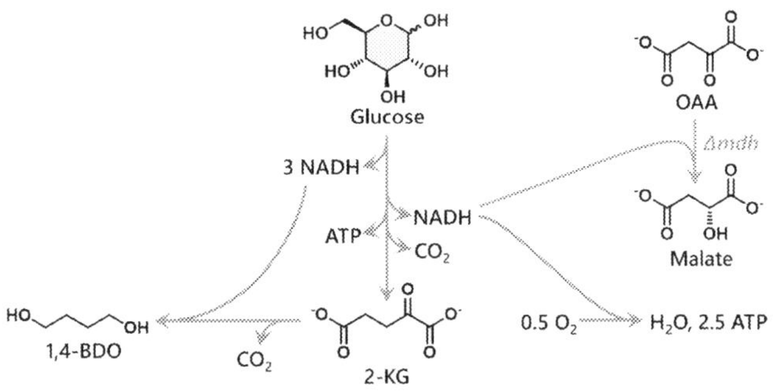

FIG. 10.4 **Redox balance in the microaerobic production of 1,4-butanediol (1,4-BDO).** The purple line with arrowhead indicates multistep pathway. The gray line with arrowhead indicates inactivated or repressed native reaction. 1,4-BDO is needed to consume the accumulated NADH under the microaerobic condition, creating a strong driving force to produce 1,4-BDO. This was made possible by controlling the rate of oxygen supply and inactivating other major NADH-consuming reactions in the cells. The reaction to reduce oxaloacetate (OAA) into malate is used as an example to illustrate such NADH-consuming reactions. The reaction is catalyzed by malate dehydrogenase, which is encoded by *mdh*. 2-KG: 2-ketoglutarate.

sensitive to NADH, which was abundant in the 1,4-BDO-producing strain. The native *lpdA* was replaced by a counterpart from *Klebsiella pneumonia* (^{Kp}*lpdA*). ^{Kp}LpdA had a mutation (Asp^{354}→Lys) and was less sensitive to NADH. Gene *gltA* was also mutated so that its protein product had the Arg^{163}→Leu mutation that could overwrite the repression by NADH. Yim et al. also deleted *arcA*, which encodes a global transcription repressor that negatively affects the transcription of many genes involved in the oxidative TCA pathway, including *gltA*, *acnAB*, and *icd* (Yim et al., 2011).

The strain containing all the modifications to the central metabolism described above (*E. coli* **ECKh-422-002C**) could produce ~13 mM (1.2 g/L) 1,4-BDO from 10 g/L glucose, which was approximately 10 times the titer of the parent strain without these modifications. However, the 1,4-BDO yield (0.12 g/g) was still far from the theoretical yield (0.5 g/g). The strain accumulated pathway intermediates (4-HB) and still produced 1.3 g/L ethanol as a byproduct, despite that *adhE* was deleted. This was because ^{Ca}AdhE used in reducing 4-HbCoA into 1,4-BDO also accepted acetyl-CoA as a substrate (acetyl-CoA would be reduced into ethanol). Yim et al. screened ADHs in their company's collection and found that an ADH from *Clostridium beijerinckii* (^{Cb}Adh) had a much better specificity (Yim et al., 2011). When ^{Ca}*adhE* in **ECKh-422-002C** was replaced by ^{Cb}*adh*, the new strain (**ECKh-422-025B**) produced much less ethanol (0.2 g/L) and maintained the same yield of 1,4-BDO. When **ECKh-422-025B** was cultivated in 2 L bioreactor with better control of pH, dissolved oxygen, and substrate levels, a higher cell density was achieved, which led to the increase of the 1,4-BDO titer to 18 g/L.

Genomatica further developed the strain, which under optimized bioreactor conditions produced >125 g/L 1,4-BDO with a yield of 0.4 g/g and a productivity of >3.5 g/L/h. The general principles of the strain developments are published in two reviews (Barton et al., 2015; Burgard et al., 2016). The company has also developed *E. coli* strains that are able to produce 1,4-BDO from xylose, which would be very useful in utilizing cellulosic waste streams as substrates to produce this useful product.

10.3   Engineering *S. cerevisiae* to utilize xylose

Xylose is the second major product in hydrolysate of many lignocellulosic wastes (after glucose). *Saccharomyces cerevisiae* is the workhorse behind industrial production of ethanol from glucose, but wild-type *S. cerevisiae* strains cannot utilize xylose. Since the 1980s, extensive research has been done to engineer *S. cerevisiae* to utilize xylose. The section of the chapter aims to explain the basic principles behind establishing the xylose utilization pathway, improving the utilization rate, and overcoming catabolite repression caused by glucose.

Cellular metabolism consists of an extensive network of reactions that utilize resources acquired by an organism to meet its needs, for example, to derive energy and to make cellular building blocks for growth. Within this network, several key modules are common across various domains of life, including glycolysis, TCA cycle, and the pentose phosphate pathway (PPP). These modules are interconnected and are collectively referred to as central carbon metabolism, due to their central roles in utilizing carbon-containing substrates. Owing to its interconnected nature, many intermediates of central carbon metabolism can support the growth of an organism when they could be converted to all other essential metabolites that the organism needs. Therefore, a common strategy for engineering microbes to utilize new feedstocks involves identifying a native metabolite of the central carbon metabolism that is closely related to the new feedstock, and establishing a pathway to transform the new feedstock into that native metabolite within the microbe. While attempting to engineer *S. cerevisiae* to utilize xylose, it was found that *S. cerevisiae* could grow slowly on xylulose (Wang and Schneider, 1980), an isomer of xylose (Fig. 10.5). *S. cerevisiae* was thought to utilize

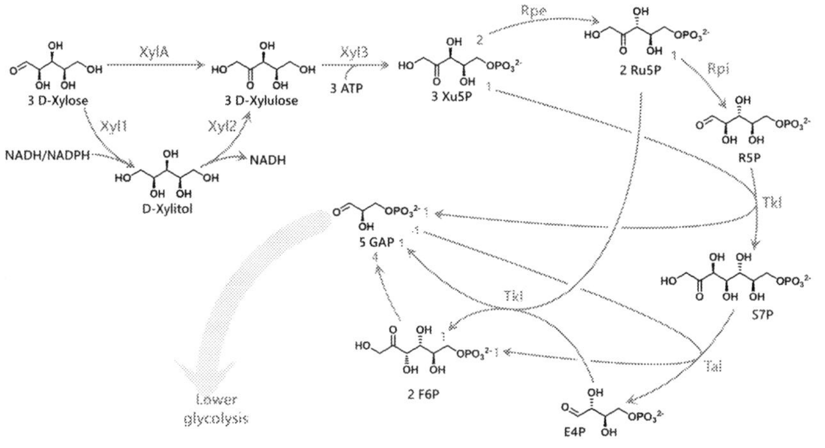

FIG. 10.5   **D-Xylose utilization pathway.** D-Xylose needs to be isomerized into D-xylulose, via a direct isomerization reaction or a two-step oxidoreductive pathway. D-Xylulose enters the nonoxidative pentose phosphate pathway via D-xylulose 5-phosphate (Xu5P). Three Xu5P can be transformed into five glyceraldehyde 3-phosphate (GAP), which enters the lower glycolysis. Blue line with arrowhead indicates new reaction to *S. cerevisiae*. Green and purple lines with arrowhead indicate single- and multiple-step pathway, respectively. The thick, transparent green line illustrates the lower glycolysis. The green numbers at a branch point indicate the desired flux distribution ratio. XylA, xylose/xylulose isomerase; Xyl1, xylose reductase; Xyl2, xylitol dehydrogenase; Xyl3, xylulose kinase; Rpe, ribulose 5-phosphate epimerase; Rpi, ribose 5-phosphate isomerase; Tkl, transketolase; Tal, transaldolase.

xylulose by converting it into xylulose 5-phosphate, which could then enter central carbon metabolism through the PPP. This was confirmed in later years by isolation of a key gene responsible for xylulose phosphorylation, *xyl3*, and its product xylulokinase whose activity was essential for the growth of S. *cerevisiae* on xylulose (Rodriguez-Peña et al., 1998). Several organisms that naturally grow on xylose utilize the same pathway for channeling the sugar into central carbon metabolism, by first isomerizing xylose into xylulose as an intermediate (Jeffries, 1983). This isomerization step, however, appears to be absent in S. *cerevisiae*. Therefore, reconstituting the pathway of isomerizing xylose into xylulose in S. *cerevisiae* was a rational strategy to enable its growth on xylose. At the same time, the slow growth of S. *cerevisiae* on xylulose was also a problem that needed to be tackled in order for xylose to be utilized efficiently. As one of the main goals in the earlier years for engineering xylose utilization in S. *cerevisiae* was to produce ethanol (a fermentative product), most of the studies during that time period were focused on improving anaerobic utilization of xylose.

10.3.1 Constructing S. *cerevisiae* strains to grow on xylose

During the initial screening of various yeast strains on xylulose utilization, Wang and Schneider reported a low specific growth rate of 0.02 h^{-1} for their laboratory S. *cerevisiae* strain **X2180-1B** grown in an aerobic condition (Wang and Schneider, 1980). Ciriacy and Porep attempted to improve this growth rate by generating S. *cerevisiae* mutants through UV-irradiation, followed by selecting for fast-growing mutants on high xylulose media. A respiratory inhibitor antimycin A was used to force fermentative growth. Through this process, they isolated a mutant strain **PUA1** which had an improved specific growth rate of 0.07 h^{-1} with respiratory inhibition (Porep and Ciriacy, 1986). Enzyme profiling of the mutant revealed elevated activity of xylulose kinase (Xyl3), suggesting that the first phosphorylation step in xylulose utilization may be a limiting step in wild-type S. *cerevisiae* stains. **PUA1** was subsequently put through further rounds of selection and the new mutants generated had proven to be helpful in the construction of xylose utilization pathway in S. *cerevisiae*.

To reconstitute the xylose isomerization pathway in S. *cerevisiae*, two different routes could be considered. The first route is the direct isomerization between xylose and xylulose catalyzed by xylulose isomerase (XylA, Fig. 10.5), which is commonly employed by various xylose-utilizing bacterial species. Xylose-utilizing yeast and fungal species usually employ the second route which consists of two steps: xylose is first reduced into xylitol with xylose reductase (Xyl1) as the catalyst, and xylitol is subsequently oxidized into xylulose with xylitol dehydrogenase (Xyl2). Although the direct isomerization route appears simpler, earlier attempts to express XylA of various bacterial origins in S. *cerevisiae* did not result in functional proteins (Kuyper et al., 2003a). In contrast, pioneering work with Xyl1 and Xyl2 expression in S. *cerevisiae* produced promising results (Kotter et al., 1990), and therefore the second route was primarily considered for establishing xylose utilization in S. *cerevisiae* before the 2000s.

In the pioneering work, Kotter et al. simultaneously expressed *xyl1* and *xyl2* from *Pichia stipites* (^{Ps}*xyl1* and ^{Ps}*xyl2*) in a laboratory S. *cerevisiae* strain (not specified in the original study) (Kotter et al., 1990). The transformant grew on xylose at aerobic condition with a specific growth rate of 0.07 h^{-1}. As the team had previously worked on improving xylulose utilization in S. *cerevisiae* (Porep and Ciriacy, 1986), they were aware of the possible rate-limiting step of xylulose phosphorylation by Xyl3, and had since selected for a strain (**PUA3-1**) with fivefold higher Xyl3 activity. When $^{Ps}xyl1$ and $^{Ps}xyl2$ were coexpressed in **PUA3-1**, the new strain had a substantially higher specific growth rate ($0.17\,h^{-1}$) at aerobic condition. The rate was, however, still half of that of growing on glucose.

In a subsequent experiment, the team found that when another strain with even higher XylA activity (**PUA6-9**) was transformed to overexpress $^{Ps}xyl1$ and $^{Ps}xyl2$, the transformant could not grow on xylose when its respiration was inhibited by antimycin A (Kötter and Ciriacy, 1993). The parent strain **PUA6-9**, however, could grow on xylulose with respiratory inhibition. This difference in respiratory requirements between xylose and xylulose utilization suggested that an inhibitory factor may have been generated during the isomerization of xylose to xylulose, which is likely related to cellular redox imbalance as it could be partially resolved through respiration.

10.3.2 Redox-imbalance limited the ethanol yield through the oxidoreductive pathway

Both steps of the newly reconstituted xylose isomerization pathway—xylose reduction and xylitol oxidation—require redox cofactors to function. There are two main types of redox cofactor in *S. cerevisiae*: $NAD^+/NADH$, which are usually involved in catabolism and respiration, and $NADP^+/NADPH$, which are more commonly used for biomolecule synthesis. Xylose reduction catalyzed by Xyl1 from *P. stipites* could use either NADH or NADPH as reducing factors (Amore et al., 1991), while *P. stipites* Xyl2 strictly requires NAD^+ for oxidation of xylitol to xylulose (Kötter et al., 1990). Normally, oxidation and reduction steps within a pathway could be coupled to achieve redox balance, such as in the case of glucose fermentation into ethanol. However, this was not the case for the two-step xylose isomerization: Xyl1 generates a mixture of NAD^+ and $NADP^+$ during xylose reduction; the generated NAD^+ could not support the oxidation of all the formed xylitol. $NADP^+$ could not be easily used to regenerate NAD^+ as this reaction requires a transhydrogenase which is apparently absent in yeasts (Bruinenberg et al., 1985). Additional NAD^+ is frequently used to oxidize the remaining fraction of xylitol, producing excess NADH in the process. This results in the depletion of NAD^+ and accumulation of NADH, creating a state of redox imbalance that is unfavorable for cellular processes. To achieve redox balance, the excess NADH could donate their electrons to the electron transport chain during respiration. During anaerobic growth or when respiration is inhibited; however, redox balance would likely be achieved by the reduction of xylulose back into xylitol. This view was supported by the observation that **PUA6-9** expressing $^{Ps}xyl1$ and $^{Ps}xyl2$ secreted about half of the xylitol it derived from xylose during anaerobic fermentation of xylose (Kötter and Ciriacy, 1993). The inability for this strain to grow on xylose during respiratory inhibition could be explained by the poor xylose utilization efficiency: reduction of xylose into xylitol consumed cellular resources in the form of NADPH, but half of the xylitol generated could not be further utilized and was secreted. The reduced carbon flux into central carbon metabolism coupled with rapid depletion of NADPH may have resulted in a bottleneck restricting cell growth. In a parallel experiment where xylose fermentation was carried out aerobically, **PUA6-9** expressing $^{Ps}xyl1$ and $^{Ps}xyl2$ still secreted about a quarter of the xylitol it derived from xylose. This showed that respiration alone was not sufficient to sink the excess NADH that was being produced by Xyl2. From these results, it was evident that a problem of redox cofactor imbalance arises during the two-step conversion of xylose to xylulose and this affects xylose utilization in both aerobic and anaerobic conditions.

10.3.3 Xylulose isomerase–based xylose utilization pathway

Using the direct isomerization route for converting xylose into xylulose would bypass the problem of cofactor imbalance. As stated earlier, attempts to express functional XylA from various bacteria including *E. coli*, *Bacillus subtilis* (Amore et al., 1989), and *Clostridium thermosulfurogenes* (Moes et al., 1996) in *S. cerevisiae* had not been successful. In *S. cerevisiae* expressing *E. coli* or *B. subtilis* XylA, the full-length proteins could be detected by western blot analysis, but almost no enzyme activity was detected from the cell lysate. The main hypothesis for the lack of enzyme activity was the improper folding of these proteins in a non-native environment. Protein chaperones are molecular catalysts that aid in protein folding. The type and molecular mechanism of chaperones differ across different organisms (Saibil, 2013). In a recent study, Xia et al. demonstrated that coexpression of an *E. coli* chaperone GroEL/ES could rescue the inactive *E. coli* XylA expressed in *S. cerevisiae* (Xia et al., 2016). A similar approach was used to recover the activity of *Propionibacterium acidipropionici* XylA in *S. cerevisiae* (Temer et al., 2017). These results support the hypothesis that the XylAs expressed in the earlier studies were not folded well in *S. cerevisiae*, likely due to the dependence of these proteins on a bacterial chaperone system for proper folding. It was thus not surprising that the first example of a highly active XylA that could be expressed independently in *S. cerevisiae* originated from a fungal genus *Piromyces* (Kuyper et al., 2003b). Kuyper et al. isolated the *xylA* gene from *Piromyces* sp. **E2** and constructed a plasmid for its expression in *S. cerevisiae* **CEN.PK113-7D**. The resulting *S. cerevisiae* strain **RWB202** displayed high xylose isomerase activity (1.1 U/ mg protein) and could grow slowly on xylose in aerobic conditions (specific growth rate of 0.005 h^{-1}). The slow growth of **RWB202** on xylose was expected as its parent strain **CEN.PK113-7D** had not been selected or engineered for xylulose utilization.

10.3.4 Improving xylose utilization based on the xylulose isomerase pathway

The specific growth rate of **RWB202** at aerobic condition was improved to 0.12 h^{-1} after 30 serial transfers of the culture in chemically defined medium with xylose as a sole carbon source. **RWB202** and the evolved strain could not grow under anaerobic condition. To enable this phenotype, the evolved strain went through another 10 cycles of serial transfers under oxygen-limited condition in bioreactor. The new strain was able to grow at anaerobic condition, and was then further evolved through 20 cycles of serial transfers under strict anaerobic condition. The adaptive laboratory evolution (ALE, also known as evolutionary engineering) resulted in *S. cerevisiae* **RWB202-AFX**, which had a specific growth rate of 0.03 h^{-1} at anaerobic condition and could convert 20 g/L xylose into 8.4 g/L ethanol (82% of the theoretical yield). Only a small amount of xylitol (0.4 g/L) accumulated, which made the ethanol yield to be much better than the highest yield obtained prior to the study (47% of theoretical yield, based on the xylitol route). This fact highlighted the redox imbalance issue caused by the xylitol route and the clear advantage of using the xylulose isomerization route. As with this study, many well-known works of improving *S. cerevisiae* growth on xylose were based on XylA.

ALE is an effective tool that is fortunately available for improving utilization of most new carbon sources, because it is straightforward to associate the carbon source utilization with cell growth. Through ALE, the host's genes may be mutated, recombined, deleted, and/or duplicated. These genetic operations are very effective in improving a phenotype that has been enabled by rational genetic engineering (RGE). But ALE alone may be insufficient in enabling a difficult-to-gain phenotype. For example, ALE so far has failed to enable a wild-type *S. cerevisiae* strain to gain the function of xylose isomerase. Thus, ALE is frequently combined with RGE—RGE is used to create new strains and ALE is applied to improve these strains. Generally, a better starting strain resulted in superior fermentation performance after ALE.

As an example, **RWB202** was rationally engineered to overexpress Xyl3 and the four enzymes involved in the nonoxidative PPP (Rpi, Rpe, Tkl, and Tal, Fig. 10.5), with the hypothesis that some of these enzymes was limiting the anaerobic growth on xylose. A gene encoding aldose reductase (*gre3*) was also deleted to reduce xylitol accumulation. Without ALE, the resulting strain (**RWB217**) was able to grow at anaerobic condition, with a specific growth rate of 0.09 h^{-1}, which was three times that of **RWB202-AFX**. The Pronk laboratory (Delft U. of Tech.) which developed **RWB217** later focused on evolving the strain for utilizing mixed sugars, because hydrolysates of lignocellulosic biomass usually contain a mixture of glucose, xylose, and arabinose. As it is difficult to compare fermentation performance between using xylose alone and sugar mixture, we will first discuss a subsequent work done by the Stephanopoulos laboratory (MIT), which focused on evolving a strain with a similar genetic background (**H131-A3**) for anaerobic growth on xylose (Zhou et al., 2012).

Similar to **RWB217**, **H131-A3** overexpressed the four enzymes in the nonoxidative PPP from its genome and overexpressed XylA and Xyl3 from replicative plasmids. After **H131-A3** was evolved in a batch operation mode (through serial transfers) under aerobic condition, the obtained strain was characterized at anaerobic condition and the specific growth rate was 0.05 h^{-1}. The specific growth rate was improved to 0.07 h^{-1} (the strain was **H131-A3^{SB-3}**) after further evolution in a batch operation mode under microaerobic and anaerobic conditions (the oxygen supply was gradually reduced to challenge the strain). The strain was further challenged through evolution in a continuous mode, in which the dilution rate was gradually increased to selectively retain faster growing cells. At the end of this evolution, one resulting strain (**H131-A3^{CS}**) had a specific growth rate of 0.12 h^{-1}, which superseded **RWB217**. After two minor genetic modifications to remove the auxotrophic markers in **H131-A3^{CS}** and medium optimization (addition of ergosterol and Tween 80), the new strain (**H131-A3-AL^{CS}**) had a much higher specific growth rate (0.2 h^{-1}) at anaerobic condition and was able to produce ethanol at 80% of the theoretical yield. Each auxotrophic marker required supplementation of an amino acid, which would increase fermentation cost when deployed in large scale. Ergosterol was added based on the hypothesis that its synthesis limited cell growth under the anaerobic condition. In the future, ergosterol biosynthesis may be manipulated to overcome the limitation without ergosterol supplementation. Tween 80 was a surfactant that may reduce cell aggregation and improve cell growth.

10.3.5   Understanding how ALE improved xylose utilization

An important part of the study done by Zhou et al. was that extensive works were done to understand the altered genotypes responsible for the obtained phenotype (high specific growth rate on xylose under anaerobic condition) (Zhou et al., 2012). Fragments of the genomic DNA of **H131-A3^{SB-3}** were introduced back on replicative plasmid to its ancestor strain **H131-A3**, producing a library of strains. The strain library was screened by using a water-in-oil droplet-based microfluidic device, which did not need serial transfers and thus minimized introducing new genetic changes. Genotype of the faster growing cells was characterized. Because the fragments were carried on replicative plasmids, they can be easily analyzed through Sanger sequencing and restriction enzyme digestion. It was found that many genomic DNA fragments contained two to three copies of the XylA expression cassette, suggesting that multiple copies of the cassette were integrated into the yeast genome. Further quantitative polymerase chain reaction (PCR) analysis revealed that **H131-A3**^{CS} had as high as 50 copies of *xylA* in its chromosomes per cell (the location of the multicopy *xylA* integration was determined to be in Chromosome VI). The transcription level and enzymatic activity of XylA in **H131-A3**^{CS} were also found to be approximately 10 times those in the ancestor strain **H131-A3**. All the data suggest that XylA catalyzed the rate-limiting step. Lastly, a global transcriptome analysis done with a microarray technology showed that transcription level of many genes (involved in PPP, glucogenesis, and production of ethanol and acetate) was fine-tuned through the ALE.

It is important to understand the genotype and mechanism behind a phenotype obtained through ALE. The generated knowledge would not only deepen our fundamental understanding of cellular metabolism but also identify potential rate-limiting steps for subsequent research efforts. The fact that the *xylA* gene was replicated in yeast chromosome to a level as high as 50 copies per cell suggested that specific activity of XylA needs improvement. This could be achieved by using DE of XylA. Lee et al. randomly mutated *xylA* by using error-prone PCR (Lee et al., 2012) and introduced a pool of the mutated genes into a strain that overexpressed ^{Ss}*tal1* and had *gre3* deleted. The gene mutants were carried by replicative plasmids. Because activity of XylA was associated with cell growth, XylA mutants with higher expression level or specific activity could be selected through growth competition in a way that is similar to ALE (5–7 serial transfers). The mutations introduced into *xylA* could be easily determined through Sanger sequencing, and this DE of XylA could be iterated to accumulate beneficial mutations. After three iterations, the maximal velocity (V_{max}) of XylA was almost doubled, which could be attributed to elevated enzyme expression level ([E]) and/or specific enzymatic activity (k_{cat}). It was demonstrated that the XylA mutant could substantially improve cell growth on xylose under microaerobic condition compared with the wild-type XylA. It would be useful to test the obtained XylA mutant in a sophisticatedly engineered xylose utilization strain, such as **RWB217** or **H131-A3-AL**^{CS}.

10.3.6 Alleviating catabolite repression in S. *cerevisiae* through adaptive laboratory evolution

When a species is exposed to multiple carbon substrates, the species may choose one substrate and only start to utilize a second substrate after the depletion of the first substrate. Such traits were selected through natural evolution to maximize a species' survival. For example, the substrate being utilized first may enable the species to grow faster (related to specific growth rate) and/or more efficiently (related to biomass formation yield) than the other substrates. Such phenomenon is known as catabolite repression. The molecular mechanism enabling catabolite repression is complex, including selective substrate transporters and intracellular signaling. Catabolite repression is often undesired in industrial biotechnology. For example, if a yeast utilized glucose and xylose in hydrolysate of lignocellulosic biomass sequentially, the productivity would be lower, because the yeast would need to have a second lag phase between growth on glucose and xylose (such growth behavior is termed as diauxic growth).

RWB217 was an excellent *S. cerevisiae* strain developed in the Pronk laboratory for utilizing xylose, but it was found to exhibit diauxic growth when growing on a mixture of glucose and xylose. ALE was employed to address the problem (Kuyper et al., 2005). **RWB217** was first evolved on xylose by using an anaerobic continuous culture mode. The steady state shifted over evolution time with xylose concentration decreased to 30% of that at the initially reached steady state. The evolved strain grew better on the sugar mixture when characterized. The strain was subjected to another round of evolution on a mixture of glucose and xylose. The new strain (**RWB218**) still exhibited diauxic growth but with a minimal lag phase. Within 24 h, **RWB218** was able to completely consume 100 g/L glucose and 25 g/L xylose with a low inoculation density (1.1 g/L dry cell weight). As a comparison, **RWB217** needed approximately 40 h to consume 25 g/L glucose and 25 g/L xylose. The authors attributed the success of the ALE to how they set up the second round of evolution. The ratio of glucose to xylose was found to be the key to the successful evolution—a large amount of xylose must be provided to allow substantial growth on xylose after depletion of glucose.

Through growth kinetics characterization done by using 14C-labelled sugars in a continuous culture setup, **RWB218** was found to have a lower value (15 g/L) of the Michaelis–Menten constant (K_m) for xylose than **RWB217** (20 g/L). K_m is the concentration of a substrate at which the reaction rate is half of the maximum reaction rate. A lower value indicates a better affinity of the enzyme with the substrate and results in higher reaction rate when the substrate binding sites are not saturated. Despite **RWB218** also had a higher K_m value (0.9 g/L) for glucose than **RWB217** (0.2 g/L), its K_m value for xylose was still higher than that for glucose by more than one order of magnitude, explaining why the growth behavior was still diauxic. The molecular mechanisms behind the altered growth kinetics of **RWB218** were not elucidated in the study, and could involve mutations introduced to the sugar transporters.

10.3.7   Alleviating catabolite repression in *S. cerevisiae* through sugar transporter engineering

More recently, the Alper laboratory (UT Austin) and the Boles laboratory (Goethe U.) focused on addressing the glucose/xylose co-utilization challenge through DE of sugar transporters.

Young et al. from the Alper laboratory first screened 26 sugar transporters in *S. cerevisiae* **EY12**, which harbored replicative plasmids to overexpress *xyl1* and *xyl2* from *Scheffersomyces stipitis* and did not have any sugar transporter gene (Young et al., 2011). **EY12** was derived from *S. cerevisiae* **EBY.VW4000**, a transporter-null strain (Wiedemann and Boles, 2008). **EY12** could not grow on glucose or xylose. Each of the 26 transporter-encoding genes was expressed in **EY12** by using a replicative plasmid, and the transformants were tested in a medium containing xylose as the sole carbon source. Seven transporters were found to enable growth on xylose. The yeast strains expressing these seven transporters (one transporter per strain) were further characterized in a medium containing 20 g/L glucose and 20 g/L xylose. All the strains were still found to prefer glucose over xylose. *S. stipitis* Xut1 ([Ss]Xut1) and *Cattleya intermedia* Gxs1 ([Ci]Gxs1) were the top two in terms of the xylose/glucose preference ratio, which was defined as the amount of consumed xylose divided by that of consumed glucose within the fermentation period (the ratio of [Ss]Xut1: 0.69; [Ci]Gxs1: 0.51). These two transporters were subjected to DE in their subsequent study (Young et al., 2012).

Similar to the workflow described above for DE of *xylA*, each of the transporter-coding gene (^{Ss}*xut1* or ^{Ci}*gxs1*) was mutated via error-prone PCR and the resulting yeasts carrying the gene mutants were selected by using growth in a medium with xylose as the sole carbon source. After a few rounds of mutagenesis and selection, improved mutants were identified for both transporters in terms of K_m and/or V_{max}, but none of the mutants enabled simultaneous utilization of glucose and xylose—the cell growth was still diauxic. When the sequence of the mutants was analyzed, a residue (Phe^{40}) of ^{Ci}Gxs1 caught the authors' attention because three substitutions at this position arose independently in one iteration of the evolution. By using bioinformatics to re-analyze the sequence of the 26 transporter they previously studied (Young et al., 2011), the same team found that Phe^{40} was part of a highly conserved motif (G-G/F-XXXG, where each X is a variable residue and Phe^{40} was mapped to the underlined X) (Young et al., 2014). They further used site-saturation mutagenesis to study each variable residue (the amino acid at each position was replaced by using each of the other 19 amino acids). The obtained mutants were characterized in terms of cell growth rate in a medium containing xylose as the sole carbon source. The best substitutions for each of the three variable residues were combined to create a triple mutant, which disabled growth on glucose and resulted in better growth on xylose than the wild-type ^{Ci}Gxs1. However, the growth was still diauxic when both glucose and xylose were present. This was because glucose could still bind the mutated ^{Ci}Gxs1 with higher affinity than xylose.

The Boles laboratory subsequently made a breakthrough toward the glucose-xylose co-utilization when they used a new evolution strategy (Farwick et al., 2014). They genetically modified **EBY.VW4000** so that it could not utilize glucose even when it is imported into the cell—all the genes encoding hexokinases were deleted. The xylose isomerization pathway was further established in the new strain by overexpressing the relevant genes, including those in nonoxidative PPP. The obtained strain (**AFY10X**) was a suitable host for selecting xylose transporters that are not repressed by glucose.

When **AFY10X** was engineered to overexpress an endogenous promiscuous transporter (Gal2) that can import xylose, the resulting strain could grow on xylose but not on glucose (due to lack of hexokinase activity); the strain could not grow on xylose when the glucose concentration was >5 g/L because Gal2 could be still occupied by glucose. Under these constraints, it became possible to select Gal2 mutants that could import xylose in the presence of glucose. Error-prone PCR was used to mutate *gal2*, and the mutant genes were expressed in **AFY10X**. The resulting strains were cultured in a medium containing both xylose and glucose. The glucose concentration was gradually increased from 2 g/L to 20 g/L to select the mutants that were less sensitive to the glucose inhibition through serial transfers. Most of the isolated yeast clones had a mutation to Asn^{376}, which led Farwick et al. to study the residue by using site-saturation mutagenesis and find the mutation Asn^{376}→Phe (N376F). Gal2^{N376F} had a much better affinity with xylose than Gal2 (K_m value decreased from 33 g/L to 14 g/L), and did not support cell growth on glucose (Farwick et al., 2014).

The Alper laboratory used an approach similar to what was used by Farwick et al., and identified the residue in ^{Ci}Gxs1 that corresponded to Asn^{376} of Gal2 (Li et al., 2016). Li et al. continued to iteratively improve ^{Ci}Gxs1 through DE and the final variant of ^{Ci}Gxs1 they obtained enabled simultaneous utilization of glucose and xylose when it was introduced into an **EBY.VW4000** derivative that carried essential xylose utilization genes.

These recent developments in improving yeast growth on xylose provide additional examples to discuss the relationships between ALE and RGE. New constraints were introduced by RGE for more effective selection of desired mutants during ALE (Farwick et al., 2014). Once a mutation lead was identified after ALE, it could be further studied through site-saturation mutagenesis, leading to discovery of better mutations (Farwick et al., 2014; Li et al., 2016). Multiple beneficial mutations can be easily combined through modern molecular cloning techniques.

10.4   Conclusions and perspectives

In this chapter, we provided a few examples to discuss how genetic engineering of microbes could be used to produce new product and to enable utilization of new feedstock. 1,4-BDO is a very useful product but no wild-type organism is known to produce it. It was thus difficult to develop a microbial strain to make 1,4-BDO without genetic engineering. The example we discussed involved establishment of segments of the pathway, integration of the segments, and optimization of the entire pathway. As all the carbon atoms, energy, reducing equivalents, and catalysts needed to synthesize 1,4-BDO were all provided by the host metabolism, multiple modifications to the central metabolism of the host were needed to reduce byproduct formation and to create a stronger force driving the desired product synthesis. Engineering Baker's yeast to grow fast and efficiently on xylose at anaerobic condition was another difficult-to-gain phenotype that was made possible by genetic engineering. Thanks to the easy association between xylose utilization and cell growth, ALE played a critical role in developing the yeast strains. ALE could be used to select beneficial spontaneous mutations across the entire genome and mutations introduced by genetic tools (e.g. error-prone PCR) to a much smaller genetic region. The synergy between ALE and RGE has been observed in a few studies covered in this chapter.

Currently, there is no generic method to associate a product formation with cell growth. Innovations enabling such association would be instrumental in rapid improvement of desired product formation through growth selection. New tools in sequencing, synthesizing, and manipulating genetic materials would also be crucial to improving our ability of producing new products and using new feedstocks. In the xylose utilization story, we have already observed how much the technological advancement has transformed the methodologies used to approach similar problems. For example, while it was a major advance to clone and sequence a gene (e.g., $^{Ps}xyl1$) in early 1990s, more than 40 transporter genes can be easily synthesized in early 2010s based on the information in public genetic sequence database and screened in highly engineered host strains. Illumina sequencing technologies are already frequently used in the field to rapidly elucidate changes in genome and transcriptome, and based on our prediction would be routinely used in the future. Nanopore sequencing technologies will provide long reads that are very useful in revealing highly repetitive sequences such as the region in which 50 copies of *xylA* were integrated. There are many other emerging technologies that are changing the field, including microfluidic screening, robot-based lab operations and artificial intelligence-powered *in-silico* screening and optimization. With these new tools, we are confident that more industrial processes based on genetically engineered microbes will be established to transform wastes into valuable molecules.

Acknowledgment

KZ acknowledges the following research grants that supported research in his laboratory on upcycling waste streams by using genetically engineered microbes: a research grant from Ministry of Education Singapore (R-279-000-594-112); a research grant from Singapore Millennium Foundation (R-279-000-516-592). These projects established the expertise in the laboratory that is critical to completing this chapter. JZ is financially supported by a PhD scholarship from National University of Singapore.

References

Amore, R., Kötter, P., Küster, C., Ciriacy, M., Hollenberg, C.P., 1991. Cloning and expression in *Saccharomyces cerevisiae* of the NAD(P)H-dependent xylose reductase-encoding gene (XYL1) from the xylose-assimilating yeast *Pichia stipitis*. Gene 109, 89–97.

Amore, R., Wilhelm, M., Hollenberg, C.P., 1989. The fermentation of xylose —an analysis of the expression of Bacillus and Actinoplanes xylose isomerase genes in yeast. Appl. Microbiol. Biotechnol. 30, 351–357.

Barton, N.R., Burgard, A.P., Burk, M.J., Crater, J.S., Osterhout, R.E., Pharkya, P., Steer, B.A., Sun, J., Trawick, J.D., Van Dien, S.J., Yang, T.H., Yim, H., 2015. An integrated biotechnology platform for developing sustainable chemical processes. J. Ind. Microbiol. Biotechnol. 42, 349–360.

Bruinenberg, P.M., Jonker, R., van Dijken, J.P., Scheffers, W.A., 1985. Utilization of formate as an additional energy source by glucose-limited chemostat cultures of *Candida utilis* CBS 621 and *Saccharomyces cerevisiae* CBS 8066. Arch. Microbiol. 142, 302–306.

Burgard, A., Burk, M.J., Osterhout, R., Van Dien, S., Yim, H., 2016. Development of a commercial scale process for production of 1,4-butanediol from sugar. Curr. Opin. Biotechnol. 42, 118–125.

Chavali, A.K., Rhee, S.Y., 2018. Bioinformatics tools for the identification of gene clusters that biosynthesize specialized metabolites. Brief. Bioinform. 19, 1022–1034.

Dale, J.M., Popescu, L., Karp, P.D., 2010. Machine learning methods for metabolic pathway prediction. BMC Bioinform 11, 15.

Farwick, A., Bruder, S., Schadeweg, V., Oreb, M., Boles, E., 2014. Engineering of yeast hexose transporters to transport D-xylose without inhibition by D-glucose. Proc. Natl. Acad. Sci. USA 111, 5159–5164.

Gleizer, S., Ben-Nissan, R., Bar-On, Y.M., Antonovsky, N., Noor, E., Zohar, Y., Jona, G., Krieger, E., Shamshoum, M., Bar-Even, A., Milo, R., 2019. Conversion of *Escherichia coli* to generate all biomass carbon from CO_2. Cell 179, 1255–1263 e12.

Grove, T.Z., Magliery, T.J., 2020. Editorial overview: From powerful tools to useful products: protein engineering after 35 years of directed evolution. Curr. Opin. Struct. Biol. 63, vi–viii.

Jeffries, T.W., et al., 1983. Utilization of xylose by bacteria, yeasts, and fungi. In: Chan, Y.K. et al (Ed.), Pentoses and Lignin. Springer Berlin Heidelberg, Berlin, Heidelberg, pp. 1–32.

Jewell, J.B., Coutinho, J.B., Kropinski, A.M., 1986. Bioconversion of propionic, valeric, and 4-hydroxybutyric acids into the corresponding alcohols by *Clostridium acetobutylicum* NRRL 527. Curr. Microbiol. 13, 215–219.

Kötter, P., Amore, R., Hollenberg, C.P., Ciriacy, M., 1990. Isolation and characterization of the *Pichia stipitis* xylitol dehydrogenase gene, XYL2, and construction of a xylose-utilizing *Saccharomyces cerevisiae* transformant. Curr. Genet. 18, 493–500.

Kötter, P., Ciriacy, M., 1993. Xylose fermentation by *Saccharomyces cerevisiae*. Appl. Microbiol. Biotechnol. 38, 776–783.

Kuyper, M., Harhangi, H.R., Stave, A.K., Winkler, A.A., Jetten, M.S., de Laat, W.T., den Ridder, J.J., Op den Camp, H.J., van Dijken, J.P., Pronk, J.T., 2003a. High-level functional expression of a fungal xylose isomerase: the key to efficient ethanolic fermentation of xylose by *Saccharomyces cerevisiae*? FEMS Yeast Res. 4, 69–78.

Kuyper, M., Toirkens, M.J., Diderich, J.A., Winkler, A.A., van Dijken, J.P., Pronk, J.T., 2005. Evolutionary engineering of mixed-sugar utilization by a xylose-fermenting *Saccharomyces cerevisiae* strain. FEMS Yeast Res. 5, 925–934.

Lee, S.M., Jellison, T., Alper, H.S., 2012. Directed evolution of xylose isomerase for improved xylose catabolism and fermentation in the yeast *Saccharomyces cerevisiae*. Appl. Environ. Microbiol. 78, 5708–5716.

Li, H., Schmitz, O., Alper, H.S., 2016. Enabling glucose/xylose co-transport in yeast through the directed evolution of a sugar transporter. Appl. Microbiol. Biotechnol. 100, 10215–10223.

Liu, H., Lu, T., 2015. Autonomous production of 1,4-butanediol via a de novo biosynthesis pathway in engineered *Escherichia coli*. Metab. Eng. 29, 135–141.

Liu, Y., Esen, O., Pronk, J.T., van Gulik, W.M., 2021. Uncoupling growth and succinic acid production in an industrial *Saccharomyces cerevisiae* strain. Biotechnol. Bioeng. 118, 1576–1586.

Ma, X., Gozaydin, G., Yang, H., Ning, W., Han, X., Poon, N.Y., Liang, H., Yan, N., Zhou, K., 2020. Upcycling chitin-containing waste into organonitrogen chemicals via an integrated process. Proc. Natl. Acad. Sci. USA 117, 7719–7728.

Moes, C.J., Pretorius, I.S., van Zyl, W.H., 1996. Cloning and expression of the *Clostridium thermosulfurogenes* D-xylose isomerase gene (xyLA) in *Saccharomyces cerevisiae*. Biotechnol. Lett. 18, 269–274.

Ni, Z., Stine, A.E., Tyo, K.E.J., Broadbelt, L.J., 2021. Curating a comprehensive set of enzymatic reaction rules for efficient novel biosynthetic pathway design. Metab. Eng. 65, 79–87.

Panda, S., Fung, V.Y.K., Zhou, J.F.J., Liang, H., Zhou, K., 2021. Improving ethylene glycol utilization in *Escherichia coli* fermentation. Biochem. Eng. J. 168, 107957.

Porep, H., Ciriacy, M., 1986. Conversion of pentoses to ethanol by baker's yeast. In: Magnien, E. (Ed.), Biomolecular Engineering in the European Community: Achievements of the Research Programme (1982 –1986) — Final Report. Springer Netherlands, Dordrecht, pp. 675–681.

Primrose, S.B., Twyman, R., 2013. Principles of Gene Manipulation and Genomics. John Wiley & Sons, NJ.

Rodriguez-Peña, J.M., Cid, V.J., Arroyo, J., Nombela, C., 1998. The YGR194c (XKS1) gene encodes the xylulokinase from the budding yeast *Saccharomyces cerevisiae*. FEMS Microbiol. Lett. 162, 155–160.

Saibil, H., 2013. Chaperone machines for protein folding, unfolding and disaggregation. Nat. Rev. Mol. Cell Biol. 14, 630–642.

Shelp, B.J., Zarei, A., 2017. Subcellular compartmentation of 4-aminobutyrate (GABA) metabolism in arabidopsis: an update. Plant Signal. Behav. 12, e1322244.

Temer, B., dos Santos, L.V., Negri, V.A., Galhardo, J.P., Magalhães, P.H.M., José, J., Marschalk, C., Corrêa, T.L.R., Carazzolle, M.F., Pereira, G.A.G., 2017. Conversion of an inactive xylose isomerase into a functional enzyme by co-expression of GroEL-GroES chaperonins in *Saccharomyces cerevisiae*. BMC Biotech. 17, 71.

Wang, F., Zhao, J., Li, Q., Yang, J., Li, R., Min, J., Yu, X., Zheng, G.W., Yu, H.L., Zhai, C., Acevedo-Rocha, C.G., Ma, L., Li, A., 2020. One-pot biocatalytic route from cycloalkanes to alpha,omega-dicarboxylic acids by designed *Escherichia coli* consortia. Nat. Commun. 11, 5035.

Wang, P.Y., Schneider, H., 1980. Growth of yeasts on D-xylulose. Can. J. Microbiol. 26, 1165–1168.

Wiedemann, B., Boles, E., 2008. Codon-optimized bacterial genes improve L-Arabinose fermentation in recombinant *Saccharomyces cerevisiae*. Appl. Environ. Microbiol. 74, 2043–2050.

Xia, P.-F., Zhang, G.-C., Liu, J.-J., Kwak, S., Tsai, C.-S., Kong, I.I., Sung, B.H., Sohn, J.-H., Wang, S.-G., Jin, Y.-S., 2016. GroE chaperonins assisted functional expression of bacterial enzymes in *Saccharomyces cerevisiae*. Biotechnol. Bioeng. 113, 2149–2155.

Yim, H., Haselbeck, R., Niu, W., Pujol-Baxley, C., Burgard, A., Boldt, J., Khandurina, J., Trawick, J.D., Osterhout, R.E., Stephen, R., Estadilla, J., Teisan, S., Schreyer, H.B., Andrae, S., Yang, T.H., Lee, S.Y., Burk, M.J., Van Dien, S., 2011. Metabolic engineering of *Escherichia coli* for direct production of 1,4-butanediol. Nat. Chem. Biol. 7, 445–452.

Young, E.M., Comer, A.D., Huang, H., Alper, H.S., 2012. A molecular transporter engineering approach to improving xylose catabolism in *Saccharomyces cerevisiae*. Metab. Eng. 14, 401–411.

Young, E.M., Tong, A., Bui, H., Spofford, C., Alper, H.S., 2014. Rewiring yeast sugar transporter preference through modifying a conserved protein motif. Proc. Natl. Acad. Sci. USA 111, 131–136.

Young, E., Poucher, A., Comer, A., Bailey, A., Alper, H., 2011. Functional survey for heterologous sugar transport proteins, using *Saccharomyces cerevisiae* as a host. Appl. Environ. Microbiol. 77, 3311–3319.

Zhang, Y., Li, Z., Liu, Y., Cen, X., Liu, D., Chen, Z., 2021. Systems metabolic engineering of *Vibrio natriegens* for the production of 1,3-propanediol. Metab. Eng. 65, 52–65.

Zhou, H., Cheng, J.S., Wang, B.L., Fink, G.R., Stephanopoulos, G., 2012. Xylose isomerase overexpression along with engineering of the pentose phosphate pathway and evolutionary engineering enable rapid xylose utilization and ethanol production by *Saccharomyces cerevisiae*. Metab. Eng. 14, 611–622.

Zhou, K., Ng, W., Cortes-Pena, Y., Wang, X., 2020. Increasing metabolic pathway flux by using machine learning models. Curr. Opin. Biotechnol. 66, 179–185.

Anaerobic digestion *via* codigestion strategies for production of bioenergy

Wangliang Li

CAS Key Laboratory of Green Process and Engineering, Institute of Process Engineering, University of Chinese Academy of Sciences, Beijing, China

11.1 Introduction

Anaerobic digestion (AD) is a useful method for producing biogas from renewable biomass feedstocks, using specific microorganisms in the absence of oxygen. AD is an environmentally friendly and green way for disposing of organic solid waste, for its high efficiency in energy recovery in the form of biogas (55%–65% CH_4 and 35%–45% CO_2) and low emission of secondary pollutants (Wang & Lee, 2021). A wide range of organic biomass materials including municipal, agricultural, and industrial wastes, plant residues, animal manure, etc. can be decomposed in an AD process. Considering the energy recovery efficiency, AD processes show obvious advantages for the treatment of wet wastes, which are problematic for thermochemical methods such as combustion, pyrolysis, and gasification (Zamri et al., 2021).

A typical AD process involves a series of metabolic reactions including hydrolysis, acidogenesis, acetogenesis, and methanogenesis. Hydrolysis is conversion of insoluble complex organic matter (carbohydrates, proteins, and lipids) into soluble molecules such as sugars, amino acids, and long-chain fatty acids (LCFAs). Hydrolysis reactions are carried out by extracellular enzymes called hydrolase that are produced by hydrolytic bacteria. In the acidogenesis step, LCFAs are converted to short-chain volatile fatty acids (VFAs) mainly including C1-C5 fatty acid, that is, formic, acetic, propionic, butyric, and valeric acids. Meanwhile, hydrogen, carbon dioxide, and acetic acid are produced. Acetate is produced by a group of bacteria called acetate-forming fermentative bacteria such as *Acetobacterium*, *Clostridium*, *Eubacterium*, and *Sporomusa*. In the acetogenesis step, the VFAs and other degraded products are fermented to form acetate, carbon dioxide, and/or hydrogen. At last, methanogens utilize acetate, carbon dioxide, and/or hydrogen to produce methane (Paul & Dutta, 2018).

Biomass, Biofuels, Biochemicals.
DOI: https://doi.org/10.1016/B978-0-323-90633-3.00016-X

According to the configuration of digester and the operation parameters, AD processes can be divided into (1) wet and dry (high-solid) processes, (2) monodigestion and codigestion, and (3) one-, two-, and multistage processes (Franca & Bassin, 2020).

11.2   Composition of organic wastes and their monodigestion performances

AD can process a variety of biomass materials such as organic fractions of municipal solid waste (MSW), sewage sludge, agricultural wastes, manure, and industrial wastes. However, the AD of these materials is facing up with various challenges such as materials recalcitrance and low volatile solids (VS) conversion ratio and the presence of inhibitors.

11.2.1   Lignocellulosic waste

Horticultural waste, agricultural residues, energy crops, and forest residues are most reported lignocellulosic substrate suitable for AD process, which is composed of three major components namely cellulose, hemicellulose, and lignin. Average values of these components of biomass are listed in Table 11.1 (Kumar et al., 2020b). It can be clearly noticed that the C/N ratio of most lignocellulosic biomass is much higher than the optimal range of 20–30, resulting in serious inhibition from the organic acids (Li, 2016). The methane potential of lignocellulosic biomass is high, but the rigid complex structure seriously inhibits the microbial conversion process. During the AD process, cellulose and hemicellulose are easily converted and degraded, both of which provide feedstocks for biogas production. However, in the lignocellulosic biomass cellulose is stabilized laterally and stiffened by hydrogen bonds, forming a rigid crystalline structure. The lignin content determines the biodegradation of cellulose and hemicellulose, and is negatively related to the biogas yield of lignocellulosic biomass (Shrestha et al., 2017). Compared with the soft wood, the content of cellulosic is higher, and the lignin content is much lower in hard wood (Paul & Dutta, 2018). Therefore, biogas yield of hard wood is much higher than that of soft wood. To improve the AD process and to increase biogas yield, pretreatments are efficient methods to obtain a higher biogas yield by breaking the rigid structure, expanding the surface area of biomass, and reducing the particle sizes of biomass. The hydrolysis of cellulose improves by increasing the accessibilities of enzymes to the surface of cellulose (Akhlisah et al., 2021; Kumari & Singh, 2018; Rezania et al., 2020). The pretreatment methods can be physical (mechanical, microwave, and ultrasound methods), chemical (using acids, alkalis, organic solvents, and ionic liquids), physicochemical (hydrothermal pretreatment/steam explosion, ammonia freeze/fiber explosion, wet oxidation, liquid hot water (LHW)), and biological method (using bacteria and fungi) (shown in Fig. 11.1) (Kumar et al., 2020a; Paul & Dutta, 2018).

11.2.2   Animal manure

The composition (e.g., moisture, protein and lignin) of manure vary considerably, which is closely related to livestock diets. The main feature of animal manure is the high nitrogen contents ranging from 35.42% to 87.44%. The average manure nitrogen contents for swine, poultry, and small ruminants (sheep and goats) were reported to be 19.49%, 9.75%, and 5.91%, respectively (Yao et al., 2020).

TABLE 11.1 Composition of lignocellulosic biomass (Kumar et al., 2020b).

Biomass name	Composition (%, dry basis)			C/N ratio
	Cellulose	Hemicellulose	lignin	
Rice straw	35–44	27–34	12–13	47–67
Cotton straw	42	12	15	–
Cotton stalk	31	11	28	41
Wheat straw	38–42	20–27	20–22	50–60
Soybean straw	38	16	16	50
Barley straw	38–48	21–25	11–26	71
Oats straw	33	23	21	95
Alfalfa straw	31	10	10	47
Rape seed straw	37	25	17	28
Corn stover	40	25–31	14–17	50–63
Corn cobs	45	25	15	123
Sugarcane bagasse	40–45	20–24	25–30	118–150
Poplar	45	21	24	103
Pine	25–44	26–32	28–48	140
Red maple/oak	39	33	23	250
Spruce	28–43	20–30	28–35	170
Soft wood stem	40	30	30	511
Switchgrass	36–45	28–30	12–26	90
Miscanthus	38–48	19–30	12–25	77
Pea vines	40	10	9	120
Green bean	17	16	8	11
Tomato pomace	39	5	11	–
Tomato plant	39	29	12	35
Cucumber plant	17	17	3	11
Pepper plant	18	12	8	13

The high nitrogen content significantly affects the AD process by ammonia inhibition, leading to the reduction in biogas yield. During the AD process, the organic nitrogen compounds are converted into ammonia, which leads to the fluctuation of pH and serious inhibition on AD process, especially for high-solid process. It is important to control the pH level to reduce the free ammonia and to mitigate its associated toxicity. One of the effective methods is the codigestion of the mixture of different biomass with high C/N ratio with manure to alleviate the inhibition of ammonia. Codigestion processes generally have higher pH buffering capacities and optimal C/N ratio in addition to the condition of increased biodegradable organic fraction.

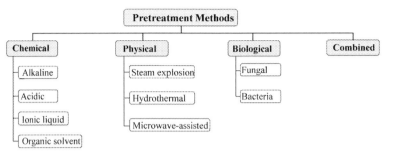

FIG. 11.1   Different pretreatment methods.

11.2.3   Food waste

Food waste (FW) is an important organic fraction in MSW, coming from household, food-processing, canteen, and restaurant. Globally, the annual production of MSW is around 2 billion tons, 32%–56% of which is organic biodegradable waste. It is estimated that the total amount of FW in Asian countries would increase to 4.16 billion tons by 2025. Especially in China, the growth rate was higher more than 10% due to the acceleration of urbanization (Ma et al., 2018). The total FW amount was about 91.1 million tons in 2015 in China. China's total annual production of FW is 3 times, 4.5 times, and 20 times that of the United States, Japan, and South Korea, respectively (Ma et al., 2018). The composition of FW may vary depending upon the country, region, seasons, cultural habits, as well as climate and economic level. Generally, FW moisture content is of 70%–80% (wet basis) and TS (total solids) are around 20%–30%, 90% of which are VS (Zhang, 2014). FW is composed mostly of easily degradable carbohydrates, proteins, and lipids. The ingredients of the Chinese household FW include vegetables (54%), rice (13%), pork, legumes, and fruit (13%), and the items of wheat, beverages, fish, and dairy products (Negri, 2020).

   The TS content of FW has a wide range from a dilute liquid to high-solids (up to 90%). The obvious characteristics of FW are easily biodegradable, rapid hydrolysis, and high methane yield. Unlike the lignocellulosic waste, the rate-limiting step of the AD processes of FW is the methanogenesis step, instead of the hydrolysis step. AD process has to be designed and adjusted to obtain optimal treatment efficiency based on the characteristics of the ingredients such as the contents of carbohydrates, proteins, and lipids (Xu et al., 2018a). The characteristics and methane yield of common FWs are summarized in Table 11.2 (Xu, 2019). As it can be seen, the FW with high lipids content such as fat, oil, and grease (FOG), and dairy waste have higher methane yields, while the waste with high fiber such as brewery, vegetable, and fruit wastes have relatively lower methane yields.

   Some of the ingredients in FW are rich in nitrogen, such as protein. The digestion of these ingredients with low C/N ratio produces excessive ammonia, leading to an increase of pH, inhibitory effects, and even process deterioration. The inhibitory concentrations of free ammonia nitrogen (FAN) and total ammonia nitrogen (TAN) were in a quite wide range, from 53 mg/L to 1450 mg/L and 1500–7000 mg/L (Akindele & Sartaj, 2018). On the other hand, FW is easily biodegraded and its rapid hydrolysis leads to the accumulation of VFAs

TABLE 11.2 Methane yield (MY) of common food waste (Xu, 2019).

Food wastes	Main components	TS%	C/N ratio	MY (m³/kg/VS)
FOG	Lipids, water, and other impurities.	1.3–3.2	22.1	0.4–1.1
Dairy industry wastewater	Water, milk solids, detergents, sanitizers, milk wastes, etc.	0.1–7	11.4–13.6	0.1–0.85
Household and restaurant food waste	Nonedible portions of food and uneaten food.	4.0–41.5	11.4–36.4	0.46–0.53
Slaughter house waste	Blood, manure, hair, fat, feathers, and bones.	2.0–28.3	3–6	0.20–0.50
Waste pet food	Smashed meat, starch, dietary fiber.	86–93	10–25	0.15–0.50
Fruit and vegetable waste	Leaves, peels, pomace, skins, rinds, cores, pits, pulp, stems, seeds, twigs, and spoiled fruits and vegetables.	7.4–17.9	15.2–18.9	0.16–0.35
Brewery waste	Spent brewer's grain, yeast, hops and trub.	23.0–29.2	18.8–54.9	0.22–0.31

and process of acidification because the rate of hydrolysis and acidogenesis is higher than that of methanogenesis. Thus, the enrichment and proliferation of acid-producing bacteria inhibited the activity of methanogens. How to improve the buffering capacity without VFA and NH_4^+ accumulation has become a hot topic in AD of FW (Ren et al., 2018).

11.3 Anaerobic codigestion

11.3.1 Definition of codigestion

Anaerobic codigestion (ACoD) (Fig. 11.2) is defined as a biological process that treats different types of substrates with the aim of enhancing methane production, increasing process efficiency, and improving process stability. This strategy could serve to balance the content of nutrients and reduced the negative effects of substrate inhibitors in the AD process. Agricultural residues, animal manures, MSW, and industrial wastes can be utilized as cosubstrates for biogas production. The combination of cosubstrates offers a better C/N ratio balance, dilute the inhibitory toxic compounds, and supple trace nutrient elements. Thus, codigestion is a promising way to solve the limitations of monodigestion by stabilizing process, increasing biogas production, improving the economy of AD (Mata-Alvarez et al., 2014).

11.3.2 Bioenergy recovery

As mentioned above, inhibition and instability always occur when one type of organic waste is used as a substrate alone due to lack of buffering capacity, lack of trace elements, or high TS content. Higher bioenergy recovery from organic waste can be achieved through

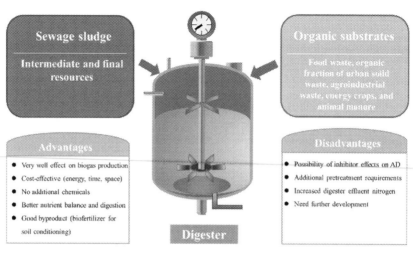

FIG. 11.2   ACoD process scheme, advantages, and drawbacks.

codigestion by means of the improved biogas production and enhanced process stability. For example, the methane yield of the codigestion of animal manure and FW was reported to be 249 L/kg VS, significantly higher than that of mono-AD of animal manure, 171 L/kg VS. In continuous digestion system, the cattle manure codigestion had the higher methane yield improvement of 124 L/kg VS, swine manure codigestion ranked second with 110 L/kg VS, and poultry manure ranked third with 70 L/kg VS. Improved methane yield was obtained at the C/N ratio in the range of 26–34 (Ma et al., 2020).

For different systems, the codigestion process shows higher methane or biogas yield than monodigestion. The methane yields of ACoD processes were associated with several factors such as substrate type and dose, VS, C/N ratio, organic loading rate (OLR), and hydraulic residence time (HRT). The AcoD of sugarcane rind (SR) or pretreated SR slurry (PSRS) with kitchen waste (KW) not only increased biogas yield but also was beneficial to the process stability. The codigestion of the mixture of PSRS and KW (mass ratio 1:3) had the highest methane production of 430.4 mL CH_4/g VS, which was 86% higher than 100% SR, respectively (Sun et al., 2020). The methane yield of the codigestion of FW and tylosin fermentation dreg was 14.8%–55.5% higher than that of monodigestion of FW (256 mL/gVS), indicating that codigestion is an effective method to increase biomethane yield and to remove tylosin (Gao et al., 2021). Compared with the AD of sewage sludge alone, codigestion of organic fraction of MSW (OFMSW) and grease trap sludge with sewage sludge resulted in an increase of methane yield and VS removal of up to 82% and 29%, respectively (Grosser et al., 2017). An increase of biogas yields up to 135.5% was obtained for codigestion of OFMSW with pretreated thickened waste activated sludge and pretreated rice straw (Abudi et al., 2016).

The highest cumulative specific methane yield of 225.7 mL/gVS was achieved by the codigestion of pretreated wheat straw and pretreated sludge, indicating that the adding sludge improved the buffer capacity of the system. The codigestion of sludge and wheat straw both pretreated at 175 °C obtained the maximum energy production of 7901.1 MJ/t,

52% higher than the monodigestion without pretreatment. The economic analysis showed that the monodigestion of wheat straw obtained relatively low net profits and the monodigestion of sludge pretreated at 175 °C achieved the highest net profit of US$31.44 per ton. The codigestion of both pretreated wheat straw and sludge can achieve the highest biogas yield and energy conversion efficiency (Tian et al., 2020).

Monodigestion of pig manure decreases direct greenhouse gas (GHG) emissions by 48% than direct land application. Codigestion of pig manure and grass silage increases the total energy recovery by 226% than monodigestion. Considering the utilization of nitrogen, the nitrogen in digestate (41.8%) is a little lower than that in raw pig manure (43.2%) due to the ammonia emissions (Zhang et al., 2021b).

The waste cardboard was used to improve the C/N in chicken manure (CM) and to enhance monodigestion system stability. ACoD of CM and cellulose-rich substances can significantly improve methane production. Codigestion of CM with cardboard produced 319.62 mL CH_4/g VS due to well-controlled VFAs concentration, 14.2% higher than monodigestion. Compared with the monodigestion process, codigestion enhanced system stability due to the match of metabolism between hydrolytic bacteria and methanogens. The energy balance assessment indicated that codigestion boosted the total energy output by nearly 38% and improved the energy recovery to 46.4% (Zhao et al., 2021a)

11.3.3 Process stability

11.3.3.1 *Influence of operating parameters on stability*

Typical process parameters of AD include temperature, C/N ratio, pH, OLR, HRT, alkalinity, and VFA concentrations. The pH level critically affects the activities of methanogens and the preferred pH range for AD is generally between 6.5 and 7.6. The AD process requires the joint work of several groups of microorganisms, among which the methanogens are the most sensitive to pH variations. Changes of operating parameters or production of inhibitors may result in process imbalance, performance decline, or even lead to the failure of process. On the contrary, in a well-operated system, a slight increase of pH is expected, because microorganisms produce alkalinity as they consume nitrogen-rich organic matter (Labatut and Pronto, 2018). VFAs such as acetic acid, butyric acid, and propionic acid are the intermediate products of methane production and the accumulation of the VFAs could lower the pH level to below 6, due to the inactivation of methanogenic microorganisms and the pathway changes. Alkali compounds or high nitrogen materials such as manure, sludge, cheese whey (CW) with high alkaline capacity could be used to neutralize the acids and mitigate the impact of the accumulation of VFAs (Neshat et al., 2017).

11.3.3.2 *Improvement of codigestion process stability*

Codigestion has the advantages of improving the robustness of AD system and preventing instability due to inhibition from VFAs and ammonia, etc. The TAN content was above 3000 mg/L in monodigestion of swine manure, which was greatly higher than that found in codigestion system of swine manure and rice straw (Lauterboeck et al., 2012; Rajagopal et al., 2013; Riya et al., 2016; Ryue et al., 2020; Yeniguen & Demirel, 2013). The codigestion process can dilute ammonia levels and stabilize the production of methane. Wang et al. (2014)

reported an increase in the C/N ratio caused the decrease of pH which diluted the TAN and FAN levels; while higher C/N ratios were attributed to lower protein solubilization rates that led to lower ammonia levels (Wang et al., 2014) .

In semicontinuous tests, monodigestion of FW could not support a high load and even failed (Zhang et al., 2015). FW codigestion with activated sludge can provide a high pH buffering capacity from ammonia and support high OLR. With regard to economic and environmental consideration, codigestion of FW and activated sludge is an efficient method to increase AD efficiency. The use of animal manure for codigestion of FW is also typical and widely reported because of its high pH buffering capacity from alkalinity, trace elements supplement, and rich micronutrients (Ebner et al., 2016). The concentration of VFAs in both monodigestion of cotton stalk and its codigestion with manure was 320 and 3000 mg/L, respectively. For both cases, the pH was about 6.5 without any significant change. The high VFA concentration can seriously inhibit the AD process or even lead to the process failure if the buffering capacity of the medium was not enough (Cheng & Zhong, 2014; Ma et al., 2018).

The ACoD of olive mill waste water (OMWW) has been proposed to enhance the AD performances, by mixing OMWW with other nitrogen-rich substrates such as animal manure and agro-industrial residues. The ACoD of OMWW and laying hen litter (LHL) and CW showed that the biogas yield was enhanced by 90% when OMWW was mixed with LHL, and 22% more was achieved with CW (Azbar et al., 2008). Agro-industrial residues such as slaughterhouse wastewater were also used to improve the AD of OMWW in terms of methane production and process stability (Gannoun et al., 2016).

For the horticultural waste with the high C/N ratio up to 35, high-solid anaerobic digestion (HSAD) process was seriously inhibited due to the accumulation of VFAs. The reasons of the rapid decrease of pH value and the accumulation of VFAs are that the growth rate of acidogenic bacteria was higher than that of the highly sensitive methanogenic archaea. During the semicontinuous codigestion of grass and horse manure (HM) process, the pH value was in the range of 7.33–8.43, slightly higher than the optimal pH for AD (from 6.8 to 7.5), which was due to a high nitrogen content in HM and sludge cake. The semicontinuous codigestion process had good stability with OLR in the range of 1.25–1.88 gVS/L day (Li, 2016). Adding alkali (ADAA) and codigestion are two methods to maintain the pH value and the process stability of HSAD of horticultural waste. The codigestion system would produce more methane than ADAA (CaO as alkali), because of more trace elements and richer microbial community in the codigestion process (Li et al., 2017a).

11.4 Microbial community in codigestion system

11.4.1 Microbials in codigestion system

The activities of microorganisms directly impact the AD efficiency, biogas yield, and process stability. The microbial community is the main driving force behind AD process, which could strongly affect the overall AD performance (Muratçobanoğlu et al., 2020). Meanwhile, the microbial population and distribution can be affected by various process conditions such as C/N ratio, pH level, trace elements nutrition, and pretreatment of substrates (Wang et al., 2018; Wei et al., 2021). The microbial community structure can be studied using various analytical techniques such as polymerase chain reaction, DNA extraction, high-throughput DNA sequencing, etc.

11.4.2 Effect of C/N ratio on microbial community

In a codigestion process, the C/N ratio can be adjusted by mixing and changing the mass ratio of different types of organic biomass. Appropriate C/N ratio in a range of 20–30 is one of the key factors for the stability of ACoD, which affects the population, community structures, and functions of bacteria and archaea (Li et al., 2018; Sung & Tao, 2003). Bacterial communities would change in the codigestion system with C/N ratio 20 and 30, which consequently influence the archaeal communities. A large decrease in *Halothermothrix* would lead to the decrease in *Methanothermobacter* because the latter produces hydrogen, and also the decreases in *Thermotogae*, which can degrade acetate syntrophically with hydrogenotrophic methanogens. In contrast, increases in *Methanosarcina* could be related to increases in bacteria capable of producing acetate, such as *Moorella*, which produces acetate from hydrogen and carbon dioxide (Turker et al., 2016).

In the monodigestion of cardboard, the accumulation of VFAs induced the proliferation of *Clostridia*. As one kind of *Clostridia*, the abundance of *Syntrophomonas* increased from 0.34% to 5.01%, which can convert butyrate and propionate to acetate and H_2. The excessive ammonia could inhibit the growth and metabiotic activity of hydrogenotrophic methanogens. The abundance of *Methanobacterium* decreased from 20.90% to 14.41% and caused the mismatch of H_2 production and utilization. The codigestion of CM and cardboard also showed higher acetoclastic *Methanosaeta* abundance (Zhao et al., 2021b). The codigestion of potato pulp and diary manure showed that the microbial diversity increased and enhanced the relative abundance of the bacterial community. Compared to inoculum, the total gene concentrations at different potato pulp and diary manure ratios increased (Chen et al., 2021).

Compared with monodigestion, codigestion of wastewater biosolids (WWB) and FW enhanced biogas production by 13% and chemical oxygen demand (COD) degradation rates by up to 101%. *Syntrophomonas* was the dominant genus in FW digesters in contrast to the dominance of *Clostridium* in WWB digesters. The predominant methanogen was *Methanosarcina* in FW digesters in contrast to *Methanosaeta* in WWB digesters. Codigestion systems showed a higher abundance of *Methanosarcina* over *Methanosaeta*, compared with WWB monodigestion (Kim et al., 2019). The codigestion of FW, cattle manure, and corn straw showed that VFAs concentrations in codigestion systems were lower than those in monodigestion system. Under the stress of accumulative VFAs inhibition, the enrichment of syntrophic communities of *Syntrophomonadaceae* with hydrogenotrophic methanogens was beneficial to system stability and biogas production. The synergy depended on great methanogenic community richness and diversity and the efficient methanogenic pathway owing to the cooperation of diverse methanogens (Zhang et al., 2021a).

In the dry ACoD of swine manure and rice straw, the abundance of *Halothermothrix* and *Halocella* were reduced, while *Moorella*, *Clostridium*, and *Selenomonas* were increased. Archaeal communities were drastically altered at different C/N ratios. *Methanothermobacter* and *Methanoculleus* were dominant in the inoculum. During the codigestion, the predominant archaea were shifted from *Methanothermobacter* and *Methanoculleus* to *Methanosarcina* (Riya et al., 2016).

11.4.3   Effect of pH on microbial community

pH is a commonly used indicator to control the stability of AD process. The change of pH has significant impact on the functions and activity of microorganisms (Li et al., 2018). For the substrates with high C/N ratio, the rapid acidification leads to a sharp decrease in pH and serious inhibition of growth and activity of methanogens. The possible reason was the acid resistance of some *Firmicutes* species, which can grow at low pH. In the codigestion process, *Anaerophaera* and *Aminomonas* played an important role in maintaining a neutral pH by utilizing ammonia, acetate, and propionate. *Chloroflexi* can degrade monosaccharides and polysaccharides to acetic acid. However, *Chloroflexi* abundance was sensitive to pH fluctuation. A stable pH is urgent for retaining and enriching of *Chloroflexi* (Li et al., 2017b). In ACoD of pig manure and FW, the methane yield did not drop extensively despite elevated VFA concentrations, due to the high pH and extensive buffering capacity (Dennehy et al., 2017).

Carbonaceous adsorbents can remove inhibitors and improve the process efficiency of AD. The microbial community structure could be affected by the presence of porous materials such as biochar (Shakourifar et al., 2020). During codigestion processes, biochar can enhance the abundance of the bacterial community. For example, *Methanomicrobiales* and *Methanosaetaceae* were the main methanogens in the AD of potato pulp waste. Biochar addition increased the abundance of *Bacteroidetes* and decreased the abundance of *Firmicutes*, increasing the population of methanogenic microorganisms (Chen et al., 2021). Similarly, *Methanosaeta* was the most abundant methanogen in the graphite-added FW digester, which is favorable for the growth of *Aminiphilus* and can enhance the biogas production (Muratçobanoğlu et al., 2020).

The community structure of bacterial and archaeal was affected by the nutritional balance of the cosubstrates. Methane yield can be enhanced by clay addition because it counteracts the high content of ammonium and nitrogen in swine manure. The phylum *Firmicutes* and *Methanosarcina* species dominate the AD of swine manure. However, in the codigestion of rice straw and swine manure, the *Bacteroidetes* phylum becomes dominant and *Methanosaeta* species are dominant in archaeal community (Jimenez et al., 2016).

11.4.4   Effect of trace elements on microbial community

Trace elements addition had positive impacts on AD processes and could improve process stability, degradation rate, and biogas yield. Appropriate amount of trace elements is needed to support the metabolism and growth of microorganisms. For example, adding Fe is beneficial to stimulating the formation of cytochromes and ferroxins which are vital for cellular energy metabolism (Xu et al., 2018b). Trace elements enriched by mineral waste supplementation are favorable to accelerating bacterial growth (Shamurad et al., 2019).

In codigestion system of yard waste (YW), FW, and sewage sludge, sewage sludge is important to maintain rich microbes' abundance and distributions. The addition of FW is favorable for the high level of microbes' abundance and high biogas productivity. Adding YW was beneficial for broader metabolic diversity. Therefore, ACoD of YW, FW, and sewage sludge could support a more diverse bacteria and archaea, thus ensuring process stability and high digestion performance (Mu et al., 2020). In ACoD of CM and corn stover, the composition and diversity of archaea and bacteria varied at genus level by pretreatment methods and trace elements supplementation. *Methanospirillum* was most sensitive to Mn and Co

(Wei et al., 2021). Addition of only Fe resulted in a more stable process than the combined addition of Fe and Co, indicating an efficient acidogenesis and/or homoacetogenesis in relation to a Ni-deprived methanogenic population (Moestedt et al., 2016).

11.4.5 Effect of pretreatment on microbial community

Pretreatment could substantially influence hydrolysis, biogas production, and archaeal communities of AD. For example, *Methanomassiliicoccus* was the predominant archaeal populations in the pretreated sludges (Li et al., 2019). The communities of bacteria (*Firmicutes, Proteobacteria*, and *Bacteroidetes)* and archaea (*Methanosaeta, Methanobacterium*, and *Methanosarcina*) were enriched by the enhanced hydrolysis and acidification process, and their interactions contributed to the improved digestion performance (Wei et al., 2020).

11.5 Life-cycle assessment of codigestion process

Life-cycle assessment (LCA) is a practical method to evaluate the impact of a product, a system, or an activity on the environment, which is a technique for assessing the environmental aspects associated with a product over its life cycle. The most important applications of LCA include analysis of the contribution of the life cycle stages to the overall environmental load, prioritizing improvements on products or processes, comparison between products for internal use.

11.5.1 Methodology of LCA study

The LCA was conducted according to ISO 14040-44 (ISO 14040: 2006; ISO 14044: 2006) in four phases: the goal and scope definition phase, the inventory analysis phase, the impact assessment phase, and the interpretation phase (Muralikrishna & Manickam, 2017). In details, as shown in Fig.11.3, the content of the four stages is as follows (Yang et al., 2020):

Stage 1: Goal and scope aims to define how big a part of the life cycle will be taken in assessment and to what end will assessment be serving. The criteria serving to system comparison and specific times are described in this step.

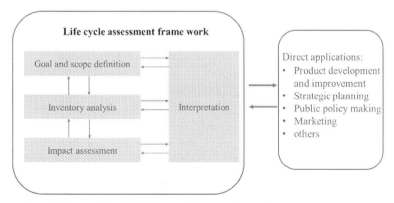

FIG. 11.3   Stages of an LCA according EN ISO 14040 (Yang et al., 2020).

Stage 2: Inventory analysis gives a description of material and energy flows in the system and its interaction with environment, consumed raw materials, and emissions to the environment. Important processes and subsidiary energy and material flows are described later.

Stage 3: Details from inventory analysis serve for impact assessment. The indicator results of all impact categories are detailed in this step; the importance of every impact category is assessed by normalization and eventually by weighting.

Stage 4: Interpretation of a life cycle involves critical review, determination of data sensitivity, and result presentation.

11.5.2 LCA of the ACoD process

To date, LCA associated with codigestion systems has been widely used to evaluate the environmental impact and economy by comparison with current waste management system or conventional handling method, monodigestion, the effects of pretreatment, posttreatment or end use of products, waste transport, and so on. LCA conclusions can be significantly affected by multiple factors such as codigestion system boundary, life cycle inventory, and the methodologies or software for calculation. For different wastes, it is difficult to compare the LCA results and to reach common conclusions, even achieving contradictory conclusions. The goal, scope, and methodology of LCA vary widely; therefore, sensitivity analysis is required to show the robustness of the LCA results and the potential for improvement (Wang et al., 2020).

11.5.2.1 *Comparison of codigestion with current management system*

The environmental influential variables including global warming potential (GWP), acidification potential (AP), eutrophication potential (EP), and photochemical ozone creation potential (POCP) were considered. The conventional handling method was land application and stockpiling. The codigestion of cow manure, feed waste, sludge, and returned dairy products was conducted on a large-scale dairy farm (>10,000 cows). The codigestion showed better environmental performances over the conventional practice by reducing GWP, AP, EP, and POCP by 25.7%, 49.5%, 18.1%, and 16.1%, respectively. In both monodigestion and conventional scenarios, accumulation of manure on sand beds offers the highest contribution to GWP and POCP. The sensitivity analysis indicated that increasing the manure collection efficiency is a good method to reduce the environmental footprint in the AD scenario (Adghim et al., 2020).

The codigestion of algae and agricultural biological waste was studied on a real pilot plant to figure out whether cultivation of macroalgae in near-shore open ponds be beneficial for biogas production. Three scenarios were considered, namely A (algae-based codigestion scenario), B (agricultural mix feedstock scenario), and C (conventional nonrenewable-based technologies). Scenario A represents a feasible solution to replace classical biomass for biofuel production. The improvement of the environmental performances is quantifiable on 10% respect to scenario B, and 38 times higher than scenario C (Cappelli et al., 2015).

High-solids ACoD (HS-ACoD) of sewage sludge with the OFMSW is a promising waste management way due to high methane yields, lower reactor volume requirements, lower energy inputs, and less leachate production than liquid AD. The environmental and economic benefits of HS-ACoD of sludge, FW, and YW were evaluated using LCA and life cycle cost analysis methods. Compared with the traditional methods such as incineration, composting, and landfilling, the HS-ACoD process had the lowest environmental impacts in terms of GWP, acidification, eutrophication, and ecotoxicity, and also had the lowest life cycle cost. HS-ACoD is the best choice to manage sewage sludge and organic waste due to both environmental and economic sustainability (Lee et al., 2020). In conclusion, ACoD had a smaller GWP than the current waste service (Edwards et al., 2017).

11.5.2.2 Comparison of codigestion and monodigestion process

Monodigestion of manure showed a potential for an improved environmental performance for global warming and phosphorus-eutrophication in comparison to conventional management. Codigestion of manure and reed canary grass produced two times higher energy than monodigestion process. The codigestion of grass with manure leads to an enhanced performance of the global warming and phosphorus-induced eutrophication impacts, which can be affected by the choice of the indirect land use change factor for modeling and the energy source displaced (Pehme et al., 2017).

One of the advantages of codigestion is the higher yields and higher quality of biogas than those of the monodigestion process. During the AD process, biomass production, digester operation, and digestate emission are the main contributors to environmental impacts (Bacenetti et al., 2016). From the aspect of bioenergy recovery, the codigestion of livestock manure and organic waste has less environmental impact than the use of energy crops due to the high environmental impacts of energy crop production. Therefore, a higher proportion of agricultural or organic waste in the codigestion process is favorable to maximize the environmental benefit (Lijó et al., 2015).

The performances of monodigestion and codigestion systems using cattle manure, maize silage, and grass silage as substrates were evaluated from the aspect of energy production and CO_2 release. The CO_2 released from the generation of electricity and the burning of methane represented 3% and 97% of the total CO_2 avoided. The codigestion process presents better performance in terms of size, electricity consumption, and heat consumption compared to monodigestion systems (Jean Agustin Velásquez Piñas et al., 2018). While the research of Tong indicated that codigestion of sewage sludge and FW may be neutral, synergistic and antagonistic based on the biogas production equivalent, higher or lower than that of monodigestion. The key efforts of the plant manager should be focused on the maximization of biogas output (Tong et al., 2019).

11.5.2.3 Effects of end-products utilization on LCA conclusions

Biogas is the target main product of AD process, and the byproduct digestate often is used as organic fertilizer, the utilization of which can influence the results of environmental, energetic, and economic analyses of codigestion systems. The codigestion of sewage sludge and OFMSW has great potential for reducing the environmental impacts and increasing the economic and energetic value of the substances by producing biomethane, electricity, and fertilizer (Morero et al., 2017). In the codigestion of manure and food industry wastes, biogas is upgraded and utilized as a vehicle fuel, distributed via the natural gas grid. The codigestion plant could reduce GHG emissions by approximately 90% (Lantz & Boerjesson, 2014).

The results of environmental impacts are affected by soil nutrient conditions, which determine the organic fertilizer application and the required land areas and transport distances. When considering the land application of digestate in terms of nutrient profiles, soil nutrient status, and environmental regulations, monodigestion of pig manure decreases GHG emissions by 48% than direct land application. The bioenergy recovery of codigestion of pig manure and grass silage is 226% higher than monodigestion. However, considering the land application, the nitrogen available for plants in codigestion digestate (41.8%) is slightly lower than that in raw pig manure (43.2%) because of the ammonia emission. The LCA study of codigestion provides practical insights to farmers, gas industry, and policy makers to conduct effective organic fertilizer application, AD plant location, and maximize environmental benefits (Zhang et al., 2021b).

Transport is one of the most important factors in all scenarios by considering SO_2 and NO_x emission. The environmental impact of a decentralized ACoD system using OFMSW and sewage sludge was investigated in small plants (capacity: 3000 t/A coupled at a 75 kW combined heat and power (CHP) unit). Compost or/and power production allows a significant cutdown on SO_2 and NO_x emission. The AcoD process in small plants combined with compost was an environmentally sustainable option in small communities because of the obvious reduction in transport, low energy input, energy saving from CHP unit, and high resources utilization as an organic fertilizer (Righi et al., 2013).

11.5.2.4   Effects of pretreatment on LCA conclusions

Pretreatments using physical, chemical, and biological methods are efficient technologies to improve AD process and enhance biogas yield by breaking the rigid structure of lignocellulosic biomass and enlarge the accessibility of microbials.

The economic, energetic, and environmental impacts of the pretreatments of LHW and thermo-alkaline (TA) were studied in the scenarios of monodigestion and codigestion. The LHW and TA pretreatments are self-sufficient in terms of thermal requirements as they can recover heat from the biogas engine, but the maximum electric and thermal net energy (64 MWh/d and 95 MWh/d) was obtained during codigestion with vinasse. The LCA showed that the AD of pretreated press mud by LHW was the most viable for the scenario of a sugar mill without distillery, while the codigestion with the vinasse of the press mud without pretreatment was the most viable for the scenario of a sugar mill with distillery. The environmental and energetic profiles and the profitability of methane production can be improved when the pretreatment and codigestion of these wastes from the sugar-alcohol production process are considered (González et al., 2020).

11.5.2.5   Effects of policy on LCA conclusions

AD growth has been driven by the government policies associated with climate change and the energy security nexus, waste management, and regional development (Edwards, 2015). AD systems on dairy farms in New York State rely on gate-fee revenues from codigestion to ensure economic viability. When the gate fee is codigested at rates exceeding the design capacity of the digester, potentially significant technical, environmental, and economic consequences may arise. To better understand these tradeoffs, a combined LCA and economic assessment with uncertainty analysis was performed. AD Model 1 was used to simulate the codigestion process for 10 cosubstrates that were hypothetically mixed with dairy manure. Higher loading rates were more economically favorable and caused considerable reductions in the degree of waste stabilization, which dramatically increased downstream methane emissions (e.g., >450%) on the farm, compared to monodigestion of manure-only. Regardless, most codigestion processes led to a net reduction in total lifecycle emissions compared to monodigestion of manure only and other method due to more electric power production and synthetic fertilizer replacement. Economically, gate-fee revenue was the most important contributor to profitability, while also compensating for the increased handling costs of the added waste volume. The model demonstrated the important environmental and economic implications arising from current AD implementation practices and policy in New York State. In addition, the model highlighted key stages in the system life-cycle, which was used to instruct and recommend immediately actionable policy changes (Usack et al., 2018).

11.6   Conclusions and perspectives

Codigestion process is an economically and environmentally feasible method for bioenergy production at laboratory and industrial scale. This chapter reviewed the advantages of codigestion process over monodigestion process and traditional solid wastes disposal approaches in the aspects of the bioenergy recovery, process stability, microbial community, and LCA. The digestion performances of codigestion are affected by various parameters, such as the composition and mass ratio of feedstocks, C/N ratio, pH, trace elements supplement, pretreatment, etc. During the codigestion process, microbial community structure may be changed, depending on the C/N ratio, pH, trace elements supplement, pretreatment, etc. The sustainability of codigestion is measured by LCA method in consideration of ecological impact, feedstock pretreatment, end-products utilization, economy and policy making, etc. The present studies have revealed that codigestion could efficiently convert organic wastes to the sustainable energy and mitigate carbon emission.

However, the performances of codigestion process are determined by the properties of feedstocks. The availability of year-round feedstock and their complexities are big challenges for codigestion process, due to different types, compositions, trace elements, and digestion performances. The in-depth analysis of the microbial communities and population and their function with feedstocks and process efficiency are needed to be conducted. The utilization of digestate as biofertilizer in agriculture and forestry should be deeply investigated for safety concerns, especially the possible contamination of heavy metals, pathogen, virus, etc. Finally, considering the source, type and production costs and logistics of feedstocks and energy cost, carbon emission, the technoeconomic analysis and LCA are crucial, which would be used to propose the suitable scenario for codigestion process and policy making.

Acknowledgements

This work was financially supported by the National Natural Science Foundation of China (No. 21878313), Science and Technology Program of Guiyang City ([2020]-18-8).

References

Abudi, Z.N., Hu, Z., Sun, N., Xiao, B., Rajaa, N., Liu, C., Guo, D., 2016. Batch anaerobic co-digestion of OFMSW (organic fraction of municipal solid waste), TWAS (thickened waste activated sludge) and RS (rice straw): influence of TWAS and RS pretreatment and mixing ratio. Energy107, 131–140.

Adghim, M., Abdallah, M., Saad, S.A., Shanableh, A., Mansouri, A., 2020. Comparative life cycle assessment of anaerobic co-digestion for dairy waste management in large-scale farms. J. Clean. Prod. 256, 120320.

Akhlisah, Z.N., Yunus, R., Abidin, Z.Z., Lim, B.Y., Kania, D., 2021. Pretreatment methods for an effective conversion of oil palm biomass into sugars and high-value chemicals. Biomass Bioenergy 144, 105901.

Akindele, A.A., Sartaj, M., 2018. The toxicity effects of ammonia on anaerobic digestion of organic fraction of municipal solid waste. Waste Manage. (Oxford) 71, 757–766.

Azbar, N., Keskin, T., Yuruyen, A., 2008. Enhancement of biogas production from olive mill effluent (OME) by co-digestion. Biomass Bioenergy 32 (12), 1195–1201.

Bacenetti, J., Sala, C., Fusi, A., Fiala, M., 2016. Agricultural anaerobic digestion plants: what LCA studies pointed out and what can be done to make them more environmentally sustainable. Appl. Energy 179, 669–686.

Cappelli, A., Gigli, E., Romagnoli, F., Simoni, S., Guerriero, E., 2015. Co-digestion of macroalgae for biogas production: an LCA-based environmental evaluation. Energy Procedia 72, 3–10.

Chen, M., Liu, S., Yuan, X., Li, Q.X., Wang, F., Xin, F., Wen, B., 2021. Methane production and characteristics of the microbial community in the co-digestion of potato pulp waste and dairy manure amended with biochar. Renew. Energy 163, 357–367.

Cheng, X.Y., Zhong, C., 2014. Effects of feed to inoculum ratio, co-digestion, and pretreatment on biogas production from anaerobic digestion of cotton stalk. Energy Fuels 28 (5), 3157–3166.

Dennehy, C., Lawlor, P.G., Gardiner, G.E., Jiang, Y., Cormican, P., Mccabe, M.S., Zhan, X., 2017. Process stability and microbial community composition in pig manure and food waste anaerobic co-digesters operated at low HRTs. Front. Environ. Sci. Eng. 11 (3), 53–66.

Ebner, J.H., Labatut, R.A., Lodge, J.S., Williamson, A.A., Trabold, T.A., 2016. Anaerobic co-digestion of commercial food waste and dairy manure: characterizing biochemical parameters and synergistic effects. Waste Manage. 52, 286–294.

Edwards, J., Othman, M., Burn, S., 2015. A review of policy drivers and barriers for the use of anaerobic digestion in Europe, the United States and Australia. Renew. Sustain. Energy Rev. 52, 815–828.

Edwards, J., Othman, M., Crossin, E., Burn, S., 2017. Anaerobic co-digestion of municipal food waste and sewage sludge: a comparative life cycle assessment in the context of a waste service provision. Bioresour. Technol. 223, 237–249.

Franca, L.S., Bassin, J.P., 2020. The role of dry anaerobic digestion in the treatment of the organic fraction of municipal solid waste: a systematic review. Biomass Bioenergy 143, 105866.

Gannoun, H., Omri, I., Chouari, R., Khelifi, E., Keskes, S., Godon, J.J., Hamdi, M., Sghir, A., Bouallagui, H., 2016. Microbial community structure associated with the high loading anaerobic codigestion of olive mill and abattoir wastewaters. Bioresour. Technol. 201, 337–346.

Gao, M., Yang, M., Ma, X., Xie, D., Wang, Q., 2021. Effect of co-digestion of tylosin fermentation dreg and food waste on anaerobic digestion performance. Bioresour. Technol. 325 (36), 124693.

González, L.M.L., Reyes, I.P., Gárciga, J.P., Barrera, E.L., Romero, O.R., 2020. Energetic, economic and environmental assessment for the anaerobic digestion of pretreated and codigested press mud. Waste Manage. (Oxford) 102, 249–259.

Grosser, A., Neczaj, E., Singh, B.R., Almas, A.R., Brattebø, H., Kacprzak, M., 2017. Anaerobic digestion of sewage sludge with grease trap sludge and municipal solid waste as co-substrates. Environ. Res. 155, 249.

Jimenez, J., Theuerl, S., Bergmann, I., Klocke, M., Guerra, G., Romero-Romero, O., 2016. Prokaryote community dynamics in anaerobic co-digestion of swine manure, rice straw and industrial clay residuals. Water Sci. Technol. 74 (4), 824–835.

Kim, M., Abdulazeez, M., Haroun, B.M., Nakhla, G., Keleman, M., 2019. Microbial communities in co-digestion of food wastes and wastewater biosolids. Bioresour. Technol. 289, 121580.

Kumar, A., Rapoport, A., Kunze, G., Kumar, S., Singh, D., Singh, B., 2020a. Multifarious pretreatment strategies for the lignocellulosic substrates for the generation of renewable and sustainable biofuels: a review. Renew. Energy 160, 1228–1252.

Kumar, B., Bhardwaj, N., Agrawal, K., Chaturvedi, V., Verma, P., 2020b. Current perspective on pretreatment technologies using lignocellulosic biomass: an emerging biorefinery concept. Fuel Process. Technol. 199, 106244.

Kumari, D., Singh, R., 2018. Pretreatment of lignocellulosic wastes for biofuel production: a critical review. Renew. Sustain. Energy Rev. 90, 877–891.

Labatut, R.A., Pronto, J.L., 2018. CHAPTER 4-Sustainable waste-to-energy technologies: anaerobic digestion. In: Sustainable Food Waste-to-Energy Systems, Elsevier, pp. 47–67.

Lantz, M., Boerjesson, P., 2014. Greenhouse gas and energy assessment of the biogas from co-digestion injected into the natural gas grid: a Swedish case-study including effects on soil properties. Renew. Energy 71, 387–395.

Lauterboeck, B., Ortner, M., Haider, R., Fuchs, W., 2012. Counteracting ammonia inhibition in anaerobic digestion by removal with a hollow fiber membrane contactor. Water Res. 46 (15), 4861–4869.

Lee, E., Oliveira, D., Oliveira, L., Jimenez, E., Kim, Y., Wang, M., Ergas, S.J., Zhang, Q., 2020. Comparative environmental and economic life cycle assessment of high solids anaerobic co-digestion for biosolids and organic waste management. Water Res. 171, 115443.

Li, W., 2016. High-solid anaerobic codigestion of horse manure and grass in batch and semi-continuous systems. Energy Fuels 30, 6419–6424.

Li, W., Fang, A., Liu, B., Xie, G., Lou, Y., 2019. Effect of different co-treatments of waste activated sludge on biogas production and shaping microbial community in subsequent anaerobic digestion. Chem. Eng. J. 378, 122098.

Li, W., Loh, K.C., Zhang, J., Tong, Y.W., Dai, Y., 2018. Two-stage anaerobic digestion of food waste and horticultural waste in high-solid system. Appl. Energy 209, 400–408.

Li, W., Lu, C., An, G., Chang, S., 2017a. Comparison of alkali-buffering effects and co-digestion on high-solid anaerobic digestion of horticultural waste. Energy Fuels 31 (10), 10990–10997.

Li, Y., Zhang, Y., Sun, Y., Wu, S., Kong, X., Yuan, Z., Dong, R., 2017b. The performance efficiency of bioaugmentation to prevent anaerobic digestion failure from ammonia and propionate inhibition. Bioresour. Technol. 231, 94–100.

Lijó, L., González-García, S., Bacenetti, J., Negri, M., Fiala, M., Feijoo, G., Moreira, M.T., 2015. Environmental assessment of farm-scaled anaerobic co-digestion for bioenergy production. Waste Manage. (Oxford) 41 (7), 50–59.

Ma, C., Liu, J., Ye, M., Zou, L., Qian, G., Li, Y.Y., 2018. Towards utmost bioenergy conversion efficiency of food waste: pretreatment, co-digestion, and reactor type. Renew. Sustain. Energy Rev. 90, 700–709.

Ma, G., Ndegwa, P., Harrison, J.H., Chen, Y., 2020. Methane yields during anaerobic co-digestion of animal manure with other feedstocks: a meta-analysis. Sci. Total Environ. 728, 138224.

Mata-Alvarez, J., Dosta, J., Romero-Gueiza, M.S., Fonoll, X., Peces, M., Astals, S., 2014. A critical review on anaerobic co-digestion achievements between 2010 and 2013. Renew. Sustain. Energy Rev. 36, 412–427.

Moestedt, J., Nordell, E., Yekta, S.S., Lundgren, J., Martí, M., Sundberg, C., Ejlertsson, J., Svensson, B.H., Björn, A., 2016. Effects of trace element addition on process stability during anaerobic co-digestion of OFMSW and slaughterhouse waste. Waste Manage. (Oxford) 47, 11–20.

Morero, B., Vicentin, R., Campanella, E.A., 2017. Assessment of biogas production in Argentina from co-digestion of sludge and municipal solid waste. Waste Manage. (Oxford) 61, 195–205.

Mu, L., Zhang, L., Zhu, K., Ma, J., Ifran, M., Li, A., 2020. Anaerobic co-digestion of sewage sludge, food waste and yard waste: synergistic enhancement on process stability and biogas production. Sci. Total Environ. 704, 135429.

Muralikrishna, I.V., Manickam, V., 2017. Life cycle assessment. In: Muralikrishna, I.V., Manickam, V. (Eds.), Environmental Management. Butterworth-Heinemann, UK, pp. 57–75.

Muratçobanoğlu, H., Gökçek, Ö.B., Mert, R.A., Zan, R., Demirel, S., 2020. Simultaneous synergistic effects of graphite addition and co-digestion of food waste and cow manure: biogas production and microbial community. Bioresour. Technol. 309, 123365.

Negri, C., Ricci, M., Zilio, M., et al., 2020. Anaerobic digestion of food waste for bio-energy production in China and Southeast Asia: a review. Renew. Sustain. Energy Rev. 133, 110138.

Neshat, S.A., Mohammadi, M., Najafpour, G.D., Lahijani, P., 2017. Anaerobic co-digestion of animal manures and lignocellulosic residues as a potent approach for sustainable biogas production. Renew. Sustain. Energy Rev. 79, 308–322.

Paul, S., Dutta, A., 2018. Challenges and opportunities of lignocellulosic biomass for anaerobic digestion. Resour. Conserv. Recycl. 130, 164–174.

Pehme, S., Veromann, E., Hamelin, L., 2017. Environmental performance of manure co-digestion with natural and cultivated grass—a consequential life cycle assessment. J. Clean. Prod. 162, 1135–1143.

Piñas, J.A.V., Venturini, O.J., Lora, E.E.S., Roalcaba, O.D.C, 2018. Technical assessment of mono-digestion and co-digestion systems for the production of biogas from anaerobic digestion in Brazil. Renew. Energy 117, 447–458.

Rajagopal, R., Massé, D., Singh, G., 2013. A critical review on inhibition of anaerobic digestion process by excess ammonia. Bioresour. Technol. 143 (17), 632–641.

Ren, Y., Yu, M., Wu, C., Wang, Q., Gao, M., 2018. A comprehensive review on food waste anaerobic digestion: research updates and tendencies. Bioresour. Technol. 247, 1069–1076.

Rezania, S., Oryani, B., Cho, J., Talaiekhozani, A., Sabbagh, F., Hashemi, B., Rupani, P.F., Mohammadi, A.A., 2020. Different pretreatment technologies of lignocellulosic biomass for bioethanol production: an overview. Energy 199, 117457.

Righi, S., Oliviero, L., Pedrini, M., Buscaroli, A., Casa, C.D., 2013. Life cycle assessment of management systems for sewage sludge and food waste: centralized and decentralized approaches. J. Clean. Prod. 44, 8–17.

Riya, S., Suzuki, K., Terada, A., Hosomi, M., Sheng, Z., 2016. Influence of C/N ratio on performance and microbial community structure of dry-thermophilic anaerobic co-digestion of swine manure and rice straw. J. Med. Bioeng. 5 (1), 11–14.

Ryue, J., Lin, L., Kakar, F.L., Elbeshbishy, E., Al Mamun, A., Dhar, B.R., 2020. A critical review of conventional and emerging methods for improving process stability in thermophilic anaerobic digestion. Energy Sustain. Dev. 54, 72–84.

Shakourifar, N., Krisa, D., Eskicioglu, C., 2020. Anaerobic co-digestion of municipal waste sludge with grease trap waste mixture: point of process failure determination. Renew. Energy 54, 117–127.

Shamurad, B., Gray, N., Petropoulos, E., Tabraiz, S., Acharya, K., Quintela-Baluja, M., Sallis, P., 2019. Co-digestion of organic and mineral wastes for enhanced biogas production: reactor performance and evolution of microbial community and function. Waste Manage. (Oxford) 87, 313–325.

Shrestha, S., Fonoll, X., Khanal, S.K., Raskin, L., 2017. Biological strategies for enhanced hydrolysis of lignocellulosic biomass during anaerobic digestion: current status and future perspectives. Bioresour. Technol. 245, 1245–1257.

Sun, C., Xie, Y., Hou, F., Yu, Q., Zhao, Y., 2020. Enhancement on methane production and anaerobic digestion stability via co-digestion of microwave-Ca(OH)$_2$ pretreated sugarcane rind slurry and kitchen waste. J. Clean. Prod. 264, 121731.

Sung, S., Tao, L., 2003. Ammonia inhibition on thermophilic anaerobic digestion. Chemosphere 53 (1), 43–52.

Tian, W., Chen, Y., Shen, Y., Zhong, C., Gu, L., 2020. Effects of hydrothermal pretreatment on the mono- and co-digestion of waste activated sludge and wheat straw. Sci. Total Environ. 732, 139312.

Tong, H., Tong, Y.W., Peng, Y.H., 2019. A comparative life cycle assessment on mono- and co-digestion of food waste and sewage sludge. Energy Procedia 158, 4166–4171.

Turker, G., Aydin, S., Akyol, A., Yenigun, O., Ince, O., Ince, B., 2016. Changes in microbial community structures due to varying operational conditions in the anaerobic digestion of oxytetracycline-medicated cow manure. Appl. Microbiol. Biotechnol. 100 (14), 6469–6479.

Usack, J.G., Gerber, V.D., Posmanik, R., Labatut, R.A., Tester, J.W., Angenent, L.T., 2018. An evaluation of anaerobic co-digestion implementation on New York State dairy farms using an environmental and economic life-cycle framework. Appl. Energy 211, 28–40.

Wang, D., He, J., Tang, Y., Higgitt, D., Robinson, D., 2020. Life cycle assessment of municipal solid waste management in Nottingham, England: past and future perspectives. J. Clean. Prod. 251, 119636.

Wang, P., Wang, H., Qiu, Y., Ren, L., Jiang, B., 2018. Microbial characteristics in anaerobic digestion process of food waste for methane production—a review. Bioresour. Technol. 248, 29–36.

Wang, W., Lee, D.J., 2021. Valorization of anaerobic digestion digestate: a prospect review. Bioresour. Technol. 323 (113218), 124626.

Wang, X., Lu, X., Fang, L., Yang, G., 2014. Effects of temperature and carbon-nitrogen (C/N) ratio on the performance of anaerobic co-digestion of dairy manure, chicken manure and rice straw: focusing on ammonia inhibition. PLoS One 9, 1–7.

Wei, Y., Li, Z., Ran, W., Yuan, H., Li, X., 2021. Performance and microbial community dynamics in anaerobic co-digestion of chicken manure and corn stover with different modification methods and trace element supplementation strategy. Bioresour. Technol. 325, 12471.

Wei, Y., Wachemo, A.C., Yuan, H.R., Li, X.J., 2020. Enhanced hydrolysis and acidification strategy for efficient co-digestion of pretreated corn stover with chicken manure: digestion performance and microbial community structure. Sci. Total Environ. 720, 137401.

Xu, F., Li, Y., Ge, X., Yang, L., Li, Y., 2018a. Anaerobic digestion of food waste—challenges and opportunities. Bioresour. Technol. 247, 1047–1058.

Xu, F., Li, Y., Wicks, M., Li, Y., Keener, H., 2019. Anaerobic digestion of food waste for bioenergy production. In: Ferranti, P., Berry, E.M., Anderson, J.R. (Eds.). Elsevier, Amsterdam, pp. 530–537.

Xu, R., Zhang, K., Liu, P., Khan, A., Xiong, J., 2018b. A critical review on the interaction of substrate nutrient balance and microbial community structure and function in anaerobic co-digestion. Bioresour. Technol. 247, 1119–1127.

Yang, S., Ma, K., Liu, Z., Ren, J., Man, Y., 2020. Development and applicability of life cycle impact assessment methodologies. In: Ren, J., Toniolo, S. (Eds.), Life Cycle Sustainability Assessment for Decision-Making. Elsevier, Amsterdam, pp. 95–124.

Yao, Y., Huang, G., An, C., Chen, X., Zhang, P., Xin, X., Shen, J., Agnew, J., 2020. Anaerobic digestion of livestock manure in cold regions: technological advancements and global impacts. Renew. Sustain. Energy Rev. 119, 109494.

Yeniguen, O., Demirel, B., 2013. Ammonia inhibition in anaerobic digestion: a review. Process Biochem. 48 (5), 901–911.

Zamri, M., Hasmady, S., Akhiar, A., Ideris, F., Mahlia, T., 2021. A comprehensive review on anaerobic digestion of organic fraction of municipal solid waste. Renew. Sustain. Energy Rev. 137, 110637.

Zhang, C., Su, H., Baeyens, J., Tan, T., 2014. Reviewing the anaerobic digestion of food waste for biogas production. Renew. Sustain. Energy Rev. 38, 383–392.

Zhang, W., Wang, X., Xing, W., Li, R., Yang, T., Yao, N., Lv, D., 2021a. Links between synergistic effects and microbial community characteristics of anaerobic co-digestion of food waste, cattle manure and corn straw. Bioresour. Technol. 329 (2), 124919.

Zhang, W., Zhang, L., Li, A., 2015. Anaerobic co-digestion of food waste with MSW incineration plant fresh leachate: process performance and synergistic effects. Chem. Eng. J. 259, 795–805.

Zhang, Y., Jiang, Y., Wang, S., Wang, Z., Zhan, X., 2021b. Environmental sustainability assessment of pig manure mono- and co-digestion and dynamic land application of the digestate. Renew. Sustain. Energy Rev. 137, 110476.

Zhao, S., Chen, W., Luo, W., Fang, H., Lv, H., Liu, R., Niu, Q., 2021b. Anaerobic co-digestion of chicken manure and cardboard waste: focusing on methane production, microbial community analysis and energy evaluation. Bioresour. Technol. 321, 124429.

Feedstock pretreatment for enhanced anaerobic digestion of lignocellulosic residues for bioenergy production

Xihui Kang[a,b], Chao Xu[c], Richen Lin[a,b], Bing Song[d], David Wall[a,b], Jerry D Murphy[a,b]

[a]MaREI Centre, Environmental Research Institute, University College Cork, Cork, Ireland
[b]Civil, Structural and Environmental Engineering, School of Engineering and Architecture, University College Cork, Cork, Ireland
[c]Institute of Bast Fiber Crops, Chinese Academy of Agricultural Sciences, Changsha, China
[d]Scion, Te Papa Tipu Innovation Park, Rotorua, New Zealand

12.1   Introduction

The estimated global energy demand by 2030 will increase by 50% based on the predicted increases in population size; with approximately 80% of worldwide energy consumption provided by fossil fuels, this continued use of nonsustainable resources would result in severe environmental issues such as climate change, pollution, and resource depletion (IEA, 2020; O'Neill & Oppenheimer, 2016). This necessitates the development of sustainable alternatives for fossil fuels to satisfy the global world energy demand. The large-scale production of carbon-neutral bioenergy from biomass for renewable heat, electricity, and transport fuel is of great importance in many climate change mitigation and energy supply scenarios (O'Neill et al., 2017). In 2018, bioenergy accounted for 11.3% (around 49.6 exajoule (EJ)) of total world energy demand (WBA, 2020). The estimated bioenergy production potential from key sources, namely energy crops, agricultural residues, forestry residues, wastes, and forestry are 22–1272 EJ, 10–66 EJ, 3–35 EJ, 12–120 EJ, and 60–230 EJ by 2050, respectively (Fig. 12.1) (Slade et al., 2014), with a market share potential of up to 50%. A more detailed study estimated that the primary bioenergy availability in 2050 will be in the range of 112–794 EJ/year, including 634–2807 Mha of land for biomass cultivation in low, mid, and high bioenergy availability scenarios (Table 12.1). The use of abundant and inexpensive

Biomass, Biofuels, Biochemicals.
DOI: https://doi.org/10.1016/B978-0-323-90633-3.00004-3

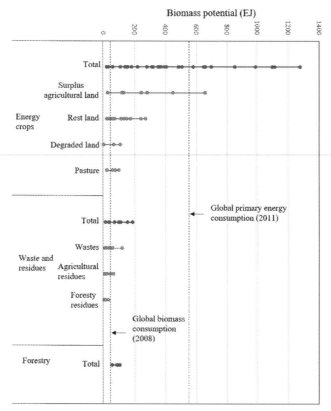

FIG. 12.1   Estimation of the future energy potential from energy crops, waste and residues, and forestry (adapted from Slade et al. (2014)).

TABLE 12.1   Primary bioenergy availability potential with land demand in 2050 under low, mid, and high availability scenarios (Staples et al., 2017).

	Low		Mid		High	
Bioenergy availability scenario	**Primary energy (EJ/year)**	**Land (Mha)**	**Primary energy (EJ/year)**	**Land (Mha)**	**Primary energy (EJ/year)**	**Land (Mha)**
Vegetable oil energy crops	17	216	43	500	32	688
Sugary and starchy energy crops	33	204	60	498	100	802
Lignocellulosic energy crops	43	214	210	664	561	1317
Energy crop subtotal	93	634	312	1662	693	2807
Agricultural residues	15	-	46	-	81	-
Forestry residues	4	-	9	-	19	-
Waste fats, oils, and greases	1	-	1	-	1	-
Residue and waste subtotal	19	-	56	-	101	-
Total	112	634	368	1662	794	2807

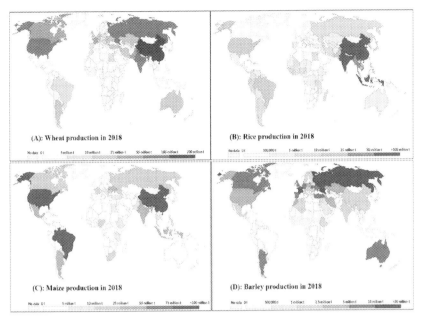

FIG. 12.2   Worldwide crop production of wheat (A), rice (B), maize (C), and barley (D) in 2018 adapted from Ritchie and Roser (2020). (Unit = tons, with data sourced from the UN Food and Agricultural Organization)

lignocellulosic residues such as agricultural residues and forest residues for large-scale bioenergy production has become common practice in many countries including China, India, Germany, Denmark, Sweden, and Italy (Bentsen et al., 2018; Kang et al., 2020d; Kumar et al., 2018).

Agricultural crops such as wheat, rice, corn (maize), and barley are typical lignocellulosic biomass (LCB). The worldwide production of major crops is presented in Fig. 12.2 (data from the UN Food and Agriculture Organization). Along with crop production, a significant number of lignocellulosic residues are also generated during crop harvesting and processing. Compared with other waste treatment methods (such as landfill) and biomass conversion technologies, anaerobic digestion (AD) of lignocellulose has the dual effects of managing waste (or resource) in a sustainable and environmentally friendly way and producing renewable gaseous biofuel—biogas (50%–70% of CH_4 and 30%–50% CO_2), which can be used for heat supply and power generation, transport fuel, or as a sustainable raw material for the production of energy-dense liquid biofuels.

AD is a successive biological process consisting of hydrolysis, acidogenesis, acetogenesis, and methanogenesis, in which microorganisms break down organic matter in the absence of oxygen (Lin et al., 2018; Lin et al., 2015). Naturally, AD of LCB is less favored compared with that of food wastes and other easily biodegradable substrates due to the inherent complex polymeric structure of lignocellulose. Table 12.2 summarizes the annual global yield, chemical composition, and theoretical energy potential of several key LCB such as rice straw, wheat straw, and maize straw. Generally, lignocellulose contains 25%–45%

TABLE 12.2   Average annual global yield, chemical properties, and energy potential of common lignocellulosic waste.

Lignocellulosic feedstock	Global yield in 2019 (million tons)[a]	Chemical composition (based on dry matter)			Higher heating value (MJ/kg)[b]	Reference
		Cellulose (%)	Hemicellulose (%)	Lignin (%)		
Wheat straw	589	35.1–39.2	25.6–26.1	7.5–15	17.00–18.91	(Schroyen et al., 2015)
Rice straw	755	30.0–41.9	29.8-36.4	3.1-6.5	15.09–15.61	(Peng et al., 2019; Zhang et al., 2015)
Corn stover (maize straw)	876	32.6-43.1	13.7-31.2	5.0-23.6	17.68	(Li et al., 2019; Schroyen et al., 2015; Zhang et al., 2015)
Sorghum straw	36	47.5-48.9	27.4-35.1	4.1-7.0	15.67–16.99	(Sambusiti et al., 2013)

[a] The global yield of each type of straw is calculated using the equation: $Y_{straw} = Y_{grain} / (grain\text{-}to\text{-}straw\ ratio)$, where the grain-to-straw ratio of wheat, rice, corn and sorghum is 1.3, 1, 1.32 and 1.6, respectively (Song et al., 2013); the yield of each type of crop in 2019 is sourced from the UN Food and Agricultural Organization (http://www.fao.org/faostat/en/#data/QC).
[b] the higher heating value of wheat straw, rice straw, and corn stover is sourced from Uzun et al. (2017), and that of sorghum straw is from Zhang et al. (2017).

cellulose, 15%–25% hemicellulose, and 10%–25% lignin (Zheng et al., 2014). Cellulose, the most abundant biopolymer in nature, is polymer of glucose, which is linked by a β-1,4- glycosidic bond. Hemicellulose, mainly consisting of glucose, xylose, mannose, and arabinose, is a heterogeneous branched polysaccharide that binds tightly to the surface of each cellulose microfibril. Lignin, an amorphous aromatic polymer, is primarily composed of syringyl, coniferyl, and coumary, linked by β-O-4 and other bonds (Xu et al., 2021a). The abundant composition of carbohydrates (hemicellulose and cellulose) is ideal for AD treatment. However, these compositions are intertwined in a complex matrix structure with a considerable amount of carbohydrates covered by or interlinked to lignin (Fig. 12.3); as such a considerable amount of carbohydrates may remain undegraded, resulting in low digestion efficiency and low biogas/biomethane production (Song et al., 2021). To fully exploit the biomethane potential from these substrates, pretreatment prior to AD is required to overcome the recalcitrance and thereby enhance the hydrolysis rate, digestion efficiency, and subsequent biomethane yield.

The objectives of this chapter are to (1) review the state-of-the-art pretreatment technologies for improving the digestion efficiency and biomethane production of LCB; and (2) discuss the opportunities for AD contributing to a sustainable low-carbon emissions circular economy by integration with other biorefinery technologies and CO_2 utilization processes (such as algae cultivation through photosynthetic microalgae upgrading).

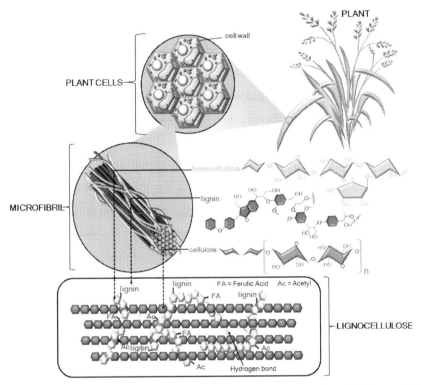

FIG. 12.3   A simplified cell wall structure of lignocellulosic biomass adapted from Ribeiro et al. (2016) with permission. This shows the linkages between crystalline cellulose (green), hemicellulose (yellow), and lignin (red).

12.2   Biomass pretreatment for enhanced AD

Pretreatment of lignocellulose is the process where complex lignocellulose (carbohydrates-lignin complexity) is fractionated into its major components, that is, cellulose, hemicellulose, and lignin. Typically, pretreatment methods can be categorized into physical, chemical, and biological methods and combinations of each (Lin et al., 2017). The goal of a pretreatment is to increase the conversion efficiency and ensuing biogas/biomethane production by preserving the hemicellulose, removing lignin, decreasing the cellulose crystallinity, and increasing the porosity of the feedstock. An ideal pretreatment should be inexpensive, with the majority of carbohydrates retained, minimum fermentative inhibitors formed, and minimum energy required (Kumari & Singh, 2018).

The effects of different pretreatment methods on the structure and composition of lignocellulose are listed in Table 12.3. It can be concluded that all pretreatment methods can increase the accessible area of lignocellulose to microorganisms or enzymes. However, as the biodegradability property of each substrate is different, pretreatment for AD must be properly selected on the basis of the technological specificity of the substrate (Paudel et al., 2017). Fundamental and technical characteristics of various pretreatments for promoting AD of lignocellulosic feedstocks are presented as follows.

TABLE 12.3   The effects of pretreatment on the chemical and structural properties of lignocellulosic material (adapted from Zheng et al. (2014)).

Pretreatment	Physical	Chemical	Biological
Increased accessible surface area	●●	●●	●●
Decreased cellulose crystallinity	●●	●●	NS
Dissolution of hemicellulose	●	●●	●●
Lignin removal	○	●●	●●
Lignin structure alteration	○	●●	●●
Formation of inhibitors	○	●●	○

Note:
●● means intensive effect;
● means moderate effect;
○ represents no effect and NS means not sure.

12.2.1   Physical pretreatment

Physical pretreatment refers to the methods without the addition of chemicals, microorganisms, or enzymes during the pretreatment process. Over the last decade, well-developed physical pretreatment techniques used to improve the biogas/biomethane production from LCB include extrusion, comminution (such as milling and grinding), and irradiation (Zheng et al., 2014). In general, physical pretreatment improves the AD performance of lignocellulose by reducing the particle size, increasing the accessible surface area, and decreasing cellulose crystallinity. In particular, particle size reduction may be an essential step to ensure a high-efficiency AD process as larger particle sizes may lead to mechanical problems within the digester mixing or pumping systems (Wall et al., 2015).

12.2.1.1   Mechanical pretreatment

Mechanical pretreatment including chipping, milling (such as ball milling, disk milling, and hammer milling) and grinding are efficient methods in reducing the particle size of lignocellulosic materials and thereby lead to increased accessible surface area and hydrolysis rate of cellulose and hemicellulose. Typically, chipping, and milling/grinding generate final particle sizes of 10–30 mm and 0.2–2 mm, respectively (Kratky & Jirout, 2011; Kumari & Singh, 2018). Chopping is usually employed to treat lignocellulosic materials with a large size (>5 cm) such as corn stalk and forest residues at an industrial scale, while milling and grinding pretreatment have received much interest in lab-scale research. Dahunsi et al. (2019) reduced the particle size of different lignocellulosic materials including elephant grass, sunflower, and siam weed with mechanical pretreatment, and found that the methane yield was enhanced by 57%, 48%–63% and 56%–76% for elephant grass, sunflower, and siam weed, respectively. With similar pretreatment apparatus, Tsapekos et al. (2018) reduced the particle size of meadow grass and obtained 27% more methane yield from pretreated meadow grass than that from untreated meadow grass. Size reduction, however, may not be linearly related to augmented methane yield as excessive size reduction can result in the build-up of volatile fatty acids, which inhibits methanogenesis in AD (Izumi et al., 2010). Studies have reported that further particle size reduction over a threshold (0.4 um) would not increase the methane yield from lignocellulosic material (Kang et al., 2019). Despite the achieved progress, the high energy demand in mechanical pretreatment negates its commercially feasible; however, when combined with chemical pretreatment, mechanical pretreatment may be more efficient, which may enhance the economic feasibility (Zheng et al., 2014).

12.2.1.2 Extrusion

Extrusion is a process whereby moistened biomass material is put through an extruder barrel that applies pressure with a screw (Ravindran & Jaiswal, 2016). Consequently, extrusion pretreatment of lignocellulosic materials will lead to larger specific areas, a higher water-holding capacity, and a lower bulk density that facilitates access for microorganisms and enzymes (Duque et al., 2017; Zheng & Rehmann, 2014). Similar to milling and grinding, extrusion pretreatment has been reported as an effective method for biomass size reduction when screws with shear damage were applied, which enhances the AD of lignocellulosic materials, and thereby the biogas production (Hjorth et al., 2011). After extrusion pretreatment, the methane yield from rice straw increased from 153 to 177 L/kg volatile solid (VS), with an improvement of 16%. Wahid et al. (2015) compared the effects of extrusion pretreatment (600 rpm for 30–120 s) with different types of screw configuration on the sugar availability and methane yield of wheat straw and deep litter. Results showed that the sugar availability after pretreatment was increased by 7%–42% for both wheat straw and deep litter, while the extrusion pretreatment was more efficient in enhancing the methane yield (by 16%) for wheat straw than that of deep litter (5% increase). The authors attributed this to the soft texture of deep litter and possible sugar hydrolysis during storage, which makes deep litter more readily degradable. Panepinto and Genon (2016) examined the digestion performance of extrusion-pretreated maize silage harvested at different times and found that the methane yield from treated maize silage was 15% higher than that from raw samples, which led to a 6.5% increase in the final electric energy. Coupled with other pretreatment, extrusion pretreatment can be more efficient in improving the digestion efficiency of lignocellulosic materials under mild condition and lower chemical demands. Gu et al. (2015) investigated the impact of combined extrusion and $Ca(OH)_2$ pretreatment on the AD of rice straw and found that the methane yield from extrusion and 8% $Ca(OH)_2$ treated rice straw was 30% higher than that from extrusion of rice straw alone.

12.2.1.3 Microwave pretreatment

Microwave pretreatment is the most tested irradiation pretreatment, during which microwave energy directly penetrates the biomass. Such a process enables the expeditious heating of the substrate with a minimum thermal gradient, altering the rigid polymeric structure of lignocellulosic material and thereby increasing its accessibility and degradation (Abraham et al., 2020; Kumari & Singh, 2018). Compared with traditional thermal pretreatment in a water/oil bath, the advantages of microwave pretreatment are high uniformity and selectivity, short process time, and lower energy requirement. Feng et al. (2019) reported that the methane yield from microwave treated green algae increased by 30% compared to that from untreated algae. The methane yield from microwave pretreated water hyacinth was reported to be 38% higher than that from water-heating pretreated samples (Zhao et al., 2017).

Chemical pretreatment with the assistant of microwave was shown to be superior to solely chemical pretreatment or solely microwave pretreatment in terms of pretreatment efficiency, methane yield augmentation, and energy and chemicals saving (Bundhoo, 2018; Kavitha et al., 2018). Kaur and Phutela (2016) compared the methane enhancement from paddy straw pretreated with conventional NaOH (soaking for 24 h) and microwave-assisted NaOH for 30 min and found that under the same concentration of NaOH (4%), the methane yield from microwave-assisted NaOH treated paddy straw was 8% higher. Kan et al. (2018) also employed microwave-assisted NaOH pretreatment to improve the methane yield from brewery spent grain. Under the optimal pretreatment conditions, 46% of lignin and 38% of hemicellulose were removed and the specific surface area of the substrate increased from 9.55 to 161.98 m^2/g, which led to a 52% increment in methane yield from the pretreated brewery spent grain.

12.2.1.4 Steam explosion

Steam explosion, also called autohydrolysis pretreatment, works on the basis of steam injection to the substrate followed by an abrupt pressure reduction, which causes the biomass to undergo an explosive decompression resulting in the hydrolysis of the hemicellulose into individual sugars (Rodriguez et al., 2017). This method is recognized as a cost-effective method for LCB pretreatment as there is no additional external catalyst needed. The major process parameters that affect the efficiency of steam explosion pretreatment are reaction temperature (T), time (t), and particle size of the substrate (Negro et al., 2003). The pretreatment severity (S) can be expressed by a function of temperature and reaction time $[S = \log\{t \times \exp((T - 100)/14.75)\}]$ to simultaneously demonstrate the process conditions (Yang et al., 2018). Although increased pretreatment severity significantly deconstructs the rigid polymerization structure of lignocellulosic material, it is not always associated with an improvement in methane yield. With an effort to increase the methane yield from late-harvested hay, Bauer et al. (2014) carried out steam explosion pretreatment with elevated severity from 160 to 200 °C for 5–15 min and reported that pretreatment at 200 °C induced almost 100% removal of hemicellulose with insignificant changes in the cellulose content. This result was attributed to the excessive degradation of hemicellulose and formation of inhibitors under harsher conditions (over 200 °C). The maximum methane yield was obtained from pretreated hay at 175 °C for 10 min, which corresponded to an improvement of 16%. Lizasoain et al. (2016) screened the optimal process conditions (160–220 °C, 5–20 min) of steam explosion pretreatment for boosting the methane yield from reed biomass and observed similar results with the hemicellulose almost completely removed with process temperatures over 200 °C. The highest methane yield from pretreated reed was 89% higher than that from untreated reed under the optimal condition of 200 °C for 15 min. As steam explosion pretreatment has a limited effect on the lignin fraction, and lignin plays a negative role in bioconversion of lignocellulosic materials (Zeng et al., 2014), steam explosion pretreatment is usually integrated with alkaline reagents such as ammonia (known as ammonia fiber expansion (AFEX)) to simultaneously dissolve hemicellulose and remove lignin under mild pretreatment conditions. Mokomele et al. (2019) compared the promotional effects of steam explosion pretreatment and AFEX pretreatment on the methane yield of sugarcane residues and found that AFEX pretreatment led to higher methane yields from the substrate than steam explosion pretreatment alone.

12.2.1.5 Liquid hot water pretreatment

Liquid hot water (LHW) pretreatment, using only pressure to maintain water in the liquid state at elevated temperatures (160–240 °C), is an effective and attractive method for LCB deconstruction. During LHW pretreatment, water is cleaved into hydroxide anions and hydrogen cations at elevated temperature, which hydrolyses cellulose and hemicellulose into correspondent monosaccharides (see the stoichiometric reactions shown in Eqs. 12.1–12.5). LHW pretreatment recovers the majority of pentoses from hemicellulose in the liquid fraction and produces few digestion inhibitors (such as furfural) (Yu et al., 2011). Without the involvement of catalysts or organic solvents, LHW pretreatment is considered as a clean, sustainable, and efficient method for deconstructing lignocellulosic substrates by dissolving hemicellulose and segregating lignin (Zhuang et al., 2016). LHW pretreatment below 240 °C is highly efficient in dissolving hemicellulose (up to 100%); the removal of hemicellulose increases the accessibility of cellulose to the enzymes or microorganisms, enhancing the biodegradability and biofuel potential from lignocellulosic substrates (Li et al., 2017; Yang et al., 2018). Previous studies have shown that under LHW pretreatment of 175 °C for 35 min, the removal of 40% hemicellulose from an energy crop led to 33% more methane yield than that from the raw energy crop (Kang et al., 2020c). Under similar pretreatment conditions (175 °C for 30 min), up to 89% removal of hemicellulose was achieved from rice straw, which resulted in 63% more methane yield than that from the untreated rice straw (Shang et al., 2019). Forest woody biomass is generally more recalcitrant than grass biomass. For example, to achieve 89% removal of hemicellulose from *Tamarix ramosissima* (a kind of woody biomass), Xiao et al. (2011) employed LHW pretreatment at 200°C for 3h. The higher treatment temperature and time required led to greater energy consumption and increased cost of the pretreatment.

$$nC_6H_{10}O_5(\text{Cellulose}) + nH_2O \rightarrow nC_6H_{12}O_6(\text{Glucose}) \tag{12.1}$$

$$nC_5H_8O_4(\text{Xylan}) + nH_2O \rightarrow nC_5H_{10}O_5(\text{Xylose}) \tag{12.2}$$

$$\text{Xylan} \rightarrow (\text{Xylan})aq \rightarrow \text{Xylose} \rightarrow \text{furfural} \tag{12.3}$$

$$\text{Arabinan} \rightarrow (\text{Arabinan})aq \rightarrow \text{Arabinose} \rightarrow \text{furfural} \tag{12.4}$$

$$\text{Glucan} \rightarrow (\text{Glucan})aq \rightarrow \text{Glucose} \rightarrow \text{Hydroxy methyl furfural(HMF)} \tag{12.5}$$

12.2.2 Chemical pretreatment

Chemical pretreatment techniques are generally divided into alkaline, acid, oxidation, and organic solvents pretreatment according to the type of chemical used in the process. NaOH, KOH, $Ca(OH)_2$, and ammonia are commonly used agents during alkaline pretreatment. The major acidic chemicals used in acid pretreatment are H_2SO_4 and HCl (inorganic acids), and CH_3COOH (organic acids). The function of these pretreatment is to dissolve hemicellulose and/or remove lignin, depending on the type of chemical usage and the process conditions in the pretreatment. An effective chemical pretreatment can loosen the carbohydrate-lignin complex structures, and increase the accessibility of cellulose, which can facilitate the subsequent AD process and increase the achievable biogas/biomethane yield. Table 12.4 summarizes the promoting effects of chemical pretreatment on the methane yield from AD of various lignocellulosic materials such as straw, and agrifood processing byproducts.

TABLE 12.4 The effects of chemical pretreatment on the digestion of lignocellulosic feedstocks.

Chemical pretreatment	Reagent	Feedstock	Pretreatment conditions	Chemical content changes	Digestion conditions	Biogas/methane improvement	Reference
Alkaline	NaOH	Rice straw	1% NaOH at room temperature for 3 h	Lignin content decreased from 10.2% to 3.5%; xylan content decreased from 12.2% to 4.8%	Semicontinuous mode at 37 °C with an HRT of 15 days for 92 days	34% increase in methane yield	(Shetty et al., 2017)
		Rice straw	1.6% NaOH at 30 °C for 24 h	Not mentioned	Batch mode at 37 °C for 40 days	21% increase in methane yield	(Mancini et al., 2018c)
		Wheat straw	1.6% NaOH at 30 °C for 24 h	Lignin content decreased from 18.2% to 11.6%; xylan content decreased from 15.5% to 9.7%	Batch mode at 37 °C for 40 days	15% increase in methane yield	(Mancini et al., 2018a)
		Maize straw	6% NaOH at 37 °C for 120 h	Lignin content decreased from 3.8% to 3.1%; xylan content decreased from 29.8% to 8.2%	Batch mode at 37 °C for 52 days	22.5% increase in methane yield	(Khatri et al., 2015)
		Corn stover	1.2% NaOH at room temperature for 24 h	Xylan content decreased by 22%	Batch mode at 38 °C for 50 days	59% increase in methane yield	(Xu et al., 2020)
	KOH	Wheat straw	6% KOH at room temperature for 72 h	Lignin content decreased from 12.5% to 9.1%; xylan content decreased from 28.3% to 17.6%	Batch mode at 35 °C for 40 days	40% increase in methane yield	(Jaffar et al., 2016)
		Corn stover	2.5% KOH at 20 °C for 24 h	Not mentioned	Batch mode at 37 °C for 35 days	96% increase in methane yield	(Li et al., 2015a)
	Ca(OH)$_2$	Corn stover	2.5% Ca(OH)$_2$ at 20 °C for 24 h	Not mentioned	Batch mode at 37 °C for 35 days	40% increase in methane yield	(Cayetano et al., 2019)
	Ammonia	Maize bran	15 wt% aqueous ammonia at 70 °C for 7 h	Lignin content reduced by 69%	Batch mode at 37 °C for 35 days	315% increase in methane yield	(Cayetano et al., 2019)
		Sugarcane bagasse	10% aqueous ammonia at 70 °C for 24 h	Lignin content reduced by 65%	Batch mode at 37 °C for 45 days	140% increase in methane yield	(Hashemi et al., 2019)

(continued)

Chemical pretreatment	Reagent	Feedstock	Pretreatment conditions	Chemical content changes	Digestion conditions	Biogas/methane improvement	Reference
Acid	H_2SO_4	Wheat plant	1% H_2SO_4 at 121 °C for 2 h	Lignin content reduced by 15%; hemicellulose content reduced by 91.5%	Batch mode at 37 °C for 30 days	16% increase in methane yield	(Taherdanak et al., 2016)
		Salvinia molesta	4% H_2SO_4 at 30 °C for 48 h	Lignin content reduced by 63%; hemicellulose content reduced by 40%	Batch mode at 30 °C for 30 days	82% increase in biogas yield	(Syaichurrozi et al., 2019)
		Water hyacinth	5% H_2SO_4 at 121 °C for 1 h	Lignin content reduced by 19%; cellulose content reduced by 65%	Batch mode at 30 °C for 90 days	132% increase in biogas yield	(Sarto et al., 2019)
	HCl	Dairy cow manure	2% HCl at 37 °C for 12 h	Not mentioned	Batch mode at 37 °C for 39 days	21% increase in methane yield	(Passos et al., 2017)
	H_3PO_4	Wheat straw	1.2% H_3PO_4 at 195 °C for 7 min	Hemicellulose content reduced by 96%	Batch mode at 55 °C for 35 days	150% increase in methane yield	(Nair et al., 2018)
	Acetic acid	Fruit waste	0.2 mol/L CH_3COOH at 62.5 °C for 0.5 h	Crystallinity index increased by 56%	Batch mode at 37 °C for 86 days	10% increase in methane yield	(Saha et al., 2018)
		Rice straw	2 % CH_3COOH at 62.5 °C for 24 h	Cellulose content reduced by 9%; hemicellulose content reduced by 18%	Batch mode at 37 °C for 28 days	24% increase in methane yield	(Peng et al., 2019)
Oxidation	$NaClO_2$	Energy grass	0.93% $NaClO_2$ and 0.31% CH_3COOH (per gram sample) at 80 °C for 200 min	Lignin content reduced by 80%	Batch mode at 37 °C for 30 days	38% increase in methane yield	(Kang et al., 2020b)
	H_2O_2	Maize straw	12% H_2O_2 at 58 °C for 58 min	Lignin content reduced by 72%; hemicellulose content reduced by 19%	Batch mode at 37 °C for 90 days	22% increase in biogas yield	(Venturin et al., 2018)
		Sorghum bicolor stalk	6.8% H_2O_2 at 28 °C for 85 min	Lignin content reduced by 73%; hemicellulose content reduced by 42%	Batch mode at 37 °C for 28 days	65% increase in biogas yield	(Dahunsi et al., 2019)
		Oil palm empty fruit bunches	6% H_2O_2 at 180 °C for 45 min	Hemicellulose decreased from 24.7% to 19.1%	Batch mode at 37 °C for 50 days	43% increase in methane yield	(Lee et al., 2020)
	Fenton	Oliver mill waste	$H_2O_2/[Fe^{2+}]$ ratio of 1000, $[Fe^{2+}]$ of 1.5 mM, pH 3 at 25 °C for 2 h	Lignin content reduced by 43%; hemicellulose content reduced by 20%	Batch mode at 37 °C for 30 days	24% increase in methane yield	(Maamir et al., 2017)

TABLE 12.4 (Cont'd)

Chemical pretreatment	Reagent	Feedstock	Pretreatment conditions	Chemical content changes	Digestion conditions	Biogas/methane improvement	Reference
Organic solvent	Ethanol	Forest residues	50% ethanol at 190 °C for 1 h	Lignin content reduced from 41% to 38%; hemicellulose content reduced from 20% to 14%	Batch mode at 55 °C for 40 days	500% increase in methane yield	(Kabir et al., 2015)
		Rice straw	50% ethanol at 180 °C for 1 h	Lignin content reduced by 15%	Batch mode at 37 °C for 43 days	42% increase in methane yield	(Mancini et al., 2018b)
		Rubber wood waste	75% ethanol at 210 °C for 1 h	Lignin content reduced by 72%; hemicellulose content reduced by 38%	Batch mode at 35 °C for 48 days	175% increase in methane yield	(Tongbuekeaw et al., 2020)
Ionic liquids	Ionic liquid	*Agave tequilana* bagasse	Cholinium lysinate at 124 °C for 205 min	Lignin content reduced by 45%	Batch mode at 35 °C for 40 days	500% increase in methane yield	(Perez-Pimienta et al., 2020)
Deep eutectic solvent	H_2O and Choline Chloride	Garden waste	H_2O-ChCl mass ratio 1:2 at 210 °C for 0.5 h	Lignin content reduced by 35%; hemicellulose content reduced by 26%	Batch mode at 37 °C for 30 days	309% increase in methane yield	(Yu et al., 2019b)

HRT, hydraulic retention time.

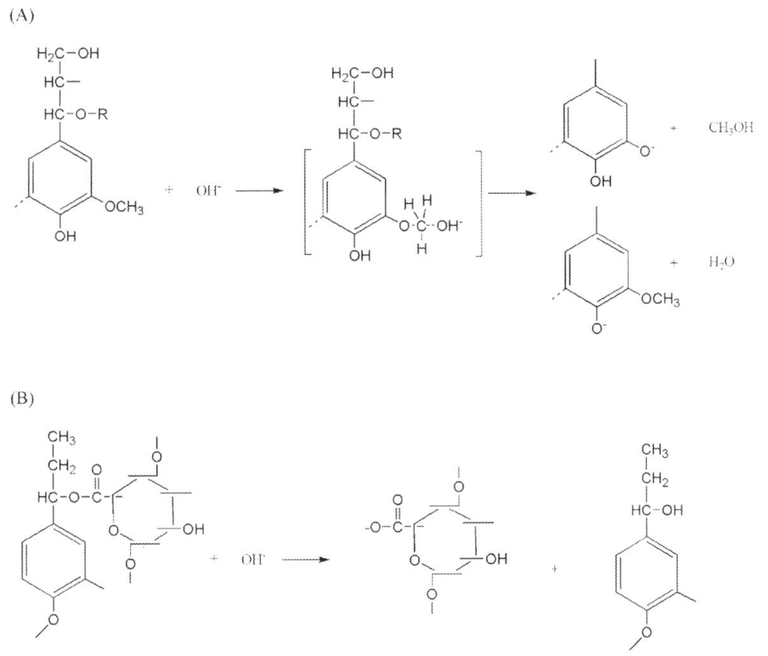

(A)

(B)

FIG. 12.4 Schematic diagram of reaction between lignin and OH⁻ (adapted from Yu et al. (2019a)).

12.2.2.1 Alkaline pretreatment

Compared to other chemical pretreatment (such as acid pretreatment), alkaline pretreatment involves lower temperatures and pressures (typically atmospheric pressure and under 100 °C) (Song et al., 2021). Alkaline pretreatment can be carried out at ambient conditions, but the reaction time would extend to hours or days. Alkaline pretreatment are reported to effectively remove lignin and partially remove hemicellulose due to the function of the hydroxide ion (OH⁻), which can cleave the ester bonds between hemicellulose and lignin, and weaken the hydrogen bond between cellulose and hemicellulose (Fig. 12.4). This increases the porosity of the feedstock, which can contribute to enhanced digestion efficiency and biogas/biomethane production. The additional advantage of alkaline pretreatment includes that the residual alkali could function as a pH buffer, stabilizing the pH in acidogenesis step of AD (Rodriguez et al., 2017). Recent studies regarding the promotional effect of alkaline pretreatment on digestion efficiency and biogas/biomethane production improvement are shown in Table 12.4.

NaOH is the most used reagent in alkaline pretreatment due to its low cost, high efficiency in removing lignin and dissolving hemicellulose, and its ability to reduce the cellulose crystallinity. As the structural and compositional characteristics vary in different types of lignocellulose, it may be necessary to optimize the process condition for a specific feedstock. For example, with xylan content decreased by 22% under the pretreatment condition of 1.2% NaOH at room temperature (ca. 30 °C) for 24 h, a 59% improvement in methane production from corn stover was achieved (Xu et al., 2020). Mancini et al. (2018c) reported that the methane production from rice straw increased by 21%, under the conditions of 1.6% NaOH at 30 °C for 24 h. Under the same pretreatment condition, the lignin content in wheat straw decreased from 18.2% to 11.6%, and the xylan content decreased from 15.5% to 9.7%; consequently, the methane yield from pretreated wheat straw was 15% higher than that from untreated wheat straw (Mancini et al., 2018a). Khatri et al. (2015) applied harsher process conditions to treat maize straw, which resulted in the xylan content decreasing from 29.8% to 8.2%, and the methane yield increasing by 22.5%. Although NaOH pretreatment can boost the methane production from different lignocellulose biomasses, negative results have also been observed associated with NaOH pretreatment at high concentrations and temperatures (Kang et al., 2018). This can be attributed to the inhibition of high sodium ions concentration (Na^+ concentration > 5 g/L) to methanogenesis due to increased osmotic pressure or dehydration of microbes (Hierholtzer & Akunna, 2012), and the formation of inhibitors (such as 5-hydroxymethylfurfural (HMF) and phenolic compounds) from the degradation of hemicellulose and lignin (Koyama et al., 2017). To increase the economic feasibility of NaOH pretreatment and make the process more environmentally friendly, recycling of pretreatment black liquor (liquid fraction after NaOH pretreatment) has recently received much attention as this enables less consumption of NaOH and water (Fan et al., 2020; Liu et al., 2015; Wang et al., 2016). However, as lignin-derived phenols accumulate in the hydrolysate, careful attention should be paid to its inhibitory effects on methanogenesis.

Compared with NaOH, KOH modifies the lignocellulose structure with a similar degree of delignification but can improve the digestion efficiency with higher concentrations. In addition, KOH can be used as agriculture fertilizer, increasing the value of the digestate. Jaffar et al. (2016) employed KOH to treat wheat straw at conditions of 6% KOH at room temperature for 72 h; they found that both hemicellulose and lignin content were reduced, and the methane yield was 40% higher than that from the untreated sample. Li et al. (2015a) also found that with pretreatment conditions of 2.5% KOH at room temperature for 1 day, the methane yield from AD of pretreated corn stover increased from 150 mL/g VS to 295 mL/g VS (Li et al., 2015a). The main concern with KOH as a pretreatment is the cost of the chemical which is considerably higher than NaOH; however, this may be offset by the added value in the digestate in biofertilizer applications.

Ammonia solution is also a good reagent for chemical pretreatment of lignocellulose. Li et al. (2015b) found that ammonia pretreatment is helpful for the AD of corn stover and the biogas production could reach 427.1 mL/g VS with 4% concentration of ammonia. The main mechanism is the saponification and cleavage of lignin-carbohydrates linkages (Wang et al., 2015). One issue with ammonia pretreatment is that the transformation between NH_4OH and NH_3 with varying pH values, which may negatively impact the activity of methane-producing microorganisms if the concentration of NH_3 is over a specific value (100–460 mg/L) (Capson-Tojo et al., 2020). Ammonia inhibition is a common issue when high nitrogen-containing material (such as distillery spent grain, pig slurry, and chicken manure) is used for biogas/biomethane production through AD (Bougrier et al., 2018; Capson-Tojo et al., 2020). Therefore, studies on ammonia pretreatment should take this into consideration when optimizing the process conditions.

12.2.2.2 *Acid pretreatment*

Acid pretreatment uses inorganic acids (such as H_2SO_4 and HCl) and organic acids (such as acetic acid and maleic acid) to alter the physicochemical characteristics of lignocellulose. The mechanism of acid pretreatment is the breakdown of intermolecular and/or intramolecular glycol bonds in cellulose and hemicellulose. This hydrolysis process leads to the release of respective monosaccharides (such as glucose and xylose) from hemicellulose and amorphous cellulose, decreases of cellulose polymerization, and increases cellulose accessibility. Acid pretreatment can also remove lignin from the biomass to some degree depending on the process conditions.

Sulfuric acid (H_2SO_4) is the most commonly used reagent in acid pretreatment. Taherdanak et al. (2016) improved the methane yield from wheat plant by 16%, using a pretreatment consisting of 1% H_2SO_4 at 121 °C for 2 h, where 91.5% of xylan content was dissolved and 15% of lignin was removed. Syaichurrozi et al. (2019) reported that pretreatment with 4% H_2SO_4 at 30 °C for 48 h led to 63% lignin removal and 40% xylan degradation, and the biogas yield from *Salvinia molesta* improved by 82%. Employing more severe pretreatment conditions (5% H_2SO_4 at 121 °C for 1 h) to treat water hyacinth, Sarto et al. (2019) found that the methane yield was 1.4 times that of the untreated sample. Although acid pretreatment with dilute H_2SO_4 reagent can significantly modify the lignocellulose structure by dissolving hemicellulose by up to 100% and hydrolyze cellulose, improvement in the methane yield may not arise. Deng et al. (2019) observed a methane yield reduction of 9% from grass silage, pretreated with 2% H_2SO_4 at 135 °C for 15 min, even with a 100% reduction in hemicellulose. Dahunsi et al. (2019) also reported a methane yield reduction of 13% from *Sorghum bicolor* stalk, with 65% hemicellulose dissolved after pretreatment at lower severity (0.75% H_2SO_4 at 118 °C for 52 min). These lower methane production yields from pretreated substrates might be attributed to the Maillard reaction, which takes place between an amino group (–NH_2) in ammonia acids and carbonyl group (–C=O) in sugars during the degradation of complex substrates (with high carbohydrate and protein contents) (Lin et al., 2017). As such, the nitrogen-containing reaction products are difficult to degrade in AD (Li et al., 2016). Therefore, future research may apply lower acid pretreatment severity to improve the methane production from various lignocellulosic materials.

Recently, organic acids such as acetic acid ($CH_3COOH \rightarrow CH_3COO^- + H^+$) have been used as acid catalysts for biomass pretreatment, which mainly alter the lignocellulosic structure by cleaving the lignin-hemicellulose bonds. Compared with inorganic acids and bases, using organic acids has some advantages such as high sugar recovery, easier handling, and less toxicity/causticity (Choi et al., 2019). In addition, organic acids such as acetic acid and propionic acid can be produced from dark fermentation of biomass (Lin et al., 2016); thus, using these acids can contribute to a green and sustainable pretreatment process. Peng et al. (2019) investigated the effects of elevated concentrations of acetic acid (2%–10%) in pretreatment on the digestion performance of rice straw and found that pretreatment with 2% acetic acid at 80 °C for 24 h led to the reduction of cellulose and hemicellulose content by 9% and 18%, respectively; under this pretreatment condition, the maximum methane yield from rice straw was obtained, corresponding to an increase of 24% as compared with that from the untreated sample. Saha et al. (2018) also observed the promotional impact of acetic acid pretreatment (0.2 M CH_3COOH at 62.5 °C for 0.5 h) on methane production from fruit waste, which was 10% higher than that from raw fruit waste. Coupled with sodium chlorite, acetic acid pretreatment can selectively remove lignin from lignocellulose through an oxidation reaction, with little impact on the carbohydrate content. With almost 80% lignin removed, the methane yield from an acetic acid and sodium chlorite pretreated energy crop was shown to increase by 38% (Kang et al., 2020b).

12.2.2.3 Wet oxidation pretreatment

Wet oxidation pretreatment can effectively degrade lignin while having limited effects on the carbohydrate content of lignocellulose. Lignin has been well highlighted as the main hurdle in the efficient conversion of lignocellulose, thus its removal could enhance the methane production in AD (Kang et al., 2020b). H_2O_2 is the most used oxidant in oxidation pretreatment due to the advantage that its only hydrolysis products are water and oxygen. Venturin et al. (2018) enhanced the methane yield from maize straw by 22% under the optimal pretreatment condition of 12% H_2O_2 at 58 °C for 58 min, where 72% of the lignin content was removed. Dahunsi et al. (2019) also found that the lignin content in *Sorghum* stalk was removed by up to 73% under the pretreatment condition of 6.8% H_2O_2 at 28 °C for 85 min, and the hemicellulose content was also reduced by 42%; this resulted in the increase of biogas yield from *Sorghum* stalk by 65% (Dahunsi et al., 2019). Treated at higher reaction temperatures (6% H_2O_2 at 180 °C for 45 min), the methane yield from oil palm fruit bunches was promoted by 43% (Lee et al., 2020). H_2O_2 can be combined with Fe^{2+} to work as Fenton pretreatment (Michalska et al., 2012). The mechanism of Fenton pretreatment is that Fe^{2+} triggers and catalyzes the rapid decomposition of H_2O_2 and leads to the generation of •OH. The •OH are highly destructive oxygen-derived free radicals, which can disrupt the recalcitrant structure of LCB and facilitates its subsequent degradation (Zhang et al., 2018). However, the Fenton oxidation tends to nonselectively degrade organic compounds (Mert et al., 2010). Maamir et al. (2017) found that 43% of the lignin content in olive mill waste was removed through Fenton pretreatment with a $H_2O_2/[Fe^{2+}]$ ratio of 1000, $[Fe^{2+}]$ of 1.5 mM, pH 3 at 25 °C for 2 h, and the methane yield was enhanced by 24%.

12.2.2.4 Organosolv pretreatment

Similar to organosolv pulping that uses organic solvents to extract lignin from LCB, organosolv pretreatment techniques involve the use of organic solvents but the degree of delignification is not required to be as high as that of pulping. The advantages of organosolv pretreatment are (1) easy recycle and reuse of the organic solvent through distillation, and (2) the production of carbohydrate-rich residue and pure lignin after chemical recovery, which creates more value in the process chain. Mancini et al. (2018b) pretreated rice straw with ethanol under the condition of 50% ethanol at 180 °C for 1 h, which led to an increase of 42% in the methane yield, even with only 15% lignin removal. Organosolv pretreatment with ethanol had a notable impact on the methane yield from woody biomass due to its high recalcitrance and low degradability. Kabir et al. (2015) found that the methane yield from ethanol-treated forest residues increased fivefold compared to that of the untreated sample. Tongbuekeaw et al. (2020) also enhanced the methane yield from rubber wood waste by 175%, using a pretreatment with 75% ethanol at 210 °C for 1 h. Although organosolv pretreatment has the aforementioned merits and enhances methane production from various lignocellulosic feedstocks, there are inherent drawbacks. To prevent the reprecipitation of dissolved lignin, the solid residues after pretreatment need to be washed with other organic solvents prior to water washing process, which is cumbersome (Zhao et al., 2009). Organic solvents are expensive and harmful to the bacteria in AD, and as such should be recovered as much as possible; this, however, increases energy consumption during the recovery process. In addition, due to the high volatility of organic solvents, organosolv pretreatment must be performed extremely carefully. Reactor leakage could lead to intrinsic fire and explosion hazards involving the organic solvents (Zheng et al., 2014). In general, organosolv pretreatment is still being questioned with regard to the technology readiness in the recovery of solvents and products and the cost-effectiveness of treatment systems. A more recent study (Wu et al., 2020b) has provided a novel combination of gamma-valerolactone pretreatment and AD for the coproduction of biogas and lignin nanoparticles. This integration of biomaterials and biogas is likely to increase the competitiveness of conventional AD processes.

12.2.2.5 *Ionic liquid and deep eutectic solvent pretreatment*

Recently developed ionic liquids (ILs) pretreatment of lignocellulosic material have attracted much interest in the research community. ILs, a mixture of salts, are liquid at room temperatures or melt at slightly elevated temperatures, with the ability to dissolve large amounts of lignin or cellulose under moderate conditions and furthermore have the feasibility to be recovered (Tadesse & Luque, 2011). Most ILs pretreatment are developed to enhance the enzymatic hydrolysis efficiency for sugar and ethanol production (Elgharbawy et al., 2016; Elgharbawy et al., 2020), with few studies applying such methods to improve the methane yield from lignocellulosic feedstocks. Using ILs synthesized from bio-based cholinium lysinate, Perez-Pimienta et al. (2020) optimized the pretreatment conditions for improving the methane yield from *Agave tequilana* bagasse; with 45% of the lignin removed under the optimum condition of 124 °C for 205 min. The authors observed that the methane yield increased fivefold compared to that from untreated *A. tequilana* bagasse.

With physical and chemical properties similar to ILs, deep eutectic solvents (DESs), a low-cost eutectic mixture synthesized from hydrogen bond donors (HBD, such as acetic acid, oxalic acid, glycerol, and urea) and hydrogen bond acceptors (mainly choline chloride (ChCl)), have been extensively used to treat lignocellulosic substrates for efficient bioconversion of carbohydrates into chemicals and bioenergy production (Abbott et al., 2004; Satlewal et al., 2018). DESs outpace conventional ILs due to the following advantages: easy to synthesize, stable, cost-competitive, biodegradable, and environmentally friendly (Mbous et al., 2017). It was estimated that synthesizing a DES costs only 20% of that of an IL (Xu et al., 2016). To date, studies in terms of employing DESs pretreatment to improve the methane production of lignocellulosic materials are scarce. However one study investigated the pretreatment conditions of H_2O-ChCl with a mass ratio of 1:2 at 210 °C for 0.5 h, whereby the results showed that the lignin content in garden waste could be removed by 35% and the hemicellulose content could be reduced by 26% (Yu et al., 2019b). As such, this pretreatment led to the methane yield increasing from 0.8 mL/g substrate to 261.8 mL/g substrate. Recent studies demonstrated that DESs pretreatment of lignocellulose with acetic acid, propionic acid, and lactic acid as HBD exhibited great potential in delignification or lignin extraction, producing a carbohydrate-enriched solid residue (Alvarez-Vasco et al., 2016; Yee Tong et al., 2019). Therefore, given the fact that the HBD (1) such as acetic acid and propionic acid in DESs can be derived from acetogenesis process (the first stage of a two-stage AD process), and (2) such as lactic acid can be obtained by bacterial fermentation of glucose, applying DESs pretreatment to enhance the AD of LCB can be considered a promising method. As such, sustainable and cost-effective DESs pretreatment of lignocellulose could be achievable.

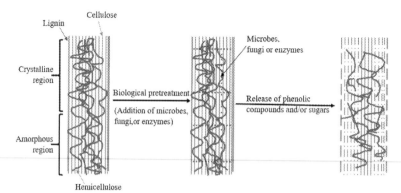

FIG. 12.5   Schematic flow of degradation of lignocellulosic material during biological pretreatment (adapted from Kainthola et al. (2021)).

12.2.3   Biological pretreatment

Biological pretreatment refers to techniques using microorganisms or enzymes to destroy the cell wall structure of lignocellulosic materials, as shown in Fig. 12.5. The generic advantages of biological pretreatment outcompeting chemical or thermochemical pretreatment are that it can take place under low temperatures without chemicals, consume less energy, and produce fewer inhibitors. The process parameters such as pH value, pretreatment temperature and time, and types of microbes and enzymes play important roles in the effects of biological pretreatment on the digestion performance of LCB. The main drawback of biological pretreatment are the longer reaction times required compared to that of other pretreatment methods.

12.2.3.1   Microbial pretreatment

Microbial pretreatment involves the process of selecting and application of suitable microorganisms isolated from the natural environment that will particularly assist in the hydrolysis of biomass, the first step in the AD process (Mishra et al., 2018). Microbial pretreatment can take place under anaerobic or aerobic conditions depending on the bacteria used in the pretreatment process.

The AD process can be classified into single-stage or two-stage digestion systems. In a two-stage system, the first stage is termed dark fermentation or preacidification, which is recognized as a pretreatment process (anaerobic microbial pretreatment) as a consortium of hydrolytic bacteria are used to produce cellulose-degrading and hemicellulose-degrading enzymes (Lin et al., 2016). The resultant products are short-chain carboxylic acids such as acetic acid, propionic acid, and butyric acid. The conversion of these acids to methane in the subsequent methanogenesis process is an exergonic reaction (Eqs. 12.6–12.8), which can accelerate the degradation rate of the substrate and increase the methane production rate.

$$CH_3COOH \rightarrow CH_4 + CO_2 \Delta G^0 = -54.95 kJ/reaction \tag{12.6}$$

$$CH_3CH_2COOH + 1/2H_2O \rightarrow 7/4CH_4 + 5/4CO_2 \Delta G^0 = -102.15 kJ/reaction \tag{12.7}$$

$$CH_3CH_2CH_2COOH + H_2O \rightarrow 5/2CH_4 + 3/2CO_2 \Delta G^0 = -128.68 kJ/reaction \tag{12.8}$$

Akobi et al. (2016) compared the methane yield from single-stage and two-stage AD of extruded woody biomass and reported that the methane yield from two-stage AD was 14% higher than single stage. Deng et al. (2019) reported that the methane yield from the two-stage AD of grass silage increased to 393 mL/g VS, an improvement of 50.5% when compared to that from single-stage AD (261 mL/g VS). Barua et al. (2018) treated water hyacinth with three different bacterial strains isolated from soil (*Bordetella muralis VKVVG5*), silver fish (*Citrobacter werkmanii VKVVG4*), and millipede (*Paenibacillus* sp. *VKVVG1*). Results showed that pretreatment with *C. werkmanii VKVVG4* led to the highest solubilization of the substrate and the biogas yield was increased by 23% under the optimal pretreatment condition dosage of 10^9 colony-forming unit (CFU)/mL for 4 days.

Microbial pretreatment can be carried out under aerobic conditions with pure strain or mixed microbes. These aerobic microbes can rapidly secrete cellulase, hemicellulase, and/or enzymes degrading lignin in large quantities, which hydrolyses the specific chemical component of the substrate and thereby facilitates the conversion efficiency of lignocellulosic feedstock. Xu et al. (2018) treated corn straw with a pure strain, *Bacillus subtilis* at 37 °C for 24 h; the results showed that microbial pretreatment led to 18% removal of hemicellulose and 23% removal of lignin, which resulted in 17% more methane yield from the pretreated corn straw. Yuan et al. (2016) treated cotton stalk with thermophilic mixed bacteria under microaerobic conditions and reported that the pretreatment led to significant changes in cellulose and hemicellulose contents; consequently, the methane yield from pretreated cotton stalk within 8 days was increased by 136%. An improvement of 17% in methane yield from corn stover was also obtained with microaerobic pretreatment at a pH of 8 for 12 h (Xu et al., 2021b).

12.2.3.2 Fungal pretreatment

Brown, white, and soft rot fungi are commonly employed microorganisms in fungal pretreatment used to destroy the rigid structure of lignocellulosic material. This is achieved through depolymerization and solubilization of the lignin by applying secreted various redox enzymes such as laccases and peroxidases (Kainthola et al., 2021; Mishra et al., 2018). After assessing the effects of fungal pretreatment with *Leiotrametes menziesii* and *Abortiporus biennis* on the methane yield of willow sawdust for 31 days, Alexandropoulou et al. (2017) found that the former strain effectively removed lignin by 31%, cellulose by 27% and hemicellulose by 42% from the substrate, while the latter strain removed 17% lignin, 7% cellulose, and 19% hemicellulose. Mustafa et al. (2016) investigated the effects of fungal pretreatment with *Pleurotus ostreatus* and *Trichoderma reesei* (different strains of white rot fungi) on the compositional structure and digestion performance of rice straw; under the optimal pretreatment conditions (75% moisture content and 20 days incubation), pretreatment with *P. ostreatus* resulted in the removal of 33% of the lignin and the methane yield was 120% higher than that from the untreated sample; while pretreatment with *T. reesei* removed 24% of the lignin in rice straw and increased the methane yield by 78% (Mustafa et al., 2016).

12.2.3.3 Enzymatic pretreatment

In the AD of lignocellulosic feedstock, the hydrolysis of holocellulose (cellulose and hemicellulose) is one of the rate-limiting steps causing low methane yields. Hydrolytic enzymes such as cellulases and hemicellulases or lignin-degrading enzymes such as laccases and peroxidases can be used to treat lignocellulosic feedstock for improved methane yields. Compared with microbial and fungal pretreatment, enzymatic pretreatment of lignocellulose is less attractive due to the substrate selectivity and the high cost of enzymes. The cost, however, could be notably reduced with the development of enzyme immobilization. Perez-Rodriguez et al. (2017) reported that the methane yield from enzymatically pretreated corn cob was increased by 16% using Ultraflo L, a mixture of endoglucanase, xylanases, and cellulase. Schroyen et al. (2015) used mixed laccase and versatile peroxidase to treat different lignocellulosic substrates including *Miscanthus,* willow, corn stover, and wheat straw; results showed that the methane yield from pretreated corn stover was increased by 24% after 6 h incubation, which was 12% for wheat straw under the same pretreatment condition.

12.2.4 Combinational pretreatment

Combined pretreatment of physical, chemical, and biological methods have advantages of shorter reaction time, moderate pretreatment conditions, fewer chemical demands, and improved economic feasibility. This is because combinational pretreatment integrates the merits of different types of mono-pretreatment method, which may successively destroy the recalcitrate lignocellulosic structure by reducing cellulose crystallinity, breaking the linkage between lignin and carbohydrates, removing lignin, and dissolving hemicellulose. For example, Guan et al. (2018) treated rice straw with the combination of CaO and liquid fraction of digestate (LFD) (a byproduct of AD) and found that when compared with the pretreatment of CaO or LFD alone, the combined pretreatment removed more lignin from the substrate. A 58% higher methane yield than that from the untreated sample could be achieved (Guan et al., 2018). Li et al. (2019) combined ammonia and a freeze-thaw pretreatment to enhance the methane yield from corn stover and found that the methane yield increased by ca. 30% in continuous AD experiments. Furthermore, Alexandropoulou et al. (2017) carried out the combined pretreatment of willow sawdust with white rot fungi and NaOH. The methane yield from the pretreated substrate dramatically increased by 115% compared to the control. These results verify the potential beneficial effects of integrated pretreatment techniques on increasing the methane yield and improving the conversion efficiency from lignocellulosic material.

12.3 Opportunities for AD in a circular bioeconomy

As opposed to the traditional linear economy, a circular economy aims at maintaining the value of end-of-life materials (and/or products, resources) within the economy as long as possible and reducing the waste to a minimum to develop a resource-efficient, competitive, and low carbon economy. This approach can not only boost economic growth but reduce greenhouse gas (GHG) emissions and further facilitate climate mitigation (Leipold & Petit-Boix, 2018). The circular bioeconomy is defined as the intersection of bioeconomy and circular economy, which facilitates the cascading use of biomass (Zabaniotou, 2018). Among the different bioconversion processes, AD plays a critical role in developing a circular bioeconomy (Fagerströ et al., 2018). By closing the loops on the formerly linear processes (take-make-consume-throw away), AD can simultaneously address waste, energy, nutrient recycling, and sustainable food production challenges in a sustainable and circular manner (Wall et al., 2017). In many European countries, AD plays an important role in the agroindustrial sector, displacing emission-intensive waste management strategies such as landfilling (Carlsson et al., 2015; Murphy & Power, 2009; Rekleitis et al., 2020). AD of agricultural residues (such as cereal straws) is a well-established technique, that not only aids in reducing the GHG emissions from the agricultural sector, but also produces a renewable gaseous biofuel—biogas. Biogas can be used to produce heat electricity, or, after biogas upgrading (removal of carbon dioxide) advanced transport fuel; upgraded biogas can also can be injected to the natural gas grid for storage and transportation to the site of demand (Liebetrau et al., 2020; Murphy & Power, 2009; Wu et al., 2020a). It can also be used as a carbon source replacing natural gas for methanotrophic bacterial for single-cell protein production (Strong et al., 2016). Furthermore, the digestate can be used as a biofertilizer to support the growth of plants for food, feed, and other uses, which reduces the consumption of synthetic fertilizer and related GHG emissions (Kang et al., 2020a). Previous studies have shown that the installation of an AD plant at a distillery (with the distillation byproducts used as AD feedstocks) could reduce the energy-related GHG emissions by approximately 50% (Kang et al., 2020a; O'Shea et al., 2020). Furthermore, the emissions could be further reduced with the introduction of a pretreatment for the spent grains and the use of digestate as a fertilizer substituting synthetic fertilizers.

Other innovative technologies could further complement the use of AD in a circular economy. For example, the integration of AD and pyrolysis, hydrothermal carbonization, and algae cultivation has recently received much attention as a means of approaching a negative GHG emissions system (Bose et al., 2020; Deng et al., 2020a; Deng et al., 2020b). In a representative cascading circular bioenergy production system (shown in Fig. 12.6),

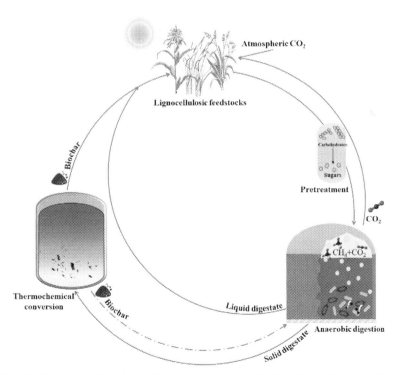

FIG. 12.6   Simplified schematic flow for the integration of anaerobic digestion with thermochemical conversion technologies toward low/negative carbon emission circular economy. Dashed line represents biochar addition to improve direct interspecies electron transfer and biogas production rate in AD system.

the liquid digestate from the AD of pretreated/nonpretreated organic lignocellulosic residues (such as cattle manure, straws, and agricultural and forestry residues) can be used for algae cultivation, and the produced algae can be used as feedstocks for animals or AD. Microalgae cultivation may also be generated in a circular economy system whereby microalgae are used in a photosynthetic upgrading system including for a carbonate/bicarbonate cycle (Bose et al., 2020). Solid digestate after AD can be used in thermo-chemical processes, namely pyrolysis or hydrothermal carbonization, to produce pyro-oil, syngas, and biochar; biochar when applied to land serves as a negative emission technology in adding to soil organic content and sequestering CO_2 from the atmosphere through increased photosynthesis, but also shows significant potential as a carbonaceous conductive material for boosting the digestion efficiency of several organic wastes (Deng et al., 2020b; Stefaniuk & Oleszczuk, 2015). A recent study exhibited that the addition of biochar (from the pyrolysis of wood biomass) in AD improved the methane yield from seaweed by 17%, with the digestate mass flow decreased by 26% (Deng et al., 2020b). Monlau et al. (2015) reported that the combination of AD and pyrolysis conversion of agricultural residues generated 42% more electricity than that from stand-alone AD treatment, where the energy needed for drying the solid digestate was fully covered by the thermal energy produced from biogas. However, the variation of feedstocks in AD may lead to various biochar properties, which diversify the effects of biochar addition on AD performance (Chiappero et al., 2020; Stefaniuk & Oleszczuk, 2015). Future studies may be needed to glean more insights into the correlations between feedstock properties, biochar characteristics, and digestion efficiency.

12.4   Conclusions and perspectives

AD of lignocellulosic residues such as crop residues, cattle manure, food processing byproducts (husk, spent grains) offers not only a sustainable method of waste treatment and nutrient recycling, but also produces renewable low-carbon energy in the form of biogas. Pretreatments are efficient in breaking down the recalcitrant polymer structure of lignocellulosic residues, increasing its accessibility via size reduction, hemicellulose dissolution, and/or lignin removal, resulting in the improved methane yield and production rate. The combined pretreatment of biological, physical, and chemical methods may outpace the stand-alone pretreatment method in terms of energy efficiency, inhibitor formation, and economic feasibility. The implementation of AD in food and beverage industries such as distilleries could reduce fossil fuel consumption, aiding in the transition to a green and sustainable production process. The integration of AD and other thermochemical conversion processes such as pyrolysis and hydrothermal carbonization is of great importance in contributing to a negative emission circular economy. Future studies are needed to optimize the process conditions of pretreatment, AD, and pyrolysis/carbonization to fully realize the cascading utilization and add value to lignocellulosic residues. The overall sustainability of cascading circular bioenergy production systems needs to be demonstrated through life cycle assessments and technoeconomic analysis before such systems can be deployed.

References

Abbott, A.P., Boothby, D., Capper, G., Davies, D.L., Rasheed, R.K., 2004. Deep eutectic solvents formed between choline chloride and carboxylic acids: versatile alternatives to ionic liquids. J. Am. Chem. Soc. 126 (29), 9142–9147.

Abraham, A., Mathew, A.K., Park, H., Choi, O., Sindhu, R., Parameswaran, B., Pandey, A., Park, J.H., Sang, B.-I., 2020. Pretreatment strategies for enhanced biogas production from lignocellulosic biomass. Bioresour. Technol. 301, 122725.

Akobi, C., Yeo, H., Hafez, H., Nakhla, G., 2016. Single-stage and two-stage anaerobic digestion of extruded lignocellulosic biomass. Appl. Energy 184, 548–559.

Alexandropoulou, M., Antonopoulou, G., Fragkou, E., Ntaikou, I., Lyberatos, G., 2017. Fungal pretreatment of willow sawdust and its combination with alkaline treatment for enhancing biogas production. J. Environ. Manage. 203, 704–713.

Alvarez-Vasco, C., Ma, R., Quintero, M., Guo, M., Geleynse, S., Ramasamy, K.K., Wolcott, M., Zhang, X., 2016. Unique low-molecular-weight lignin with high purity extracted from wood by deep eutectic solvents (DES): a source of lignin for valorization. Green Chem. 18 (19), 5133–5141.

Barua, V.B., Goud, V.V., Kalamdhad, A.S., 2018. Microbial pretreatment of water hyacinth for enhanced hydrolysis followed by biogas production. Renew. Energy 126, 21–29.

Bauer, A., Lizasoain, J., Theuretzbacher, F., Agger, J.W., Rincon, M., Menardo, S., Saylor, M.K., Enguidanos, R., Nielsen, P.J., Potthast, A., Zweckmair, T., Gronauer, A., Horn, S.J., 2014. Steam explosion pretreatment for enhancing biogas production of late harvested hay. Bioresour. Technol. 166, 403–410.

Bentsen, N.S., Nilsson, D., Larsen, S., 2018. Agricultural residues for energy-A case study on the influence of resource availability, economy and policy on the use of straw for energy in Denmark and Sweden. Biomass Bioenergy 108, 278–288.

Bose, A., O'Shea, R., Lin, R., Murphy, J.D., 2020. A perspective on novel cascading algal biomethane biorefinery systems. Bioresour. Technol. 304, 123027.

Bougrier, C., Dognin, D., Laroche, C., Rivero, J.A.C., 2018. Use of trace elements addition for anaerobic digestion of brewer's spent grains. J. Environ. Manage. 223, 101–107.

Bundhoo, Z.M.A., 2018. Microwave-assisted conversion of biomass and waste materials to biofuels. Renew. Sustain. Energy Rev. 82, 1149–1177.

Capson-Tojo, G., Moscoviz, R., Astals, S., Robles, Á., Steyer, J.-P., 2020. Unraveling the literature chaos around free ammonia inhibition in anaerobic digestion. Renew. Sustain. Energy Rev. 117, 109487.

Carlsson, M., Holmström, D., Bohn, I., Bisaillon, M., Morgan-Sagastume, F., Lagerkvist, A., 2015. Impact of physical pre-treatment of source-sorted organic fraction of municipal solid waste on greenhouse-gas emissions and the economy in a Swedish anaerobic digestion system. Waste Manage. (Oxford) 38, 117–125.

Cayetano, R.D.A., Oliwit, A.T., Kumar, G., Kim, J.S., Kim, S.-H., 2019. Optimization of soaking in aqueous ammonia pretreatment for anaerobic digestion of African maize bran. Fuel 253, 552–560.

Chiappero, M., Norouzi, O., Hu, M.Y., Demichelis, F., Berruti, F., Di Maria, F., Masek, O., Fiore, S., 2020. Review of biochar role as additive in anaerobic digestion processes. Renew. Sustain. Energy Rev. 131, 110037.

Choi, J.-H., Jang, S.-K., Kim, J.-H., Park, S.-Y., Kim, J.-C., Jeong, H., Kim, H.-Y., Choi, I.-G., 2019. Simultaneous production of glucose, furfural, and ethanol organosolv lignin for total utilization of high recalcitrant biomass by organosolv pretreatment. Renew. Energy 130, 952–960.

Dahunsi, S., Adesulu-Dahunsi, A., Osueke, C., Lawal, A., Olayanju, T., Ojediran, J., Izebere, J., 2019. Biogas generation from Sorghum bicolor stalk: effect of pretreatment methods and economic feasibility. Energy Rep. 5, 584–593.

Dahunsi, S.O., 2019. Mechanical pretreatment of lignocelluloses for enhanced biogas production: methane yield prediction from biomass structural components. Bioresour. Technol. 280, 18–26.

Deng, C., Kang, X., Lin, R., Murphy, J.D., 2020a. Microwave assisted low-temperature hydrothermal treatment of solid anaerobic digestate for optimising hydrochar and energy recovery. Chem. Eng. J. 395, 124999.

Deng, C., Lin, R., Cheng, J., Murphy, J.D., 2019. Can acid pre-treatment enhance biohydrogen and biomethane production from grass silage in single-stage and two-stage fermentation processes? Energy Convers. Manage. 195, 738–747.

Deng, C., Lin, R., Kang, X., Wu, B., O'Shea, R., Murphy, J.D., 2020b. Improving gaseous biofuel yield from seaweed through a cascading circular bioenergy system integrating anaerobic digestion and pyrolysis. Renew. Sustain. Energy Rev. 128, 109895.

Duque, A., Manzanares, P., Ballesteros, M., 2017. Extrusion as a pretreatment for lignocellulosic biomass: fundamentals and applications. Renew. Energy 114, 1427–1441.

Elgharbawy, A.A., Alam, M.Z., Moniruzzaman, M., Goto, M., 2016. Ionic liquid pretreatment as emerging approaches for enhanced enzymatic hydrolysis of lignocellulosic biomass. Biochem. Eng. J. 109, 252–267.

Elgharbawy, A.A., Moniruzzaman, M., Goto, M., 2020. Facilitating enzymatic reactions by using ionic liquids: a mini review. Curr. Opin. Green Sustain. Chem. 27, 100406.

Fagerströ, A., Al Seadi, T., Rasi, S., Briseid, T., 2018. The role of anaerobic digestion and biogas in the circular economy. IEA Bioenergy Task 37.

Fan, Z., Lin, J., Wu, J., Zhang, L., Lyu, X., Xiao, W., Gong, Y., Xu, Y., Liu, Z., 2020. Vacuum-assisted black liquor-recycling enhances the sugar yield of sugarcane bagasse and decreases water and alkali consumption. Bioresour. Technol. 309, 123349.

Feng, R., Zaidi, A.A., Zhang, K., Shi, Y., 2019. Optimisation of microwave pretreatment for biogas enhancement through anaerobic digestion of microalgal biomass. Periodica Polytechnica Chem. Eng. 63 (1), 65–72.

Gu, Y., Zhang, Y.L., Zhou, X.F., 2015. Effect of Ca(OH)(2) pretreatment on extruded rice straw anaerobic digestion. Bioresour. Technol. 196, 116–122.

Guan, R.L., Li, X.J., Wachemo, A.C., Yuan, H.R., Liu, Y.P., Zou, D.X., Zuo, X.Y., Gu, J.Y., 2018. Enhancing anaerobic digestion performance and degradation of lignocellulosic components of rice straw by combined biological and chemical pretreatment. Sci. Total Environ. 637, 9–17.

Hashemi, S.S., Karimi, K., Karimi, A.M., 2019. Ethanolic ammonia pretreatment for efficient biogas production from sugarcane bagasse. Fuel 248, 196–204.

Hierholtzer, A., Akunna, J.C., 2012. Modelling sodium inhibition on the anaerobic digestion process. Water Sci. Technol. 66 (7), 1565–1573.

Hjorth, M., Granitz, K., Adamsen, A.P.S., Moller, H.B., 2011. Extrusion as a pretreatment to increase biogas production. Bioresour. Technol. 102 (8), 4989–4994.

IEA, 2020. World Energy Outlook 2020. IEA, Paris. https://www.iea.org/reports/world-energy-outlook-2020. (Accessed on 16th April 2021).

Izumi, K., Okishio, Y.K., Nagao, N., Niwa, C., Yamamoto, S., Toda, T., 2010. Effects of particle size on anaerobic digestion of food waste. Int. Biodeterior. Biodegrad. 64 (7), 601–608.

Jaffar, M., Pang, Y., Yuan, H., Zou, D., Liu, Y., Zhu, B., Korai, R.M., Li, X., 2016. Wheat straw pretreatment with KOH for enhancing biomethane production and fertilizer value in anaerobic digestion. Chin. J. Chem. Eng. 24 (3), 404–409.

Kabir, M.M., Rajendran, K., Taherzadeh, M.J., Horváth, I.S., 2015. Experimental and economical evaluation of bioconversion of forest residues to biogas using organosolv pretreatment. Bioresour. Technol. 178, 201–208.

Kainthola, J., Podder, A., Fechner, M., Goel, R., 2021. An overview of fungal pretreatment processes for anaerobic digestion: applications, bottlenecks and future needs. Bioresour. Technol. 321, 124397.

Kan, X., Zhang, J.X., Tong, Y.W., Wang, C.H., 2018. Overall evaluation of microwave-assisted alkali pretreatment for enhancement of biomethane production from brewers' spent grain. Energy Convers. Manage. 158, 315–326.

Kang, X., Lin, R., O'Shea, R., Deng, C., Li, L., Sun, Y., Murphy, J.D., 2020a. A perspective on decarbonizing whiskey using renewable gaseous biofuel in a circular bioeconomy process. J. Cleaner Prod. 255, 120211.

Kang, X., Zhang, Y., Li, L., Sun, Y., Kong, X., Yuan, Z., 2020b. Enhanced methane production from anaerobic digestion of hybrid Pennisetum by selectively removing lignin with sodium chlorite. Bioresour. Technol. 295, 122289.

Kang, X.H., Sun, Y.M., Li, L.H., Kong, X.Y., Yuan, Z.H., 2018. Improving methane production from anaerobic digestion of Pennisetum hybrid by alkaline pretreatment. Bioresour. Technol. 255, 205–212.

Kang, X.H., Zhang, Y., Lin, R.C., Li, L.H., Zhen, F., Kong, X.Y., Sun, Y.M., Yuan, Z.H., 2020c. Optimization of liquid hot water pretreatment on hybrid Pennisetum anaerobic digestion and its effect on energy efficiency. Energy Convers. Manage. 210, 112718.

Kang, X.H., Zhang, Y., Song, B., Sun, Y.M., Li, L.H., He, Y., Kong, X.Y., Luo, X.J., Yuan, Z.H., 2019. The effect of mechanical pretreatment on the anaerobic digestion of hybrid Pennisetum. Fuel 252, 469–474.

Kang, Y., Yang, Q., Bartocci, P., Wei, H., Liu, S.S., Wu, Z., Zhou, H., Yang, H., Fantozzi, F., Chen, H., 2020d. Bioenergy in China: evaluation of domestic biomass resources and the associated greenhouse gas mitigation potentials. Renew. Sustain. Energy Rev. 127, 109842.

Kaur, K., Phutela, U.G., 2016. Enhancement of paddy straw digestibility and biogas production by sodium hydroxide-microwave pretreatment. Renew. Energy 92, 178–184.

Kavitha, S., Banu, J.R., Kumar, G., Kaliappan, S., Yeom, I.T., 2018. Profitable ultrasonic assisted microwave disintegration of sludge biomass: modelling of biomethanation and energy parameter analysis. Bioresour. Technol. 254, 203–213.

Khatri, S., Wu, S., Kizito, S., Zhang, W., Li, J., Dong, R., 2015. Synergistic effect of alkaline pretreatment and Fe dosing on batch anaerobic digestion of maize straw. Appl. Energy 158, 55–64.

Koyama, M., Watanabe, K., Kurosawa, N., Ishikawa, K., Ban, S., Toda, T., 2017. Effect of alkaline pretreatment on mesophilic and thermophilic anaerobic digestion of a submerged macrophyte: inhibition and recovery against dissolved lignin during semi-continuous operation. Bioresour. Technol. 238, 666–674.

Kratky, L., Jirout, T., 2011. Biomass size reduction machines for enhancing biogas production. Chem. Eng. Technol. 34 (3), 391–399.

Kumar, S., Paritosh, K., Pareek, N., Chawade, A., Vivekanand, V., 2018. De-construction of major Indian cereal crop residues through chemical pretreatment for improved biogas production: an overview. Renew. Sustain. Energy Rev. 90, 160–170.

Kumari, D., Singh, R., 2018. Pretreatment of lignocellulosic wastes for biofuel production: a critical review. Renew. Sustain. Energy Rev. 90, 877–891.

Lee, J.T.E., Khan, M.U., Tian, H.L., Ee, A.W.L., Lim, E.Y., Dai, Y.J., Tong, Y.W., Ahring, B.K., 2020. Improving methane yield of oil palm empty fruit bunches by wet oxidation pretreatment: mesophilic and thermophilic anaerobic digestion conditions and the associated global warming potential effects. Energy Convers. Manage. 225, 113438.

Leipold, S., Petit-Boix, A., 2018. The circular economy and the bio-based sector—perspectives of European and German stakeholders. J. Cleaner Prod. 201, 1125–1137.

Li, J., Wachemo, A.C., Yuan, H.R., Zuo, X.Y., Li, X.J., 2019. Evaluation of system stability and anaerobic conversion performance for corn stover using combined pretreatment. Waste Manage. (Oxford) 97, 52–62.

Li, L., Chen, C., Zhang, R., He, Y., Wang, W., Liu, G., 2015a. Pretreatment of corn stover for methane production with the combination of potassium hydroxide and calcium hydroxide. Energy Fuels 29 (9), 5841–5846.

Li, M., Cao, S.L., Meng, X.Z., Studer, M., Wyman, C.E., Ragauskas, A.J., Pu, Y.Q., 2017. The effect of liquid hot water pretreatment on the chemical-structural alteration and the reduced recalcitrance in poplar. Biotechnol. Biofuels 10, 237.

Li, X.J., Dang, F., Zhang, Y.T., Zou, D.X., Yuan, H.R., 2015b. Anaerobic digestion performance and mechanism of ammoniation pretreatment of corn stover. Bioresources 10 (3), 5777–5790.

Li, Y.Y., Jin, Y.Y., Li, J.H., Li, H.L., Yu, Z.X., 2016. Effects of thermal pretreatment on the biomethane yield and hydrolysis rate of kitchen waste. Appl. Energy 172, 47–58.

Liebetrau, J., Kornatz, P., Baier, U., Wall, D., Murphy, J., 2020. Integration of biogas systems into the energy system: technical aspects of flexible plant operation. IEA Bioenergy Task 37.

Lin, R., Cheng, J., Ding, L., Murphy, J.D., 2018. Improved efficiency of anaerobic digestion through direct interspecies electron transfer at mesophilic and thermophilic temperature ranges. Chem. Eng. J. 350, 681–691.

Lin, R., Cheng, J., Song, W., Ding, L., Xie, B., Zhou, J., Cen, K., 2015. Characterisation of water hyacinth with microwave-heated alkali pretreatment for enhanced enzymatic digestibility and hydrogen/methane fermentation. Bioresour. Technol. 182, 1–7.

Lin, R., Cheng, J., Yang, Z., Ding, L., Zhang, J., Zhou, J., Cen, K., 2016. Enhanced energy recovery from cassava ethanol wastewater through sequential dark hydrogen, photo hydrogen and methane fermentation combined with ammonium removal. Bioresour. Technol. 214, 686–691.

Lin, R.C., Cheng, J., Murphy, J.D., 2017. Unexpectedly low biohydrogen yields in co-fermentation of acid pretreated cassava residue and swine manure. Energy Convers. Manage. 151, 553–561.

Liu, X., Zicari, S.M., Liu, G., Li, Y., Zhang, R., 2015. Improving the bioenergy production from wheat straw with alkaline pretreatment. Biosystems Eng. 140, 59–66.

Lizasoain, J., Rincon, M., Theuretzbacher, F., Enguidanos, R., Nielsen, P.J., Potthast, A., Zweckmair, T., Gronauer, A., Bauer, A., 2016. Biogas production from reed biomass: effect of pretreatment using different steam explosion conditions. Biomass Bioenergy 95, 84–91.

Maamir, W., Ouahabi, Y., Poncin, S., Li, H.Z., Bensadok, K., 2017. Effect of Fenton pretreatment on anaerobic digestion of olive mill wastewater and olive mill solid waste in mesophilic conditions. Int. J. Green Energy 14 (6), 555–560.

Mancini, G., Papirio, S., Lens, P.N., Esposito, G., 2018a. Increased biogas production from wheat straw by chemical pretreatments. Renew. Energy 119, 608–614.

Mancini, G., Papirio, S., Lens, P.N.L., Esposito, G., 2018b. Anaerobic digestion of lignocellulosic materials using ethanol-organosolv pretreatment. Environ. Eng. Sci. 35 (9), 953–960.

Mancini, G., Papirio, S., Riccardelli, G., Lens, P.N., Esposito, G., 2018c. Trace elements dosing and alkaline pretreatment in the anaerobic digestion of rice straw. Bioresour. Technol. 247, 897–903.

Mbous, Y.P., Hayyan, M., Hayyan, A., Wong, W.F., Hashim, M.A., Looi, C.Y., 2017. Applications of deep eutectic solvents in biotechnology and bioengineering—promises and challenges. Biotechnol. Adv. 35 (2), 105–134.

Mert, B.K., Yonar, T., Kiliç, M.Y., Kestioğlu, K., 2010. Pre-treatment studies on olive oil mill effluent using physico-chemical, Fenton and Fenton-like oxidations processes. J. Hazard. Mater. 174 (1-3), 122–128.

Michalska, K., Miazek, K., Krzystek, L., Ledakowicz, S., 2012. Influence of pretreatment with Fenton's reagent on biogas production and methane yield from lignocellulosic biomass. Bioresour. Technol. 119, 72–78.

Mishra, S., Singh, P.K., Dash, S., Pattnaik, R., 2018. Microbial pretreatment of lignocellulosic biomass for enhanced biomethanation and waste management. 3 Biotech 8 (11), 458.

Mokomele, T., Sousa, L.D., Balan, V., van Rensburg, E., Dale, B.E., Gorgens, J.F., 2019. Incorporating anaerobic co-digestion of steam exploded or ammonia fiber expansion pretreated sugarcane residues with manure into a sugarcane-based bioenergy-livestock nexus. Bioresour. Technol. 272, 326–336.

Monlau, F., Sambusiti, C., Antoniou, N., Barakat, A., Zabaniotou, A., 2015. A new concept for enhancing energy recovery from agricultural residues by coupling anaerobic digestion and pyrolysis process. Appl. Energy 148, 32–38.

Murphy, J., Power, N., 2009. Technical and economic analysis of biogas production in Ireland utilising three different crop rotations. Appl. Energy 86 (1), 25–36.

Mustafa, A.M., Poulsen, T.G., Sheng, K.C., 2016. Fungal pretreatment of rice straw with Pleurotus ostreatus and Trichoderma reesei to enhance methane production under solid-state anaerobic digestion. Appl. Energy 180, 661–671.

Nair, R.B., Kabir, M.M., Lennartsson, P.R., Taherzadeh, M.J., Horvath, I.S., 2018. Integrated process for ethanol, biogas, and edible filamentous fungi-based animal feed production from dilute phosphoric acid-pretreated wheat straw. Appl. Biochem. Biotechnol. 184 (1), 48–62.

Negro, M.J., Manzanares, P., Oliva, J.M., Ballesteros, I., Ballesteros, M., 2003. Changes in various physical/chemical parameters of Pinus pinaster wood after steam explosion pretreatment. Biomass Bioenergy 25 (3), 301–308.

O'Neill, B., Oppenheimer, M., 2016, IPCC Reasons for Concern Regarding Climate Change Risks: Implications for 1.5 and 2 C Targets. AGU Fall Meeting Abstracts GC24D-07.

O'Neill, B.C., Oppenheimer, M., Warren, R., Hallegatte, S., Kopp, R.E., Pörtner, H.O., Scholes, R., Birkmann, J., Foden, W., Licker, R., 2017. IPCC reasons for concern regarding climate change risks. Nat. Clim. Change 7 (1), 28–37.

O'Shea, R., Lin, R., Wall, D.M., Browne, J.D., Murphy, J.D., 2020. Using biogas to reduce natural gas consumption and greenhouse gas emissions at a large distillery. Appl. Energy 279, 115812.

Panepinto, D., Genon, G., 2016. Analysis of the extrusion as a pretreatment for the anaerobic digestion process. Ind. Crops Prod. 83, 206–212.

Passos, F., Ortega, V., Donoso-Bravo, A., 2017. Thermochemical pretreatment and anaerobic digestion of dairy cow manure: experimental and economic evaluation. Bioresour. Technol. 227, 239–246.

Paudel, S.R., Banjara, S.P., Choi, O.K., Park, K.Y., Kim, Y.M., Lee, J.W., 2017. Pretreatment of agricultural biomass for anaerobic digestion: current state and challenges. Bioresour. Technol. 245, 1194–1205.

Peng, J.J., Abomohra, A., Elsayed, M., Zhang, X.Z., Fan, Q.Z., Ai, P., 2019. Compositional changes of rice straw fibers after pretreatment with diluted acetic acid: towards enhanced biomethane production. J. Cleaner Prod. 230, 775–782.

Perez-Pimienta, J.A., Icaza-Herrera, J.P.A., Mendez-Acosta, H.O., Gonzalez-Alvarez, V., Mendoza-Perez, J.A., Arreola-Vargas, J., 2020. Bioderived ionic liquid-based pretreatment enhances methane production from Agave tequilana bagasse. RSC Adv. 10 (24), 14025–14032.

Perez-Rodriguez, N., Garcia-Bernet, D., Dominguez, J.M., 2017. Extrusion and enzymatic hydrolysis as pretreatments on corn cob for biogas production. Renew. Energy 107, 597–603.

Ravindran, R., Jaiswal, A.K., 2016. A comprehensive review on pre-treatment strategy for lignocellulosic food industry waste: challenges and opportunities. Bioresour. Technol. 199, 92–102.

Rekleitis, G., Haralambous, K.-J., Loizidou, M., Aravossis, K., 2020. Utilization of agricultural and livestock waste in anaerobic digestion (AD): applying the biorefinery concept in a circular economy. Energies 13 (17), 4428.

Ribeiro, G., Gruninger, R., Badhan, A., McAllister, T., 2016. Mining the rumen for fibrolytic feed enzymes. Animal Front. 6 (2), 20–26.

Ritchie, H., Roser, M., 2020. Agricultural Production. OurWorldInData.org. Published online at. https://ourworldin-data.org/agricultural-production [Online Resource]. (Accessed on 23rd March 2021).

Rodriguez, C., Alaswad, A., Benyounis, K.Y., Olabi, A.G., 2017. Pretreatment techniques used in biogas production from grass. Renew. Sustain. Energy Rev. 68, 1193–1204.

Saha, S., Jeon, B.H., Kurade, M.B., Jadhav, S.B., Chatterjee, P.K., Chang, S.W., Govindwar, S.P., Kim, S.J., 2018. Opti-mization of dilute acetic acid pretreatment of mixed fruit waste for increased methane production. J. Cleaner Prod. 190, 411–421.

Sambusiti, C., Monlau, F., Ficara, E., Carrère, H., Malpei, F., 2013. A comparison of different pre-treatments to increase methane production from two agricultural substrates. Appl. Energy 104, 62–70.

Sarto, S., Hildayati, R., Syaichurrozi, I., 2019. Effect of chemical pretreatment using sulfuric acid on biogas produc-tion from water hyacinth and kinetics. Renew. Energy 132, 335–350.

Satlewal, A., Agrawal, R., Bhagia, S., Sangoro, J., Ragauskas, A.J., 2018. Natural deep eutectic solvents for lignocel-lulosic biomass pretreatment: recent developments, challenges and novel opportunities. Biotechnol. Adv. 36 (8), 2032–2050.

Schroyen, M., Vervaeren, H., Vandepitte, H., Van Hulle, S.W.H., Raes, K., 2015. Effect of enzymatic pretreatment of various lignocellulosic substrates on production of phenolic compounds and biomethane potential. Bioresour. Technol. 192, 696–702.

Shang, G.Y., Zhang, C.G., Wang, F., Qiu, L., Guo, X.H., Xu, F.Q., 2019. Liquid hot water pretreatment to enhance the anaerobic digestion of wheat straw-effects of temperature and retention time. Environ. Sci. Pollut. Res. 26 (28), 29424–29434.

Shetty, D.J., Ks hirsagar, P., Tapadia-Maheshwari, S., Lanjekar, V., Singh, S.K., Dhakephalkar, P.K., 2017. Alkali pre-treatment at ambient temperature: a promising method to enhance biomethanation of rice straw. Bioresour. Technol. 226, 80–88.

Slade, R., Bauen, A., Gross, R., 2014. Global bioenergy resources. Nat. Clim. Change 4 (2), 99–105.

Song, B., Lin, R., Lam, C.H., Wu, H., Tsui, T.-H., Yu, Y., 2021. Recent advances and challenges of inter-disciplinary biomass valorization by integrating hydrothermal and biological techniques. Renew. Sustain. Energy Rev. 135, 110370.

Song, J., Yang, W., Yabar, H., Higano, Y., 2013. Quantitative estimation of biomass energy and evaluation of biomass utilization-a case study of Jilin Province, China. J. Sustain. Dev. 6 (6), 137.

Staples, M.D., Malina, R., Barrett, S.R., 2017. The limits of bioenergy for mitigating global life-cycle greenhouse gas emissions from fossil fuels. Nat. Energy 2 (2), 1–8.

Stefaniuk, M., Oleszczuk, P., 2015. Characterization of biochars produced from residues from biogas production. J. Anal. Appl. Pyrolysis 115, 157–165.

Strong, P.J., Kalyuzhnaya, M., Silverman, J., Clarke, W.P., 2016. A methanotroph-based biorefinery: potential sce-narios for generating multiple products from a single fermentation. Bioresour. Technol. 215, 314–323.

Syaichurrozi, I., Villta, P.K., Nabilah, N., Rusdi, R., 2019. Effect of sulfuric acid pretreatment on biogas production from *Salvinia molesta*. J.Environ. Chem. Eng. 7 (1), 102857.

Tadesse, H., Luque, R., 2011. Advances on biomass pretreatment using ionic liquids: an overview. Energy Environ. Sci. 4 (10), 3913–3929.

Taherdanak, M., Zilouei, H., Karimi, K., 2016. The influence of dilute sulfuric acid pretreatment on biogas production from wheat plant. Int. J. Green Energy 13 (11), 1129–1134.

Tongbuekeaw, T., Sawangkeaw, R., Chaiprapat, S., Charnnok, B., 2020. Conversion of rubber wood waste to methane by ethanol organosolv pretreatment. Biomass Convers.Biorefin. 11, 999–1011.

Tsapekos, P., Kougias, P.G., Angelidaki, I., 2018. Mechanical pretreatment for increased biogas production from lignocellulosic biomass; predicting the methane yield from structural plant components. Waste Manage. (Oxford) 78, 903–910.

Uzun, H., Yıldız, Z., Goldfarb, J.L., Ceylan, S., 2017. Improved prediction of higher heating value of biomass using an artificial neural network model based on proximate analysis. Bioresour. Technol. 234, 122–130.

Venturin, B., Camargo, A.F., Scapini, T., Mulinari, J., Bonatto, C., Bazoti, S., Siqueira, D.P., Colla, L.M., Alves Jr, S.L., Bender, J.P., 2018. Effect of pretreatments on corn stalk chemical properties for biogas production purposes. Bioresour. Technol. 266, 116–124.

Wahid, R., Hjorth, M., Kristensen, S., Moller, H.B., 2015. Extrusion as pretreatment for boosting methane production: effect of screw configurations. Energy Fuels 29 (7), 4030–4037.

Wall, D.M., McDonagh, S., Murphy, J.D., 2017. Cascading biomethane energy systems for sustainable green gas production in a circular economy. Bioresour. Technol. 243, 1207–1215.

Wall, D.M., Straccialini, B., Allen, E., Nolan, P., Herrmann, C., O'Kiely, P., Murphy, J.D., 2015. Investigation of effect of particle size and rumen fluid addition on specific methane yields of high lignocellulose grass silage. Bioresour. Technol. 192, 266–271.

Wang, F., Niu, W.S., Zhang, A.D., Yi, W.M., 2015. Enhanced anaerobic digestion of corn stover by thermo-chemical pretreatment. Int. J. Agric. Biol. Eng. 8 (1), 84–90.

Wang, W., Wang, Q., Tan, X.S., Qi, W., Yu, Q., Zhou, G.X., Zhuang, X.S., Yuan, Z.H., 2016. High conversion of sugarcane bagasse into monosaccharides based on sodium hydroxide pretreatment at low water consumption and wastewater generation. Bioresour. Technol. 218, 1230–1236.

WBA, 2020. Global bioenergy statistics 2020. https://www.worldbioenergy.org/global-bioenergy-statistics/. Global Bioenergy Association. (Accessed on 29th March 2021).

Wu, B., Lin, R., Kang, X., Deng, C., Xia, A., Dobson, A.D., Murphy, J.D., 2020a. Graphene addition to digestion of thin stillage can alleviate acidic shock and improve biomethane production. ACS Sustain. Chem. Eng. 8 (35), 13248–13260.

Wu, P., Li, L., Sun, Y., Song, B., Yu, Y., Liu, H., 2020b. Near complete valorisation of Hybrid pennisetum to biomethane and lignin nanoparticles based on gamma-valerolactone/water pretreatment. Bioresour. Technol. 305, 123040.

Xiao, L.-P., Sun, Z.-J., Shi, Z.-J., Xu, F., Sun, R.-C., 2011. Impact of hot compressed water pretreatment on the structural changes of woody biomass for bioethanol production. BioResources 6 (2), 1576–1598.

Xu, C.C., Dessbessell, L., Zhang, Y., Yuan, Z., 2021a. Lignin valorization beyond energy use: has lignin's time finally come? Biofuels Bioprod. Biorefin. 15 (1), 32–36.

Xu, G.-C., Ding, J.-C., Han, R.-Z., Dong, J.-J., Ni, Y., 2016. Enhancing cellulose accessibility of corn stover by deep eutectic solvent pretreatment for butanol fermentation. Bioresour. Technol. 203, 364–369.

Xu, H., Li, Y., Hua, D., Zhao, Y., Chen, L., Zhou, L., Chen, G., 2021b. Effect of microaerobic microbial pretreatment on anaerobic digestion of a lignocellulosic substrate under controlled pH conditions. Bioresour. Technol. 328, 124852.

Xu, H., Li, Y., Hua, D., Zhao, Y., Mu, H., Chen, H., Chen, G., 2020. Enhancing the anaerobic digestion of corn stover by chemical pretreatment with the black liquor from the paper industry. Bioresour. Technol. 306, 123090.

Xu, W.Y., Fu, S.F., Yang, Z.M., Lu, J., Guo, R.B., 2018. Improved methane production from corn straw by microaerobic pretreatment with a pure bacteria system. Bioresour. Technol. 259, 18–23.

Yang, B., Tao, L., Wyman, C.E., 2018. Strengths, challenges, and opportunities for hydrothermal pretreatment in lignocellulosic biorefineries. Biofuels Bioprod. Biorefin. 12 (1), 125–138.

Yee Tong, T., Gek Cheng, N., Adeline Seak May, C., 2019. Effect of functional groups in acid constituent of deep eutectic solvent for extraction of reactive lignin. Bioresour. Technol. 281, 359–366.

Yu, Q., Liu, R., Li, K., Ma, R., 2019a. A review of crop straw pretreatment methods for biogas production by anaerobic digestion in China. Renew. Sustain. Energy Rev. 107, 51–58.

Yu, Q., Qin, L., Liu, Y., Sun, Y., Xu, H., Wang, Z., Yuan, Z., 2019b. In situ deep eutectic solvent pretreatment to improve lignin removal from garden wastes and enhance production of bio-methane and microbial lipids. Bioresour. Technol. 271, 210–217.

Yu, Q.A., Zhuang, X.S., Yuan, Z.H., Wang, W., Qi, W., Wang, Q.O., Tan, X.S., 2011. Step-change flow rate liquid hot water pretreatment of sweet sorghum bagasse for enhancement of total sugars recovery. Appl. Energy 88 (7), 2472–2479.

Yuan, X.F., Ma, L., Wen, B.T., Zhou, D.Y., Kuang, M., Yang, W.H., Cui, Z.J., 2016. Enhancing anaerobic digestion of cotton stalk by pretreatment with a microbial consortium (MC1). Bioresour. Technol. 207, 293–301.

Zabaniotou, A., 2018. Redesigning a bioenergy sector in EU in the transition to circular waste-based bioeconomy—a multidisciplinary review. J. Cleaner Prod. 177, 197–206.

Zeng, Y., Zhao, S., Yang, S., Ding, S.-Y., 2014. Lignin plays a negative role in the biochemical process for producing lignocellulosic biofuels. Curr. Opin. Biotechnol. 27, 38–45.

Zhang, K., Si, M., Liu, D., Zhuo, S., Liu, M., Liu, H., Yan, X., Shi, Y., 2018. A bionic system with Fenton reaction and bacteria as a model for bioprocessing lignocellulosic biomass. Biotechnol. Biofuels 11 (1), 1–14.

Zhang, K., Zhou, L., Brady, M., Xu, F., Yu, J., Wang, D., 2017. Fast analysis of high heating value and elemental compositions of sorghum biomass using near-infrared spectroscopy. Energy 118, 1353–1360.

Zhang, Y., Chen, X., Gu, Y., Zhou, X., 2015. A physicochemical method for increasing methane production from rice straw: extrusion combined with alkali pretreatment. Appl. Energy 160, 39–48.

Zhao, B.-H., Chen, J., Yu, H.-Q., Hu, Z.-H., Yue, Z.-B., Li, J., 2017. Optimization of microwave pretreatment of lignocellulosic waste for enhancing methane production: hyacinth as an example. Front. Environ. Sci. Eng. 11 (6), 1–9.

Zhao, X.B., Cheng, K.K., Liu, D.H., 2009. Organosolv pretreatment of lignocellulosic biomass for enzymatic hydrolysis. Appl. Microbiol. Biotechnol. 82 (5), 815–827.

Zheng, J., Rehmann, L., 2014. Extrusion pretreatment of lignocellulosic biomass: a review. Int. J. Mol. Sci. 15 (10), 18967–18984.

Zheng, Y., Zhao, J., Xu, F.Q., Li, Y.B., 2014. Pretreatment of lignocellulosic biomass for enhanced biogas production. Prog. Energy Combust. Sci. 42, 35–53.

Zhuang, X.S., Wang, W., Yu, Q., Qi, W., Wang, Q., Tan, X.S., Zhou, G.X., Yuan, Z.H., 2016. Liquid hot water pretreatment of lignocellulosic biomass for bioethanol production accompanying with high valuable products. Bioresour. Technol. 199, 68–75.

CHAPTER

13

Application of enzymes in microbial fermentation of biomass wastes for biofuels and biochemicals production

Luciana Porto de Souza Vandenberghe, Gustavo Amaro Bittencourt, Kim Kley Valladares-Diestra, Nelson Libardi Junior, Luiz Alberto Junior Letti, Zulma Sarmiento Vásquez, Ariane Fátima Murawski de Mello, Susan Grace Karp, Maria Giovana Binder Pagnoncelli, Cristine Rodrigues, Adenise Lorenci Woiciechowski, Júlio César de Carvalho, Carlos Ricardo Soccol

Department of Bioprocess Engineering and Biotechnology, Federal University of Paraná, Centro Politécnico, Curitiba, Paraná, Brazil

13.1 Introduction

The utilization of lignocellulosic or starchy biomass wastes as feedstock to produce biofuels and value-added products has been studied as an alternative to the depletion of fossil fuels to develop more cost-effective processes, decreasing the price of the final bioproduct (Gutiérrez-García et al., 2020; Sitaraman et al., 2019).

However, many of the current technologies employ starchy and/or lignocellulosic biomasses as substrates that are abundantly generated in agricultural countries. These materials must pass through different treatments including thermal, chemical, and enzymatic procedures to liberate easily assimilable and high concentration of sugars for biofuels production. The involved degrading enzymes have been intensively studied, being the main challenges of biofuels viability due to their high production cost. In recent years, many research efforts have been made to enhance the production and stability of enzymes (Gutiérrez-García et al., 2020).

There have been several governmental investments and incentives, together with the development of new powerful biocatalyzers, to turn biofuels processes economic and

Biomass, Biofuels, Biochemicals.
DOI: https://doi.org/10.1016/B978-0-323-90633-3.00012-2

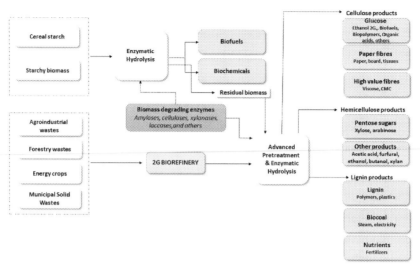

FIG. 13.1   **Biomass-degrading enzymes in the biorefinery context.**

sustainable. In this context, the biorefinery concept provides a wide perspective to reach these goals, with concomitant use of biomass conversion to new value-added products. Sugarcane, corn, maize, cotton, soy, wheat, and rice are the main produced crops in the world and their processing generates large amounts of wastes/byproducts. Beyond the traditional use for electricity and thermal generation, there is the possibility of a more efficient exploitation of these byproducts to produce different products such as biofuels, organic acids, biopolymers, enzymes, biofertilizers, and others (Fig. 13.1).

Different starchy and/or lignocellulosic degrading enzymes including amylases, cellulases, hemicellulases, and laccases will be presented with examples of their efficiency, employed substrate, conditions, and type of operation (separate or simultaneous processes). The involved pretreatment will also be described as a previous condition to prepare the substrate for sequential enzymatic treatment. The complexity and heterogeneity of biomass leads to the use of not only one enzyme, but an enzymatic complex to achieve efficient hydrolysis. The previous analysis of substrate composition is then a *sine qua non* condition for the correct choice of the employed enzyme or enzymatic complex.

13.2   Biomass-degrading enzymes

Annually, large amounts of biomass waste are generated worldwide. Biomass wastes include a variety of forestry residues, agricultural wastes, and food processing waste. In the current global context, the bioeconomy perspective is to reduce the environmental pollution and guarantee the sustainability (Chen and Zhang, 2015). In this regard, an efficient reuse of this biomass waste is a strategy to improve the economy. In addition, the generated wastes are rich in nutrients and have potential to be used as sources of biofuels or in the production value-added products (Cho et al., 2020; Clauser et al., 2021). The chemical composition of these wastes is generally rich in polysaccharides, consisting of a lignocellulosic structure where cellulose, hemicellulose, and lignin are the three main constituents. Cellulose appears as the most abundant polymer composed of D-glucose units (Bhatia et al., 2021).

Biofuels and biochemical products could be obtained using these lignocellulosic and/or starchy residues through the action of microorganisms. However, microbial cells cannot easily metabolize these complex polysaccharides (Bonechi et al., 2017; Khaire et al., 2021). The structure of the lignocellulosic biomass is extremely stable and recalcitrant due to some characteristics of the components and the interactions between them, requiring some steps, called pretreatment. Different pretreatments have been researched and developed to allow the efficient use of biomass fractions, maximizing the release of fermentable sugars from enzymatic hydrolysis (Ravindran and Jaiswal, 2016; Zanirun et al., 2015), with lower concentration of inhibitors, and less degradation of carbohydrates (Chen and Fu, 2016). Treatments can be physical, chemical, biological, or a combination of them. The effectiveness of the application of each of these treatments depends on the characteristics of the biomass and treatment conditions (Sindhu et al., 2016).

The enzymatic conversion of biomass into its primary constituents is considered an environment-friendly technology. Lignocellulose can be transformed through enzymatic hydrolysis into monosaccharides (glucose, fructose, mannose, xylose, and others), oligosaccharides (xylo-oligo, fructo-oligo, galacto-oligo, and others), nanocellulose (nanofiber, nanocrystal, and bacterial nanocellulose), lignin (carbon fiber, activated carbon, benzene, toluene, binders, and dispersants), and bioactive molecules (flavonoids, phenolic acids, terpenoids, and carotenoids). The main enzymes employed in biomass hydrolysis to monomeric sugars, such glucose and xylose for further production of biofuels and biochemicals are glycosyl hydrolases (cellulases, β-glucosidase, hemicellulases) and α-amylases, for starchy nature biomass. In addition, other enzymes such as laccases and peroxidases have been highlighted in the conversion of lignin, from lignocellulosic materials, into different products of commercial value (Cho et al., 2020).

13.2.1 Cellulases

Cellulolytic enzymes are employed in the enzymatic hydrolysis of cellulose for its efficient conversion to glucose (Gatt et al., 2019). These enzymes' complexes consist of endoglucanases (EC 3.2.1.4), also known as 1,4-β-D-glucan-4-glucanohydrolase or carboxymethyl cellulase; exoglucanases (EC 3.2.1.176), such as cellodextrinase or 1,4-β-D-glucanglucano hydrolase and cellobiohydrolase (EC 3.2.1.91) or 1,4-β-D-glucan cellobiohydrolase, and β-glucosidases (EC 3.2.1.21) or cellobiases (Fig. 13.2). Each enzyme has a mechanism of action and is responsible for one stage during the entire hydrolytic process (Juturu and Wu, 2014). Hydrolysis begins at the amorphous regions of cellulose, which are easier for enzymes to access. According to their structure and function in the cellulolytic complex, cellulases act by catalyzing different reactions that lead from the breakdown of the amorphous structure of cellulose to the conversion of cellobiose into glucose units (Lynd et al., 2002).

Endoglucanases are responsible for the cleavage of the amorphous regions of cellulose, releasing cell-oligosaccharide products with different amounts of monomers and nonreducing terminals. Acting in synergy with endoglucanases, exoglucanases cleave the released oligosaccharides, in their reducing or nonreducing ends, generating cellobioses. The reaction

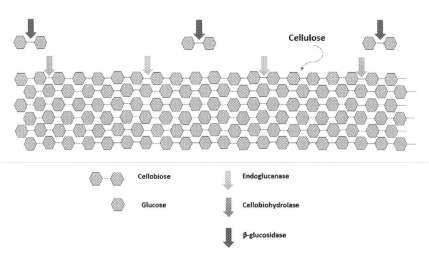

FIG. 13.2   Cellulose-degrading enzymes.

occurs when the substrate is inserted into the enzyme structure, which is spiraled, resembling a tunnel. Cellobiohydrolases can be of two types, those that cleave chains with reducing terminals and those that cleave nonreducing terminals. The released cellobioses are then cleaved into glucose by cellobiases or β-glucosidases (Juturu and Wu, 2014). Cellobiases or β-glucosidases act on the glycosidic links between the glucose units of cellobiose and other low polymerization saccharides, generating the final reaction product, which is glucose (Lynd et al., 2002).

Industrially, cellulases are produced by filamentous fungi of the genera *Aspergillus* and *Trichoderma*. However, other genera have also been studied, such as *Penicillium, Clostridium, Cellulomonas,* and *Thermomonospora* (Manisha and Yadav, 2017).

The use of cellulases in the hydrolysis of biomass waste has been intensively developed. In biorefineries, cellulases are used to process the second-generation substrates, such as agroindustrial residues releasing free sugars for bioethanol production. They also have been employed in the extractions of bioactive products such as organic acids, lipids, phenolic compounds, polysaccharides, organic solvents, among others (Saldarriaga-Hernández et al., 2020). Bambusa bambos was enzymatically treated using a commercial cellulose with a production of 132.67 mg/dL of glucose (Chavan and Gaikwad, 2021). Bioextrusion of corn subproducts was carried out with the use of cellulases to produce glucose, reaching a conversion of 94% (Gatt et al., 2019). Hu et al. (2021) used cellulase to obtain H_2 in a two-stage fermentation using unpretreated sorghum. Polyhydroxyalkanoates (PHA) were produced by solid-state enzymatic hydrolysis using spent grain, grape pomace, and olive-mill solid wastes with the addition of cellulases and xylanases with a productivity of 0.33 g/kg/h (Martínez-Avila et al., 2021).

13.2.2   Hemicellulases

Hemicellulose is one of the most abundant polysaccharides. It consists of a polymeric carbohydrate complex with an amorphous structure, which is found in plant's cell walls. This polysaccharide is associated with cellulose by hydrogen bonds and with lignin through covalent bonds. Structurally, they are constituted mainly by units of pentoses (xylose and arabinose) and hexoses (glucose, galactose, mannose, and rhamnose). Hemicelluloses can be classified as xylan, glucuronoxylan, arabinoxylan, mannan, glucomannan, and galactomannan (Bonechi et al., 2017).

Xylans are polysaccharides of hardwood hemicellulose, which are commonly found in crop residues and in paper production wastes. Xylans have a central linear structure composed by D-xylose residues that are linked by β1→4 bonds, and sometimes they are branched to L-arabinofuranoside, 4-o-methylglucuronic acid, mannose, galactose, and rhamnose. After xylans, mannans are the main constituents of hemicelluloses, which are found in softwoods. Mannans have a central linear structure that is mainly composed by D-mannose residues linked by β1→4 bonds that are occasionally linked with D-glucose or branched with D-galactose. Based on the prevalence of the monomers, mannans can be classified into four families (linear mannans, glucomannans, galactomannans, and galactoglucomannans). The great diversity of mannans guarantees an important interest of the food industry (Bajpai, 2009; Bonechi et al., 2017).

Hemicellulose, as a heteropolymer, needs an enzymatic pool for its complete degradation. Usually, hemicellulases refer to enzyme mixtures and they are divided into depolymerases and debranching enzymes. The xylanolytic system consists of endoxylanases (EC 3.2.1.8) and β-xylosidases (EC 3.2.1.37). The endoxylanases catalyze the hydrolysis of internal 1,4-beta-xylosidic linkages in xylan, while the xylosidases act on the terminal linkages (Fig. 13.3) (Vandenberghe et al., 2020).

Xylanolitic enzymes can be produced by a high variety of organisms; however xylanases from filamentous fungi are the most used due to the higher yields and the production of extracellular enzymes. The main producers are *Aspergillus, Disporotrichum, Penicillium, Neurospora, Fusarium, Neocallimastix, Trichoderma, Coniothyrium,* and *Streptomyces.* Some extreme bacteria can also produce xylanases. Xylanase production is induced by the presence of xylan in the medium, but it is easily repressed by metabolized carbon sources (glucose or xylose),

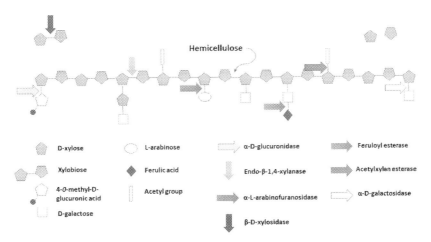

FIG. 13.3   Hemicellulose-degrading enzymes.

which impact its productivity. (Bajpai, 2009). *Myceliophthora thermophila* produced 51.70 U/mL of endoxylanase using rice straw as a substrate (Dahiya et al., 2020). Agroforestry wastes were used as substrates for endoxylanases production using *Aspergillus niger* (Díaz et al., 2019). Recombinant DNA techniques have been studied to improve xylanase production (Mander et al., 2014).

Xylose, which is the main product of xylan hydrolysis, can be used for the production of several compounds; the most important is xylitol that is obtained through catalytic hydrogenation. Xylitol is a high value product with sweetening power, noncariogenic properties, and insulin-independent metabolism. While the controlled hydrolysis of xylan by endoxylanases produces xylooligosaccharides (XOS), which can be used as prebiotic agents for promoting the growth of beneficial bacteria in humans' colon (Blibech et al., 2011; Cho et al., 2002; Lee et al., 2021).

Mannans can be degraded by mannanases that are identified as β-glucosidases (EC 3.2.1.21), β-mannosidases (EC 3.2.1.25), and β-mannanases (EC 3.2.1.78). β-mannanases hydrolyze internal glycosidic bonds of mannan backbone chain, while β-mannosidases act on the terminal ends of mannose oligosaccharides, which are released by the action of mannanases. β-glucosidases remove 1,4-glucopyranose units from glucomannan and galactoglucomannan oligomers (Dimarogona and Topakas, 2016). Many microorganisms have been reported as mannanase producers. β-Mannanases are produced by *Aspergillus* species, *Sclerotium rolfsii, Trichoderma reesei,* and *Penicillium occitanis.* Fungal mannanases are generally produced with mannan-rich substrates, including locust bean gum, guar gum, konjac flour, and copra meal. Mannanases' production was carried out by *P. occitanis* in fed-batch using acacia seed powder as inducer, reaching an activity of 76.00 U/mL (Blibech et al., 2011). β-mannanase from *Bacillus subtilis* was produced using palm kernel cake as a substrate in semisolid fermentation, reaching an activity of 805.12 U/mL (Norizan et al., 2020). β-mannosidases can be produced by fungi, for example, *Aspergillus* spp., *T. reesei, S. rolfsii* and bacteria, including *Thermobifida fusca, Thermotoga neapolitana,* and *Streptomyces.* β-mannosidase production by *A. niger,* which was sequenced, and expressed in the yeast *Pichia pastoris,* reached a specific activity of 4 U/mg after 4 days of cultivation (Fliedrová et al., 2012).

Enzymes that act on sugar chain backbones are considered major enzymes and their action is complemented by debranching enzymes (auxiliary or accessory enzymes). The debranching enzymes act on ester and glycoside bonds such as α-arabinofuranosidase (EC 3.2.1.55), α-glucuronidase (EC 3.2.1.139), α-D-galactosidase (EC 3.2.1.22), acetyl xylan esterase (EC 3.1.1.72), β-feruloyl esterase (EC 3.1.1.73) (Fig. 13.3). α-D-galactosidase removes -1,6-links of d-galactopyranosyl, and acetyl xylan esterase removes acetyl group side chains of galactoglucomannan (Dimarogona and Topakas, 2016).

13.2.3   Amylases

Starch is a reserve polysaccharide in higher plants and appears as a promising biopolymer for being a rich natural source of carbon. It is produced abundantly by photosynthesis and it is composed exclusively by D-glucose residues. The molecule consists of amylose and amylopectin. Amylose accounts for about 15%–25% of starch, and amylopectin represents 75%–85% of starch. Structurally, amylose consists of linear chains of D-glucose units linked by α1→4 glycoside bonds and, eventually, some branches of α1→6 bonds (0.3%–0.5%). Amylopectin has a similar backbone to amylose; however, it is a highly branched molecule consisting of glucose α-1→4 and α-1→6 linkages. Due to the chemical composition, the starch biomass represents an abundant sugar source for biofuels and biochemicals production through the action of microorganisms. Sugars can be efficiently obtained through enzymatic hydrolysis, using starch enzymes (Bonechi et al., 2017).

α-amylases (EC 3.2.1.1) are endoamylases that randomly hydrolyze the α-1,4 glycosidic bonds in the interior of starch with the production of low molecular weight molecules. Exoamylases, β-amylases (EC 3.2.1.2), and glucoamylases (EC 3.2.1.3), hydrolyze starch from its nonreducing end, generating short products. β-amylases cleave nonreducing chain ends of the starch molecule, while glucoamylases hydrolyze single glucose units from nonreducing ends of amylose and amylopectin. α-glucosidases (EC 3.2.1.20) cleave both α-1,4- and α-1,6-bonds on the external glucose residues of amylopectin or amylose from the nonreducing end of the molecule. Pullulanases (EC 3.2.1.41) catalyze the hydrolysis of α-1,6 linkages of pullulan that is a linear α-glucan polymer composed of maltotriosyl units connected by α-1,6-glycosidic bonds. Maltotetraohydrolases, (EC 3.2.1.60) hydrolyze α-1,4-glucosidic linkages to remove successive maltotetraose residues from the nonreducing chain ends. Isoamylases (EC 3.2.1.68), similar to pullulanases, are capable of hydrolyzing α-1,6-glycosidic bonds in amylopectins and are the only known enzymes that completely debranch glycogen (Marques et al., 2017; Vandenberghe et al., 2020).

Amylases can be synthesized by plants, animals, and microorganisms and they have a wide spectrum of industrial applications due to their stability and specificity. Commercial α-amylases are produced by fungus *Aspergillus* spp. and bacterium *Bacillus* spp. Sweet potatoes and plant seed (wheat, barley, soybeans) are primary sources of β-amylases, which are produced by *Bacillus* spp. or *Pseudomonas* spp. Amyloglucosidases are obtained from *Aspergillus* spp. and *Rhizopus* spp. The main glucosidases' producers are *Aspergillus* spp., *Mucor* spp., *Bacillus* spp., or *Pseudomonas* spp. Pullulanases are derived from different microorganisms such as *Bacillus acidopullulyticus*, *Klebsiella planticola*, *Bacillus* spp., and *Geobacillus stearothermophilus* (Marques et al., 2017).

13.2.4 Laccases and peroxidases

Laccases belong to the group of the oxireductases, or more, phenol oxidizers, EC 1.10.3.2, described as multicopper-containing enzymes that act by removing electrons, leading to the oxidation of many substrates. This property, due to the presence of four atoms of Cu in its structure, is useful in many industrial processes. The oxidation mechanism is very well described and explained due to the redox potential of the three copper atoms presented at the laccases structure, named T1 (acceptor of the substrate electron), T2 (responsible for the redox catalytic activity), and two catalytic sites T3 (responsible for the O_2 transportation and substrate oxygenation) characteristic of the copper-oxidize enzymes. T1 copper transfers electrons to T2 and T3 sites, named electrons receptors, where O_2 is reduced to water (Rivera-Hoyos et al., 2013; Shleev et al., 2006).

Laccases are extracellular enzymes that have several industrial applications. Besides, this generic oxidation mechanism gives laccases an important property, a low specificity to their reducing substrates, meaning that these enzymes can catalyze oxidation reactions of phenolic and aromatic compounds, amine, ester, and ether compounds. They are present in toxic and recalcitrant compounds, such as pesticides, herbicides, and dyes. This is useful in many industrial applications, wastewater treatment plants, environmental remediation, soil decontamination, among others. Laccases can oxidize substances such as anilines, promoting the decolorization of dyes, which is useful in wastewater treatment of textile, cellulose, and paper industries (Aljeboree et al., 2017; Dükkancı et al., 2014; Fernández-Fernández et al., 2013; Guimarães et al., 2017; Niebisch et al., 2014; Torres-duarte et al., 2009).

Lignin is an amorphous and aromatic vegetable polymer with many ether and esters links, tightly united to cellulose and hemicelluloses that is responsible for the recalcitrance, chemical and mechanical stability of lignocellulosic biomass. Laccases are important alternative for delignification, allowing the recovering of biomass saccharide fraction, including wheat straw, oil palm empty fruit bunches, soy hulls, and corn processing residues. Laccases are currently employed in the biobleaching of cellulose pulp in the pulp and paper industries, instead of chemical products (Boruah et al., 2019; Fonseca et al., 2014; Martín-Sampedro et al., 2012; Sharma et al., 2014).

The delignification of lignocellulosic biomass is being intensively studied all over the world to expose biomass saccharidic fraction, mainly agroresidues, to the action of cellulolytic enzymes to get a sugary broth that can be used as substrate for many fermentable processes to produce different biomolecules (Agrawal et al., 2019; Liu et al., 2020).

Laccases are found in many natural sources such as plants, insects, and microorganisms, mainly superior fungi known as wood-decomposing fungi named white and brown rot fungi, among them are the genus *Trametis, Pleurotus, Phanerochaete, Ganoderma, Lentinus*, and others, which are employed in the industrial production.

Peroxidases (EC.1.11.1.X) are oxidoreductases enzymes that catalyze the oxidation of many substrates, using hydrogen peroxides as acceptor of hydrogen donated by another compound, defined as heme proteins containing iron (III) in its structure and a protoporphyrin as a prosthetic group. Lignin peroxidase, LiP, (EC.1.11.1.14) and manganese peroxidase, MnP, (EC.1.11.1.13) are enzymes that present a high-redox potential, described as ligninases that degrade lignin and other compounds with similar linkages. Peroxidases act on chemical groups such as CH-OH, C=O, HC=CH, CH-NH-, NADH, and NADPH, structures that are present in many substrates. Similar, to laccases, these enzymes can cause the oxidation of a large variety of substrates, with the potential to be used in bioremediation, biopulping, and wastewater treatment to improve the quality of the plant effluent, phenolic compounds degradation, synthetic textile dyes decolorization; in food industry preserving products, development of medical kits, and others. The combination of lignin peroxidase, Mn peroxidase, and laccase is an efficient agent for biopulping, and also to delignify plant biomasses, agro-industrial residues aiming the second-generation ethanol production, under the biorefinery concept (Daoutidis et al., 2013; Li et al., 2020; Magalhães et al., 2017; Santhi et al., 2014; Wang et al., 2013).

Peroxidases are produced from many microorganisms, animals, and plants with a very important role in their energetic metabolism. Among bacteria, peroxidases (MnP and LiP) are produced by *Pseudomonas, Serratia, Bacillus,* and others; fungal peroxidases, both, MnP and LiP, are found in *Penicillium, Lentinus, Pleurotus, Phanerochaete, Fomitopsis, Auricularia,* and others.

13.3   Separated enzymatic hydrolysis and fermentation processes

Many bioproducts from different biomass waste have been produced by numerous microorganisms that have the capacity to convert simple sugar monomers into strategic platform chemicals by means of fermentation (Bardhan et al., 2015; Jin et al., 2018). One of the critical steps of biomass waste conversion is the saccharification, in which polymeric sugars become bioaccessible for the microbial community (Guo et al., 2018; Liu et al., 2020).

Concentrated acid hydrolysis, which employs H_2SO_4 or HCl, is a traditionally used process for fermentable sugar recovery from biomass that can reach high conversions (>90%). However, it has unsuitable characteristics that can add extra cost to the process, such as: end products contamination with toxic compounds, use of large quantities of neutralization agent with subsequent step for calcium sulfate separation, and use of expansive corrosion-resistant equipment (Cho et al., 2020). With the growing need for sustainable processes in industry, ecofriendly bio-based approaches, such as enzymatic hydrolysis, are replacing chemical-based concepts of manufacturing for fermentable sugars production. Different concepts for saccharification and fermentation have been developed. The separated hydrolysis and fermentation (SHF) system involves two reactors with different optimized configurations with efficient use of raw materials and reduced environmental impacts (Pino et al., 2018). In the first bioreactor, biomass waste is degraded to monomeric sugar, and in the second microbial fermentation occurs (Sudiyani et al., 2019), with their own individual optimal conditions. These procedures are conducted after a pretreatment step.

13.3.1   Pretreatments

Enzymatic hydrolysis is not only affected by enzymatic reaction conditions itself, but also by morphological characteristics of the involved substrates. Cellulose crystallinity, which as result of the well-ordered hydrogen bonding networks between hydroxyl groups along cellulose backbone, limits the efficient accessibility of cellulases to the substrate (Sun et al., 2016). Due to the size of cellulases (3–18 nm), regions of high crystallinity are difficultly digested, resulting in low conversion yields. The presence of lignin in the substrates hinders a proper contact of enzymes, causing nonproductive binding to it. Particle size and porosity also have influence on the accessible specific surface area, where larger biomass pores enhance conversion yields. Due to the strong recalcitrance of biomass, a suitable step of pretreatment is necessary to enhance bioaccessibility of the substrates, with economic viability and without causing sugar losses (Kumar et al., 2020). In the last few decades, many pretreatment methods were developed with different mechanisms, such as decrystallization, solubilization of hemicellulose, and alteration of lignin structure (Bychkov et al., 2019).

Physical pretreatments, such as grinding and milling, aim to increase the substrates specific surface area by reducing their particle size and disrupting their structure. However, as it cannot effectively improve hydrolysis, it is commonly used as a prestep to other pretreatment processes (Sheng et al., 2021). Irradiation methods, such as microwave, ultrasound and gamma ray, can be combined with other methods to reach higher sugar recoveries (Yu et al., 2018). Yin and Wang (2016) evaluated gamma irradiation combined with alkaline pretreatment, reaching 72.7% of reducing sugar yield after irradiation at 100 kGy and 2% NaOH for 1 h.

Chemical pretreatment, such as alkaline, acid, and Organosolv, are the most widely used methods. Acid pretreatment acts through the solubilization of hemicellulose into pentoses by glucosides cleavage, partially degrading lignin to open biomass fibers. In alkaline pretreatment, ether and ester linkages can be broken down, swelling cellulose and exposing it to enzymes due to reduced crystallinity (Woiciechowski et al., 2020; Sun et al., 2016). Organosolv pretreatment employs organic solvents, such as ethanol, methanol, glycerol, ethylene glycol, and others, and acts on the cleavage of α- and β-aryl ether linkages in lignin and hemicellulose hydrolysis (Zhang et al., 2016). The potential of solvent recycles and relative pure lignin byproducts recovery makes this method a promising approach for integrated use of biomass (Ferreira and Taherzadeh, 2020). Recent studies targeted the optimization of chemicals concentration, temperature, and time reaction, and reduction of the costs and laboriousness of post treatment steps, trying to evaluate effects on hydrolysis (Gomes et al., 2020; Lee and Yu, 2020; Saulnier et al., 2020). Hernández-Guzmán et al. (2020) studied different biomass loadings and particle sizes in an alkaline oxidative pretreatment with batch and semibatch operations. Delignification extents were around 60%, with semibatch process time significantly lower with 72% yield of reducing sugars conversion.

Ionic liquids (ILs) are nonmolecular components, which are liquid salts at temperatures lower than 100 °C, known as green solvents, with low vapor pressure and good thermal stability (Usmani et al., 2020). Depending on the applied IL, the lignin fraction can be depolymerized and condensed, and cellulose and hemicellulose fractions are dissolved into liquid fractions. After biomass regeneration in antisolvents of IL, cellulose, and hemicellulose can be efficiently hydrolyzed by enzymes, improving considerably the yield of enzymatic hydrolysis. Wei et al. (2020) evaluated enzymatic hydrolysis of rice straw with 1-butyl-3-methylimidazolium chloride pretreatment. The reducing sugar yields reached 98.9% even after 8 IL reuse cycles, and the hydrolysates were fermented to ethanol and succinic acid.

Hydrothermal pretreatment, steam explosion, and ammonia fiber expansion (AFEX) are examples of methods that apply physical and chemical procedures to treat biomass. Hydrothermal depolymerizes biomass fibers through acid medium originated by self-hydrolysis of hemicellulose under high temperatures, while steam explosion adds into it the shearing action of pressure, with a sudden decompression (Kumar et al., 2020; Ruiz et al., 2020). Pimienta et al. (2019) evaluated a continuous pilot-scale steam explosion pretreatment at 180 °C during 20 min to pretreat four different biomass feedstocks. The subsequential hydrolysis reached 82.5%, 23.8%, 49.8%, and 72.4% of glucose recovery from wheat straw, corn stover, sugarcane bagasse, and agave bagasse, respectively. Wang et al. (2014) applied 2.0% sulfuric acid in a steam explosion pretreatment of corn stover at 160 °C during 5 min, reaching 95.0% of glucose recovery. In the case of AFEX, nitrogen acts as a reactant, saving water (Da Costa Sousa et al., 2016). Flores-Gómez et al. (2018) applied a low-severity AFEX pretreatment at 120 °C on agave residues, reaching >85% of sugar conversion and 90% of ethanol production efficiency without any washing step.

As each single pretreatment has its different disadvantages, the alternative that is employed to fractionate biomass, combining two or more methods, can improve the integrated use of biomass with best reaction conditions and economic feasibility (Xing et al., 2018; H. Zhang et al., 2021). Also, coproduction is an alternative to improve integrated biomass exploration. Ebrahimian et al. (2020) evaluated an integrated exploitation of organic fraction of municipal solid waste (OF-MSW) by *Enterobacter aerogenes*, through pretreatment, enzymatic hydrolysis (applying cellulases and amylases) and fermentation steps. The authors reached 139.1 g 2,3-butanediol, 98.3 g ethanol, 28.6 g acetic acid, 71.4 L biohydrogen, and 40 g PHA production from 1 kg of OF-MSW, showing the potential of this biomass waste. Zhang et al. (2020) evaluated the coproduction of XOS and glucose from seawater hydrothermal pretreatment of sugarcane bagasse, yielding 67.1% of XOS from liquid pretreatment stream, and 94.7% enzymatic digestibility of cellulosic solid residue. Other works evaluated different strategies, as well as coproduction processes, to explore the obtaining of sugar fractions from biomass, showing results that may indicate a possibility on the next generation of biorefineries (Kuglarz et al., 2018; Lai et al., 2019; Liu et al., 2017; Qian et al., 2021; Valladares-Diestra et al., 2020).

13.3.2 Enzymatic hydrolysis

Due to the complex nature of lignocellulosic materials, the use of different hydrolytic and accessory enzymes, which act synergistically in the degradation of polymers, is required. Commonly, filamentous fungi, which are the main degraders of lignocellulosic biomass, are used in the production of enzyme complexes at laboratory and industrial scale (Kunitake and Kobayashi, 2017; Siqueira et al., 2020).

An important factor in the enzymatic hydrolysis process is the control of the physicochemical conditions, due to the biochemical and structural nature of enzymes, which are sensitive to abrupt changes or extreme conditions. For this reason, adequate conditions are necessary to avoid denaturation and loss of enzymatic activity to maintain a high performance in its hydrolytic action. Among the most important physicochemical variables that affect the performance of enzymes are the reaction temperature and the pH of the system in which the enzyme is suspended. These two factors directly influence the hydrolytic capacity of each enzyme; that is why they are the most studied variables in processes' optimization.

In SHF each process is carried out in its individual optimized reaction conditions that can lead to production maximization, allowing the application of enzymes and yeasts fermentation recycle techniques (Su et al., 2020). The development of the parameters of enzymatic hydrolysis, as well as process conditions optimization, has tremendous importance in biorefineries. In the case of the second-generation bioethanol production, one of its most common raw material is cellulose (Rastogi and Shrivastava, 2017). Cellulose is hydrolyzed by synergistic reactions of endocellulase, exocellulase, and β-glucosidase under milder conditions, typically at 40–50 °C and at pH 4.5–5, generating glucose hydrolysates with low toxicity (Maitan-Alfenas et al., 2015). In the case of xylanase-mediated hemicellulose hydrolysis, the generation of xylose and XOS commonly occurs at 50 °C and pH 4.8-5 (Álvarez et al., 2017; Chen et al., 2018). After enzymatic hydrolysis, ethanolic fermentation is carried out between 30 and 36 °C (Mohd Azhar et al., 2017).

López-Gutiérrez et al. (2020) optimized a binary mixture of commercial enzymes for the saccharification of agave bagasse. The optimum temperature for saccharification using Cellic CTec2/Cellic HTec2 was 41.4 °C, while with the use of Celluclast 1.5 and Viscozyme L higher efficiency was obtained at a maximum pH of 3.5. Rawat et al. (2014) analyzed different filamentous fungi, among them three strains of the genus *Aspergillus* were selected with a high potential for cellulolytic enzyme production. Cellulases of the three strains showed an optimal pH of 5, while the optimum temperature for cellulases produced by *A. niger* and *Aspergillus fumigatus* was 55 °C and for *Aspergillus oryzae* was 50 °C. The majority of commercial enzymes derived from mesophilic filamentous fungi have an optimal reaction temperature of 45–55 °C, while the pH varies from 3.5 to 5.5, always under low acid conditions (Siqueira et al., 2020).

Other well-studied examples of SHF and further production of biomolecules, especially bioethanol, were conducted with starch sources. Martín et al., (2017) evaluated the benefits of a previous enzymatic hydrolysis application in cassava stems, which are composed of starch, and are a novel source in the bioethanol industry. The results showed that the previous use of α-amylase and amyloglucosidase generated an increase in glucan conversion, decreasing the degradation of sugars by other pretreatment, showing a high potential to produce ethanol from cassava stalks. Pradyawong et al., (2018) reported that starch from cassava has a high potential as a source for the alcohol industry. The hydrolysis and fermentation process of cassava starch was compared with three different sources of corn starch. The results showed a fermentation profile of cassava starch, in terms of ethanol production (124 g/L) and glycerol formation (7 g/L), which was similar to that of corn starch, showing a high potential of starch as a carbon source in the alcohol production at industrial scale.

Finding the optimal reaction conditions for enzymatic hydrolysis ensures the maximum hydrolytic performance and therefore a greater release of sugars. However, despite the optimal results, applying enzymes within bioprocesses for lignocellulosic and/or starchy biomass valorization is still a bottleneck due to its costs that hinder its massive use, greatly affecting the economic viability of their application in industrial biorefineries. Therefore, it is necessary to find the best conditions for the maximum degradation of lignocellulosic and/or starchy biomass, thus reducing the total costs of the process. This parameter, however, is strongly related to the nature of biomass, hydrolysis time, composition, and characteristics of the enzyme complex. Xing et al., (2018) observed that in the enzymatic hydrolysis of pretreated rice straw, the application of 50 filter paper units (FPU) or 70 FPU did not show significant differences in the release of sugars. In addition, after 48 h of hydrolysis, over 40 g/L of sugars were released with the use of 40 FPU, where higher enzyme loading did not influence the process. On the other hand, Bu et al. (2019) used commercial Cellic Ctec2 for the hydrolysis of pretreated sugarcane bagasse, using enzymatic loads of 5, 10, and 15 FPU, which allowed a final ethanol production of 104.8, 115.9, and 120.8 g/L, respectively. As it can be seen, the final ethanol production is similar for the different employed enzymatic concentrations, which could lead to enzyme economy with the same high production efficiencies.

Other alternatives to improve the enzymatic hydrolysis with minimum enzymatic loading is the use of additives, such as chemical surfactants, which have showed good results in the improvement of enzymatic conversion efficiency with different cellulolytic complexes. Zhang et al. (2021) showed that cellulase complex supplemented with Tween 80 improved the yield of released glucose during enzymatic hydrolysis of pretreated sugarcane bagasse. They observed that an enzyme loading of 10 FPU supplemented with Tween 80 presented the same efficiency as using an enzyme loading of 20 FPU, demonstrating that the use of Tween 80 not only reduced 50% the use of enzymes, but also reduced the hydrolysis time. In addition, the use of a surfactant with polyethylene glycol of 34 degrees of polymerization showed an increase in enzymatic hydrolysis efficiency (24%–72%) of treated bamboo biomass using Cellic Ctec2 (Lin et al., 2019).

The addition of accessory enzymes is also an alternative to increase the performance of enzymatic hydrolysis. Xu et al. (2019) evaluated the addition of hemicellulase, laccase, pectinase, and β-glucosidase for the enzymatic hydrolysis of pretreated sugarcane bagasse using the Cellic Ctec3. Their results showed an increase in glucose release performance with the addition of hemicellulose and β-glucosidase, while the use of pectinases and laccases did not show a significant influence. Finally, with an addition of 150 U of hemicellulase and 60 mg of β-glucosidase, a glucose release yield of 80% was achieved with the use of only 4 FPU of the cellulolytic complex. Patel et al. (2017) showed that the use of β-xylosidase together with a commercial cellulase significantly increased the hydrolysis of treated sugarcane bagasse than using each enzyme separately. A total sugar concentration of 540 mg/g of treated biomass was obtained with the use of 8 FPU of cellulase and the addition of 23 U g of β-xylosidase, increasing twofold the yield of sugars released after 24 h. Furthermore, authors also showed that the increase of cellulase loading to 12 FPU, or doubling the amount of β-xylosidase, did not show significant differences in the release of sugars.

Another strategy to reduce process' costs is the reuse of enzymes, through its recovery and subsequent reuse. It is interesting to note that the recovery of enzymes is possible if the hydrolysis and fermentation stage are carried out separately, thus achieving partial recovery according to the employed method. Xing et al. (2018) applied the reuse of enzymes that bind to the solid residues during enzymatic hydrolysis. They managed to reduce the enzymatic quantity by 10% for each cycle, maintaining almost the same sugars' release rates. This methodology allowed saving almost 40% of enzyme use in the fifth hydrolysis cycle, which means a substantial economy. On the other hand, Yuan et al. (2018) showed that the recyclability degree was related to the pretreated material. Authors evaluated acid and alkali pretreatment of corn stover. In the case of alkaline pretreatment, an enzymatic loading of 60% was required in the second cycle, maintaining the same concentration of sugars was released after the first cycle. With the acid pretreatment, an enzymatic loading of 50% was required in the second cycle. In both cases, the sugar release profile remained stable after at least six cycles.

Finally, it is also important to study different strategies of biomass loading for pretreatment. Strategies such as enzymatic batch hydrolysis, fed batch, or continuous are being studied. Although the most studied and used strategy is the batch process, the fed batch system has also been studied with a high release of sugars (Godoy et al., 2019; Hernández-Beltrán and Hernández-Escoto, 2018). Xu et al. (2019) evaluated different loadings of pretreated sugarcane bagasse with progressive addition of biomass, at different times, which allowed finding the optimal biomass feeding condition for maximal sugar release. The optimal conditions were defined with an initial hydrolysis of 10% w/v of biomass for 8 h, and addition of 5% of biomass, then 4% and 3% of biomass were added after 12 h and 16 h, respectively. With this sequence, an enzymatic conversion of 80.1% was obtained, reaching a release of 122.2 g of glucose with the use of 4 FPU of cellulase with a final biomass loading of 22% w/v. As the pretreated biomass is normally recovered in the solid state and is poorly soluble in different buffered enzymatic solutions, it is difficult to employ continuous enzymatic hydrolysis processes. For this reason, there are few studies of continuous enzymatic hydrolysis systems. Stickel et al. (2018) analyzed a continuous process of enzymatic hydrolysis of pretreated corn stover with a feeding of 2.5% or 5% of biomass, and sugar recovery by ultrafiltration membrane. Their results showed that 50% of enzyme loss occurred in the continuous sugar clarification process, so a continuous enzyme feeding was also necessary.

In addition, the implementation of a continuous system with two reactors in line was proposed, which would increase the retention time of the biomass and, consequently, a higher enzymatic efficiency would provide at least a double of the volumetric production. Among the main challenges, it was found that the increase in biomass loading would lead to technical problems in the feeding the system. Table 13.1 summarizes some examples of SHF processes.

13.4   Simultaneous enzymatic hydrolysis and fermentation processes

Simultaneous enzymatic hydrolysis and fermentation, also called as simultaneous saccharification and fermentation (SSF) is a method widely used for lignocellulosic biofuel production; whereas the transformation of lignocellulosic or starchy biomass into sugars, which are employed for ethanol production, takes place in the same reactor vessel. It is an alternative to the SHF where the enzymatic pretreatment is performed previously to the fermentation step.

The main reason behind the development of the SSF processes is the end-product inhibition of the hydrolysis mechanism caused by cellobiose and glucose. In fact, this is one of the most problematic aspects concerning the application of cellulases for biomass hydrolysis and further ethanol production. Besides preventing product inhibition, other preeminent advantages of the SSF include hydrolysis and the fermentation steps performed in the same vessel, resulting in the reduction of sugar losses that is common for SHF during the transfer of the hydrolysate from the hydrolysis reactor to the fermentation reactor. Also, the use of less unit operations results in lower residence times for each step and the coconsumption of hexoses and pentoses, which represent economic advantages for the global ethanol productivity.

However, as the optimum temperature of enzymatic hydrolysis is typically different from that of yeasts' fermentation, lower metabolic conversions can be achieved in SSF processes. In addition, the fermentation product, that is, ethanol, could inhibit the hydrolysis step, but to a lesser extent when compared to glucose and cellobiose inhibition. Another issue is that the recycling of enzymes back to the process is unpractical for SSF, as many impurities of the fermented broth imply in technical difficulties for separation processes.

Another point to be observed is the incomplete hydrolysis of the solid lignocellulosic fraction that is observed for SSF processes due to enzyme denaturation and unproductive enzyme adsorption to the solid particles, decreasing availability of chain ends, and increasing crystallinity with conversion of pretreated cellulose. According to Olofsson et al. (2008), the enzymes applied for SSF at industrial level should have a balanced composition including endoglucanases, cellobiohydrolases and β-glucosidases to minimize costs for yeast and enzyme production. The traditional SSF and proposed process modifications including preliquefaction and consolidation of the enzyme production and hydrolysis are presented in Fig. 13.4.

Liu et al. (2020) classified the lignocellulose bioconversion strategies in two categories named off-site and on-site. The off-site approach considers that the enzymes used for the hydrolysis step are previously produced and added to the saccharification reactor. On the other hand, the on-site saccharification considers the production of enzymes together with the hydrolysis and fermentation steps.

TABLE 1 Some recent SHF processes applied to biomass wastes.

Source	Pretreatment	Enzymes applied	Enzymatic condition	Biomolecules	Microorganism	Reference
Rice Straw	Two stage deep eutectic solvent by combination of foraceline and Na$_2$CO$_3$	Cellulase cocktail by Vland Biotech Co. Ltd.	10% substrate loading, 50 FPU cellulase recycle saving 10% each cycle, 50 °C, 150 rpm, 24 h	Butanol	C. saccharobu-tylicum DSM 13864	(Xing et al., 2018)
Sugarcane bagasse	15% ammonia solution; biomass ratio of 1:4.5 (w/v), autoclaved at 121 °C, 15 lbs for 1 h.	Cellulases by MAPs 450 and hemicellulases by Aspergillus niger ADH-1	2.5% substrate loading, 50 °C, 8 FPU/g of cellulase with 23 U/g of b-xylosidase	Ethanol	Pichia stipitis NCIM 3506	(Patel et al., 2017)
Acacia wood	0.05% sulfuric acid at 200 °C for 5 min	Cellic®CTec3 supplemented by soy protein	2.5% substrate loading, 10 FPU/g glucan, 50 °C, 200 rpm, 96 h	Ethanol	Saccharomyces cerevisiae (ATCC 24858)	(Lee and Yu, 2020)
Organic fraction of municipal solid waste	Acetic acid-catalyzed ethanol organosolv pretreatment, with 10% solid loading, at 120–160 °C, for 30–60 min	Cellic® CTec2 cellulase, α-amylase and glucoamylase (Novozymes A/S)	20 FPU/g of Cellic® CTec2, at 5% substrate loading, 45 °C, 72 h. Subsequently, 0.1 g/g starch of α-amylase and glucoamylase, at 65–90 °C for 26 h	Biohydrogen, 2,3-butanediol, ethanol, acetic acid and biogas	Enterobacter aerogenes PTCC 1221	(Ebrahimian et al., 2020)
Sugarcane bagasse	2 w/mL Ultrasound for 30 min, and then 2% NaOH, at 121 °C for 15 min, with 5% biomass loading	Trichoderma reesei cellulase (Celluclast 1.5 L)	2% substrate loading, 40 FPU/g, at 50 °C for 24 h	Polyhydroxybutyrate	Lysinibacillus sp. RGS	(Saratale et al., 2021)
Rapeseed straw	1% sulfuric acid, 20% biomass loading, 180 °C for 10 min	Cellic® CTec2 supplemented by endoxylanases (Cellic® HTec2)	5% substrate loading, 64.6 FPU/g (10.3 FPU/g for xylanases), 50 °C, 48 h	Succinic acid and ethanol	S. cerevisiae and Actinobacillus succinogenes 130Z	(Kuglarz et al., 2018)
Wheat straw	Subcritical water at 220.5 °C, 22 min, 2.5% substrate loading	β-glucosidase by SPE-007AL and cellulases by five different commercial cocktails	5% substrate loading, 20 FPU/g of cellulase and 1 U/g of β-glucosidase, 50 °C, 120 rpm, 96 h	Ethanol and biogas	S. cerevisiae NX11424	(Chen et al., 2021)

(continued)

TABLE 1   (Cont'd)

Source	Pretreatment	Enzymes applied	Enzymatic condition	Biomolecules	Microorganism	Reference
Cassava starch	-	α-amylase and glucoamylase	Liquefaction by 0.9 mg/g of α-amylase at 85 °C for 180min, and then saccharification by 1.5 mg/g of glucoamylase at 60 °C for 90 min	Ethanol	*S. cerevisiae* and *Zymomonas mobilis*	(Wangpor et al., 2017)
Sorghum grain	-	Termamyl SC DS α-amylase (or Amylex BT2) and San Extra L glucoamylase (or Diazyme SSF)	Liquefaction by 0.2 mL/g of α-amylase for 1 h at 100 °C, and then saccharification by 0.6 mL/g of glucoamylase at 60 °C for 100 min	Ethanol	*S. cerevisiae* (Fermiol and Ethanol Red)	(Szambelan et al., 2020)
Solanum lycocarpum starch	0.8% citric acid starch extraction	α-amylase and glucoamylase from *Aspergillus niger* L119	Liquefaction and subsequent saccharification with immobilized enzymes onto polyaniline, at 40 °C for 60 min, 20% substrate loading,	Ethanol	*Saccharomyces bayanus* INCQS 40235	(Morais et al., 2019)

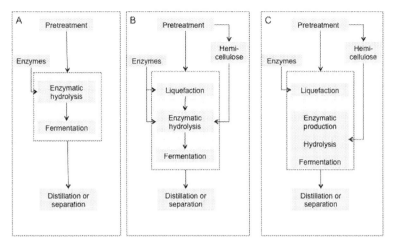

FIG. 13.4 **Lignocellulose to ethanol process concepts.** (A) SSF; (B) SSF with preliquefaction; (C) Consolidated bioprocess.

The simultaneous saccharification and cofermentation (SSCF) process is an SSF process that includes the fermentation of pentoses together with hexoses. This aspect offers the advantage of using more sugars that are released in the hydrolysis step. In addition, the consumption of pentoses such as xyloses reduces its natural inhibitory effect. The simultaneous saccharification, filtration, and fermentation process is characterized by the cross-flow filtration of the fermented broth for the separation of the retentate, which is pumped back to the hydrolysis vessel, while the permeate with high sugar concentration is maintained in the fermentation vessel. This process prevents glucose accumulation and inhibition, as well as the use of different temperatures for fermentation yeast growth and enzymatic activity. Also, this process allows the recycling of yeast cells, so it is a hybrid process between SHF and SSF. In the consolidated bioprocessing (CBP) the enzyme production CBP is performed in the same vessel where the saccharification and fermentation takes place, simultaneously. In this aspect, not only yeasts are employed, but also other microbial strains responsible for enzyme production. Although this process strategy has been reported as cost effective, enzyme and ethanol productivities are lower. The cotreatment (CT) and the consolidated biosaccharification (CBS) are CBP-derived strategies with other modifications to solve CBP limitations. CT involves mechanical treatment and CBS separates the fermentation broth from the CBP to avoid the effect of side products in the fermentation process (Liu et al., 2020; Olofsson et al., 2008). Various SSF processes for ethanol or other biochemicals' production from lignocellulosic biomass are presented in Table 13.2.

13.4.1 Product inhibition

The mechanism of cellulose hydrolysis consists of the synergistic action of endoglucanases, cellobiohydrolases, and β-glucosidases. Cellobiose is the end-product of the enzymatic depolymerization performed by endoglucanases and cellobiohydrolases, which is frequently reported as the rate-limiting step of cellulose hydrolysis. Cellobiose causes a noncompetitive

TABLE 2 Bioethanol and other biochemicals production by SSF using different lignocellulosic biomass substrate.

Strategy	Substrate	Enzyme	Fermentation microorganism	Reactor type and mode of operation	Reactor conditions	Product	Reference
SSF	Cactus pear (*Opuntia ficus Indica* and *Nopalea cochenilifera*)	Enzymatic cocktail ACCELERASE® 1500 (provided by GENECOR®)	*Saccharomyces cerevisiae* LNFCA-11 and PE-2	250 mL Erlenmeyer flasks	Working volume: 50 mL Temperature: 35–45 °C pH: not informed Agitation: 150 rpm Length: 48 h	Bioethanol	(Souza Filho et al., 2016)
SSF	Sugarcane straw	Enzymatic complex (Cellic® CTec2, from Novozymes Latin America)	*Saccharomyces cerevisiae* Y-904	200 mL bench scale reactors – batch mode	Utile volume: 50 mL Temperature: 40 °C pH: 5.5 Agitation: 250 rpm Length: 24 h	Bioethanol	(Pratto et al., 2020)
SSF	Rice straw	Commercial enzyme blend (Cellic® CTec2)	*Clostridium beijerinckii* NRRL B-592	50 mL serum bottle – batch fermentation	Utile volume: 40 mL Temperature: 37 °C pH: 5.2 to 6.2 Agitation: 150 rpm Optimum Length: 48 h	Butanol	(Valles et al., 2020)
SSF	Oil palm front	Commercial cellulase (Advanced Enzymes, Thane, India)	*Enterobacter cloacae* sp. SG1	500 mL bench scale bioreactors (batch mode)	Working volume: 300 mL Temperature: 37 °C pH: absence of control Agitation: 200 rpm Length: 120 h	2,3-butanediol	(Hazeena et al., 2019)
SSF	Wheat and rice straw, aspen and pine sawdust, Jerusalem artichoke stems and tubers, biomass of macro and microalgae	Celluclast 1.5 L, Glucosidase, Inulinase, Glucanex, α-amylase, Viscozyme L, Laminarinase, Alginate lyase, Agarase (from Sigma, St. Louis, MO, USA)	*Rhizopus oryzae* (F-814, F-1127) and *Actinobacillus succinogenes* (B-10111)	Use of immobilized biocatalysts (lack of data regarding the bioreactors)	Temperature: 40–50 °C pH: 5.0 Agitation: 140-250 rpm Length: 24 h	Lactid, succinic and fumaric acids	(Maslova et al., 2019)
SSF	Corn stover	Cellulase complex (Cellic® CTec 2.0, from Novozymes, China)	*Aspergillus niger* SIIM M288	5 L bioreactor (batch mode)	Working volume: not informed Temperature: 50 °C pH: 4.8 Aearation rate: 1 vvm Length: 48 h	Citric acid	(Hou and Bao, 2018)

(continued)

Strategy	Substrate	Enzyme	Fermentation microorganism	Reactor type and mode of operation	Reactor conditions	Product	Reference
SSF	Rice straw	Cellulase complex and xylanase (Siam Victory Chemicals Co., Ltd., Bangkok, Thailand)	*Escherichia coli* KJ122	2 L bioreactor (batch and fed-batch modes)	Working volume (batch): 1.2 L Initial Working volume (fed-batch): 0.7 L Temperature: 39 °C pH: 6.5 Agitation: 250–300 rpm Length: 5 days	Succinate	(Sawisit et al., 2018)
SSF	Rice straw and vegetable waste	Cellulase and B-glucosidase	Consortium (hydrogen producing bacteria domesticated via acclimated repetition. Source: municipal sewage sludge)	Dual-chamber Microbial electrolysis cell (batch mode)	Volume (anode chamber): 98 mL Temperature: 35 °C pH: 7.0 Agitation: none Length: 196 h	Hydrogen	(Zhang et al., 2019)
SSF	Fossil fuels (DBT, 2-HBP and DMF)	Invertase produced by *Zaigosaccharomyces baiii* Talf1	*Gordonia alkanivorans* 1B	500 mL shake flasks (batch mode)	Working volume: 150 mL Temperature: 30 °C pH: 5.5–7.5 Agitation: 150 rpm Length: 48 h	(objective: biodesulfurization)	(Paixão et al., 2016)
SSF, Fed-SSF and Fed-non-isother-mal-SSF	Paper mulberry (*Broussonetia papyrifera*)	Cellulase (Sigma C8546) and B-glucosidase (Sigma G0395)	*Saccharomyces cerevisiae*	Erlenmeyers (working volume of 30 mL) – tested both batch and fed-batch modes	Working volume: 30 mL Temperature: 43 °C (SSF and Fed-SSF) and switching between 30 °C and 43 °C (Fed-nonisothermal-SSF) pH: non adjusted Agitation: 200 rpm Length: 48 h	Bioethanol	(Wang et al., 2020)
Semi-SSF with pre-hydrolysis	Pulp and paper sludge	Crude cellulose (SUPERCUT) from Zytex India Private Limited	*Pichia stiptis* and Baker's yeast (*Saccharomyces cerevisiae*)	3 L Hygene Plus, Lark autoclavble fermenter – fed batch	Working volume: 2 L Temperature: 50 °C (pre-hydroliysis) and 30 °C (fermentation) pH: 4.8 Length: 48 h (pre-hydrolysis) plus 60 h (fermentation)	Bioethanol	(Dey et al., 2021)

TABLE 2 (Cont'd)

Strategy	Substrate	Enzyme	Fermentation microorganism	Reactor type and mode of operation	Reactor conditions	Product	Reference
SSF with pre-liquefaction	Waste paper and kitchen waste	Alpha amylase and glucoamylase (Sichuan Shanye Co., Ltd., China) and cellulose (Cellic® CTec2, from Novozymes, Denmark)	Saccharomyces cerevisiae KF-7	Pilot scale kneader (working volume of 22 L – batch mode)	Liquefaction (with amylase) at 90 °C, 20 rpm, 1 h Pre-saccharification (with glucoamylase and cellulase) at 50 °C, 20 rpm, 16 h SSF at 35 °C, 20 rpm, 96 h	Bioethanol	(Nishimura et al., 2017)
SSF and solid-state fermentation	Bread waste	Amyloglucosidase from Aspergillus sp.	Lactobacillus paracasei	Solid state fermentation in small tubes (batch mode)	Temperature: 37 °C pH: 6.0 Length: 24–92 h	Lactic acid	(Sadaf et al., 2021)
SSF and co-fermentation	Sugarcane bagasse	Cellulase complex (Cellic® CTec2, from Novozymes Latin America)	Saccharomyces cerevisiae SHY07-1	50 mL serum bottles (batch mode)	Working volume: 15 mL Temperature: 30 °C pH: not informed Agitation: 150 rpm Length: 240 h	Bioethanol and succinic acid	(Bu et al., 2019)
SSF and co-fermentation	Cassava bagasse	Glucoamylase (provided by handong Sukahan Bio-Technology Co., Ltd.) and cellulose (provided by Vland Biotech Inc.)	Bacillus coagulans and lactobacillus rhamnosus	5 L bioreactor – batch mode	Utile volume: 3 L Temperature: 42–50 °C pH: 5.0–6.5 Agitation: 200 rpm Length: 40 h	Lactic acid	(Chen et al., 2020)
CBP	Food wastes	Endogenous amylase	Clostridium sp. HN4	3 L bioreactor (Bio-flo110, USA) (batch and fed-batch)	Working volume: not informed Temperature: 35 °C Initial pH: 6.2 (in some experiments not controlled, in others adjusted to stay in the range from 5.2–5.5) Agitation: 150 rpm Length: 144 h	Biobutanol	(Qin et al., 2018)
CBP	Liquefied corn starch	Enzymes secreted by the fermenter strain (cellulase, mannanase, amylase and xylanase)	Aureobasidium pullulans GXL-1	10 L fermenter – batch and repeated batch modes	Utile volume: 6.9 L Temperature: 30 °C pH: not informed Agitation: 200–400 rpm Length: 196 h	Polymalic acid	(Zeng et al., 2020)

SSF – simultaneous saccharification and fermentation.

inhibition in endoglucanases and cellobiohydrolases, reducing the overall conversion of cellulose to glucose. Cellobiose is the natural substrate of β-glucosidases, which catalyzes its cleavage to glucose. Glucose causes a feedback inhibition to β-glucosidases, blocking the enzyme's active site or the release of the hydrolyzed substrate (Libardi et al., 2020; Payne et al., 2015).

The natural behavior of the cellulosic biomass enzymatic depolymerization pushed the development of some strategies to minimize product inhibition and increase of glucose productivities. The use of an enzymatic cocktail with β-glucosidases is the most commonly used strategy to prevent the cellobiohydrolases and endoglucanases inhibition. In fact, the presence of β-glucosidases has a double benefit, it cleaves cellobiose to glucose, getting rid of the inhibitor and transforming it into the desired biomolecule for further application. Other strategies have been proposed to overcome product inhibition in industrial processes, such as the removal of the product through membrane filtration, with the retention and recycling of enzymes and the permeation of the product. Engineered strains are another way to prevent the feedback inhibition. The gene *cre1* is responsible for the catabolic repression regulation and the inhibitory effect that glucose has in cellulase production. The deletion of this gene in the mutant strain RUT C-30 of the fungus *Trichoderma*, well known as industrial cellulase producer, promoted the ability of the strain to grow in glucose media for cellulase production, in contrast to the parental strain QM6a (Payne et al., 2015; Singhania et al., 2017).

Another class of inhibiting compounds are related to the byproducts released during pretreatment, impacting the hydrolysis step in SHF processes. These compounds include furaldehydes such as 5-hydroxymethyl-2-furaldehyde and furfural (2-furalde-hyde); organic acids such as acetic, formic, and levulinic acids, which are originated from lignocellulosic biomass; and diverse aromatic phenolic compounds derived from lignin degradation (Olofsson et al., 2008; Sjulander and Kikas, 2020). So, in the SSF process these inhibitory byproducts are consumed by the fermenting yeast, reducing their negative impact in the enzymatic hydrolysis. Yeast strains with high tolerance to inhibitors have been developed through genetic engineering tools (Sjulander and Kikas, 2020).

13.4.2 Strain thermotolerance in SSF processes

As previously presented, one of the disadvantages of SSF processes is the different optimum conditions of saccharification and fermentation. The enzymatic saccharification requires an optimum temperature to enhance enzyme activity. Fermentation usually requires the optimum temperature for yeast grown and metabolic activities, ranging between 28 and 37 °C for mesophilic strains. In this way, thermotolerant strains may undergo the fermentation at temperatures above 40 °C to overcome the limitations of SHF, meeting the requirements for SSF (Choudhary et al., 2016). Some traditional thermotolerant strains such as *Saccharomyces cerevisiae*, other strains of *Candida* spp., *Kluyveromyces* spp., *Pichia* spp., and *Zymomonas* have been adapted or modified to tolerate high temperatures. Different strategies have been carried for the development of thermotolerant strains, such as physiological adaptation, site-directed mutagenesis, genome shuffling, metabolic engineering, or even yeast cell encapsulation in hydrogel supports for protection against high temperatures (Choudhary et al., 2016).

Thermotolerant strains allow the use of higher fermentation temperatures, which reduce the overall ethanol production costs. Also, higher temperatures improve ethanol evaporation that helps to maintain ethanol concentrations below toxic levels for the fermenting yeasts (Choudhary et al., 2016). In addition to thermotolerant fermentation yeast, thermotolerant enzymes have also been proposed to guarantee efficient enzymatic conversions even under high temperatures. Although enzymes have their activity improved until around 50 °C, their exposure to high temperatures for longer residence times may favor denaturation processes, with the loss of enzymatic activity.

13.5 Commercial enzymes and enzymes' costs

The enzyme costs are the most important challenge that limit their application. The majority of commercial biomass-degrading enzymes are produced by submerged fermentation, over-expressing selected genes in either native or heterologous microbial hosts. According to Klein-Marcuschamer et al. (2012), the basis cost for cellulase production from fungal strains is considered to be about $10 per kg protein. The authors proposed the off-site saccharification costs based on sugar production using cellulases. The authors found a cellulase cost of $250 per ton sugar, considering cellulase load of 10 FPU/g. The cellulase loading, which is presented in terms of enzymatic units (international units—IU or FPU per gram of substrate), is the key parameter that affects hydrolysis costs. According to Chandel et al. (2019), sugar production should not overcome $100 per ton, otherwise the production process is not feasible.

The world commercial cellulase producers are the Danish Novozymes and the North American Genecor, recently taken over by DuPont. Available commercial cellulase complex such as Accelerase 1500, Accelerase XY, and Accelerase TRIO from DuPont as well as Cel-luclast, Novozyme 188L, Cellic Ctec series from Novozymes have been commercially available as off-site enzymatic formulations for biomass hydrolysis. Other important players in the cellulase market are the Japanese Amano Enzyme, Inc., and the Indian Advanced Enzymes (Chandel et al., 2019).

13.6 Advancements and innovation in biomass-degrading enzymes

As shown throughout this chapter, enzymes can be applied in several processes of complex biomass saccharification and biomolecules production. To complement this vision and to reveal the state of the art and main topics of innovation toward enzyme applications in the industry, a patent search was conducted in the Lens (worldwide) and Latipat (Latin American) databases, in the period between 2010 and 2021. The main objective of this search was to understand that enzymes were applied to which main processes. Therefore, data were retrieved from the Lens and Latipat databases on 24 February 2021 using Boolean language and the following search criteria:

- in the title field, the following words were used: simultaneous fermentation and saccharification; CBP; fermentation, bioprocessing; lignocellulosic biomass, biomass waste; biomass residue; agricultural waste; agricultural residue; kitchen waste; kitchen residue; fruit residue; forestry biomass;
- the following terms were excluded: apparatus, device, machine;

* the International Patent Classification field, the subclassification code C12-P as code was used (C12P is the code for fermentation or enzyme-using processes to synthesis a desired chemical compound or composition or to separate optical isomers from a racemic mixture).

After manual classification and sorting in Microsoft Excel Software, about 700 patents, which are related to the discussed topic in this chapter, were analyzed. During this period, an average of 70 patents were published per year with a constant growth throughout the period of 2010 to 2017, revealing great interest of industries and universities for the development of biotechnological processes with the application of enzymes. The low number of published documents reported in the last 2 years does not reveal a lack of interest, but a period of secrecy of 18 months between the dates of filing and publishing of the document granted by patents offices. The time evolution of patent filings is represented in Fig. 13.5A. The countries with highest participation in patent filings were China (38%), the United States (25%), Australia (6%), and Brazil (6%). China, the United States, and Brazil are countries that have a great agro-industrial activity and, consequently, generate numerous subproducts that can be used as biomass for biotechnological processes involving enzymes and the production of high added value biomolecules. On the other hand, Australia is developing a strong biotechnology field with 470 companies in the country with the focus of therapeutics and agricultural technology (Austrade, 2021).

Consulted patents contain information that is mainly related to general processes of biomass processing (33%), production of biofuels (29%) and biomolecules (22%), development of new microbial strains (7%), and downstream processes (3%) (categories developed by the authors) (Fig. 13.5B). The technological development related to processes will be described using the following patent documents as examples (these documents were sorted by the authors based on the number of patents' citations and on how they fitted in the theme of the chapter). Biofuels' production technological development (i.e., ethanol, butanol, hydrogen, and biogas) from forest and agricultural residues and kitchen waste, through the application of enzymes, has been a widely studied subject, which was applied in academic and

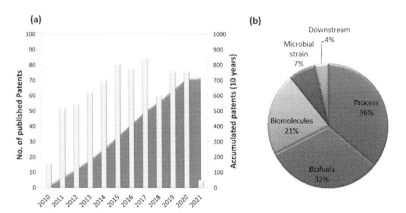

FIG. 13.5 (A) Number of published patent documents related to enzymes in microbial fermentation of biomass wastes to produce biofuels and biomolecules, per year and accumulated number in 10 years. (B) Classification of patents by field of application.

industrial environments. As it was mentioned in the previous sections, the SHF, SSF, SSCF, and CBP processes have been implemented to obtain biofuel and chemical products from lignocellulosic biomass, with good perspectives of cost effectiveness. However, industries still require the implementation of upgraded technologies, improvements in productivity, and cost minimization to produce these components at a large scale. A brief description of recently patented technologies that address those requirements can be found as follows.

The invention developed by Quanguo et al. (2017), from the Henan Agricultural University, refers to a reactor and a method of SSF for hydrogen production, with enzyme recycling. The reactor is formed by four compartments, one for each operation, that is, enzymatic hydrolysis, hydrogen production, enzyme desorption, and enzyme adsorption. The enzymatic hydrolysis unit is located inside the hydrogen production unit. The units are interconnected so that the compartments of enzymatic hydrolysis and hydrogen production communicate, and a stream flows through pipelines from the enzymatic hydrolysis unit to the enzyme desorption unit; also, a stream flows from the hydrogen production unit to the enzyme adsorption unit; and a discharge stream flows from the enzyme desorption unit to the enzyme adsorption unit. The method involves the conversion of straw biomass to hydrogen using cellulase as a saccharification agent.

The patent document authored by Jun and Gahin (2011) and applied by Oji Paper Co. proposes an SSF of cellulosic biomass to produce ethanol. Cellulolytic enzymes and the alcohol-producing yeast *Issatchenkia orientalis* are added to the cellulosic biomass in aqueous suspension, and after the simultaneous reactions the alcohol-containing stream is separated. The residues of the SSF process, that is, enzyme, yeast, and residual biomass are recirculated.

The granted patent filed by Sharma et al. (2019) from the Indian Oil Corporation Limited along with the Indian Department of Biotechnology describes a process of ethanol production from acid pretreated lignocellulosic biomass (straw, wheat straw, sugarcane bagasse, cotton stalk, barley stalk, bamboo, or any agriculture residues that contain cellulose or hemicellulose, or both) via modified SSCF using a single yeast strain *S. cerevisiae*. They proposed a two-stage SSCF, in which mainly xylose fermentation (30–35 °C/17 h) and limited enzymatic hydrolysis take place. Free xylose and other oligosaccharides are targeted in the initial stage of fermentation, which ultimately reduce enzymatic inhibition. This stage is followed by enzymatic hydrolysis (48–55 °C/23 h), and then glucose fermentation is promoted at different temperatures (35–37 °C/6 h). These steps result in a 50% reduction of the commercial enzyme dose and the processing time takes 26 h less than the conventional SSCF process, with an increase in ethanol yield (71%).

In another example, to address the need to increase the efficiency of the currently known techniques, Zhiliang and Edyta (2013) from California University (United States) filed a patent document related to two novel and previously unreported processes. The first one is the integrated bioprocessing and separation (IBS) process and the second one is the one-step bioconversion and separation (OBS) process. The novel IBS process can consolidate cellulase and hemicellulase production, enzymatic hydrolysis, pentose fermentation, hexose fermentation, and product recovery into one single step. This process implements a coculture under aerobic conditions. One culture is formed by an aerobic mutant population of lignocellulolytic cells (*Neurospora crassa*) containing a mutation in one or more beta glucosidase genes, which preferentially utilize monosaccharides for growth and hydrolytic enzymes production; the other is a culture of anaerobic fermentative engineered microorganisms (*Klebsiella oxytoca*) that preferentially consume oligosaccharides to increase biofuel production. This last group of microorganisms is immobilized on a gel matrix to establish a local microaerobic or anaerobic environment. Finally, biofuels or chemicals are preferably extracted from the fermentation broth by gas stripping.

The OBS process involves engineered lignocellulolytic microorganisms to aerobically produce lignin-solubilizing enzymes that efficiently degrade that component. In this sense, the OBS is fully compatible with the IBS process, and the IBS process could be improved including the lignocellulosic pretreatment step proposed in the OBS process to create a consolidated one-step process. Strategically, the IBS and OBS processes have higher levels of consolidation than known CBP processes. Thus, these processes achieve an even greater reduction in lignocellulosic biomass processing costs for biofuels and chemical components production.

Besides biofuels, several biomolecules can be produced through fermentation routes applying lignocellulosic material and enzymes for their saccharification. Among these molecules are organic acids (e.g., lactic acid, propionic acid, citric acid), bioplastics (e.g., PHA, poly-lactic acid, nanocellulose), and antibiotics (e.g., nisin). As previously shown, 22% of the filed patent documents referred to the production of biomolecules and some examples will be explored next.

Ren et al. (2014) described a process for SSF of lignocellulosic material for coproduction of nisin and lactic acid. The substrate (wheat straw, corn straw, sorghum straw, reed, straw, wood chip, or bamboo powder) was pretreated (process not described) and washed until most part of glucose, furfural, and other inhibitory components are degraded to enhance enzymatic hydrolysis. This material is then added to a reactor at a rate of 4% with other fermentation components: nitrogen source (peptone and yeast extract), trace elements, and the enzymes (cellulase, β-glucosidase, xylanases, and pectinase). There is a phase of prehydrolysis for the initial liberation of reducing sugars (5 h) and then the microorganism is added (*Lactococcus lactis* subsp. *lactis*, 10%) and the process is carried until 168 h at 37 °C and 160 rpm. Lactic acid and nisin were quantified and reached concentrations of 39 g/L and 8230 IU/mL, respectively.

Another group of biomolecules that can be produced by fermentation and enzymatic hydrolysis is represented by isoflavones. These compounds can be naturally found in grains such as soy, peanuts, and chickpeas and have been reported to improve blood vessels and cellular health and increase the level of antioxidants in the body, being applied in several pharmaceutical applications (Thrane et al., 2017). In the process described by Macedo et al. (2016), isoflavones and equol are produced by biotransformation processes of soy extracts applying fermentation, enzymatic hydrolysis, or a combination of both. The soy extract is obtained through macerating and cooking soybeans. Three biotransformation processes are proposed: fermentation with lactic acid bacteria—with soy extract inoculated at 1% and cultivated at 37 °C for 12 h; the same fermentation was carried out followed by enzymatic hydrolysis with tannase (1.8 U/mL) at 40 °C for 30 min, which was interrupted by submitting the mixture to 0 °C (ice bath) for 15 min; and only enzymatic hydrolysis with tannase using the same described process. Tannase is an enzyme that hydrolyzes ester bonds and demonstrates great potential for breaking phenolic compounds and has been reported for the production of isoflavones (Macedo et al., 2016). The produced biomolecules were detected by high-performance liquid chromatography and the best results were obtained with enzymatic hydrolysis of soy extract (944.5 mg/mL of daidzein and 321.6 mg/mL of genistein) and fermentation with enzymatic hydrolysis (717.2 mg/mL of daidzein and 251.4 mg/mL of genistein) proving that tannase enhances isoflavones production and recovery from soy extract.

Finally, the invention related to new microbial strains developed by Clarkson et al. (2013) and applied by Danisco US Inc. claims a conversion process to transform a starchy substrate into alcohol (ethanol or butanol) using cocultured microorganisms expressing a variety of enzymes, for example, yeasts and *A. niger* expressing endogenous glucoamylase and alpha-amylase, in an SSCF process. The enzymes produced by *A. niger* do not need to be added as exogenous enzymes.

13.7   Conclusions and perspectives

Biomass-degrading enzymes are powerful tools in the new bioindustry where the production of different biofuels and biochemicals depends more and more on these biocatalyzers to guarantee a sustainable process. The economic use of enzymes in the context of biorefineries depends not only on the biocatalyzers' costs, but also on the new defined and studied strategies of SHF or SSF processes. The development of new and highly efficient enzymes through advanced technologies is also the object of numerous research and investments by the academy and bioenergy industrial sector. One possible alternative to ameliorate enzyme performance is enzyme immobilization through different techniques, which allow the reuse of the biocatalysts in different cycles, which is an economic strategy. New enzymes can also be produced by genetic engineered microorganism that are designed especially for each process purpose. These approaches contribute to the bioeconomy and definition of sustainable processes.

References

Agrawal, K., Bhardwaj, N., Kumar, B., Chaturvedi, V., Verma, P., 2019. Process optimization, purification and characterization of alkaline stable white laccase from Myrothecium verrucaria ITCC-8447 and its application in delignification of agroresidues. Int. J. Biol. Macromol. 125, 1042–1055. https://doi.org/10.1016/j.ijbiomac.2018.12.108.

Aljeboree, A.M., Alshirifi, A.N., Alkaim, A.F., 2017. Kinetics and equilibrium study for the adsorption of textile dyes on coconut shell activated carbon. Arab. J. Chem. 10, S3381–S3393. https://doi.org/10.1016/j.arabjc.2014.01.020.

Álvarez, C., González, A., Negro, M.J., Ballesteros, I., Oliva, J.M., Sáez, F., 2017. Optimized use of hemicellulose within a biorefinery for processing high value-added xylooligosaccharides. Ind. Crops Prod. 99, 41–48. https://doi.org/10.1016/j.indcrop.2017.01.034.

Austrade. Biotechnology, 2021. A powerhouse for science and innovation. https://www.austrade.gov.au/international/buy/australian-industry-capabilities/biotechnology (Accessed on 09 May 2021).

Bajpai, P., 2009. Xylanases. Encyclopedia of Microbiology. Editor(s): Schaechter, M. Academic Press, 3rd Edition, 600–612. https://doi.org/10.1016/B978-012373944-5.00165-6.

Bardhan, S.K., Gupta, S., Gorman, M.E., Haider, M.A., 2015. Biorenewable chemicals: feedstocks, technologies and the conflict with food production. Renew. Sustain. Energy Rev. 51, 506–520. https://doi.org/10.1016/j.rser.2015.06.013.

Bhatia, S.K., Jagtap, S.S., Bedekar, A.A., Bhatia, R.K., Rajendran, K., Pugazhendhi, A., Rao, C.V., Atabani, A.E., Kumar, G., Yang, Y.H., 2021. Renewable biohydrogen production from lignocellulosic biomass using fermentation and integration of systems with other energy generation technologies. Sci. Total Environ. 765, 144429. https://doi.org/10.1016/j.scitotenv.2020.144429.

Blibech, M., Ellouz Ghorbel, R., Chaari, F., Dammak, I., Bhiri, F., Neifar, M., Ellouz Chaabouni, S., 2011. Improved mannanase production from *Penicillium occitanis* by fed-batch fermentation using acacia seeds. ISRN Microbiol 2011, 938347. https://doi.org/10.5402/2011/938347.

Bonechi, C., Consumi, M., Donati, A., Leone, G., Magnani, A., Tamasi, G., Rossi, C., 2017. Biomass: An overview, Bioenergy Systems for the Future: Prospects for Biofuels and Biohydrogen. Elsevier Ltd, Cham. https://doi.org/10.1016/B978-0-08-101031-0.00001-6.

Boruah, P., Sarmah, P., Das, P.K., Goswami, T., 2019. Exploring the lignolytic potential of a new laccase producing strain Kocuria sp. PBS-1 and its application in bamboo pulp bleaching. Int. Biodeterior. Biodegrad. 143, 104726. https://doi.org/10.1016/j.ibiod.2019.104726.

Bu, J., Yan, X., Wang, Y.T., Zhu, S.M., Zhu, M.J., 2019. Co-production of high-gravity bioethanol and succinic acid from potassium peroxymonosulfate and deacetylation sequentially pretreated sugarcane bagasse by simultaneous saccharification and co-fermentation. Energy Convers. Manage. 186, 131–139. https://doi.org/10.1016/j.enconman.2019.02.038.

Bychkov, A., Podgorbunskikh, E., Bychkova, E., 2019. Current achievements in the mechanically pretreated conversion of plant biomass. Biotechnol. Bioeng. 116 (5), 1231–1244. https://doi.org/10.1002/bit.26925.

Chandel, A.K., Albarelli, J.Q., Santos, D.T., Chundawat, S.P., Puri, M., Meireles, M.A.A., 2019. Comparative analysis of key technologies for cellulosic ethanol production from Brazilian sugarcane bagasse at a commercial scale. Biofuels, Bioprod. Biorefining 13, 994–1014. https://doi.org/10.1002/bbb.1990.

Chavan, S., Gaikwad, A., 2021. Optimization of enzymatic hydrolysis of bamboo biomass for enhanced saccharification of cellulose through Taguchi orthogonal design. J. Environ. Chem. Eng. 9, 104807. https://doi.org/10.1016/j.jece.2020.104807.

Chen, H., Chen, B., Su, Z., Wang, K., Wang, B., Wang, Y., Si, Z., Wu, Y., Cai, D., Qin, P., 2020. Efficient lactic acid production from cassava bagasse by mixed culture of Bacillus coagulans and lactobacillus rhamnosus using stepwise pH controlled simultaneous saccharification and co-fermentation. Ind. Crops Prod. 146, 112175. https://doi.org/10.1016/j.indcrop.2020.112175.

Chen, H., Fu, X., 2016. Industrial technologies for bioethanol production from lignocellulosic biomass. Renew. Sustain. Energy Rev. 57, 468–478. https://doi.org/10.1016/j.rser.2015.12.069.

Chen, H.G., Zhang, Y.H.P., 2015. New biorefineries and sustainable agriculture: increased food, biofuels, and ecosystem security. Renew. Sustain. Energy Rev. 47, 117–132. https://doi.org/10.1016/j.rser.2015.02.048.

Chen, J., Wang, X., Zhang, B., Yang, Y., Song, Y., Zhang, F., Liu, B., Zhou, Y., Yi, Y., Shan, Y., Lü, X., 2021. Integrating enzymatic hydrolysis into subcritical water pretreatment optimization for bioethanol production from wheat straw. Sci. Total Environ. 770, 145321. https://doi.org/10.1016/j.scitotenv.2021.145321.

Chen, X., Cao, X., Sun, S., Yuan, T., Shi, Q., Zheng, L., Sun, R., 2018. Evaluating the production of monosaccharides and xylooligosaccharides from the pre-hydrolysis liquor of kraft pulping process by acid and enzymatic hydrolysis. Ind. Crops Prod. 124, 906–911. https://doi.org/10.1016/j.indcrop.2018.08.071.

Cho, C.H., Hatsu, M., Takamizawa, K., 2002. The production of D-xylose by enzymatic hydrolysis of agricultural wastes. Water Sci. Technol. 45, 97–102. https://doi.org/10.2166/wst.2002.0414.

Cho, E.J., Trinh, L.T.P., Song, Y., Lee, Y.G., Bae, H.J., 2020. Bioconversion of biomass waste into high value chemicals. Bioresour. Technol. 298, 122386. https://doi.org/10.1016/j.biortech.2019.122386.

Choudhary, J., Singh, S., Nain, L., 2016. Thermotolerant fermenting yeasts for simultaneous saccharification fermentation of lignocellulosic biomass. Electron. J. Biotechnol. 21, 82–92. https://doi.org/10.1016/j.ejbt.2016.02.007.

Clarkson, K.A., Reboli, M.T., Huitink, J.A., Teunissen, P.J.M., Chotani, G.K., Shetty, J.K., 2013. Simultaneous saccharification and co-fermentation of glucoamylase-expressing fungal strains with an ethanologen to produce alcohol from corn. WO 2015/164058 A1.

Clauser, N.M., González, G., Mendieta, C.M., Kruyeniski, J., Area, M.C., Vallejos, M.E., 2021. Biomass waste as sustainable raw material for energy and fuels. Sustainability 13, 794. https://doi.org/10.3390/su13020794.

Da Costa Sousa, L., Jin, M., Chundawat, S.P.S., Bokade, V., Tang, X., Azarpira, A., Lu, F., Avci, U., Humpula, J., Uppugundla, N., Gunawan, C., Pattathil, S., Cheh, A.M., Kothari, N., Kumar, R., Ralph, J., Hahn, M.G., Wyman, C.E., Singh, S., Simmons, B.A., Dale, B.E., Balan, V., 2016. Next-generation ammonia pretreatment enhances cellulosic biofuel production. Energy Environ. Sci. 9, 1215–1223. https://doi.org/10.1039/c5ee03051j.

Dahiya, S., Kumar, A., Singh, B., 2020. Enhanced endoxylanase production by *Myceliophthora thermophila* using rice straw and its synergism with phytase in improving nutrition. Process Biochem. 94, 235–242. https://doi.org/10.1016/j.procbio.2020.04.032.

Daoutidis, P., Kelloway, A., Marvin, W.A., Rangarajan, S., Torres, A.I., 2013. Process systems engineering for biorefineries: new research vistas. Curr. Opin. Chem. Eng. 2, 442–447. https://doi.org/10.1016/j.coche.2013.09.006.

Dey, P., Rangarajan, V., Nayak, J., Das, D.B., Wood, S.B., 2021. An improved enzymatic pre-hydrolysis strategy for efficient bioconversion of industrial pulp and paper sludge waste to bioethanol using a semi-simultaneous saccharification and fermentation process. Fuel 294, 120581. https://doi.org/10.1016/j.fuel.2021.120581.

Díaz, G.V., Coniglio, R.O., Velazquez, J.E., Zapata, P.D., Villalba, L., Fonseca, M.I., 2019. Adding value to lignocellulosic wastes via their use for endoxylanase production by Aspergillus fungi. Mycologia 111, 195–205. https://doi.org/10.1080/00275514.2018.1556557.

Dimarogona, M., Topakas, E., 2016. Regulation and heterologous expression of lignocellulosic enzymes in aspergillus. In: Gupta, V.K. (Ed.), New and Future Developments in Microbial Biotechnology and Bioengineering: Aspergillus System Properties and Applications. Elsevier B.V, Amsterdam, pp. 171–190. https://doi.org/10.1016/B978-0-444-63505-1.00012-9.

Dükkancı, M., Vinatoru, M., Mason, T.J., 2014. The sonochemical decolourisation of textile azo dye Orange II: effects of Fenton type reagents and UV light. Ultrason. Sonochem. 21, 846–853. https://doi.org/10.1016/j.ultsonch.2013.08.020.

Ebrahimian, F., Karimi, K., Kumar, R., 2020. Sustainable biofuels and bioplastic production from the organic fraction of municipal solid waste. Waste Manage. 116, 40–48. https://doi.org/10.1016/j.wasman.2020.07.049.

Fernández-Fernández, M., Sanromán, M.Á., Moldes, D., 2013. Recent developments and applications of immobilized laccase. Biotechnol. Adv. 31, 1808–1825. https://doi.org/10.1016/j.biotechadv.2012.02.013.

Ferreira, J.A., Taherzadeh, M.J., 2020. Improving the economy of lignocellulose-based biorefineries with organosolv pretreatment. Bioresour. Technol. 299, 122695. https://doi.org/10.1016/j.biortech.2019.122695.

Fliedrová, B., Gerstorferová, D., Křen, V., Weignerová, L., 2012. Production of Aspergillus niger β-mannosidase in *Pichia pastoris*. Protein Expr. Purif. 85, 159–164. https://doi.org/10.1016/j.pep.2012.07.012.

Flores-Gómez, C.A., Escamilla Silva, E.M., Zhong, C., Dale, B.E., Da Costa Sousa, L., Balan, V., 2018. Conversion of lignocellulosic agave residues into liquid biofuels using an AFEX-based biorefinery. Biotechnol. Biofuels 11, 1–18. https://doi.org/10.1186/s13068-017-0995-6.

Fonseca, M.I., Fariña, J.I., Castrillo, M.L., Rodríguez, M.D., Nuñez, C.E., Villalba, L.L., Zapata, P.D., 2014. Biopulping of wood chips with *Phlebia brevispora* BAFC 633 reduces lignin content and improves pulp quality. Int. Biodeterior. Biodegrad. 90, 29–35. https://doi.org/10.1016/j.ibiod.2013.11.018.

Gatt, E., Khatri, V., Bley, J., Barnabé, S., Vandenbossche, V., Beauregard, M., 2019. Enzymatic hydrolysis of corn crop residues with high solid loadings: new insights into the impact of bioextrusion on biomass deconstruction using carbohydrate-binding modules. Bioresour. Technol. 282, 398–406. https://doi.org/10.1016/j.biortech.2019.03.045.

de Godoy, C.M., Machado, D.L., da Costa, A.C., 2019. Batch and fed-batch enzymatic hydrolysis of pretreated sugarcane bagasse—assays and modeling. Fuel 253, 392–399. https://doi.org/10.1016/j.fuel.2019.05.038.

Gomes, M.G., Gurgel, L.V.A., Baffi, M.A., Pasquini, D., 2020. Pretreatment of sugarcane bagasse using citric acid and its use in enzymatic hydrolysis. Renew. Energy 157, 332–341. https://doi.org/10.1016/j.renene.2020.05.002.

Guimarães, L.R.C., Woiciechowski, A.L., Karp, S.G., Coral, J.D., Zandoná Filho, A., Soccol, C.R., 2017. Laccases. In: Pandey, A., Negi, S., Soccol, C.R. (Eds.), Laccases. Current Developments in Biotechnology and Bioengineering: Production, Isolation and Purification of Industrial Products, 199–216. https://doi.org/10.1016/B978-0-444-63662-1.00009-9.

Guo, H., Chang, Y., Lee, D.J., 2018. Enzymatic saccharification of lignocellulosic biorefinery: research focuses. Bioresour. Technol. 252, 198–215. https://doi.org/10.1016/j.biortech.2017.12.062.

Gutiérrez-García, A.K., Alvarez-Guzmán, C.L., De Leon-Rodriguez, A., 2020. Autodisplay of alpha amylase from Bacillus megaterium in E. coli for the bioconversion of starch into hydrogen, ethanol and succinic acid. Enzyme Microb. Technol. 134, 109477. https://doi.org/10.1016/j.enzmictec.2019.109477.

Hazeena, S.H., Nair Salini, C., Sindhu, R., Pandey, A., Binod, P., 2019. Simultaneous saccharification and fermentation of oil palm front for the production of 2,3-butanediol. Bioresour. Technol. 278, 145–149. https://doi.org/10.1016/j.biortech.2019.01.042.

Hernández-Beltrán, J.U., Hernández-Escoto, H., 2018. Enzymatic hydrolysis of biomass at high-solids loadings through fed-batch operation. Biomass Bioenergy 119, 191–197. https://doi.org/10.1016/j.biombioe.2018.09.020.

Hernández-Guzmán, A., Navarro-Gutiérrez, I.M., Meléndez-Hernández, P.A., Hernández-Beltrán, J.U., Hernández-Escoto, H., 2020. Enhancement of alkaline-oxidative delignification of wheat straw by semi-batch operation in a stirred tank reactor. Bioresour. Technol. 312, 123589. https://doi.org/10.1016/j.biortech.2020.123589.

Hou, W., Bao, J., 2018. Simultaneous saccharification and aerobic fermentation of high titer cellulosic citric acid by filamentous fungus *Aspergillus niger*. Bioresour. Technol. 253, 72–78. https://doi.org/10.1016/j.biortech.2018.01.011.

Hu, J., Cao, W., Guo, L., 2021. Directly convert lignocellulosic biomass to H_2 without pretreatment and added cellulase by two-stage fermentation in semi-continuous modes. Renew. Energy 170, 866–874. https://doi.org/10.1016/j.renene.2021.02.062.

Jin, Q., Yang, L., Poe, N., Huang, H., 2018. Integrated processing of plant-derived waste to produce value-added products based on the biorefinery concept. Trends Food Sci. Technol. 74, 119–131. https://doi.org/10.1016/j.tifs.2018.02.014.

Jun, S., Gahin, C., 2011. Saccharifying fermentation system of cellulose-based biomass. JP 2010015868 A.

Juturu, V., Wu, J.C., 2014. Microbial cellulases: engineering, production and applications. Renew. Sustain. Energy Rev. 33, 188–203. https://doi.org/10.1016/j.rser.2014.01.077.

Khaire, K.C., Moholkar, V.S., Goyal, A., 2021. Bioconversion of sugarcane tops to bioethanol and other value added products: an overview. Mater. Sci. Energy Technol. 4, 54–68. https://doi.org/10.1016/j.mset.2020.12.004.

Klein-Marcuschamer, D., Oleskowicz-Popiel, P., Simmons, B.A., Blanch, H.W., 2012. The challenge of enzyme cost in the production of lignocellulosic biofuels. Biotechnol. Bioeng. 109, 1083–1087. https://doi.org/10.1002/bit.24370.

Kuglarz, M., Alvarado-Morales, M., Dąbkowska, K., Angelidaki, I., 2018. Integrated production of cellulosic bioethanol and succinic acid from rapeseed straw after dilute-acid pretreatment. Bioresour. Technol. 265, 191–199. https://doi.org/10.1016/j.biortech.2018.05.099.

Kumar, B., Bhardwaj, N., Agrawal, K., Chaturvedi, V., Verma, P., 2020. Current perspective on pretreatment technologies using lignocellulosic biomass: an emerging biorefinery concept. Fuel Process. Technol. 199, 106244. https://doi.org/10.1016/j.fuproc.2019.106244.

Kunitake, E., Kobayashi, T., 2017. Conservation and diversity of the regulators of cellulolytic enzyme genes in Ascomycete fungi. Curr. Genet. 63, 951–958. https://doi.org/10.1007/s00294-017-0695-6.

Lai, C., Jia, Y., Wang, J., Wang, R., Zhang, Q., Chen, L., Shi, H., Huang, C., Li, X., Yong, Q., 2019. Co-production of xylooligosaccharides and fermentable sugars from poplar through acetic acid pretreatment followed by poly (ethylene glycol) ether assisted alkali treatment. Bioresour. Technol. 288, 121569. https://doi.org/10.1016/j.biortech.2019.121569.

Lee, I., Yu, J.H., 2020. The production of fermentable sugar and bioethanol from acacia wood by optimizing dilute sulfuric acid pretreatment and post treatment. Fuel 275, 117943. https://doi.org/10.1016/j.fuel.2020.117943.

Lee, J.W., Yook, S., Koh, H., Rao, C.V., Jin, Y.S., 2021. Engineering xylose metabolism in yeasts to produce biofuels and chemicals. Curr. Opin. Biotechnol. 67, 15–25. https://doi.org/10.1016/j.copbio.2020.10.012.

Li, X., Li, S., Liang, X., McClements, D.J., Liu, X., Liu, F., 2020. Applications of oxidases in modification of food molecules and colloidal systems: laccase, peroxidase and tyrosinase. Trends Food Sci. Technol. 103, 78–93. https://doi.org/10.1016/j.tifs.2020.06.014.

Libardi, N., Soccol, C.R., Tanobe, V.O.A., Vandenberghe, L.P., 2020. Definition of liquid and powder cellulase formulations using domestic wastewater in bubble column reactor. Appl. Biochem. Biotechnol. 190, 113–128. https://doi.org/10.1007/s12010-019-03075-1.

Lin, W., Chen, D., Yong, Q., Huang, C., Huang, S., 2019. Improving enzymatic hydrolysis of acid-pretreated bamboo residues using amphiphilic surfactant derived from dehydroabietic acid. Bioresour. Technol. 293, 122055. https://doi.org/10.1016/j.biortech.2019.122055.

Liu, Y.-J., Li, B., Feng, Y., Cui, Q., 2020. Consolidated bio-saccharification: leading lignocellulose bioconversion into the real world. Biotechnol. Adv. 40, 107535. https://doi.org/10.1016/j.biotechadv.2020.107535.

Liu, Z.H., Olson, M.L., Shinde, S., Wang, X., Hao, N., Yoo, C.G., Bhagia, S., Dunlap, J.R., Pu, Y., Kao, K.C., Ragauskas, A.J., Jin, M., Yuan, J.S, 2017. Synergistic maximization of the carbohydrate output and lignin processability by combinatorial pretreatment. Green Chem. 19, 4939–4955. https://doi.org/10.1039/c7gc02057k.

López-Gutiérrez, I., Razo-Flores, E., Méndez-Acosta, H.O., et al., 2020. Optimization by response surface methodology of the enzymatic hydrolysis of non-pretreated agave bagasse with binary mixtures of commercial enzymatic preparations. Biomass Conv. Bioref. https://doi.org/10.1007/s13399-020-00698-x.

Lynd, L.R., Weimer, P.J., van Zyl, W.H., Pretorius, I.S., 2002. Microbial cellulose utilization: fundamentals and biotechnology. Microbiol. Mol. Biol. Rev. 66, 506–577. https://doi.org/10.1128/MMBR.66.3.506.

Macedo, G. A., Macedo, J. A., De Queirós, L. D., 2016. Biotransformation method for transforming phenolic compounds from soy extract into equol and bioactive isoflavones by fermentation and/or enzymatic application, thus obtained composition and use. WO 2016/201536 A1.

Magalhães, A.I., de Carvalho, J.C., Medina, J.D.C., Soccol, C.R., 2017. Downstream process development in biotechnological itaconic acid manufacturing. Appl. Microbiol. Biotechnol. 101, 1–12. https://doi.org/10.1007/s00253-016-7972-z.

Maitan-Alfenas, G.P., Visser, E.M., Guimarães, V., 2015. Enzymatic hydrolysis of lignocellulosic biomass: converting food waste in valuable products. Curr. Opin. Food Sci. 1, 44–49. https://doi.org/10.1016/j.cofs.2014.10.001.

Mander, P., Choi, Y.H., G.C., P., Choi, Y.S., Hong, J.H., Cho, S.S., Yoo, J.C., 2014. Biochemical characterization of xylanase produced from *Streptomyces* sp. CS624 using an agro residue substrate. Process Biochem. 49, 451–456. https://doi.org/10.1016/j.procbio.2013.12.011.

ManishaYadav, S.K., 2017. Technological advances and applications of hydrolytic enzymes for valorization of lignocellulosic biomass. Bioresour. Technol. 245, 1727–1739. https://doi.org/10.1016/j.biortech.2017.05.066.

Marques, S., Moreno, A.D., Ballesteros, M., Gírio, F., 2017. Starch biomass for biofuels, biomaterials, and chemicals. In: Vaz, S. Jr. (Ed.), Starch biomass for biofuels, biomaterials, and chemicals. Biomass and Green Chemistry, 69–94. https://doi.org/10.1007/978-3-319-66736-2_4.

Martín-Sampedro, R., Eugenio, M.E., Villar, J.C., 2012. Effect of steam explosion and enzymatic pre-treatments on pulping and bleaching of *Hesperaloe funifera*. Bioresour. Technol. 111, 460–467. https://doi.org/10.1016/j.biortech.2012.02.024.

Martín, C., Wei, M., Xiong, S., Jönsson, L.J., 2017. Enhancing saccharification of cassava stems by starch hydrolysis prior to pretreatment. Ind. Crops Prod. 97, 21–31. https://doi.org/10.1016/j.indcrop.2016.11.067.

Martínez-Avila, O., Llimós, J., Ponsá, S., 2021. Integrated solid-state enzymatic hydrolysis and solid-state fermentation for producing sustainable polyhydroxyalkanoates from low-cost agro-industrial residues. Food Bioprod. Process. 126, 334–344. https://doi.org/10.1016/j.fbp.2021.01.015.

Maslova, O., Stepanov, N., Senko, O., Efremenko, E., 2019. Production of various organic acids from different renewable sources by immobilized cells in the regimes of separate hydrolysis and fermentation (SHF) and simultaneous saccharification and fermentation (SFF). Bioresour. Technol. 272, 1–9. https://doi.org/10.1016/j.biortech.2018.09.143.

Mohd Azhar, S.H., Abdulla, R., Jambo, S.A., Marbawi, H., Gansau, J.A., Mohd Faik, A.A., Rodrigues, K.F., 2017. Yeasts in sustainable bioethanol production: a review. Biochem. Biophys. Reports 10, 52–61. https://doi.org/10.1016/j.bbrep.2017.03.003.

Morais, R.R., Pascoal, A.M., Pereira-Júnior, M.A., Batista, K.A., Rodriguez, A.G., Fernandes, K.F., 2019. Bioethanol production from *Solanum lycocarpum* starch: a sustainable non-food energy source for biofuels. Renew. Energy 140, 361–366. https://doi.org/10.1016/j.renene.2019.02.056.

Niebisch, C.H., Foltran, C., Campos, R., Domingues, S., Paba, J., 2014. Assessment of *Heteroporus biennis* secretion extracts for decolorization of textile dyes. Int. Biodeterior. Biodegradation 88, 20–28. https://doi.org/10.1016/j.ibiod.2013.11.013.

Nishimura, H., Tan, L., Kira, N., Tomiyama, S., Yamada, K., Sun, Z.Y., Tang, Y.Q., Morimura, S., Kida, K., 2017. Production of ethanol from a mixture of waste paper and kitchen waste via a process of successive liquefaction, presaccharification, and simultaneous saccharification and fermentation. Waste Manage. 67, 86–94. https://doi.org/10.1016/j.wasman.2017.04.030.

Norizan, N.A.B.M., Halim, M., Tan, J.S., Abbasiliasi, S., Sahri, M.M., Othman, F., Ari, A.Bin, 2020. Enhancement of β-mannanase production by *Bacillus subtilis* ATCC11774 through optimization of medium composition. Molecules 25, 1–15. https://doi.org/10.3390/molecules25153516.

Olofsson, K., Bertilsson, M., Lidén, G., 2008. A short review on SSF—an interesting process option for ethanol production from lignocellulosic feedstocks. Biotechnol. Biofuels 1, 7. https://doi.org/10.1186/1754-6834-1-7.

Paixão, S.M., Arez, B.F., Roseiro, J.C., Alves, L., 2016. Simultaneously saccharification and fermentation approach as a tool for enhanced fossil fuels biodesulfurization. J. Environ. Manage. 182, 397–405. https://doi.org/10.1016/j.jenvman.2016.07.099.

Patel, H., Chapla, D., Shah, A., 2017. Bioconversion of pretreated sugarcane bagasse using enzymatic and acid followed by enzymatic hydrolysis approaches for bioethanol production. Renew. Energy 109, 323–331. https://doi.org/10.1016/j.renene.2017.03.057.

Payne, C.M., Knott, B.C., Mayes, H.B., Hansson, H., Himmel, M.E., Sandgren, M., Ståhlberg, J., Beckham, G.T., 2015. Fungal cellulases. Chem. Rev. 115, 1308–1448. https://doi.org/10.1021/cr500351c.

Pimienta, J.A.P., Papa, G., Rodriguez, A., Barcelos, C.A., Liang, L., Stavila, V., Sanchez, A., Gladden, J.M., Simmons, B.A., 2019. Pilot-scale hydrothermal pretreatment and optimized saccharification enables bisabolene production from multiple feedstocks. Green Chem. 21, 3152–3164. https://doi.org/10.1039/c9gc00323a.

Pino, M.S., Rodríguez-Jasso, R.M., Michelin, M., Flores-Gallegos, A.C., Morales-Rodriguez, R., Teixeira, J.A., Ruiz, H.A., 2018. Bioreactor design for enzymatic hydrolysis of biomass under the biorefinery concept. Chem. Eng. J. 347, 119–136. https://doi.org/10.1016/j.cej.2018.04.057.

Pradyawong, S., Juneja, A., Sadiq, M., Noomhorm, A., Singh, V., 2018. Comparison of cassava starch corn as a feedstock for bioethanol production with. Energies 11, 3476. https://doi.org/10.3390/en11123476.

Pratto, B., dos Santos-Rocha, M.S.R., Longati, A.A., de Sousa Júnior, R., Cruz, A.J.G., 2020. Experimental optimization and techno-economic analysis of bioethanol production by simultaneous saccharification and fermentation process using sugarcane straw. Bioresour. Technol. 297, 122494. https://doi.org/10.1016/j.biortech.2019.122494.

Qian, M., Lei, H., Villota, E., Zhao, Y., Wang, C., Huo, E., Zhang, Q., Mateo, W., Lin, X., 2021. High yield production of nanocrystalline cellulose by microwave-assisted dilute-acid pretreatment combined with enzymatic hydrolysis. Chem. Eng. Process. Process Intensif. 160, 108292. https://doi.org/10.1016/j.cep.2020.108292.

Qin, Z., Duns, G.J., Pan, T., Xin, F., 2018. Consolidated processing of biobutanol production from food wastes by solventogenic *Clostridium* sp. strain HN4. Bioresour. Technol. 264, 148–153. https://doi.org/10.1016/j.biortech.2018.05.076.

Quanguo, Z., Shengnan, Z., Jia, Z., Chenxi, X., Yameng, L., Zhiping, Z., Huan, Z., Yang, Z., 2017. Simultaneous saccharification and fermentation hydrogen-producing reactor accompanied with enzyme recycling and experimental method thereof. CN 107488578 A.

Rastogi, M., Shrivastava, S., 2017. Recent advances in second generation bioethanol production: an insight to pretreatment, saccharification and fermentation processes. Renew. Sustain. Energy Rev. 80, 330–340. https://doi.org/10.1016/j.rser.2017.05.225.

Ravindran, R., Jaiswal, A.K., 2016. A comprehensive review on pre-treatment strategy for lignocellulosic food industry waste: challenges and opportunities. Bioresour. Technol. 199, 92–102. https://doi.org/10.1016/j.biortech.2015.07.106.

Rawat, R., Srivastava, N., Chadha, B.S., Oberoi, H.S., 2014. Generating fermentable sugars from rice straw using functionally active cellulolytic enzymes from *Aspergillus niger* HO. Energy Fuels 28, 5067–5075. https://doi.org/10.1021/ef500891g.

Ren, X., Teng, L., Liu, J., Cheng, Q., Cui, Y., Meng, M., Lu, J., 2014. Method for producing nisin and lactic acid through simultaneous saccharification and fermentation and application of nisin and lactic acid in feed. Patent CN 103993043.

Rivera-Hoyos, C.M., Morales-Álvarez, E.D., Poutou-Piñales, R.A., Pedroza-Rodríguez, A.M., Rodríguez-Vázquez, R., Delgado-Boada, J.M., 2013. Fungal laccases. Fungal Biol. Rev. 27, 67–82. https://doi.org/10.1016/j.fbr.2013.07.001.

Ruiz, H.A., Conrad, M., Sun, S.N., Sanchez, A., Rocha, G.J.M., Romaní, A., Castro, E., Torres, A., Rodríguez-Jasso, R.M., Andrade, L.P., Smirnova, I., Sun, R.C., Meyer, A.S., 2020. Engineering aspects of hydrothermal pretreatment: from batch to continuous operation, scale-up and pilot reactor under biorefinery concept. Bioresour. Technol. 299, 122685. https://doi.org/10.1016/j.biortech.2019.122685.

Sadaf, A., Kumar, S., Nain, L., Khare, S.K., 2021. Bread waste to lactic acid: applicability of simultaneous saccharification and solid state fermentation. Biocatal. Agric. Biotechnol. 32, 101934. https://doi.org/10.1016/j.bcab.2021.101934.

Saldarriaga-Hernández, S., Velasco-Ayala, C., Leal-Isla Flores, P., de Jesús Rostro-Alanis, M., Parra-Saldivar, R., Iqbal, H.M.N., Carrillo-Nieves, D., 2020. Biotransformation of lignocellulosic biomass into industrially relevant products with the aid of fungi-derived lignocellulolytic enzymes. Int. J. Biol. Macromol. 161, 1099–1116. https://doi.org/10.1016/j.ijbiomac.2020.06.047.

Santhi, V.S., Gupta, A., Saranya, S., Jebakumar, S.R.D., 2014. A novel marine bacterium Isoptericola sp. JS-C42 with the ability to saccharifying the plant biomasses for the aid in cellulosic ethanol production. Biotechnol. Rep. 1–2, 8–14. https://doi.org/10.1016/j.btre.2014.05.002.

Saratale, R.G., Cho, S.K., Saratale, G.D., Ghodake, G.S., Bharagava, R.N., Kim, D.S., Nair, S., Shin, H.S., 2021. Efficient bioconversion of sugarcane bagasse into polyhydroxybutyrate (PHB) by *Lysinibacillus* sp. and its characterization. Bioresour. Technol. 324, 124673. https://doi.org/10.1016/j.biortech.2021.124673.

Saulnier, B.K., Phongpreecha, T., Singh, S.K., Hodge, D.B., 2020. Impact of dilute acid pretreatment conditions on p-coumarate removal in diverse maize lines. Bioresour. Technol. 314, 123750. https://doi.org/10.1016/j.biortech.2020.123750.

Sawisit, A., Jampatesh, S., Jantama, S.S., Jantama, K., 2018. Optimization of sodium hydroxide pretreatment and enzyme loading for efficient hydrolysis of rice straw to improve succinate production by metabolically engineered *Escherichia coli* KJ122 under simultaneous saccharification and fermentation. Bioresour. Technol. 260, 348–356. https://doi.org/10.1016/j.biortech.2018.03.107.

Sharma, A., Thakur, V.V., Shrivastava, A., Jain, R.K., Mathur, R.M., Gupta, R., Kuhad, R.C., 2014. Xylanase and laccase based enzymatic kraft pulp bleaching reduces adsorbable organic halogen (AOX) in bleach effluents: a pilot scale study. Bioresour. Technol. 169, 96–102. https://doi.org/10.1016/j.biortech.2014.06.066.

Sharma, K., Swain, R., Singh, H., Mathur, S., Gupta, P., Tuli, D., Puri, K., Ramakumar, S.S.V., 2019. Two-stage simultaneous saccharification and co-fermentation for producing ethanol from lignocellulose. EP 3 543 343 B1.

Sheng, Y., Lam, S.S., Wu, Y., Ge, S., Wu, J., Cai, L., Huang, Z., Le, Q.Van, Sonne, C., Xia, C., 2021. Enzymatic conversion of pretreated lignocellulosic biomass: a review on influence of structural changes of lignin. Bioresour. Technol. 324, 124631. https://doi.org/10.1016/j.biortech.2020.124631.

Shleev, S., Persson, P., Shumakovich, G., Mazhugo, Y., Yaropolov, A., Ruzgas, T., Gorton, L., 2006. Interaction of fungal laccases and laccase-mediator systems with lignin. Enzyme Microb. Technol. 39, 841–847. https://doi.org/10.1016/j.enzmictec.2006.01.010.

Sindhu, R., Binod, P., Pandey, A., 2016. Biological pretreatment of lignocellulosic biomass—an overview. Bioresour. Technol. 199, 76–82. https://doi.org/10.1016/j.biortech.2015.08.030.

Singhania, R., Adsul, M., Pandey, AK, Patel, A., 2017. Cellulases. In: Pandey, A, Negi, S., Soccol, C. (Eds.), Current Developments in Biotechnology and Bioengineering -Production, Isolation and Purification of Industrial Products. Elsevier, Amsterdam, pp. 73–97.

Siqueira, J.G.W., Rodrigues, C., Vandenberghe, L.P., Woiciechowski, A.L., Soccol, C.R., 2020. Current advances in on-site cellulase production and application on lignocellulosic biomass conversion to biofuels: a review. Biomass Bioenergy. https://doi.org/10.1016/j.biombioe.2019.105419.

Sitaraman, H., Danes, N., Lischeske, J.J., Stickel, J.J., Sprague, M.A., 2019. Coupled CFD and chemical-kinetics simulations of cellulosic-biomass enzymatic hydrolysis: mathematical-model development and validation. Chem. Eng. Sci. 206, 348–360. https://doi.org/10.1016/j.ces.2019.05.025.

Sjulander, N., Kikas, T., 2020. Origin, impact and control of lignocellulosic inhibitors in bioethanol production—a review. Energies 13 (18), 4751. https://doi.org/10.3390/en13184751.

Souza Filho, P.F., de Ribeiro, V.T., dos Santos, E.S., de Macedo, G.R., 2016. Simultaneous saccharification and fermentation of cactus pear biomass-evaluation of using different pretreatments. Ind. Crops Prod. 89, 425–433. https://doi.org/10.1016/j.indcrop.2016.05.028.

Stickel, J.J., Adhikari, B., Sievers, D.A., Pellegrino, J., 2018. Continuous enzymatic hydrolysis of lignocellulosic biomass in a membrane-reactor system. J. Chem. Technol. Biotechnol. 93, 2181–2190. https://doi.org/10.1002/jctb.5559.

Su, T., Zhao, D., Khodadadi, M., Len, C., 2020. Lignocellulosic biomass for bioethanol: recent advances, technology trends, and barriers to industrial development. Curr. Opin. Green Sustain. Chem. 24, 56–60. https://doi.org/10.1016/j.cogsc.2020.04.005.

Sudiyani, Y., Dahnum, D., Burhani, D., Putri, A.M.H., 2019. Evaluation and comparison between simultaneous saccharification and fermentation and separated hydrolysis and fermentation process. In: Basile, A., Dalena, F. (Eds.), Second and Third Generation of Feedstocks. Elsevier Inc., Amsterdam, pp. 273–290. https://doi.org/10.1016/B978-0-12-815162-4.00010-0.

Sun, Shaoni, Sun, Shaolong, Cao, X., Sun, R., 2016. The role of pretreatment in improving the enzymatic hydrolysis of lignocellulosic materials. Bioresour. Technol. 199, 49–58. https://doi.org/10.1016/j.biortech.2015.08.061.

Szambelan, K., Nowak, J., Szwengiel, A., Jeleń, H., 2020. Quantitative and qualitative analysis of volatile compounds in sorghum distillates obtained under various hydrolysis and fermentation conditions. Ind. Crops Prod. 155, 113109. https://doi.org/10.1016/j.indcrop.2020.112782.

Thrane, M., Paulsen, P.V., Orcutt, M.W., Krieger, T.M., 2017. Soy Protein: Impacts, Production, and Applications, Sustainable Protein Sources. Elsevier Inc, Amsterdam. https://doi.org/10.1016/B978-0-12-802778-3.00002-0.

Torres-duarte, C., Roman, R., Tinoco, R., Vazquez-duhalt, R., 2009. Halogenated pesticide transformation by a laccase-mediator system. Chemosphere 77, 687–692. https://doi.org/10.1016/j.chemosphere.2009.07.039.

Usmani, Z., Sharma, M., Gupta, P., Karpichev, Y., Gathergood, N., Bhat, R., Gupta, V.K., 2020. Ionic liquid based pretreatment of lignocellulosic biomass for enhanced bioconversion. Bioresour. Technol. 304, 123003. https://doi.org/10.1016/j.biortech.2020.123003.

Valladares-Diestra, K.K., Porto de Souza Vandenberghe, L., Zevallos Torres, L.A., Nishida, V.S., Zandoná Filho, A., Woiciechowski, A.L., Soccol, C.R., 2020. Imidazole green solvent pre-treatment as a strategy for second-generation bioethanol production from sugarcane bagasse. Chem. Eng. J. 420 (2), 127708. https://doi.org/10.1016/j.cej.2020.127708.

Valles, A., Álvarez-Hornos, F.J., Martínez-Soria, V., Marzal, P., Gabaldón, C., 2020. Comparison of simultaneous saccharification and fermentation and separate hydrolysis and fermentation processes for butanol production from rice straw. Fuel 282, 118831. https://doi.org/10.1016/j.fuel.2020.118831.

Vandenberghe, L.P., Karp, S.G., Pagnoncelli, M.G.B., von Linsingen Tavares, M., Libardi Junior, N., Valladares Diestra, K., Viesser, J.A., Soccol, C.R., 2020. Classification of enzymes and catalytic properties. In: Singh, S.P., Pandey, A., Singhania, R.R., Larroche, C. (Eds.), Classification of enzymes and catalytic properties. Biomass, Biofuels, Biochemicals: Advances in Enzyme Catalysis and Technologies, 11–30. https://doi.org/10.1016/b978-0-12-819820-9.00002-8.

Wang, W., Chen, X., Donohoe, B.S., Ciesielski, P.N., Katahira, R., Kuhn, E.M., Kafle, K., 2014. Effect of mechanical disruption on the effectiveness of three reactors used for dilute acid pretreatment of corn stover Part 1: chemical and physical substrate analysis effect of mechanical disruption on the effectiveness of three reactors used for dilute. Biotechnol. Biofuels 7, 57.

Wang, Yanxia, Liu, Q., Yan, L., Gao, Y., Wang, Yanjie, Wang, W., 2013. A novel lignin degradation bacterial consortium for efficient pulping. Bioresour. Technol. 139, 113–119. https://doi.org/10.1016/j.biortech.2013.04.033.

Wang, Z., Ning, P., Hu, L., Nie, Q., Liu, Y., Zhou, Y., Yang, J., 2020. Efficient ethanol production from paper mulberry pretreated at high solid loading in Fed-nonisothermal-simultaneous saccharification and fermentation. Renew. Energy 160, 211–219. https://doi.org/10.1016/j.renene.2020.06.128.

Wangpor, J., Prayoonyong, P., Sakdaronnarong, C., Sungpet, A., Jonglertjunya, W., 2017. Bioethanol production from cassava starch by enzymatic hydrolysis, fermentation and ex-situ nanofiltration. Energy Procedia 138, 883–888.

Wei, H.L., Wang, Y.T., Hong, Y.Y., Zhu, M.J., 2020. Pretreatment of rice straw with recycled ionic liquids by phase-separation process for low-cost biorefinery. Biotechnol. Appl. Biochem. 68 (4), 871–880. https://doi.org/10.1002/bab.2007.

Woiciechowski, A.L., Dalmas Neto, C.J., Vandenberghe, L.P.de S., de Carvalho Neto, D.P., Novak Sydney, A.C., Letti, L.A.J., Karp, S.G., Zevallos Torres, L.A., Soccol, C.R., 2020. Lignocellulosic biomass: acid and alkaline pretreatments and their effects on biomass recalcitrance—conventional processing and recent advances. Bioresour. Technol. 304, 122848. https://doi.org/10.1016/j.biortech.2020.122848.

Xing, W., Xu, G., Dong, J., Han, R., Ni, Y., 2018. Novel dihydrogen-bonding deep eutectic solvents: pretreatment of rice straw for butanol fermentation featuring enzyme recycling and high solvent yield. Chem. Eng. J. 333, 712–720. https://doi.org/10.1016/j.cej.2017.09.176.

Xu, C., Zhang, J., Zhang, Y., Guo, Y., Xu, H., Xu, J., Wang, Z., 2019. Enhancement of high-solids enzymatic hydrolysis efficiency of alkali pretreated sugarcane bagasse at low cellulase dosage by fed-batch strategy based on optimized accessory enzymes and additives. Bioresour. Technol. 292, 121993. https://doi.org/10.1016/j.biortech.2019.121993.

Yin, Y., Wang, J., 2016. Enhancement of enzymatic hydrolysis of wheat straw by gamma irradiation—alkaline pretreatment. Radiat. Phys. Chem. 123, 63–67. https://doi.org/10.1016/j.radphyschem.2016.02.021.

Yu, X., Bao, X., Zhou, C., Zhang, L., Yagoub, A.E.A., 2018. Ultrasound-ionic liquid enhanced enzymatic and acid hydrolysis of biomass cellulose. Ultrason. Sonochem. 41, 410–418. https://doi.org/10.1016/j.ultsonch.2017.09.003.

Yuan, Y., Zhai, R., Li, Y., Chen, X., Jin, M., 2018. Developing fast enzyme recycling strategy through elucidating enzyme adsorption kinetics on alkali and acid pretreated corn stover. Biotechnol. Biofuels 11, 316. https://doi.org/10.1186/s13068-018-1315-5.

Zanirun, Z., Bahrin, E.K., Lai-Yee, P., Hassan, M.A., Abd-Aziz, S., 2015. Enhancement of fermentable sugars production from oil palm empty fruit bunch by lignolytic enzymes mediator system. Int. Biodeterior. Biodegrad. 105, 13–20. https://doi.org/10.1016/j.ibiod.2015.08.010.

Zeng, W., Zhang, B., Jiang, L., Liu, Y., Ding, S., Chen, G., Liang, Z., 2020. Poly(malic acid) production from liquefied corn starch by simultaneous saccharification and fermentation with a novel isolated *Aureobasidium pullulans* GXL-1 strain and its techno-economic analysis. Bioresour. Technol. 304, 122990. https://doi.org/10.1016/j.biortech.2020.122990.

Zhang, H., Han, L., Dong, H., 2021. An insight to pretreatment, enzyme adsorption and enzymatic hydrolysis of lignocellulosic biomass: experimental and modeling studies. Renew. Sustain. Energy Rev. 140, 110758. https://doi.org/10.1016/j.rser.2021.110758.

Zhang, J., Xie, J., Zhang, H., 2021. Sodium hydroxide catalytic ethanol pretreatment and surfactant on the enzymatic saccharification of sugarcane bagasse. Bioresour. Technol. 319, 124171. https://doi.org/10.1016/j.biortech.2020.124171.

Zhang, L., Wang, Y.Z., Zhao, T., Xu, T., 2019. Hydrogen production from simultaneous saccharification and fermentation of lignocellulosic materials in a dual-chamber microbial electrolysis cell. Int. J. Hydrogen Energy 44, 30024–30030. https://doi.org/10.1016/j.ijhydene.2019.09.191.

Zhang, X., Zhang, W., Lei, F., Yang, S., Jiang, J., 2020. Coproduction of xylooligosaccharides and fermentable sugars from sugarcane bagasse by seawater hydrothermal pretreatment. Bioresour. Technol. 309, 123385. https://doi.org/10.1016/j.biortech.2020.123385.

Zhang, Z., Harrison, M.D., Rackemann, D.W., Doherty, W.O.S., O'Hara, I.M., 2016. Organosolv pretreatment of plant biomass for enhanced enzymatic saccharification. Green Chem. 18, 360–381. https://doi.org/10.1039/c5gc02034d.

Zhiliang, F., Edyta, S., 2013. Consolidated bioprocess for biofuel and chemical production from lignocellulosic biomass WO 2013/070949 AI.

CHAPTER

14

Hybrid technologies for enhanced microbial fermentation process for production of bioenergy and biochemicals

Lingkan Ding, Bo Hu

Department of Bioproducts and Biosystems Engineering, University of Minnesota, MN, USA

14.1   Introduction

Anaerobic digestion (AD) is a mature technology that has been employed in numerous regions worldwide to dispose of numerous biomass waste (Li et al., 2019; Nag et al., 2019). Currently, under the circumstances of rapidly increasing waste generation brought by the economy development, AD is still attracting increasing global attentions because of its advantages in waste minimization, bioenergy generation in the form of methane-rich biogas, and nutrient recovery via the utilization of digestate as biofertilizer. However, some intrinsic problems undermine the development of AD. For instance, the limited economic return (due to the relatively low economic value of biogas and the environmental benefits that are hard to be monetized) of AD necessitates the diversification of value-added end products (Angenent et al., 2018). Moreover, the problems associated with the routine operation (e.g., process instability due to the accumulation of volatile fatty acids (VFAs) and ammonia, limited degradation of recalcitrant substrates, slow methane production rate and long retention time, toxic impurities in biogas, etc.) require further advancement and optimization of conventional AD systems (Appels et al., 2011; Chen et al., 2008). Hence, various hybrid systems incorporating physical, chemical, physicochemical, biological, and electrochemical technologies with current AD units have been proposed and studied (Angenent et al., 2018; De Vrieze et al., 2018; Li et al., 2019; Ye et al., 2018). Typically, based on the process where these technologies are applied to AD, they can be classified into the following categories: pretreatment of feedstock/inoculum before AD (Carrere et al., 2016), process intensification during AD (Khan et al., 2016), and posttreatment of digestate/biogas after AD (Wang & Lee, 2021).

In this chapter, the focus will be put on the process intensification during AD via hybrid technologies.

AD process can be fundamentally dissected into four stages, including hydrolysis, acidogenesis, acetogenesis, and methanogenesis, among which hydrolysis or methanogenesis is usually deemed as the rate-limiting step depending mainly on the feedstock properties (Ma et al., 2013). Thus, various pretreatment methods are always employed to improve the hydrolysis of recalcitrant feedstock (Carrere et al., 2016), while the incorporation of microbial electrochemical technologies (METs) into AD has become an increasingly attractive method in the most recent decade to mitigate the limitations of slow methanogenesis (Logan & Rabaey, 2012) in addition to the conventional AD process parameter optimization. METs, typically including microbial fuel cells (MFCs) and microbial electrolysis cells (MECs), harness the power of certain electroactive microorganisms to catalyze the electrochemical reactions converting the energy in waste biomass into versatile forms such as electricity, hydrogen, and methane (Logan & Rabaey, 2012; Wang & Ren, 2013). Because of the ability to utilize nearly any source of biodegradable organics in wastewater for the direct generation of high-grade electricity, MFCs have been enormously explored. Nonetheless, some technical constraints (e.g., expensive metal catalysts and membranes, relatively low power density, instable performance with larger wastewater treatment volumes, etc.) still hamper the practical application of MFCs (Cheng & Kaksonen, 2017; Wang & Ren, 2013). In contrast, MECs, which are supplied with electrical power (via either using an external direct current power supply or setting an electrode potential using a potentiostat) to enable the conversion of biomass into biofuels (such as hydrogen and methane) (Logan & Rabaey, 2012), share the similar target products with conventional AD systems, thus offering an exciting opportunity to establish hybrid MEC-AD systems for the enhanced microbial fermentation process.

14.2   Mechanisms in hybrid MEC-AD systems

In an MEC, the solid electrodes act as a nonsoluble electron donor (cathode) or acceptor (anode) (Moscoviz et al., 2016). The oxidation of simple organics such as acetate by microbes releases electrons to the anode, which are further transferred to the cathode via the external conductive circuit for the synthesis of target products. An external voltage is required to circumvent the thermodynamic barriers of the reactions as well as the electrode overpotentials and internal resistance for the product synthesis (Cheng & Logan, 2007), while the electroactive microorganisms in the biofilms attached to the electrodes serve as the biocatalysts and thrive via this process (Logan & Rabaey, 2012). MEC was first used to generate hydrogen (H_2) gas via electrohydrogenesis as shown in Eq. (14.1), with some advantages such as smaller voltage required than water electrolysis and higher H_2 yields than bacterial fermentation (Call & Logan, 2008; Cheng & Logan, 2007; Liu et al., 2005; Rozendal et al., 2006; Rozendal et al., 2008). Nonetheless, some drawbacks limited the development of MEC-based electrohydrogenesis, including: (1) the requirement of precious catalysts (e.g., platinum) on the cathode (Cheng & Logan, 2007); (2) the H_2 losses caused by microbial conversions to methane (CH_4) or other chemicals (Call & Logan, 2008); and (3) the problematic H_2 compression and storage (Cheng et al., 2009). In contrast to preventing H_2 losses from CH_4 formation in MECs, the generation of CH_4 as the final product of MECs via either indirect hydrogenotrophic methanogenesis or direct electromethanogenesis (Cheng et al., 2009; Clauwaert et al., 2008; Clauwaert & Verstraete, 2009), became increasingly attractive.

$$2H^+ + 2e^- \rightarrow H_2, \quad Eo' = -0.42 \text{ V vs. standard hydrogen electrode (SHE)} \quad (14.1)$$

Most MECs could be typically classified as two-chamber and one-chamber designs. In a two-chamber MEC, an ion-selective membrane is usually used to separate the anode and cathode chambers, so that the protons released from the anodic oxidation of organics can migrate to the cathode chamber for H_2 formation and further utilized for methanogenesis by hydrogenotrophic methanogens (Noori et al., 2020), or directly utilized for electromethanogenesis (Cheng et al., 2009). However, the presence of a membrane increases the internal resistance between the electrodes, thus leading to a higher ohmic voltage loss (Clauwaert & Verstraete, 2009), while the pH gradient between the two chambers would also cause potential losses (Call & Logan, 2008; Clauwaert & Verstraete, 2009). Moreover, current industrial AD units are mostly one-chamber reactors, hence the addition of membranes not only increases the cost and material input, but also requires a huge workload for system modification and maintenance, although two-chamber MECs exhibited great performances in improving the CH_4 content to over 90% in the biogas through cathodic carbon dioxide (CO_2) reduction in some studies (Liu et al., 2017; Xu et al., 2014). Therefore, one-chamber MEC without the requirement of ion-selective membranes is considered a better choice to be incorporated into the AD facility with less capital costs and more convenient and practical operations (Noori et al., 2020), and the subsequent discussion in this chapter will be mainly based on the one-chamber hybrid MEC-AD configuration.

14.2.1 Methanogenesis pathways

Among the four stages of AD, hydrolysis, acidogenesis, and acetogenesis are performed mainly by hydrolytic/fermentative bacteria, while methanogenesis is carried out by methanogenic archaea. Among the three distinct methanogenesis pathways (i.e., acetoclastic, hydrogenotrophic, and methylotrophic methanogenesis), acetoclastic methanogenesis is usually considered to dominate the methane production in anaerobic digesters and contributes about two-thirds of the global biological methane formation, while hydrogenotrophic methanogenesis accounts for most of the rest one-third (Fenchel et al., 2012; Lyu et al., 2018). In either way, the effective interspecies electron transfer plays a significant role in maintaining the functional methanogenic communities (Rotaru et al., 2014b). In conventional AD, mediated interspecies electron transfer (MIET) between fermentative bacteria and methanogens via intermediates (e.g., H_2 and formate) is deemed as the major electron transfer pathway for methane production (Kouzuma et al., 2015; Rotaru et al., 2014a), and the balance of such syntrophy provides thermodynamically favorable conditions for AD (Baek et al., 2018). As an alternative to MIET, direct interspecies electron transfer (DIET) between certain syntrophic bacteria and methanogens, via biotic (e.g., outer-surface c-type cytochromes and electrically conductive pili) and abiotic (e.g., electrodes and additive conductive materials) electric conduits, serves as another electron transfer pathway for methane production (Kouzuma et al., 2015; Lovley, 2017; Rotaru et al., 2014a). As compared to MIET, DIET is energetically more advantageous and also faster because it does not require H_2 or formate being generated and consumed as the redox mediators (Baek et al., 2018; Zhao et al., 2020).

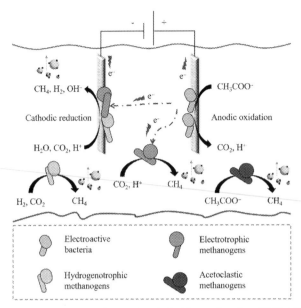

FIG. 14.1 Schematic of methanogenesis pathways in a hybrid one-chamber MEC-AD system.

With the incorporation of MEC into AD, both MIET and DIET are expected to be strengthened, thus facilitating the methanogenesis. Fig. 14.1 depicts the possible methanogenesis pathways in a hybrid one-chamber MEC-AD system. In the one-chamber MEC-AD, although the anode and cathode share the same electrolyte (namely the digestion liquid), the microbial communities attached to each electrode form distinct biofilms with syntrophic partnerships. On the anode of an MEC-AD system, the electroactive bacteria (e.g., *Geobacter* and *Shewanella*) oxidize simple organics such as acetate and release electrons to the anode electrode through a chain of c-type cytochromes and/or electrically conductive pili via DIET as shown in Eq. (14.2) (Baek et al., 2018). The existence of the solid-state conductive anode provides sufficient electrical contact between the electrode surface and electroactive bacteria (Lovley, 2017), and the anode acts as an electron acceptor and also an electron sink, thereby broadening the capacity of electron transfer of the entire system. As a result, the decomposition of organics is supposed to be facilitated as compared to the conventional anaerobic digesters without anodic oxidation. The cathode serves as an electron donor, and the pathways mainly responsible for the CH_4 production on the cathode include: first, direct electromethanogenesis catalyzed by the electrotrophic methanogens that reduce CO_2 via electron transfer from the cathode electrode through the external conductive circuit as well as the digestion liquid through biological DIET as shown in Eq. (14.3); second, indirect H_2-mediated methanogenesis by hydrogenotrophic methanogens harnessing the H_2 abiotically produced (Noori et al., 2020; Zakaria & Dhar, 2019). As shown in Eqs. (14.1) and (14.3), the cathode potential of -0.24 V versus standard hydrogen electrode (SHE) required for direct electromethanogenesis is less negative as compared to that (-0.42 V vs SHE) for abiotic electrohydrogenesis, implying less theoretical electrical energy input will be required for the CH_4 production in an MEC. Additionally, a portion of the H_2 produced on the cathode can be used by the homoacetogens in the anaerobic inoculum to synthesize acetate (Saady, 2013), which can further be either utilized by the acetoclastic methanogens to produce CH_4 or oxidized by the electroactive bacteria on the anode. Besides, in the suspension or suspended sludge in the MEC-AD reactors, the acetoclastic and hydrogenotrophic methanogens still perform their methanogenic routes as usual derived from acetate and H_2/CO_2, respectively.

$$CH_3COO^- + 2H_2O \rightarrow 2CO_2 + 7H^+ + 8e^-, \ Eo' = -0.28 \text{ V vs. SHE} \tag{14.2}$$

$$CO_2 + 8H^+ + 8e^- \rightarrow CH_4 + 2H_2O, \ Eo' = -0.24 \text{ V vs. SHE} \tag{14.3}$$

14.2.2 Microbial communities

The microorganisms are the key biocatalysts and also derive energy in a series of bioelectrochemical reactions in MECs (Logan & Rabaey, 2012). In a one-chamber MEC-AD system, the bacteria on the anode thrive catalyzing the decomposition of organics and form a biofilm attached on the anode surface for better direct electron transfer to the anode. Meanwhile, the electrotrophic methanogenic archaea capture the electrons directly transferred from the cathode, and the hydrogenotrophic methanogenic archaea utilize the abiotically produced H_2 for CH_4 synthesis, thus forming a biofilm attached on the cathode surface. Therefore, the interaction between the microbial communities in the biofilms and in the suspension or sludge, and the alteration of environmental conditions (e.g., intermediates, pH, conductivity, etc.) induced by the MEC, result in the changes of microbial communities as compared to the conventional AD systems.

The changes of main microbial communities in one-chamber MEC-AD systems in the literature are listed in Table 14.1. Despite the greatly differed microbial communities in the original inocula (e.g., AD sludge, waste activated sludge, etc.), the changes after MEC-AD show some patterns. On the anode, hydrolytic/fermentative bacteria undertake the responsibility of decomposing the large-molecular-weight organics into small-molecular-weight VFAs, while electroactive bacteria (e.g., *Geobacter* species) facilitate the oxidation of simple VFAs and release electrons. Thus, *Geobacter* species dominated the bacterial communities in MEC-AD treating simple feedstock such as acetate, while more diverse bacterial compositions were observed when complex feedstock were fed into the MEC-AD units. On the cathode, methanogenic archaea undertake CH_4 synthesis. The dominance shift from acetoclastic methanogens to hydrogenotrophic methanogens and the enrichment of latter in the cathode biofilm were widely reported, though some opposite findings were also recorded. Two reasons could possibly explain this phenomenon. First and the most important, the abiotic H_2 evolution on the cathode raises the H_2 partial pressure, thus providing more substrates for hydrogenotrophic methanogens. Second, the cathodic H_2 evolution reaction at neutral pH as shown in Eq. (14.4) releases hydroxide ions (OH^-), thus inducing localized pH elevation that favors some basophilic hydrogenotrophic methanogens (Cai et al., 2019; Cai et al., 2018). Besides, the pH elevation will lead to drastic increase in free ammonia molecules (Rajagopal et al., 2013), which exhibit a stronger toxicity to acetoclastic methanogens over hydrogenotrophic methanogens (Song et al., 2010). In addition to the dominance of hydrogenotrophic methanogens, some studies also recorded that mixotrophic *Methanosarcina* species, which were proved capable of utilizing either electrons derived from DIET via the syntrophic

TABLE 14.1 Microbial communities in one-chamber MEC-AD systems

Feedstock	Inoculum	Main bacterial communities		Main archaeal communities		Reference
		MEC-AD	Control	MEC-AD	Control	
Acetate	AD sludge, anaerobic bog sediment	*Geobacter*	Highly diverse	Hydrogenotrophic *Methanobacterium* and *Methanobrevibacter*	Acetoclastic *Methanosaeta*	(Siegert et al., 2014)
Acetate	Waste activated sludge	/	/	Hydrogenotrophic methanogens (*Methanospirillum* increased from 16.0% at 0.4 V to 68.4% at 1.0 V)	Acetoclastic and hydrogenotrophic methanogens	(Bo et al., 2014)
Acetate	Waste activated sludge	/	/	Hydrogenotrophic methanogens (56.25% in sludge, 85.01% in cathode biofilm)	Hydrogenotrophic methanogens (26.83%)	(Zhao et al., 2014)
Acetate	Waste activated sludge	*Geobacter*	/	Hydrogenotrophic *Methanobacterium alcaliphilum*	Acetoclastic *Methanosarcina mazei* LYC	(Cai et al., 2019; Cai et al., 2018)
Acetate	Waste activated sludge, *Geobacter*-containing inoculum, and *Methanosarcina* sp. culture	*Desulfuromonas* and *Geobacter*	Highly diverse	*Methanosarcina*	Acetoclastic *Methanosaeta*	(Yin et al., 2016)
Glucose	AD sludge	*Clostridium quinii*	/	Hydrogenotrophic *Methanocorpusculum bavaricum*	/	(Feng et al., 2018)
Glucose	UASB sludge	*Bifidobacterium*, *Clostridium*, and *Pectinatus*	*Bifidobacterium* and *Clostridium*	Hydrogenotrophic *Methanobacterium* and *Methanobrevibacter*	Hydrogenotrophic *Methanobacterium*	(Li et al., 2016)
Glucose and sewage sludge	AD sludge	/	/	Hydrogenotrophic *Methanomicrobiales* (6.6–12.0 times ↑) and *Methanobacteriales* (15.9–17.2 times ↑)	Acetoclastic *Methanosaeta* and *Methanosarcina*	(Gajaraj et al., 2017)
Molasses	AD sludge	*Lactobacillales* and *Clostridiales*	*Lactobacillales* and *Clostridiales*	Acetoclastic *Methanosaeta* and hydrogenotrophic *Methanobacteriales*	Acetoclastic *Methanosaeta*	(De Vrieze et al., 2014)

Feedstock	Inoculum	Main bacterial communities		Main archaeal communities		Reference
		MEC-AD	Control	MEC-AD	Control	
Mixed dextrin and peptone	AD sludge	*Bacteroidetes* (in anode biofilm)	*Firmicutes*	Hydrogenotrophic *Methanobacterium formicicum*	Acetoclastic *Methanosaeta concilii* and hydrogenotrophic *Methanobacterium formicicum*	(Dou et al., 2018)
Food waste	AD sludge	*Clostridia* (10% ↑)	*Clostridia*	*Methanosarcina thermophile* and hydrogenotrophic *Methanobacterium formicicum*	Hydrogenotrophic *Methanobacterium beijingense* and *Methanobacterium petrolearium*	(Park et al., 2018c)
Food waste leachate	AD sludge	*Clostridia* and *Bacteroidia*	*Clostridia* and *Bacteroidia*	*Methanosarcina thermophila* and hydrogenotrophic *Methanobacterium formicicum*	Hydrogenotrophic *Methanobacterium beijingense* and *Methanobacterium petrolearium*	(Lee et al., 2017)
Municipal solid waste incineration leachate	UASB sludge	*Firmicutes* and *Bacteroides* (in sludge), and *Desulfuromondales* and *Pseudomonas* (in anode biofilm)	*Firmicutes* and *Bacteroides* (in sludge)	*Methanosarcina* and hydrogenotrophic *Methanobacterium*	Hydrogenotrophic *Methanobacterium* and *Methanosphaera*	(Gao et al., 2017)
Waste activated sludge	AD sludge	*Proteobacteria* and *Firmicutes*	*Proteobacteria*, *Firmicutes*, and *Bacteroidetes*	Hydrogenotrophic *Methanoregula* (53.3%) and acetoclastic *Methanosaeta* (25.8%)	Acetoclastic *Methanosaeta* (31.6%) and hydrogenotrophic *Methanobacterium* (31.5%)	(Chen et al., 2016)
Waste activated sludge	AD sludge	*Geobacter* (9.65% in sludge and 32.18% in anode biofilm)	*Bacteroidetes*	Acetoclastic *Methanosaeta* (40.55%) and hydrogenotrophic *Methanobacterium* (10.89%)	Acetoclastic *Methanosaeta* (39.82%) and hydrogenotrophic *Methanobacterium* (9.66%)	(Zhao et al., 2015)
Waste activated sludge	AD sludge	*Clostridia* (31.5% in sludge, 38.4% in anode biofilm, *Syntrophomonas*: 3 times ↑), *Anaerolineae*, and *Geobacter*	*Clostridia* and *Anaerolineae*	Acetoclastic *Methanosaeta* (74.1% in sludge) and hydrogenotrophic *Methanobacterium* (9.4% in sludge)	Acetoclastic *Methanosaeta* (56.5% in sludge) and hydrogenotrophic *Methanobacterium* (18.8% in sludge)	(Zhao et al., 2016)
Waste activated sludge	Waste activated sludge	*Geobacteria* (21.86% in anode biofilm) and *Porphyromonadaceae* (in sludge)	*Porphyromonadaceae* (in sludge)	Hydrogenotrophic *Methanobacterium* and *Methanospirillum*, and acetoclastic *Methanosaeta* (in cathode biofilm)	Acetoclastic *Methanosaeta* (in raw sludge)	(Liu et al., 2016c)

partnerships with exoelectrogens (e.g., *Geobacter*) or H_2 for CH_4 synthesis (Rotaru et al., 2014b), were enriched in MEC-AD systems (Gao et al., 2017; Lee et al., 2017; Park et al., 2018c; Yin et al., 2016). This also indicates that the MEC incorporation can enhance the AD performance through altering the methanogenic routes via both enhanced H_2-mediated and DIET pathways.

$$2H_2O + 2e^- \rightarrow H_2 + 2OH^-, \; Eo' = -0.414 \text{ V vs. SHE} \tag{14.4}$$

With the establishment of MEC units in AD systems, the alteration of microbial communities and the enrichment of biofilms ensue. However, the start-up stages of MEC-AD systems where the biofilm enrichment and acclimation occur are usually long. Thereby, to maximize the efficiency of an MEC-AD system in utilizing the feedstock and producing methane, some strategies targeting to optimize the inoculum have been investigated. Siegert et al. (2014) compared the effects of two inocula, namely AD sludge from a digester (mainly containing acetoclastic *Methanosaeta* and no hydrogenotrophic methanogens) and sediment from a freshwater bog (more diverse and containing both acetoclastic and hydrogenotrophic methanogens originally), on the methane production in one-chamber MEC-AD systems. The results showed that the MEC-AD reactors inoculated with bog sediment outperformed the ones inoculated with AD sludge. This suggests that the presence of hydrogenotrophic methanogens in the initial inoculum could contribute to a faster enrichment of functional microbial communities, thereby benefiting the methane production in MEC-AD. In addition, precultured electrodes/inocula were also reported to be beneficial to the MEC-AD systems, and this method seems very practical to improve the MEC-AD process in future applications. De Vrieze et al. (2014) inserted the precolonized electrodes from one-chamber MEC-AD reactors that ran for 91 days into failing AD reactors, and immediate increases in CH_4 production and VFA removal ensued. Xu et al. (2019) precultured the inocula using activated carbon as the conductive additives in AD reactors and enriched the hydrogenotrophic methanogens, thus providing necessary microbiome favorable for starting up the MEC-AD systems. Similarly, in a two-chamber MEC-AD system, Dykstra & Pavlostathis (2017) inoculated the cathode using H_2/CO_2 preenriched inoculum, and found that the CH_4 production was significantly increased as compared to that inoculated with the mixed methanogenic culture. In contrast to preculturing the AD inocula, adding pure cultures of functional microbes in MEC-AD may lead to a better performance. Yin et al. (2016) directly added *Geobacter*-containing inoculum and *Methanosarcina* sp. culture in the seed sludge of a one-chamber MEC-AD reactor, resulting in a 24.1% increase in methane production as compared to the previous one without cocultivating *Geobacter* and *Methanosarcina* (Bo et al., 2014).

14.2.3   Electrode materials

The key difference between the MEC-AD and conventional AD lies in the inserted electrodes, and the interactions between the microorganisms and the electrodes fundamentally affect the system efficiency (Guo et al., 2015). Ideal electrode materials usually have some outstanding properties, including high conductivity, high specific surface area, biocompatibility, and chemical inertness (Noori et al., 2020; Zakaria & Dhar, 2019), though some requirements may change under specific scenarios (e.g., sacrificial metal anodes are preferred to chemically inert ones for sulfide removal and phosphorus recovery in MEC-AD, which will be discussed later).

To reduce the electrode overpotential of electrochemical reactions, which is negatively correlated to the electron transfer and methane production, electrode combination and modification are essential for the MEC-AD performance.

The most widely employed electrodes are made of carbon-based materials in diverse forms, such as graphite/carbon rod, graphite/carbon brush, graphite/carbon felt, carbon cloth, granular activated carbon (GAC), etc., as listed in Table 14.2. These carbon-based materials possess some fine characteristics such as high chemical stability, low cost, and rich electrocatalytic activities for a variety of redox reactions (Zhang et al., 2016). Meanwhile, the diverse structural forms of these carbon-based materials make it possible to substantially increase the specific surface areas and explore various electrode configurations. In addition to increasing the specific surface areas using brush- and felt-type electrodes, some surface modification methods were also developed to strengthen the interactions between the microbial communities and the carbon-based electrodes as well as the electron transfer. Song et al. (2016) modified the graphite fiber fabric using multiwall carbon nanotube and nickel via electrophoretic deposition as the cathode, and printed another modified graphite fiber fabric with mixed multiwall carbon nanotube and exfoliated graphite as the anode. Liu et al. (2016c) used carbon cloth covered with a Pt catalyst layer as the cathode. Tian et al. (2019) modified the carbon felt cathodes with graphene/polypyrrole and MnO_2 nanoparticles/polypyrrole. Park et al. (2018c; 2019a; 2019c) coated Ni on the surface of graphite carbon mesh as the anode, and coated a complex mixture of metal catalysts including Mn, Fe, and Cu on the graphite carbon mesh as the cathode. The coating of either metal or nonmetal materials on the electrode surface mainly targets to increase the electrode conductivity, to provide more catalysts for the electrochemical reactions and more shuttles for electron transfer, and/or to extend the surface area and room for the microbial interaction with the electrodes to facilitate the formation of dense biofilms (Noori et al., 2020). Besides, inducing ions/charge on the electrode surface could also generate electrostatic attraction that enables the acceleration of the biofilm formation on the electrodes (Guo et al., 2015; Noori et al., 2020).

Some metal-based materials with high conductivity and mechanical strength, typically including stainless steel, nickel, copper, and titanium, were also employed as the electrodes in other studies. Bo et al. (2014) and Yin et al. (2016) directly used the stainless steel reactor wall as the cathode and a piece of carbon felt was used as the anode in a one-chamber MEC-AD system. Feng et al. (2015) set an iron tube as the anode and a graphite pillar as the cathode. Under both conditions, the improved AD performances as well as the enrichment of microbial communities were observed. Three metal cathodes (i.e., stainless steel, nickel, and copper) coupled with the granular graphite anode were used in incorporated MEC and upflow anaerobic sludge blanket (UASB) reactors, and their effects on the performance were compared (Sangeetha et al., 2016). It was found that the nickel mesh cathode outperformed the other two in both the chemical oxygen demand (COD) removal and the methane production. Similar to the carbon-based materials, some modifications, such as coating with metal catalysts (Tartakovsky et al., 2011), were also investigated to improve the electrocatalytic property of the metal-based electrodes.

In addition to the selection of electrode materials, the configuration of the electrode pairs (i.e., the surface area and the distance between the anode and cathode) will also impact the MEC-AD performance. Generally, a larger electrode surface area means lower overpotential and larger room for biofilm attachment, while a smaller electrode distance means a smaller

TABLE 14.2 Summary of one-chamber MEC-AD performances in the literature.

MEC-AD type	Feedstock	Inoculum	Electrodes	Applied voltage (V)	Effects compared with control	Reference
Batch, 0.8 L working volume	Glucose and sewage sludge	AD sludge	Reticulated vitreous carbon	0.3–0.6	MEC-AD increased the CH_4 yields by 8.1%–9.4%	(Gajaraj et al., 2017)
Batch, 2.0 L working volume	Sewage sludge	UASB sludge	Anode: iron tube; cathode: graphite pillar	0.3	MEC-AD enhanced the CH_4 yield by 22.4% and the VSS removal by 11%	(Feng et al., 2015)
				0.6	MEC-AD increased the pH to over 9, generated a H_2 yield of 15 L/kg-VSS, but decreased the CH_4 yield by 30.4%	
Batch, 0.35 L working volume	Sewage sludge	AD sludge	Carbon fiber brush	1.0	MEC-AD increased the CH_4 content in biogas by 5% and the CH_4 yield by 40%	(Vu & Min, 2019)
	Glucose				MEC-AD achieved 1.4–2.4 times higher CH_4 yield with 2–4 g/L glucose dosage	(Lee et al., 2019)
Batch, 0.27 L working volume	Glucose	AD sludge	Carbon fiber brush	1.0	MEC-AD increased the CH_4 yield by 30.3%	(Choi et al., 2017)
	Acetate				MEC-AD increased the soluble COD removal by 54.8% and the CH_4 yield by 110.2%	(Flores-Rodriguez et al., 2019)
Batch, 0.5 L working volume	Waste activated sludge	Waste activated sludge	Anode: graphite brush; cathode: Pt covered carbon cloth	0.8	MEC-AD increased the CH_4 production rate by 2 times and the VSS removal by 26.3%	(Liu et al., 2016b)
Batch, 0.17 L working volume	Acetate	Waste activated sludge	Anode: carbon felt; cathode: stainless steel type 304 (as the reactor wall)	1.0	MEC-AD increased the COD removal rate by 2 times and the CH_4 yield by 1.3 times with the CH_4 content over 98% in the biogas	(Bo et al., 2014)
Batch, 0.25 L working volume	Acetate	Waste activated sludge, Geobacter-containing inoculum, and Methanosarcina sp. culture	Anode: carbon felt; cathode: stainless steel type 304 (as the reactor wall)	1.0	MEC-AD inoculated with Geobacter and Methanosarcina increased the COD removal by 79.9% and the methane yield by 59.1% with the CO_2 content of 6.9% in the biogas	(Yin et al., 2016)
Batch, 0.17 L working volume	Waste activated sludge	AD sludge	Anode: graphite brush; cathode: graphite rod	0.6	MEC-AD achieved much faster CH_4 production rates and increased the CH_4 yield by 9.2%	(Zhao et al., 2015)

MEC-AD type	Feedstock	Inoculum	Electrodes	Applied voltage (V)	Effects compared with control	Reference
Batch, 1.0 L working volume	Waste activated sludge	AD sludge	Activated carbon fiber textile	0.6	MEC-AD increased the CH_4 yield by 39.3% and the VS removal rate by 26.6%	(Chen et al., 2016)
Sequencing batch, 0.5 L working volume	Municipal solid waste incineration leachate	UASB sludge	Graphite rods	0.7	MEC-AD increased the COD removal efficiency by 8.7%, the CH_4 production by 44.3%, and the degradation rate of large macromolecules by 18.5%	(Gao et al., 2017)
Continuous, 20 L working volume	Diluted food waste	AD sludge	Anode: Ni-coated graphite carbon mesh; Cathode: Cu, Mn, and Fe coated graphite carbon mesh	0.3	MEC-AD achieved 1.7 times higher CH_4 production rate and 75% shorter stabilization time at the start-up stage, and increased the OLR by 50%–150%	(Park et al., 2019a; Park et al., 2018c; Park et al., 2019c)
Continuous, 12 L working volume	Sewage sludge	AD sludge	Anode: modified graphite fiber fabric with mixed multiwall carbon nanotube and exfoliated graphite; Cathode: modified graphite fiber fabric with multiwall carbon nanotube and Ni	0.3	MEC-AD maintained stable CH_4 production when shortening the HRT from 20 days to 5 days	(Song et al., 2016)
Continuous MEC-UASB, 0.5–3.5 L working volume	Synthetic wastewater	Anaerobic granular sludge	Anode: Ir-MMO-coated titanium mesh; cathode: stainless steel type 316 mesh	2.8–3.5	MEC-UASB increased the CH_4 production by 10%–25%, improved the COD removal efficiency, reduced H_2S, and resulted in 2%–7% H_2 in the biogas	(Tartakovsky et al., 2011)
Continuous MEC-UASB, 1.0 L working volume	Synthetic glucose wastewater	UASB sludge	Graphite	1.0	MEC-UASB increased the organic removal by 67.8% and the CH_4 production rate by 3.8 times	(Li et al., 2016)
Continuous MEC-UASB, 0.6 L working volume	Synthetic beer wastewater	MEC effluent	Anode: granular graphite; cathode: stainless steel, nickel, and copper meshes	0.8	MEC-UASB using nickel mesh cathode exhibited the highest COD removal of 85% and methane yield of 143 mL/gCOD	(Sangeetha et al., 2016)
Continuous MEC-UASB, 1.0 L working volume	Liquid dairy manure	Anaerobic granular sludge	Stainless steel type 304	0.5–1.0	MEC-UASB increased the CH_4 content by 3.0%–6.4% and the CH_4 yield by 10.9%–11.4%, and decreased the H_2S content by 20.8%–50.3%	(Ding et al., 2021b)

internal resistance with improved ionic diffusion rates (Zakaria & Dhar, 2019). Nonetheless, the optimization of these parameters would be more an empirical job and still requires efforts under each specific scenario because of the distinct interference matrix including MEC-AD configurations, medium properties, applied voltage, the cost of electrode materials, etc. Some one-chamber MEC-AD designs based on current AD reactor structures have been made with varying electrode configurations. Bo et al. (2014) designed a barrel-shaped stainless steel MEC-AD reactor, and used the reactor wall itself as the cathode and a piece of carbon felt as the anode. Similarly, Park et al. (2018a; 2018b) designed an MEC-AD reactor in which the rotating impellers (stainless steel type 304) were used as the anode and a roll of stainless steel type 304 was attached to the inner wall as the cathode. On the basis of this MEC-AD reactor with impeller anode, they further used it as an auxiliary MEC reactor and connected it with a larger AD reactor. Rapid oxidation of the accumulated VFAs and increased CH_4 production in the MEC + AD system were then observed, which were similar to that of the one-chamber MEC-AD (Park et al., 2019b). This design might offer a more practical and economical modification of existing AD facilities in industrial applications: the separation of MEC in a smaller auxiliary unit can leave the main digester unchanged, while the maintenance and optimization of the MEC unit would require less effort.

14.3   Performances of hybrid MEC-AD systems

With the enrichment and acclimation of syntrophic hydrolytic/fermentative bacteria and methanogenic archaea in the biofilms attached on the electrodes and the facilitation of interspecies electron transfer, augmented feedstock decomposition and enhanced methane production have been widely recorded as the most significant improvements of the hybrid MEC-AD systems, regardless of the varying feedstock, inocula, electrode materials, applied voltage, and digester configurations. Table 14.2 summarizes the performances of one-chamber MEC-AD systems in recent studies. Moreover, cogeneration of hydrogen and methane, biogas upgrade via CO_2 reduction to CH_4, biogas cleansing through electrochemical sulfide removal, and recovery of phosphorus in concentrated forms offer the opportunities to diversify the value-added end products for MEC-AD systems, while low-temperature MEC-AD offers a more sustainable alternative to conventional mesophilic AD with a comparable performance. Other functions that require two-chamber MEC units or even more complex configurations, such as VFA monitoring and ammonia monitoring and recovery (Yu et al., 2018; Zakaria & Dhar, 2019), are not discussed here.

14.3.1   Augmented feedstock decomposition and methane production

Typically, slow hydrolysis or slow methanogenesis is usually deemed to limit the performance of conventional AD depending mainly on the feedstock properties, while the ratio of hydrolytic/fermentative bacteria to methanogenic archaea determines which would be the rate-limiting step (Ma et al., 2013). Although the initial bacterial and archaeal communities were highly diverse in different inoculum sources, the MEC treatment could alter the distribution of microbial communities in a similar pattern: electroactive bacteria (e.g., *Geobacter* species) thrive in the biofilm of anode, while hydrogenotrophic methanogens (e.g., *Methanobacterium* species) and mixotrophic methanogens (e.g., *Methanosarcina* species) could dominate the cathode biofilm and the digestion liquid (Rotaru et al., 2014b; Siegert et al., 2014; Yin et al., 2016). Thus, the enrichment of both fermentative bacteria and methanogenic archaea offers an opportunity to simultaneously augment the feedstock decomposition and methanogenesis in MEC-AD.

Zhang et al. (2015) inserted a pair of electrodes at an external voltage of 0.3 V into an acidogenic reactor fed with synthetic wastewater, and found that the MEC contributed to the higher COD removal by 18.9% and the higher acidification efficiency by 13.8% as compared to the control. The microbial analysis revealed that some acidophilic methanogens, which were in favor of consuming accumulated VFAs and preventing excessive pH decline, were enriched in the MEC-assisted acidogenic reactor. Gao et al. (2017) used one-chamber MEC-AD at 0.7 V to treat municipal solid waste incineration leachate, and achieved significantly higher COD removal efficiencies and CH_4 production than the control AD without MEC assistance. More large macromolecules were degraded in the MEC-AD, largely due to the enrichment of exoelectrogenic bacteria on the anode which are able to degrade aromatic/complex organics and transfer electrons to the anode. Meanwhile, the cathode biofilm was enriched with methanogenic archaea that can form syntrophic partnerships with the bacteria on the anode. In this MEC-AD system, the dominance transfer from hydrogenotrophic methanogens to mixotrophic *Methanosarcina* species, which are capable of utilizing both electrons derived from DIET via the syntrophic partnerships with exoelectrogens (e.g., *Geobacter*) and H_2 for CH_4 synthesis (Rotaru et al., 2014b), indicated that the enhanced electron transfer between syntrophic microorganisms through the electrodes contributed more to the improved CH_4 production than H_2-mediated methanogenesis. Similarly, the addition of *Geobacter*-containing inoculum and *Methanosarcina* sp. culture to the seed sludge significantly improved the MEC-AD performance, leading to the increases of 79.9% in COD removal and 59.1% in methane yield as compared to the control AD (Yin et al., 2016). Gao et al. (2017) observed that the enhanced syntrophic partnerships (including both direct and indirect interspecies electron transfer) between the exoelectrogenic bacteria enriched on the anode and the aceto-clastic methanogens enriched on the cathode were responsible for the improved COD removal and CH_4 production in a sequencing batch MEC-AD system (at 0.7 V). Similarly, in another study incorporating UASB and MEC (at 0.5 V) treating acidic distillery wastewater, the facilitated DIET was suggested to be the reason for the increased CH_4 content and enhanced CH_4 yield (Feng et al., 2017). Zhao et al. (2015; 2016) calculated the current density and coulombic efficiency of one-chamber MEC-AD systems treating waste activated sludge (at 0.6 V), and found that the direct electromethanogenesis via cathodic CO_2 reduction accepting electrons from anode only contributed to a small portion of the increased CH_4 production as compared to the control AD, while the enrichment of syntrophic microbial communities (such as *Geobacter* species and *Methanosaeta* species) via DIET was supposed to play a major role in the improved CH_4 production.

In contrast, many other studies recorded that more hydrogenotrophic methanogens benefited from the MEC incorporation and dominated the archaeal communities, which were highly likely responsible for the improved methane production. Gajaraj et al. (2017) found that the increased CH_4 yields from the MEC-AD systems treating glucose and waste activated sludge at 0.3–0.6 V were accompanied by the remarkable increases in the population of hydrogenotrophic methanogens. Liu et al. (2016c) found that the graphite brush anode and

Pt-covered carbon cloth cathode provided much larger effective electrode surface areas for biofilm attachment and electron transfer, thus leading to the enrichment of exoelectrogens and hydrogenotrophic methanogens that fundamentally contributed to the significant CH_4 production improvement. Li et al. (2016) recorded that the enrichment of hydrogenotrophic methanogens in the cathode biofilm and the alleviation of increasing acidity in the MEC-UASB treating synthetic wastewater (at 1.0 V) contributed to a substantial CH_4 production rate increase from 51.3 mL/h to 248.5 mL/h.

In addition to the improved COD or volatile solids (VS) removal and CH_4 production, maintaining a stable and efficient process with higher organic loading rates (OLRs) or shorter hydraulic retention times (HRTs) in a continuous mode also represents an augmented AD performance in MEC-AD. Park et al. (2018c) established a relatively large MEC-AD system with a working volume of 20 L and ran a continuous test on it using diluted food waste as feedstock for 12 months at an OLR of 3.0 kgCOD/m^3/d and an HRT of 20 d. Six pairs of modified graphite carbon meshes were used as the electrodes to increase the surface area and the external voltage was set at 0.3 V. The MEC-AD system alleviated the VFA and ammonia accumulation during the start-up of the experiment, thus exhibiting significantly higher CH_4 production rates and shorter stabilization time as compared to the AD reactor. However, the low external voltage of 0.3 V, which probably was not high enough to overcome the overpotentials of electrodes and the internal resistance of digestion liquid, did not increase the CH_4 content and yield in the final steady state of MEC-AD. Similar results were recorded when operating the same MEC-AD and AD systems at low OLRs within 4 kgCOD/m^3/d, whereas the COD removal, pH stabilization, and CH_4 production in the MEC-AD systems started to prevail that in the AD when further increasing the OLRs to 6–10 kgCOD/m^3/d (Park et al., 2019a; Park et al., 2019c). The growing abundance of VFAs-oxidizing exoelectrogenic bacteria stimulated by the MEC contributed to the relieved VFA accumulation (Park et al., 2019a), while the enrichment of H_2-dependent methylotrophic methanogens in the bulk sludge of the MEC-AD reactor and hydrogenotrophic methanogens in the cathode biofilm explained the improvement in substrate conversion and CH_4 production (Park et al., 2019c). Song et al. (2016) also evaluated the performance of MEC-AD treating sewage sludge at 0.3 V with the shortening of HRT from 20 to 5 days (OLR from 1.44 to 5.76 kgVS/m^3/d): the maximum CH_4 yield and the highest net energy efficiency were obtained at the HRT of 10 days, while the stability of MEC-AD process was still secured at the shortest HRT of 5 days. Zhao et al. (2014) set up an MEC-UASB reactor treating synthetic acetate wastewater, and the current and coulombic efficiency calculation implied that the anodic oxidation accounted for a major part of the improved acetate removal at the start-up stage when the methanogenesis was not robust, especially under acidic conditions. The enriched hydrogenotrophic methanogens in the MEC-UASB system, which dominated the cathode biofilm and showed a much higher relative abundance in the bed sludge than the control, suggested that the improved CH_4 production was brought by the coupling of the enhanced acetate oxidation on the anode and the facilitated hydrogenotrophic methanogenesis on the cathode and in the bulky liquid medium.

Most of the one-chamber MEC-AD systems achieved facilitated feedstock decomposition and improved methane production at an applied voltage within 1.0 V, though the optimum values differed because of the distinct configurations. Generally, a high external voltage applied to the MEC-AD systems, which would probably induce excessive hydrogen and oxygen evolution, cause drastic pH rise, and generate a large current that is detrimental to many microorganisms, is not desired for improving the AD performance. For example, Feng et al. (2015) found that an external voltage of 0.3 V enhanced the volatile suspended solids (VSS) removal rate by 11% and the CH_4 yield by 22.4% in a one-chamber MEC-AD treating waste activated sludge as compared to the control, but the further increase of applied voltage to 0.6 V induced much stronger pH rise and led to apparent H_2 evolution and the inhibition of CH_4 production. Lin et al. (2016) recorded significant decreases in the accumulative biogas yields when the batch MEC-AD systems treating dairy manure were operated at an applied voltage of 3.0 V, though higher CH_4 and H_2 contents and lower CO_2 contents were shown in the biogas from the MEC-AD reactors as compared to the control without MEC. However, different findings on the high voltage were present in some other studies. Tartakovsky et al. (2011) set up MEC-UASB reactors treating synthetic wastewater at a much higher external voltage of 2.8–3.5 V. Water electrolysis was induced and the oxygen generated on the anode created micro-aerobic conditions, thus facilitating the feedstock hydrolysis and reducing the release of hydrogen sulfide due to the improved oxidation of sulfide and bisulfide. Meanwhile, the H_2 generated on the cathode contributed to the improved hydrogenotrophic methanogenesis. Through the combined effects of facilitated feedstock hydrolysis and improved hydrogenotrophic methanogenesis, the CH_4 production was increased by 10%–25% in the MEC-UASB system as compared to the conventional UASB reactor. Similarly, Dou et al. (2018) investigated the effects of external voltages ranging from 0 to 2.0 V on the MEC-AD performance and found that the highest voltage of 2.0 V led to water electrolysis, resulting in the highest CH_4 production rate and CH_4 content (up to 88.5%). The microbial community analysis also confirmed that the enrichment of exoelectrogens in the anode biofilm and hydrogenotrophic methanogens in the cathode biofilm could contribute to the enhanced CH_4 production. Nonetheless, the oxygen generation induced by the high voltage needs to be precisely controlled as its presence might inhibit the obligate anaerobic methanogens in the digestion liquid, and most importantly, the mixture of oxygen into the biogas would significantly increase the explosion risks.

14.3.2 Biogas upgrade

Biogas generated from AD of biomass waste primarily contains 40%–75% of CH_4 and 15%–60% CO_2, and the presence of high-level CO_2 significantly reduces the heating value of biogas and limits its downstream applications (Angenent et al., 2018). Thus, biogas upgrade by removing CO_2 can expand the application spectrum of the resultant biomethane, typically including injection into the natural gas grid as the renewable natural gas and compression to vehicle fuels. Microbial electrochemical CO_2 reduction to produce CH_4 is deemed as an emerging sustainable and cost-effective way to upgrade biogas (Angelidaki et al., 2018). Cheng et al. (2009) first reported electromethanogenesis via CO_2 reduction in the cathode chamber of a two-chamber MEC. In addition to direct extracellular electron transfer, H_2-mediated extracellular electron transfer strengthened by the abiotically produced H_2 on the cathode also contributed to the hydrogenotrophic methanogenesis that reduces CO_2 (Villano et al., 2010). Xu et al. (2014) further compared the performances of *in-situ* and *ex-situ* biogas upgrade using two-chamber MECs, and found that the *in-situ* biogas upgrade (the cathode chamber of MEC served as the digester) outperformed the *ex-situ* one (biogas was introduced from a separate digester to the cathode chamber of MEC), but CO_2 levels in both cases were below 10%.

The increased pH in the cathode chamber also led to the dissolution of CO_2, making the cathode chamber a carbon sink as well. These results imply that biogas upgrade or the direct generation of high-methane biogas in hybrid MEC-AD systems is very promising.

To date, most of the MEC-AD systems for biogas upgrade were operated in two-chamber configurations where ion-selective membranes were necessary, although some specific studies reported exciting results using one-chamber MEC-AD systems. Bo et al. (2014) designed a single-chamber barrel-shaped stainless steel MEC-AD reactor and observed that the CO_2 content in biogas significantly decreased from 43.2% in the control AD to 2.0% in the MEC-AD system, accompanied by the greatly increased COD removal efficiency and CH_4 yield. The enrichment of hydrogenotrophic methanogen *Methanospirillum* suggested that the facilitation of hydrogenotrophic methanogenesis in MEC-AD played a significant role in the *in-situ* CO_2 reduction to CH_4, and the large cathode surface area (the inner wall of reactor) also contributed to the biogas upgrade. Yin et al. (2016) used the similar reactor design and electrode configuration, and added *Geobacter* and *Methanosarcina* in the seed sludge, which significantly decreased the CO_2 content in the biogas from 34.8% for the control AD to 6.9% for the MEC-AD. Similarly, Vu & Min (2019) fed sewage sludge to an MEC-AD reactor at an external voltage of 1.0 V, and found that both CH_4 production (i.e., CH_4 yield and CH_4 content) and COD removal were enhanced in the MEC-AD reactor as compared to that in the control without MEC. Generally, as compared to the one-chamber systems, the two-chamber MEC-AD configuration provides a more ideal cathode environment for biogas upgrade with more alkaline catholyte for CO_2 capture and less interference. Biogas upgrade using a small MEC unit separate from the AD reactor might be a more efficient choice (Fu et al., 2020; Zeppilli et al., 2019).

14.3.3   Coproduction of hydrogen and methane

Hythane is a mixture of H_2 and CH_4 within certain ratios (typically 10%–25% of H_2). It exhibits some advantages such as wider flammability range and higher fuel efficiency over CH_4. Conventional biological coproduction of H_2 and CH_4 is through two-stage anaerobic fermentation and digestion (Liu et al., 2013). In contrast, given that the MECs were first developed to generate H_2 gas via electrohydrogenesis (Call & Logan, 2008; Cheng & Logan, 2007; Liu et al., 2005; Rozendal et al., 2006; Rozendal et al., 2008), MEC-AD systems could also be designed to coproduce H_2 and CH_4. Guo et al. (2013) applied external voltages of 1.4–1.8 V to the Ti/Ru alloy mesh electrodes in the one-chamber MEC-AD treating sewage sludge, and recovered H_2 in the first 5 days and CH_4 thereafter. The H_2 and CH_4 yields in the MEC-AD were 1.7- to 5.2-fold and 11.4- to 13.6-fold that of the control AD, respectively, while no oxygen was detected in the biogas. Prajapati & Singh (2020) further tried to employ a similar one-chamber MEC-AD system for H_2 and CH_4 coproduction via the solid-state AD of mixed wastes. However, due to the high VS content (>15%) of the feedstock and the low external voltage (20–120 mV) supplied, the effects of MEC-AD on curbing the pH drop and facilitating the biogas production were very limited as compared to that of the control. MEC deployment in the solid-state AD still requires further optimization. An anaerobic baffled reactor consisting of four compartments, of which the first one was used for H_2 fermentation and the last three were one-chamber MEC-AD units, was operated treating glucose-added wastewater (Ran et al., 2014). The H_2 content in the biogas from the first compartment was 20.1%, while the CH_4 contents from the three MEC-AD compartments operated at an external voltage of 0.9 V reached 70.1%–98.0%.

14.3.4 Low-temperature MEC-AD

Low temperatures are usually considered an inhibitor to methanogenesis because of the limited activity of microorganisms, while the incorporation of MEC into low-temperature AD (or psychrophilic AD) might be a solution to facilitate the CH_4 production (Asztalos & Kim, 2015; Liu et al., 2016a; Park et al., 2018b; Tian et al., 2019; Tian et al., 2018; Yu et al., 2019), thus significantly saving the energy input for heating the entire digester to mesophilic temperatures but still maintaining the AD performance comparable to the mesophilic one. Asztalos & Kim (2015) incorporated an MEC unit at 1.2 V into a digester treating waste activated sludge at the ambient temperature of 22.5 °C, and achieved 5%–10% higher VSS and COD removals as compared to the control without MEC. The mathematical model results indicated that the ambient MEC-AD achieved a COD removal similar to that of conventional mesophilic digesters. An MEC-AD unit was operated at 10 °C using GAC as electrodes and acetate as feedstock, and the results showed that the CH_4 yield from MEC-AD at a cathode potential of -0.90 V (vs. Ag/AgCl) was 5.3–6.6 times higher than that in the AD reactor (Liu et al., 2016a). Meanwhile, the CH_4 production rate was also close to that obtained in the AD reactor at 30 °C. More importantly, it was calculated that the energy input for heating up the reactor from 10 to 30 °C was about 10 times higher than the electrical energy required for the MEC unit at 10 °C. Similarly, Tian et al. (2018) compared the performances of MEC-AD and AD treating low-organic strength wastewater at 8, 12, and 20 °C, and concluded that the MEC-AD assisted by an external voltage of 0.4 V outperformed the AD control in both COD removal and CH_4 production at all three low temperatures. The microbial analysis indicated that the H_2-mediated syntrophy was significantly strengthened in the MEC-AD, evidenced by the enrichment of hydrogenogens and hydrogenotrophic methanogens on the cathode surface. They further replaced the carbon felt cathodes with graphene/polypyrrole and MnO_2 nanoparticles/polypyrrole modified ones, which had higher specific surface areas and favored the attachment and formation of dense biofilms on the electrode surface, and found that the COD removal and CH_4 production were all higher (Tian et al., 2019). Park et al. (2018b) found that stable COD removal and CH_4 production without VFA accumulation or pH decrease at the OLRs of 2.0–4.5 kgCOD/m^3/d were achieved in the psychrophilic MEC-AD (19.8 °C) of food waste at an applied voltage of 0.3 V, while the CH_4 production in the conventional AD reactor rapidly decreased to zero at the starting OLR of 2.0 kgCOD/m^3/d. The microbial community analysis revealed that H_2-dependent methylotrophic and hydrogenotrophic methanogens dominated the MEC-AD reactor. Yu et al. (2019) found that the MEC-AD system treating swine manure operated at the applied voltage of 0.7 V and the temperature of 25 °C exhibited competitive CH_4 production as compared to the conventional AD system at 35 °C. These results suggest that the combination of MEC and AD could offer a more sustainable alternative to conventional mesophilic AD when treating low-temperature waste streams or in cold regions.

14.3.5 Hydrogen sulfide removal

Hydrogen sulfide (H_2S) is an extremely toxic gas component commonly present in biogas. Depending on the feedstock sources (e.g., animal manure, sulfate-containing wastewater, etc.) and operating conditions of AD, the content of H_2S in biogas may range from several hundred to several thousand parts per million by volume (ppmv) (Peu et al., 2012). The toxicity to personnel and the corrosivity to facilities associated with the high H_2S content in biogas necessitate its removal prior to industrial utilization such as biogas upgrade. Conventionally employed H_2S removal methods include biofiltration and physicochemical absorption in liquids or adsorption to solids, involving biological treatment of biogas, or oxidation of sulfide by oxidants (e.g., oxygen, ferric, etc.) to elemental sulfur, or precipitation of metallic sulfides. These methods all require separate units housing the facilities to remove H_2S from the biogas stream. Meanwhile, the physicochemical treatments are usually chemical-intensive with a low rate of chemical regeneration for reuse (Muñoz et al., 2015). Hence, these requirements for H_2S removal lead to the high capital investment and operating costs. In contrast, electrochemical sulfide removal from wastewater (Pikaar et al., 2015), animal wastes (Ding et al., 2021a; Wang et al., 2019), and anaerobic digester effluent (Dutta et al., 2010), which mitigates the sulfide in the liquid phase and can be ideally combined with AD, has been investigated recently. Therefore, the hybrid one-chamber MEC-AD offers an alternative to conventional H_2S removal methods.

Theoretically, by adjusting the operating conditions such as applied voltage and electrode materials in the MEC-AD system, the microenvironment on the anode can be properly controlled via a series of anodic processes, including direct anode sulfide oxidation, anode sacrificing, and water electrolysis with associated sulfide oxidation by increased oxygen (Yu et al., 2018; Zakaria & Dhar, 2019). On the anode, under mild electrooxidation conditions such as a low applied voltage, sulfide (S^{2-}) and bisulfide (HS^-) can be oxidized to elemental sulfur and to sulfate as shown in Eqs. (14.5)–(14.8) (Lin et al., 2016; Pikaar et al., 2015). When the anode is made of sacrificial metal such as low carbon steel, ferrous ions (Fe^{2+}) would be released from the anode and further combined with S^{2-} and HS^- to form stable and insoluble precipitate (FeS) as shown in Eqs. (14.9)–(14.10) (Ding et al., 2021a; Wang et al., 2020a). Wang et al employed low carbon steel as the electrodes in the one-chamber MEC-AD systems treating swine manure (Wang et al., 2019) and sugar beet wastewater (Wang et al., 2020a), and found that the applied voltage of 0.7–1.0 V achieved over 60% of aqueous sulfide removal in the digestion liquid and over 90% of H_2S removal in the yielded biogas. Ding et al. (2021b) established MEC-UASB systems using stainless steel mesh electrodes, and recorded 20.8%–50.3% reduction in the H_2S levels in the biogas from liquid dairy manure at the external voltage of 0.5–1.0 V.

$$S^0 + H^+ + 2e^- \rightarrow HS^-, \quad E_o^{'} = -0.271 \text{ V} \tag{14.5}$$

$$S^0 + 2e^- \rightarrow S^{2-}, \quad E_o^{'} = -0.476 \text{ V} \tag{14.6}$$

$$SO_4^{2-} + 8e^- + 9H^+ \rightarrow HS^- + 4H_2O, \quad E_o^{'} = -0.213 \text{ V} \tag{14.7}$$

$$SO_4^{2-} + 6e^- + 8H^+ \rightarrow S^0 \downarrow + 4H_2O, \quad E_o^{'} = -0.194 \text{ V} \tag{14.8}$$

$$Fe^{2+} + S^{2-} \rightarrow FeS \downarrow \tag{14.9}$$

$$Fe^{2+} + HS^- \rightarrow FeS \downarrow + H^+ \tag{14.10}$$

When water electrolysis is triggered at a higher applied voltage (over 1.23 V) that overcomes the overpotentials (Mazloomi & Sulaiman, 2012), the oxygen generated on the anode not only facilitates the anodic sulfide oxidation, but also inhibits the activity of sulfate-reducing bacteria, thus minimizing the formation of gaseous H_2S and its release into biogas (Diaz et al., 2010). Tartakovsky et al. (2011) applied an external voltage of 2.8–3.5 V to the electrode pair (Ir-mixed metal oxide (MMO)-coated titanium mesh as the anode) in the one-chamber MEC-UASB treating synthetic wastewater, and found that oxygen generated on the anode created microaerobic conditions, which decreased the H_2S concentration from 0.02% to below the detection limit. When combined with a sacrificial metal anode, a higher applied voltage could contribute to a better H_2S mitigation performance. In the one-chamber MEC-AD reactors treating dairy manure, when the applied voltage increased from 1 V to 3 V, the H_2S in the biogas from the groups using stainless steel anodes decreased from 318–358 ppm to 6–7 ppm, while that from the group using carbon cloth anode only decreased from 277 ppm to 178 ppm (Lin et al., 2016). As aforementioned, however, a high applied voltage was usually considered detrimental to biogas production in lots of studies, and the oxygen level in the digesters requires strict control to eliminate the explosion risks and reduce the inhibitory effects on the obligate anaerobic methanogens.

14.3.6 Phosphorus recovery

In addition to the methane-rich biogas, the liquid digestate, which contains certain nutrients such as phosphorus (P), still requires further proper valorizations to simultaneously mitigate the adverse environmental effects and increase the nutrient recovery. Currently, the commonly adopted method for digestate utilization is direct cropland application as a biofertilizer, but this can easily lead to the over accumulation of P in the soil that can be released into water bodies with agricultural runoff and cause eutrophication. Through P separation and recovery in a concentrated form (e.g., sediments, sludge), the spread of liquid digestate with less P accumulation in nearby farmlands can reduce the environmental burdens, while the P concentrated solid digestate can be transported to farther places for more efficient nutrient management and even generate a new revenue for the AD facilities (Lin et al., 2015; Pradel & Aissani, 2019). Conventional P separation methods include biological processes such as enhanced phosphorus removal process and chemical coagulation (Lin et al., 2015). However, the biological treatment systems for P mitigation are usually inefficient, while the efficient chemical dosing is chemical- and labor-intensive (Nguyen et al., 2016). For instance, crystallization of phosphate minerals (e.g., struvite ($MgNH_4PO_4\bullet6H_2O$) and K-struvite ($KMgPO_4\bullet6H_2O$)) has been investigated for nutrient recuperation from digestate (Cerrillo et al., 2015; Moerman et al., 2009), but some important disadvantages still hinder the practical applications, including the requirement on the chemical/energy input to modify the liquid media (e.g., alkali dosing for pH adjustment, stripping, Mg^{2+} dosing, etc.) and the interference induced by the complex matrix in the digestate (e.g., various ratios of Ca to P).

    Electrochemical treatment, also known as electrocoagulation, which offers many advantages such as the elimination of excess chemical/alkali dosing, simple and continuous operation, and high efficiency, may be an alternative to chemical crystallization for P separation and recovery (Cusick & Logan, 2012; Cusick et al., 2014; Nguyen et al., 2016; Omwene et al., 2018). In the electrochemical treatment, sacrificial metal anodes (e.g., aluminum (Omwene et al., 2018), iron (Nguyen et al., 2016), magnesium (Kékedy-Nagy et al., 2020), etc.) are usually employed to release metal ions and generate hydroxides and polyhydroxides capable of participating in the precipitation and adsorption of phosphate. Meanwhile, the cathode surface conditions may locally change during the electrochemical process: the cathodic H_2 evolution reaction at neutral pH as shown in Eq. (14.4) may induce localized pH elevation. The solubilities of some phosphate minerals are very low in water and further decrease at higher pH. Thus, the crystallization of these minerals is expected to be accelerated on the cathode because of the elevated local pH, while the bulky liquid media pH hardly affects the P recovery (Lei et al., 2017).

Struvite precipitation in a two-chamber MEC was reported based on a fluidized bed configuration (Cusick et al., 2014), in which the pH increase in the cathode chamber promoted struvite formation and led to the soluble P removal ranging from 70% to 85%. Meanwhile, the scale accumulation on the cathode surface was minimized by limiting the applied voltage to 0.8–1.0 V and flushing the cathode by fluidized particles at high up-flow velocities. Similarly, in other MEC systems, local pH elevation around the cathode, rather than the pH of the bulk solution, was confirmed to be the reason for calcium phosphate precipitation (Lei et al., 2020; Lei et al., 2017; Wang et al., 2020b). Most of the studies on electrochemical P removal were conducted on wastewater, animal waste, or digested effluent, which were performed independently from AD and did not consider the CH_4 production, while very few studies simultaneously combined the electrochemical P removal with AD (Ding et al., 2021b; Hou et al., 2017). In a continuous one-chamber MEC-UASB system treating liquid dairy manure at an applied voltage of 1.0 V, 65.1% of the total P in the feeding manure was recovered mainly in the bed sludge and the cathode deposit, while the biogas production was also facilitated (Ding et al., 2021b). Different from conventional AD reactors, Hou et al. (2017) incorporated an MEC unit into an anaerobic osmotic membrane bioreactor treating synthetic wastewater. The results showed that this hybrid system recovered 41%–65% PO_4-P while maintaining efficient biogas production at an applied voltage of 0.8 V. Higher applied voltages or current densities can efficiently facilitate the P removal from wastewater (Nguyen et al., 2016; Omwene et al., 2018), but if combined with AD, these harsh conditions would certainly induce detrimental effects on the microbial communities. Besides, the accumulation of insoluble phosphate minerals on the cathode surface will form a thick passivation layer, hence reducing the electrochemical treatment efficiency (Lei et al., 2019). This will affect not only the P recovery efficiency but also the cathodic CH_4 production, thus requiring automated methods for efficient precipitate/deposit detachment in the MEC-AD configurations. Possible solutions may include dynamic brushing and polarity inversion (Takabe et al., 2020), but the corresponding adverse effects on the biofilms attached on the electrode surfaces that act as the biocatalysts for the improved AD performance also need to be investigated.

14.4 Conclusions and perspectives

MEC, as an emerging technology in the recent decade that enables the conversion of biomass waste into biofuels with the supply of electrical power and the assistance of electroactive microorganisms, can be incorporated into AD systems. By enriching the syntrophic hydrolytic/fermentative bacteria and methanogenic archaea in the biofilms attached on the electrodes and promoting the interspecies electron transfer, the hybrid MEC-AD systems can address some critical issues of conventional AD such as limited hydrolysis of feedstock and slow methanogenesis. In addition to facilitating the decomposition of feedstock and improving the CH_4 production, typical advantages of this hybrid MEC-AD technology also include but not limited to enhancing the CH_4 content by cathodic CO_2 reduction, cleansing the biogas by H_2S removal, and nutrient enrichment in sediment and sludge by phosphorus recovery.

Accompanied with the merits aforementioned, some shortcomings of the hybrid MEC-AD technology still await further scientific and technical advancement. First of all, the understandings on the underlying microbial communities that function as the key biocatalysts in MEC-AD in many previous studies are contradictory, thus still requiring more in-depth investigation, which in turn could provide orientational guidelines for further accelerating the acclimation phase, shortening the start-up stage, and pushing the methanogenesis via cathodic CO_2 reduction. Second, the selection of efficient electrode materials and the optimization of process parameters (e.g., the configuration of electrode pairs including the surface area and electrode distance, the applied voltage ranges, etc.) need to pertain more to specific feedstock (especially the actual waste streams) and existing AD facilities, thus minimizing the unreasonable consumptions in electrode materials, digester spacing, and electrical power. Third, a detailed technoeconomic analysis of such a hybrid system on a pilot scale, rather than a bench-scale reactor, is essential to assess the practical feasibility of MEC-AD and provide a reference in future industrial application.

References

Angelidaki, I., Treu, L., Tsapekos, P., Luo, G., Campanaro, S., Wenzel, H., Kougias, P.G., 2018. Biogas upgrading and utilization: current status and perspectives. Biotechnol. Adv. 36, 452–466.

Angenent, L.T., Usack, J.G., Xu, J., Hafenbradl, D., Posmanik, R., Tester, J.W., 2018. Integrating electrochemical, biological, physical, and thermochemical process units to expand the applicability of anaerobic digestion. Bioresour. Technol. 247, 1085–1094.

Appels, L., Lauwers, J., Degrève, J., Helsen, L., Lievens, B., Willems, K., Van Impe, J., Dewil, R., 2011. Anaerobic digestion in global bio-energy production: potential and research challenges. Renew. Sustain. Energy Rev. 15, 4295–4301.

Asztalos, J.R., Kim, Y., 2015. Enhanced digestion of waste activated sludge using microbial electrolysis cells at ambient temperature. Water Res. 87, 503–512.

Baek, G., Kim, J., Kim, J., Lee, C., 2018. Role and potential of direct interspecies electron transfer in anaerobic digestion. Energies 11, 107.

Bo, T., Zhu, X., Zhang, L., Tao, Y., He, X., Li, D., Yan, Z., 2014. A new upgraded biogas production process: coupling microbial electrolysis cell and anaerobic digestion in single-chamber, barrel-shape stainless steel reactor. Electrochem. Comm. 45, 67–70.

Cai, W., Liu, W., Zhang, Z., Feng, K., Ren, G., Pu, C., Li, J., Deng, Y., Wang, A., 2019. Electro-driven methanogenic microbial community diversity and variability in the electron abundant niche. Sci. Total Environ. 661, 178–186.

Cai, W., Liu, W., Zhang, Z., Feng, K., Ren, G., Pu, C., Sun, H., Li, J., Deng, Y., Wang, A., 2018. mcrA sequencing reveals the role of basophilic methanogens in a cathodic methanogenic community. Water Res. 136, 192–199.

Call, D., Logan, B.E., 2008. Hydrogen production in a single chamber microbial electrolysis cell lacking a membrane. Environ. Sci. Technol. 42, 3401–3406.

Carrere, H., Antonopoulou, G., Affes, R., Passos, F., Battimelli, A., Lyberatos, G., Ferrer, I., 2016. Review of feedstock pretreatment strategies for improved anaerobic digestion: from lab-scale research to full-scale application. Bioresour. Technol. 199, 386–397.

Cerrillo, M., Palatsi, J., Comas, J., Vicens, J., Bonmatí, A., 2015. Struvite precipitation as a technology to be integrated in a manure anaerobic digestion treatment plant— removal efficiency, crystal characterization and agricultural assessment. J. Chem. Technol. Biotechnol. 90, 1135–1143.

Chen, Y., Cheng, J.J., Creamer, K.S., 2008. Inhibition of anaerobic digestion process: a review. Bioresour. Technol. 99, 4044–4064.

Chen, Y., Yu, B., Yin, C., Zhang, C., Dai, X., Yuan, H., Zhu, N., 2016. Biostimulation by direct voltage to enhance anaerobic digestion of waste activated sludge. RSC Adv. 6, 1581–1588.

Cheng, K.Y., Kaksonen, A.H., 2017. Integrating microbial electrochemical technologies with anaerobic digestion for waste treatment. Curr. Dev. Biotechnol. Bioeng., 191–221.

Cheng, S., Logan, B.E., 2007. Sustainable and efficient biohydrogen production via electrohydrogenesis. Proc. Natl. Acad. Sci. USA 104, 18871.

Cheng, S., Xing, D., Call, D.F., Logan, B.E., 2009. Direct biological conversion of electrical current into methane by electromethanogenesis. Environ. Sci. Technol. 43, 3953–3958.

Choi, K.S., Kondaveeti, S., Min, B., 2017. Bioelectrochemical methane (CH_4) production in anaerobic digestion at different supplemental voltages. Bioresour. Technol. 245, 826–832.

Clauwaert, P., Toledo, R., van der Ha, D., Crab, R., Verstraete, W., Hu, H., Udert, K.M., Rabaey, K., 2008. Combining biocatalyzed electrolysis with anaerobic digestion. Water Sci. Technol. 57, 575–579.

Clauwaert, P., Verstraete, W., 2009. Methanogenesis in membraneless microbial electrolysis cells. Appl. Microbiol. Biotechnol. 82, 829–836.

Cusick, R.D., Logan, B.E., 2012. Phosphate recovery as struvite within a single chamber microbial electrolysis cell. Bioresour. Technol. 107, 110–115.

Cusick, R.D., Ullery, M.L., Dempsey, B.A., Logan, B.E., 2014. Electrochemical struvite precipitation from digestate with a fluidized bed cathode microbial electrolysis cell. Water Res. 54, 297–306.

De Vrieze, J., Arends, J.B.A., Verbeeck, K., Gildemyn, S., Rabaey, K., 2018. Interfacing anaerobic digestion with (bio) electrochemical systems: potentials and challenges. Water Res. 146, 244–255.

De Vrieze, J., Gildemyn, S., Arends, J.B., Vanwonterghem, I., Verbeken, K., Boon, N., Verstraete, W., Tyson, G.W., Hennebel, T., Rabaey, K., 2014. Biomass retention on electrodes rather than electrical current enhances stability in anaerobic digestion. Water Res. 54, 211–221.

Diaz, I., Lopes, A.C., Perez, S.I., Fdz-Polanco, M., 2010. Performance evaluation of oxygen, air and nitrate for the microaerobic removal of hydrogen sulphide in biogas from sludge digestion. Bioresour. Technol. 101, 7724–7730.

Ding, L., Lin, H., Hetchler, B., Wang, Y., Wei, W., Hu, B., 2021a. Electrochemical mitigation of hydrogen sulfide in deep-pit swine manure storage. Sci. Total. Environ. 777, 146048.

Ding, L., Lin, H., Zamalloa, C., Hu, B., 2021b. Simultaneous phosphorus recovery, sulfide removal, and biogas production improvement in electrochemically assisted anaerobic digestion of dairy manure. Sci. Total. Environ. 777, 146226.

Dou, Z., Dykstra, C.M., Pavlostathis, S.G., 2018. Bioelectrochemically assisted anaerobic digestion system for biogas upgrading and enhanced methane production. Sci. Total. Environ. 633, 1012–1021.

Dutta, P.K., Rabaey, K., Yuan, Z., Rozendal, R.A., Keller, J., 2010. Electrochemical sulfide removal and recovery from paper mill anaerobic treatment effluent. Water Res. 44, 2563–2571.

Dykstra, C.M., Pavlostathis, S.G., 2017. Methanogenic biocathode microbial community development and the role of bacteria. Environ. Sci. Technol. 51, 5306–5316.

Fenchel, T., King, G.M., Blackburn, T.H., 2012. Bacterial metabolism. In: Fenchel, T., King, G.M., Blackburn, T.H. (Eds.), Bacterial Biogeochemistry3rd edn. Academic Press, Boston, pp. 1–34.

Feng, Q., Song, Y.C., Ahn, Y., 2018. Electroactive microorganisms in bulk solution contribute significantly to methane production in bioelectrochemical anaerobic reactor. Bioresour. Technol. 259, 119–127.

Feng, Q., Song, Y.C., Yoo, K., Kuppanan, N., Subudhi, S., Lal, B., 2017. Bioelectrochemical enhancement of direct interspecies electron transfer in upflow anaerobic reactor with effluent recirculation for acidic distillery waste-water. Bioresour. Technol. 241, 171–180.

Feng, Y., Zhang, Y., Chen, S., Quan, X., 2015. Enhanced production of methane from waste activated sludge by the combination of high-solid anaerobic digestion and microbial electrolysis cell with iron–graphite electrode. Chem. Eng. J. 259, 787–794.

Flores-Rodriguez, C., Nagendranatha Reddy, C., Min, B., 2019. Enhanced methane production from acetate intermediate by bioelectrochemical anaerobic digestion at optimal applied voltages. Biomass Bioenergy 127, 105261.

Fu, X.-Z., Li, J., Pan, X.-R., Huang, L., Li, C.-X., Cui, S., Liu, H.-Q., Tan, Z.-L., Li, W.-W., 2020. A single microbial electrochemical system for CO_2 reduction and simultaneous biogas purification, upgrading and sulfur recovery. Bioresour. Technol. 297, 122448.

Gajaraj, S., Huang, Y., Zheng, P., Hu, Z., 2017. Methane production improvement and associated methanogenic assemblages in bioelectrochemically assisted anaerobic digestion. Biochem. Eng. J. 117, 105–112.

Gao, Y., Sun, D., Dang, Y., Lei, Y., Ji, J., Lv, T., Bian, R., Xiao, Z., Yan, L., Holmes, D.E., 2017. Enhancing biomethanogenic treatment of fresh incineration leachate using single chambered microbial electrolysis cells. Bioresour. Technol. 231, 129–137.

Guo, K., Prevoteau, A., Patil, S.A., Rabaey, K., 2015. Engineering electrodes for microbial electrocatalysis. Curr. Opin. Biotechnol. 33, 149–156.

Guo, X., Liu, J., Xiao, B., 2013. Bioelectrochemical enhancement of hydrogen and methane production from the anaerobic digestion of sewage sludge in single-chamber membrane-free microbial electrolysis cells. Int. J. Hydrog. Energy 38, 1342–1347.

Hou, D., Lu, L., Sun, D., Ge, Z., Huang, X., Cath, T.Y., Ren, Z.J., 2017. Microbial electrochemical nutrient recovery in anaerobic osmotic membrane bioreactors. Water Res. 114, 181–188.

Kékedy-Nagy, L., Teymouri, A., Herring, A.M., Greenlee, L.F., 2020. Electrochemical removal and recovery of phosphorus as struvite in an acidic environment using pure magnesium vs. the AZ31 magnesium alloy as the anode. Chem. Eng. J. 380, 122480.

Khan, M.A., Ngo, H.H., Guo, W.S., Liu, Y., Nghiem, L.D., Hai, F.I., Deng, L.J., Wang, J., Wu, Y., 2016. Optimization of process parameters for production of volatile fatty acid, biohydrogen and methane from anaerobic digestion. Bioresour. Technol. 219, 738–748.

Kouzuma, A., Kato, S., Watanabe, K., 2015. Microbial interspecies interactions: recent findings in syntrophic consortia. Front. Microbiol. 6, 477.

Lee, B., Park, J.G., Shin, W.B., Tian, D.J., Jun, H.B., 2017. Microbial communities change in an anaerobic digestion after application of microbial electrolysis cells. Bioresour. Technol. 234, 273–280.

Lee, M., Nagendranatha Reddy, C., Min, B., 2019. In situ integration of microbial electrochemical systems into anaerobic digestion to improve methane fermentation at different substrate concentrations. Int. J. Hydrog. Energy 44, 2380–2389.

Lei, Y., Remmers, J.C., Saakes, M., van der Weijden, R.D., Buisman, C.J.N., 2019. Influence of cell configuration and long-term operation on electrochemical phosphorus recovery from domestic wastewater. ACS Sustain. Chem. Eng. 7, 7362–7368.

Lei, Y., Saakes, M., van der Weijden, R.D., Buisman, C.J.N., 2020. Electrochemically mediated calcium phosphate precipitation from phosphonates: implications on phosphorus recovery from non-orthophosphate. Water Res. 169, 115206.

Lei, Y., Song, B., van der Weijden, R.D., Saakes, M., Buisman, C.J.N., 2017. Electrochemical induced calcium phosphate precipitation: importance of local pH. Environ. Sci. Technol. 51, 11156–11164.

Li, Y., Chen, Y., Wu, J., 2019. Enhancement of methane production in anaerobic digestion process: a review. Appl. Energy 240, 120–137.

Li, Y., Zhang, Y., Liu, Y., Zhao, Z., Zhao, Z., Liu, S., Zhao, H., Quan, X., 2016. Enhancement of anaerobic methanogenesis at a short hydraulic retention time via bioelectrochemical enrichment of hydrogenotrophic methanogens. Bioresour. Technol. 218, 505–511.

Lin, H., Gan, J., Rajendran, A., Reis, C.E.R., Hu, B., 2015. Phosphorus removal and recovery from digestate after biogas production. In: Biernat, K. (Ed.), Biofuels: Status and Perspective. Intechopen, London, pp. 517–546.

Lin, H., Williams, N., King, A., Hu, B., 2016. Electrochemical sulfide removal by low-cost electrode materials in anaerobic digestion. Chem. Eng. J. 297, 180–192.

Liu, D., Zhang, L., Chen, S., Buisman, C., Ter Heijne, A., 2016a. Bioelectrochemical enhancement of methane production in low temperature anaerobic digestion at 10 degrees C. Water Res. 99, 281–287.

Liu, H., Grot, S., Logan, B.E., 2005. Electrochemically assisted microbial production of hydrogen from acetate. Environ. Sci. Technol. 39, 4317–4320.

Liu, S.Y., Charles, W., Ho, G., Cord-Ruwisch, R., Cheng, K.Y., 2017. Bioelectrochemical enhancement of anaerobic digestion: comparing single- and two-chamber reactor configurations at thermophilic conditions. Bioresour. Technol. 245, 1168–1175.

Liu, W., Cai, W., Guo, Z., Wang, L., Yang, C., Varrone, C., Wang, A., 2016b. Microbial electrolysis contribution to anaerobic digestion of waste activated sludge, leading to accelerated methane production. Renew. Energy 91, 334–339.

Liu, W., Cai, W., Guo, Z., Wang, L., Yang, C., Varrone, C., Wang, A., 2016c. Microbial electrolysis contribution to anaerobic digestion of waste activated sludge, leading to accelerated methane production. Renew. Energy 91, 334–339.

Liu, Z., Zhang, C., Lu, Y., Wu, X., Wang, L., Wang, L., Han, B., Xing, X.-H., 2013. States and challenges for high-value biohythane production from waste biomass by dark fermentation technology. Bioresour. Technol. 135, 292–303.

Logan, B.E., Rabaey, K., 2012. Conversion of wastes into bioelectricity and chemicals by using microbial electrochemical technologies. Science 337, 686.

Lovley, D.R., 2017. Syntrophy goes electric: direct interspecies electron transfer. Annu. Rev. Microbiol. 71, 643–664.

Lyu, Z., Shao, N., Akinyemi, T., Whitman, W.B., 2018. Methanogenesis. Curr. Biol. 28, R727–R732.

Ma, J., Frear, C., Wang, Z.W., Yu, L., Zhao, Q., Li, X., Chen, S., 2013. A simple methodology for rate-limiting step determination for anaerobic digestion of complex substrates and effect of microbial community ratio. Bioresour. Technol. 134, 391–395.

Mazloomi, S.K., Sulaiman, N., 2012. Influencing factors of water electrolysis electrical efficiency. Renew. Sustain. Energ. Rev. 16, 4257–4263.

Moerman, W., Carballa, M., Vandekerckhove, A., Derycke, D., Verstraete, W., 2009. Phosphate removal in agro-industry: pilot- and full-scale operational considerations of struvite crystallization. Water Res. 43, 1887–1892.

Moscoviz, R., Toledo-Alarcon, J., Trably, E., Bernet, N., 2016. Electro-fermentation: how to drive fermentation using electrochemical systems. Trends Biotechnol. 34, 856–865.

Muñoz, R., Meier, L., Diaz, I., Jeison, D., 2015. A review on the state-of-the-art of physical/chemical and biological technologies for biogas upgrading. Rev. Environ. Sci. Bio/Technol. 14, 727–759.

Nag, R., Auer, A., Markey, B.K., Whyte, P., Nolan, S., O'Flaherty, V., Russell, L., Bolton, D., Fenton, O., Richards, K., Cummins, E., 2019. Anaerobic digestion of agricultural manure and biomass—critical indicators of risk and knowledge gaps. Sci. Total Environ. 690, 460–479.

Nguyen, D.D., Ngo, H.H., Guo, W., Nguyen, T.T., Chang, S.W., Jang, A., Yoon, Y.S., 2016. Can electrocoagulation process be an appropriate technology for phosphorus removal from municipal wastewater? Sci. Total Environ. 563–564, 549–556.

Noori, M.T., Vu, M.T., Ali, R.B., Min, B., 2020. Recent advances in cathode materials and configurations for upgrading methane in bioelectrochemical systems integrated with anaerobic digestion. Chem. Eng. J. 392, 123689.

Omwene, P.I., Kobya, M., Can, O.T., 2018. Phosphorus removal from domestic wastewater in electrocoagulation reactor using aluminium and iron plate hybrid anodes. Ecol. Eng. 123, 65–73.

Park, J.-G., Shin, W.-B., Shi, W.-Q., Jun, H.-B., 2019a. Changes of bacterial communities in an anaerobic digestion and a bio-electrochemical anaerobic digestion reactors according to organic load. Energies 12, 2958.

Park, J., Lee, B., Shin, W., Jo, S., Jun, H., 2018a. Application of a rotating impeller anode in a bioelectrochemical anaerobic digestion reactor for methane production from high-strength food waste. Bioresour. Technol. 259, 423–432.

Park, J., Lee, B., Shin, W., Jo, S., Jun, H., 2018b. Psychrophilic methanogenesis of food waste in a bio-electrochemical anaerobic digester with rotating impeller electrode. J. Clean. Prod. 188, 556–567.

Park, J., Lee, B., Tian, D., Jun, H., 2018c. Bioelectrochemical enhancement of methane production from highly concentrated food waste in a combined anaerobic digester and microbial electrolysis cell. Bioresour. Technol. 247, 226–233.

Park, J.G., Lee, B., Kwon, H.J., Park, H.R., Jun, H.B., 2019b. Effects of a novel auxiliary bio-electrochemical reactor on methane production from highly concentrated food waste in an anaerobic digestion reactor. Chemosphere 220, 403–411.

Park, J.G., Lee, B., Park, H.R., Jun, H.B., 2019c. Long-term evaluation of methane production in a bio-electrochemical anaerobic digestion reactor according to the organic loading rate. Bioresour. Technol. 273, 478–486.

Peu, P., Picard, S., Diara, A., Girault, R., Beline, F., Bridoux, G., Dabert, P., 2012. Prediction of hydrogen sulphide production during anaerobic digestion of organic substrates. Bioresour. Technol. 121, 419–424.

Pikaar, I., Likosova, E.M., Freguia, S., Keller, J., Rabaey, K., Yuan, Z., 2015. Electrochemical abatement of hydrogen sulfide from waste streams. Crit. Rev. Environ. Sci. Technol. 45, 1555–1578.

Pradel, M., Aissani, L., 2019. Environmental impacts of phosphorus recovery from a "product" life cycle assessment perspective: allocating burdens of wastewater treatment in the production of sludge-based phosphate fertilizers. Sci. Total Environ. 656, 55–69.

Prajapati, K.B., Singh, R., 2020. Bio-electrochemically hydrogen and methane production from co-digestion of wastes. Energy 198, 117259.

Rajagopal, R., Masse, D.I., Singh, G., 2013. A critical review on inhibition of anaerobic digestion process by excess ammonia. Bioresour. Technol. 143, 632–641.

Ran, Z., Gefu, Z., Kumar, J.A., Chaoxiang, L., Xu, H., Lin, L., 2014. Hydrogen and methane production in a bio-electrochemical system assisted anaerobic baffled reactor. Int. J. Hydrog. Energy 39, 13498–13504.

Rotaru, A.-E., Shrestha, P.M., Liu, F., Shrestha, M., Shrestha, D., Embree, M., Zengler, K., Wardman, C., Nevin, K.P., Lovley, D.R., 2014a. A new model for electron flow during anaerobic digestion: direct interspecies electron transfer to Methanosaeta for the reduction of carbon dioxide to methane. Energy Environ. Sci. 7, 408–415.

Rotaru, A.E., Shrestha, P.M., Liu, F., Markovaite, B., Chen, S., Nevin, K.P., Lovley, D.R., 2014b. Direct interspecies electron transfer between *Geobacter metallireducens* and *Methanosarcina barkeri*. Appl. Environ. Microbiol. 80, 4599–4605.

Rozendal, R., Hamelers, H., Euverink, G., Metz, S., Buisman, C., 2006. Principle and perspectives of hydrogen production through biocatalyzed electrolysis. Int. J. Hydrog. Energy 31, 1632–1640.

Rozendal, R.A., Jeremiasse, A.W., Hamelers, H.V.M., Buisman, C.J.N., 2008. Hydrogen production with a microbial biocathode. Environ. Sci. Technol. 42, 629–634.

Saady, N.M.C., 2013. Homoacetogenesis during hydrogen production by mixed cultures dark fermentation: unresolved challenge. Int. J. Hydrog. Energy 38, 13172–13191.

Sangeetha, T., Guo, Z., Liu, W., Cui, M., Yang, C., Wang, L., Wang, A., 2016. Cathode material as an influencing factor on beer wastewater treatment and methane production in a novel integrated upflow microbial electrolysis cell (Upflow-MEC). Int. J. Hydrog. Energy 41, 2189–2196.

Siegert, M., Li, X.F., Yates, M.D., Logan, B.E., 2014. The presence of hydrogenotrophic methanogens in the inoculum improves methane gas production in microbial electrolysis cells. Front. Microbiol. 5, 778.

Song, M., Shin, S.G., Hwang, S., 2010. Methanogenic population dynamics assessed by real-time quantitative PCR in sludge granule in upflow anaerobic sludge blanket treating swine wastewater. Bioresour. Technol. 101, S23–S28.

Song, Y.-C., Feng, Q., Ahn, Y., 2016. Performance of the bio-electrochemical anaerobic digestion of sewage sludge at different hydraulic retention times. Energy Fuels 30, 352–359.

Takabe, Y., Ota, N., Fujiyama, M., Okayasu, Y., Yamasaki, Y., Minamiyama, M., 2020. Utilisation of polarity inversion for phosphorus recovery in electrochemical precipitation with anaerobic digestion effluent. Sci. Total Environ. 706, 1879–1026, 136090. doi:10.1016/j.scitotenv.2019.136090. 31862599.

Tartakovsky, B., Mehta, P., Bourque, J.S., Guiot, S.R., 2011. Electrolysis-enhanced anaerobic digestion of wastewater. Bioresour. Technol. 102, 5685–5691.

Tian, T., Qiao, S., Yu, C., Yang, Y., Zhou, J., 2019. Low-temperature anaerobic digestion enhanced by bioelectrochemical systems equipped with graphene/PPy- and MnO_2 nanoparticles/PPy-modified electrodes. Chemosphere 218, 119–127.

Tian, T., Qiao, S., Yu, C., Zhou, J., 2018. Bio-electrochemically assisting low-temperature anaerobic digestion of low-organic strength wastewater. Chem. Eng. J. 335, 657–664.

Villano, M., Aulenta, F., Ciucci, C., Ferri, T., Giuliano, A., Majone, M., 2010. Bioelectrochemical reduction of CO_2 to CH_4 via direct and indirect extracellular electron transfer by a hydrogenophilic methanogenic culture. Bioresour. Technol. 101, 3085–3090.

Vu, H.T., Min, B., 2019. Enhanced methane fermentation of municipal sewage sludge by microbial electrochemical systems integrated with anaerobic digestion. Int. J. Hydrog. Energy 44, 30357–30366.

Wang, H., Ren, Z.J., 2013. A comprehensive review of microbial electrochemical systems as a platform technology. Biotechnol. Adv. 31, 1796–1807.

Wang, W., Lee, D.J., 2021. Valorization of anaerobic digestion digestate: a prospect review. Bioresour. Technol. 323, 124626.

Wang, Y., Lin, H., Ding, L., Hu, B., 2020a. Low-voltage electrochemical treatment to precipitate sulfide during anaerobic digestion of beet sugar wastewater. Sci. Total Environ. 747, 141243.

Wang, Y., Lin, H., Hu, B., 2019. Electrochemical removal of hydrogen sulfide from swine manure. Chem. Eng. J. 356, 210–218.

Wang, Z., Zhang, J., Hu, X., Bian, R., Xv, Y., Deng, R., Zhang, Z., Xiang, P., Xia, S., 2020b. Phosphorus recovery from aqueous solution via a microbial electrolysis phosphorus-recovery cell. Chemosphere 257, 127283.

Xu, H., Wang, K., Holmes, D.E., 2014. Bioelectrochemical removal of carbon dioxide (CO2): an innovative method for biogas upgrading. Bioresour. Technol. 173, 392–398.

Xu, S., Zhang, Y., Luo, L., Liu, H., 2019. Startup performance of microbial electrolysis cell assisted anaerobic digester (MEC-AD) with pre-acclimated activated carbon. Bioresour. Technol. Rep. 5, 91–98.

Ye, M., Liu, J., Ma, C., Li, Y.-Y., Zou, L., Qian, G., Xu, Z.P., 2018. Improving the stability and efficiency of anaerobic digestion of food waste using additives: a critical review. J. Clean. Prod. 192, 316–326.

Yin, Q., Zhu, X., Zhan, G., Bo, T., Yang, Y., Tao, Y., He, X., Li, D., Yan, Z., 2016. Enhanced methane production in an anaerobic digestion and microbial electrolysis cell coupled system with co-cultivation of Geobacter and Methanosarcina. J. Environ. Sci. 42, 210–214.

Yu, J., Kim, S., Kwon, O.S., 2019. Effect of applied voltage and temperature on methane production and microbial community in microbial electrochemical anaerobic digestion systems treating swine manure. J. Ind. Microbiol. Biotechnol. 46, 911–923.

Yu, Z., Leng, X., Zhao, S., Ji, J., Zhou, T., Khan, A., Kakde, A., Liu, P., Li, X., 2018. A review on the applications of microbial electrolysis cells in anaerobic digestion. Bioresour. Technol. 255, 340–348.

Zakaria, B.S., Dhar, B.R., 2019. Progress towards catalyzing electro-methanogenesis in anaerobic digestion process: fundamentals, process optimization, design and scale-up considerations. Bioresour. Technol. 289, 121738.

Zeppilli, M., Paiano, P., Villano, M., Majone, M., 2019. Anodic vs cathodic potentiostatic control of a methane producing microbial electrolysis cell aimed at biogas upgrading. Biochem. Eng. J. 152, 107393.

Zhang, J., Zhang, Y., Quan, X., Chen, S., 2015. Enhancement of anaerobic acidogenesis by integrating an electrochemical system into an acidogenic reactor: effect of hydraulic retention times (HRT) and role of bacteria and acidophilic methanogenic Archaea. Bioresour. Technol. 179, 43–49.

Zhang, W., Zhu, S., Luque, R., Han, S., Hu, L., Xu, G., 2016. Recent development of carbon electrode materials and their bioanalytical and environmental applications. Chem. Soc. Rev. 45, 715–752.

Zhao, Z., Li, Y., Zhang, Y., Lovley, D.R., 2020. Sparking anaerobic digestion: promoting direct interspecies electron transfer to enhance methane production. iScience 23, 101794.

Zhao, Z., Zhang, Y., Chen, S., Quan, X., Yu, Q., 2014. Bioelectrochemical enhancement of anaerobic methanogenesis for high organic load rate wastewater treatment in a up-flow anaerobic sludge blanket (UASB) reactor. Sci. Rep. 4, 6658.

Zhao, Z., Zhang, Y., Quan, X., Zhao, H., 2016. Evaluation on direct interspecies electron transfer in anaerobic sludge digestion of microbial electrolysis cell. Bioresour. Technol. 200, 235–244.

Zhao, Z., Zhang, Y., Wang, L., Quan, X., 2015. Potential for direct interspecies electron transfer in an electric-anaerobic system to increase methane production from sludge digestion. Sci. Rep. 5, 11094.

Acidogenic fermentation of organic wastes for production of volatile fatty acids

Le Zhang^{a,b}, To-Hung Tsui^{a,b}, Kai-Chee Loh^{b,c}, Yanjun Dai^{b,d}, Yen Wah Tong^{a,b,c}

^{a}NUS Environmental Research Institute, National University of Singapore, Singapore, ^{b}Energy and Environmental Sustainability for Megacities (E2S2) Phase II, Campus for Research Excellence and Technological Enterprise (CREATE), Singapore, ^{c}Department of Chemical and Biomolecular Engineering, National University of Singapore, Singapore, ^{d}School of Mechanical Engineering, Shanghai Jiao Tong University, Shanghai, China

15.1 Introduction

In addition to anaerobic digestion technology, microbial fermentation also includes acidogenic fermentation technology, which holds great potential in organic waste management and resource recovery. Fig. 15.1 shows the technical theory of acidogenic fermentation for the production of volatile fatty acids (VFAs). Acidogenic fermentation involves hydrolysis, acidogenesis, acetogenesis, and an arrested methanogenesis step, which produces intermediate short-chain organic acids. As most of the short-chain organic acids (e.g., acetic acid, propionic acid, and butyric acid) are volatile, they are usually termed as VFAs. Traditional anaerobic digestion shares the same first three steps with acidogenic fermentation. However, in acidogenic fermentation, a variety of approaches can be used to inhibit or primarily block methanogenesis and lead to the purposed VFAs accumulation from acidogenesis and acetogenesis steps. Common strategies to decrease or prohibit methanogenic activities include utilization of inhibitors (e.g., 2-bromoethanesulfonic acid and $CHCl_3$), adjustment of operating parameters (e.g., organic loading, pH, and temperature), bioaugmentation employing acid-forming bacteria, thermal pretreatment (thermal shock) of inoculum (Jomnonkhaow et al., 2020), feedstock pretreatment technologies, etc. The produced VFAs are potential value-added renewable carbon sources, which can be utilized to manufacture biofuels and chemicals such as cosmetics, food additives, polymers, and pharmaceutical products. After separation and purification for VFAs recovery, the produced process water can be reused in subsequent acidogenic

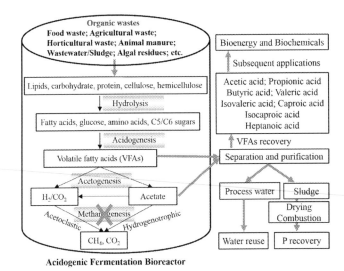

FIG. 15.1   **Technical theory of acidogenic fermentation of organic wastes for production of volatile fatty acids.**

fermentation while the remaining solid residuals (sludge) can be dried and then combusted into ash, from which valuable plant nutrients such as phosphorus can be recycled.

The most common VFAs are acetic acid, propionic acid, and butyric acid, which result from the major metabolic pathways in acidogenic fermentation. Taking food waste acidogenic fermentation as an example, the major acidogenic metabolic pathways can be categorized as acetate-ethanol type, propionate type, butyrate type, lactate type, homoacetogenic type, and mixed-acid type (Zhou et al., 2018). The major acidogenic metabolic pathways are shown in Fig. 15.2. Regarding acidogenic pathways of lignocellulosic biomass, lignocellulosic biomass was initially hydrolyzed into glucose and xylose, which are further fermented via acidogens and acetogens to produce VFAs and alcohols. Among the heterogeneous acidogenic metabolic pathways, pyruvate (CH_3COCOH) plays an essential role as an intermediate that can be converted into a wide range of metabolic products, including acetate, propionate, lactate, ethanol, etc. Acetic acid is one of the most dominant fermented compositions from various lignocellulosic substrates (Mockaitis et al., 2020). Acetate-ethanol pathway represents the pathway through which acetyl-CoA ($CH_3COSCoA$) was converted to acetate and ethanol via acetaldehyde and acetyl-P, along with the cogeneration of a small quantity of CO_2 and H_2 (Kandylis et al., 2016). Butyrate-type pathway refers to the pathway that acetyl-CoA stemming from pyruvate was converted into butyric acid along with CO_2 and H_2 through a series of biochemical reactions (Fig. 15.2). Propionate-type metabolic pathway corresponds to production of propionate from acetyl-CoA or intermediate lactate. The metabolism of each type of metabolic pathway involves unique enzymes and electron transfer which can be influenced by many factors such as pH, dominant species, feedstock (carbon source), retention time, and methanogenesis inhibitor.

The technical feasibility of acidogenic fermentation of organic wastes for production of VFAs has been demonstrated in previous studies; however, several technical limitations still remain, including high separation cost to recover VFAs of high purity, low VFAs productivity

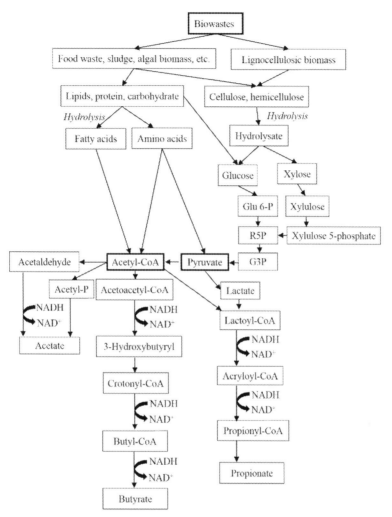

FIG. 15.2 Major acidogenic metabolic pathways. Acetyl-CoA: acetyl-coenzyme A. Modified from (Zhang et al., 2021; Zhou et al., 2018) with permission/license granted by the publisher.

from recalcitrant substrates, functional identification of key microbial consortia in various acidogenic fermentation bioreactors, unexplored applications of such VFAs, etc. To promote the technological development and potential markets of VFAs from biowastes, it is critical to consolidate insights scattered across the literature by summarizing the latest progress in acidogenic fermentation of organic wastes for production of VFAs.

Therefore, this book chapter aimed to provide the state-of-the-art technologies on acidogenic fermentation of organic wastes for production of VFAs, involving fermentation feedstock, inoculum, bioreactor and operation mode, operating parameters, intensification strategies, VFA recovery methods, and subsequent applications. The key technical bottlenecks have been summarized, and future development prospects are highlighted.

15.2 Substrates for volatile fatty acids production through acidogenic fermentation

A variety of liquid and solid biowastes as substrates have been investigated for their potential for VFAs production through acidogenic fermentation. Among them, common substrates are lignocellulosic biomass (Sawatdeenarunat et al., 2017), food waste (Strazzera et al., 2018), manure (Jomnonkhaow et al., 2020), industrial wastewater (Garcia-Aguirre et al., 2017), sludge (Chen et al., 2020), algal residue (Cho et al., 2018), and some cosubstrates (Esteban-Gutiérrez et al., 2018). Common substrates for VFAs production through acidogenic fermentation and corresponding reactor performance are summarized in Table 15.1.

Different biowastes vary in their contents of carbohydrate, protein, lipid, cellulose, and hemicellulose, which could result in complex hydrolysis steps then recovery of different of VFAs. The most predominant VFAs following acidogenic fermentation of glucose, peptone, and glycerol, representing carbohydrates, proteins, and lipids, respectively, were identified as butyric, acetic, and propionic acid, respectively (Yin et al., 2016). Instead of acidogenic fermentation of a single biowaste type, cofermentation of two or more biowastes types can help mitigate certain inhibitions due to the unbalanced feed composition while simultaneously manage a wider range of wastes (Huang et al., 2019). The synergistic effects between food and paper wastes have been confirmed for enhanced VFAs production (Soomro et al., 2020). The higher VFA yield could be due to enhanced hydrolysis of the cosubstrates (Lee et al., 2014). Hence, various cosubstrates could be further studied to enhance the VFAs production via acidogenic fermentation in future studies. Furthermore, different countries and areas may have significantly different biowastes and technical readiness of microbial fermentation. Thus, apart from the characteristics of the biowastes, other important factors such as substrate availability, environmental incentives, and economic feasibility should be taken into consideration before potential industrial application.

15.3 Inocula for acidogenic fermentation

The inoculum is an essential factor influencing the evolution of acidogenic fermentation pathways (Arras et al., 2019). Inocula quality or composition determines the acidogenic fermentation pathways that can lead to significantly varied VFAs compositions (Tang et al., 2017; van Aarle et al., 2015). Therefore, the source of inoculum should be carefully selected. Generally, three types of inocula (microbial strains) are available, including pure culture, coculture, and mixed cultures. Acidogenic fermentation by pure culture requires the sterilization process and complex operation, which regularly results in unsatisfactory reactor performance, especially for scale-up applications. In contrast, acidogenic fermentation by cocultures and mixed cultures can be performed under nonsterile conditions and able to utilize complex substrates (Bhatia & Yang, 2017; Islam et al., 2017). The advantages of pure culture fermentation are high products quality, and easier separation procedures, while fermentation with the cocultures or mixed cultures is superior in more diversified biotransformation pathways and higher flexibility of converting different substrates (Atasoy et al., 2019). Instead of using the high cost in purchasing pure cultures, the acclimatized consortia (mixed cultures) can be enriched from natural microbes, and have been successfully utilized in acidogenic fermentation to enhance the VFAs production (Chang et al., 2018; Saha et al., 2019).

TABLE 15.1 Common substrates, bioreactors, operation mode and conditions, VFA yield, and components in acidogenic fermentation of biowastes.

Substrate	Bioreactor	Operation mode/conditions	VFA yield	Main VFA components	Reference
Brewery spent grain	2.1 L batch reactor	Two-step process (hydrolysis + fermentation); pH 4.5–8.0; 37 °C; 1500 rpm	16.9 g COD/L (pH 6.0)	Acetic and butyric acids	(Castilla-Archilla et al., 2021)
Cow manure	120 mL serum glass bottles	Semicontinuous; 37 °C; 100 rpm; OLR 4.7 gVS/L/d; VFAs recovery by membrane-assisted bioreactor	0.41 g VFAs/g VS; 6.9 g/L VFAs	Acetic acids	(Jomnonkhaow et al., 2020)
Food waste	0.5–0.6 L bench scale reactors	Batch; pH 6; 28 °C; 100 rpm; salinity concentration 0–50 g/L	4.6–6.6 g/L	Acetic, propionic, butyric, and valeric acids	(Sarkar et al., 2020a)
Brewery spent grains	0.5–10 L two-stage bioreactors	Batch; initial pH 4–10; 35 °C	Stage 1: 4.87–8.94 g/L; stage 2: 4.72–10.32 g/L	Acetic, propionic, butyric, and valeric acids	(Sarkar et al., 2020b)
Food waste	15 L two-phase leachate bed bioreactor	Semicontinuous; pH 7; 35 °C	6.3 g COD/L; 0.29 g VFA/gVS	Acetic, propionic and butyric acids	(Zhang et al., 2020)
Microalgae	0.6 L batch reactors	Batch; 35 °C; 120 r/min, 112.8 mg Fe(VI)/g VS	0.73 g COD/g VS	Acetic and propionic acids	(Wang et al., 2020c)
Waste activated sludge	1 L serum bottles	Batch; additive iron and persulfate; 35 °C; 180 rpm	2.3 g COD/L	Acetic, propionic, butyric and valeric acids	(Zhang et al., 2020)
Waste activated sludge + tofu residue	1 L batch reactors	Batch; 35 °C; 100 rpm; without pH control	0.24 g COD/g VSS	Acetic, propionic, and isobutyric acids	(Huang et al., 2019)
Waste activated sludge	Glass serum bottles	Batch; additive sulfadiazine: 50 mg/kg dry sludge; 35 °C; SRT 8 days	2.0 g COD/L	Acetic, propionic and butyric acids	(Xie et al., 2019)
Food waste	0.6 L glass bottles	OLR 15 kg COD/m³-day; 35 °C; 100 rpm; buffering with Na₂CO₃, NaOH, CaCO₃	Max: 11.4 g/L	Acetic, butyric, and propionic acids	(Dahiya & Mohan, 2019)
Sewage sludge	120 mL bottles	Batch; thermal treatment: 120 °C, 15 min; HRT 10 days; OLR: 1.6 g COD/L/d	0.41 mg COD/gVS	Acetic, propionic and butyric acids	(Iglesias-Iglesias et al., 2019)

(continued)

TABLE 15.1 (Cont'd)

Substrate	Bioreactor	Operation mode/conditions	VFA yield	Main VFA components	Reference
Saccharification residue from food waste	Lab-scale reactors	Batch; 37 °C; 72 h; pH 5.5–6.5	268 mg COD/gVS	Butyric, propionic and acetic acids	(Jin et al., 2019)
Food waste	500 mL bottles	Batch; pH around 7.7; 37 °C; 84 rpm; 10 d	8.2 g/L VFAs; 10.8 g COD/L	Acetic acid (66%–80%)	(Tampio et al., 2019)
Food waste	Leachate bed reactor	Batch mode; 22 °C; 14 days	16–28.6 g COD/L	Butyric, acetic and propionic acids	(Xiong et al., 2019b)
Food waste	1 L wide-mouthed bottles	Batch; 30 °C; pH 6; NaCl 10 g/L	0.54 g/gVS	Acetic, propionic and butyric acids	(He et al., 2019)
Algal residue (Cyanobacteria)	1 L reactors	Batch; alkaline pretreatment; pH 6.5; 35 °C	12.4 g/L	Acetic, butyric, propionic and valeric acids	(Cho et al., 2018)
Sewage sludge and agrifood waste	500 mL bottles; 80 L pilot-scale reactor	Batch; pH 5.5 or 10; 35 or 55 °C; pH 9 for pilot scale	0.24–0.52 g VFA/g COD	Acetic acid (30%–65%), butyric acid (60%) when at 55 °C	(Esteban-Gutiérrez et al., 2018)
Excess sludge	2.5 L granular sludge blanket reactor	Semicontinuous; pH 10; SRT 8 days; 35 °C	0.57 g COD/gVS	Acetic and propionic acids	(Li et al., 2018)
Lignocellulosic biomass	250 mL Erlenmeyer flasks	Batch; 37 °C; 100 rpm; Micro-oxygenation	110 mg/gVS	Acetic and propionic acids	(Sawatdeenarunat et al., 2017)
Industrial wastewater	500 mL Pyrex bottles	Batch; pH 5.5; pH 10; 35 °C and 55 °C	8.3 g COD/L	Propionic, butyric and acetic acids	(Garcia-Aguirre et al., 2017)
Algal residue (*Ettlia* sp.)	–	Batch; 48 °C; pH 7	8.2 g/L	Butyric, acetic, and propionic acids	(Seo et al., 2016)
Food waste + excess sludge	500 mL reactors	Semicontinuous; 40 °C; OLR 9 g VS/L/d; FW/ES 5; SRT 7 d	0.87 g COD/gVS	Propionic and acetic acids	(Wu et al., 2016)
Glucose, peptone, and glycerol representing carbohydrates, proteins, and lipids	500 mL brown wide-mouthed bottles	Batch; pH 6.0; 30 °C; 15 days	38.2, 32.2, 31.1 g COD/L for glucose, peptone, and glycerol, respectively	Butyric, acetic, and propionic acids for glucose, peptone, and glycerol, respectively	(Yin et al., 2016)
Excess sludge	1 L Erlenmeyer flasks	Batch; 28 °C; pH 5–12; 20 days	3.2 g COD/L at optimal pH 10	Acetic and propionic acids	(Wu et al., 2016)

15.4 Bioreactors and operation modes for VFAs production *via* acidogenic fermentation

The most widely used lab-scale bioreactors (Table 15.1) for various acidogenic fermentations were single-phase bioreactors with working volumes between 0.5 L and 15 L. Only a few studies reported pilot-scale (e.g., 80 L) acidogenic fermentation for production of VFAs (Esteban-Gutiérrez et al., 2018; Garcia-Aguirre et al., 2019). Most of the bioreactors were operated in a batch mode. Thus, continuous acidogenic fermentation, especially at the pilot scale, would be worth exploring. Recent studies on bioreactors for acidogenic fermentation focused on leached-bed bioreactors and membrane bioreactors. The basic working mechanisms of the above two kinds of bioreactors are similar. Specifically, the filtration mesh in leachate bed bioreactors or the filtration membrane in the membrane bioreactors allows separation of the VFAs-containing liquid stream from the fermentation broth (Wainaina et al., 2020; Zhang et al., 2020). Therefore, compared to the traditional single-phase bioreactors, the aforementioned two kinds of bioreactors possess great advantages in the separation and recovery of VFAs. Regarding the leachate bed bioreactors, Xiong et al. (2019a; 2019b) performed systematic studies on bioreactor design and effects of inoculum to substrate ratios on reactor performance. It was found that the inoculum to substrate ratio can significantly affect the hydrolysis and fermentation kinetics. Zhang et al. (2020) conducted semicontinuous acidogenic fermentation of food waste using a leached-bed reactor and obtained an average concentration of 6.3 g chemical oxygen demand (COD)/L and yield of 0.29 g VFA/g volatile solids (VS), respectively. Moreover, an appropriate leachate recirculation rate (e.g., 50%) of the leachate bed bioreactors was capable of enhancing the hydrolysis efficiency and increasing VFA production (Hussain et al., 2017). Khan et al. (2019) optimized the organic loading rate (OLR) and the hydraulic retention time (HRT) of an anaerobic membrane bioreactor for VFA production from low-strength wastewater. Nonetheless, a potential bottleneck of the membrane bioreactor for VFA separation and recovery was the blockage/fouling of membrane pores due to substrate and sludge particles. To tackle this issue, Wainaina et al. (2019b) developed an anaerobic immersed membrane bioreactor coupled with a backwashing unit for VFA production and recovery, and achieved a high yield of 0.54 g VFA/g VS at an OLR of 2 g VS/L/d. Moreover, the effects of OLRs range between 4 and 10 g VS/L/d on the performance of immersed membrane bioreactors were tested by Wainaina et al. (2020). It attained the highest yield of 0.52 g VFA/gVS at 6 g VS/L/d. More recently, a stable long-term (114 days) acidogenic fermentation of cow manure using an immersed membrane bioreactor for *in-situ* recovery of VFAs was performed successfully (Jomnonkhaow et al., 2020). Indeed, continuous acidogenic fermentation can bring several advantages, including more consistent VFA quality, higher organic loadings, and bridging the technical gaps between laboratory research and practical application (Garcia-Aguirre et al., 2019). Furthermore, the VFA recovery efficiency can be significantly affected by the type of extractant (e.g., trioctylamine) utilized in the vapor permeation membrane contactor module (Aydin et al., 2018). Regarding pilot-scale studies, Esteban-Gutiérrez et al. (2018) conducted pilot-scale cofermentation of sewage sludge and agrifood waste, and it was found that thermophilic temperature and alkaline pH conditions boosted the VFA concentration by 1.7- to 2-fold. Da Ros et al. (2020) filtered municipal wastewater using a pilot-scale (400 m^3/day) rotating belt filter to recycle cellulosic primary sludge that was regarded as an appropriate substrate for VFA production.

15.5 Enhancing strategies for elevated VFAs yield from acidogenic fermentation

15.5.1 Optimization of operating parameters

The operating parameters of a bioreactor have direct impacts on the yield and distribution of VFA products. Hence, it is extremely essential to optimize the operating parameters during acidogenic fermentation of biowastes. Hitherto, the effects of various parameters including pH, temperature, OLR, solid retention time (SRT), HRT, carbon/nitrogen (C/N) ratio, and inoculum to substrate ratio on VFA yield and composition have been widely studied (Atasoy et al., 2018, 2019; Dahiya & Mohan, 2019; Khan et al., 2019; Strazzera et al., 2018). pH has been regarded as one of the most essential factors to affect VFA production and operation stability of acidogenic fermentation by influencing the competition between acidogens and methanogens. Indeed, acidogens can adapt to a pH range of 3 to 12, whereas the optimal pH for methanogens to survive is 6.8–7.5 (Latif et al., 2017). Meanwhile, the appropriate pH value varies depending on the type of substrate used for acidogenic fermentation (Moretto et al., 2019). For instance, a pH of 8–10 is more appropriate for the fermentation of sludge or wastewater, while a pH of 6–7 is more adaptable for acidogenic fermentation of food waste. During acidogenic fermentation, pH can be adjusted automatically or artificially to facilitate VFA accumulation and interrupt methanogenesis. Adjusting pH to below 6.5 is an ordinary approach to avoid methane generation and start the acidogenic fermentation. Afterward, the VFA accumulated during fermentation can further decrease pH value without external control. Moreover, alkaline fermentation (e.g., pH > 9) was found to able to not only inhibit the methanogenesis but also enhance the VFA yields (Cabrera et al., 2019). Reportedly, the pathway of acetoclastic methanogenesis was severely inhibited by alkaline conditions during alkaline fermentation of sludge (Ma et al., 2019). Disintegration, acidogenesis, and acetogenesis processes were enhanced by 53%, 1030%, and 30%, respectively, when the fermentation pH increased from 7 to 10 (Wang et al., 2017). Alkaline environment facilitated the hydrolysis of soluble organic matters, which contributed to subsequent acidogenesis for VFA production. However, during the alkaline fermentation without pH adjustment, a common issue is that the initial alkaline pH probably cannot be maintained continuously and would continue to drop to a neutral pH level due to the accumulation of VFA (Jankowska et al., 2017). NaOH or Ca(OH)$_2$ was usually adopted to adjust pH to maintain alkaline fermentation; however, the corresponding economic feasibility remains unclear. Furthermore, it is essential to study distribution rules of various VFA components in fermentation with different pH values because organic acids of different chain lengths have different market values.

Additionally, temperature plays a key role in microbial growth and enzyme activities, thereby being a parameter for optimization of acidogenic fermentation. Acidogenic microbes can function at different temperature ranges (e.g. 20–65 °C). The selection of the acidogenic fermenter temperature generally depends on the scope of VFA application. Within the typical range of 21–55 °C, increasing the fermentation temperature was found to be capable of facilitating hydrolysis of the substrates and enhancing the VFA yield (Cavinato et al., 2017). Specifically, Zhuo et al. (2012) investigated acidification at 10, 20, 37, and 55 °C. They demonstrated that 37 °C was the point with the highest VFA yield after a 3-day fermentation. However, an optimal temperature for VFA fermentation from the organic fraction of municipal solid waste was recently determined as 42 °C (Soomro et al., 2020). As reported by Fernández-Domínguez et al. (2020), the VFA yield was highest at 35 °C, and it was 2%, 6%, 10%, and 14% higher than that of 55, 45, 20, and 70 °C, respectively, where the VFA profile was not significantly affected by the elevated temperature.

OLR is defined as the amount of feedstock fed into the bioreactor per unit of working volume per day. A low OLR means the low use of possible treatment capacity, while the acidogenic fermentation at too high OLR can lead to unstable processes due to VFA accumulation and microbial growth inhibition. Thus, the selection of an optimal OLR is fundamental for achieving stable acidogenic fermentation for VFA production. Yet, the optimal OLR may vary greatly depending on various substrates, temperature and bioreactor, operation mode, and aeration (Nguyen et al., 2019). For instance, Magdalena et al. (2019) studied the effect of stepwise OLR increases (3, 6, 9, 12, and 15 g COD/L/d) on VFAs production from microalgae biomass. They observed that the organic matter conversion efficiency into VFAs increased as the OLR increased from 3 to 12 COD/L/d, but did not show any further increase when OLR was increased to 15 COD/L/d. Lukitawesa et al. (2020) investigated semicontinuous production of VFAs from citrus waste under an OLR from 1 to 8 g VS/L/d, and it was found that the highest yield of VFAs (i.e. 0.84 g VFA/g VS) was achieved at OLR 4 g VS/L/d. The VFA yield from acidogenic fermentation of food waste increased along with elevated OLR from 5 to 13 g/L/d, but the process stability decreased under an OLR of 13 g/L/d due to the extremely viscous fermentation broth (Lim et al., 2008). Similar instability of a bioreactor with an OLR of 11–16 gTS/L/d was observed by Jiang et al. (2013), who obtained a stable production rate by decreasing the OLR. In the case of using continuous or semicontinuous operations, HRT, representing the time of feedstock remaining in the bioreactor, has a close relationship with OLR, and can also affect the yield and profile of VFAs (Wainaina et al., 2019a). An HRT of 0.2–15 days is typical for acidogenic fermentation of biowastes for VFA production, and the exact duration strongly depends on the specific waste type (Jankowska et al., 2015; Lee et al., 2014). A relatively long HRT could be beneficial to microbes' adaptation to the targeted biowaste, thereby enhancing VFA production, especially for the lignocellulosic substrates. Moreover, alteration of the C/N ratio has been found to affect the distribution of VFA from acidogenic fermentation (Liu et al., 2008). Other factors including different substrate characteristics, pretreatments, and operation conditions should also be taken into consideration during the optimization of HRT.

15.5.2   Pretreatment of biowastes

To enhance fermentation performance and total VFA yield, a wide range of pretreatment methods have been developed, and they can be categorized into physical, chemical, and biological methods as well as their combined applications (Lee et al., 2014). The main functions of these pretreatment are to reduce the substrate size and crystallinity degrees, and augment substrate surface areas, thereby facilitating subsequent hydrolysis and acidogenic fermentation. Generally, particle sizes of raw lignocellulosic biomass substrates are not ideal at particle size of 14–28 mesh (1.168–0.589 mm in diameter) (Lynam et al., 2012). Through comminution or milling, the diameters of biomass particles can be reduced to less than 2 mm or even smaller. From the perspective of industrial application, it is acceptable to mill or cut the particle sizes of lignocellulosic biomass into 1 cm, but further size reduction may lead to much higher operational costs. Regarding physical pretreatments, hydrothermal treatment is a promising option (laqa Kakar et al., 2020). Yin et al. (2014) studied the hydrothermal treatment of food waste prior to acidogenic fermentation and demonstrated that the highest VFA yield of 0.91 g/g VS was achieved at the hydrothermal temperature of 160 °C. Xiang et al. (2021) confirmed the effectiveness of hydrothermal pretreatment (i.e., 90–130°C for 15 min) of rice straw on improving its fermentation performance. Notably, hydrothermal pretreatment at temperatures above 200 °C could inhibit the fermentation process due to the potential generation of inhibitory substances (e.g., melanoidins) (Yin et al., 2014). Microwave pretreatment is another common physical pretreatment approach, which is able to inactivate bacteria and pathogens through the high temperature generated by the applied electromagnetic field (Zhang et al., 2017). However, the improvement of VFA yield caused by microwave pretreatment cannot often cover the energy cost of the pretreatment, which greatly limits their implementation in industrial-scale plants (Strazzera et al., 2018). Additionally, sonication pretreatment with a density of 2 W/mL for 15 min can halve the fermentation time to reach the maximum VFA yield (Liu et al., 2018). Many chemical pretreatment methods including acid, alkaline, ammonia fiber explosion, ionic liquid, sulfur dioxide explosion, and others have been studied for efficient pretreatment of various biowastes (Behera et al., 2014; Yu et al., 2018). For instance, sludge biodegradability and VFA production were successfully enhanced by chemical pretreatment using tetrakis hydroxymethyl phosphonium sulfate (Wu et al., 2017). Industrial application of alkaline chemicals (e.g., NaOH and $Ca(OH)_2$) for effective chemical pretreatment has been confirmed. Biological pretreatment is another promising way to open up the recalcitrant structure of certain biowastes (e.g. lignocellulosic biomass and sludge) and enhance fermentation performance. For example, Fang et al. (2018) and Liu et al. (2016) adopted fungi species (e.g. *Pleurotus sajor-caju*) and enzymes (e.g., protease), respectively, to perform biological pretreatment of solid digestate, and successfully accelerated the carbon release and enhanced VFA yields by 117% to 124 %. Compared to the above single pretreatments, a synergistic effect could be achieved with combinations of different pretreatment methods such as microwave irradiation with sodium citrate addition for excess sewage sludge (Peng et al., 2018), ultrasound-calcium hydroxide pretreatment for grass clipping (Wang et al., 2019), calcium peroxide with free ammonia pretreatment for sludge (Wang et al., 2018), and thermo-chemical pretreatment for sludge (Yu et al., 2018), leading to enhanced fermentation performance. Although the effectiveness of various pretreatment methods and their combinations have been confirmed by many lab-scale studies, more comparative analyses are indispensable to select the suitable pretreatment method for each type of biowaste. Generally, physical and chemical pretreatments are suggested for size reduction of lignocellulosic biomass wastes, while physical pretreatments (e.g. pulping or hydrothermal treatment) are suggested for sludge and food wastes to decrease particle sizes of substrates. Regarding manure, high temperature pretreatment (e.g., microwave) is recommended to inactivate the undesired pathogens in the substrates. Moreover, previous studies mainly focus on the technical aspects of pretreatments; the corresponding economic evaluation and environmental impact of each pretreatment approach should be taken into consideration.

15.5.3   Additives

During acidogenic fermentation, complex biochemical reactions rely on hydrolytic bacteria and acidogenic bacteria (Liu et al., 2020a; Sträuber et al., 2012). Thus, strategies for enhancing the growth of these microorganisms could favor fermentation processes and enhance VFA yields. Supplementation of additives is such a useful strategy. Trace elements (e.g. cobalt, nickel, zinc, and iron) are necessary for activating and maintaining enzymatic activities in microbial metabolism. For instance, Co, Ni, and Fe are cofactors of carbon monoxide dehydrogenase, which is an essential enzyme for the production and consumption of acetate (Fang et al., 2020). Hitherto, many works of the literature reported the impact of trace element addition on the degradation efficiency of VFAs (Jiang et al., 2017), but there are very limited studies specifically focusing on the effects of trace elements on VFA yields. In this regard, further investigations on how trace elements affect acidogenic fermentation and VFA yields would be required. Biochar, as another usually employed bioamendment agent for microbial fermentation processes, was shown to immobilize the microbes, enhance bacterial growth, suppress odor generation, and increase the organic loading (Duan et al., 2019b; Kaur et al., 2020). For instance, Lu et al. (2020) performed dark fermentation with biochar-amended biological consortium for production of VFAs and found that the functional groups on biochar surface contributed to biofilm development and electron transfer, which subsequently favored the production of acetate. Duan et al. (2019a) applied algae-derived biochar in algae anaerobic fermentation for VFA production and confirmed that algae-derived biochar approximately doubled VFA yield and benefited VFA generation rate. Moreover, biochar produced under different conditions (e.g., altered temperature and carrier gas) could outcome significantly different properties in terms of surface area, particle size, pH value, pore size, bulk density, etc. (Hu et al., 2020; Rajapaksha et al., 2016). Hence, further optimization of the operating conditions for biochar with required physicochemical properties could help maximize the VFA yield in acidogenic fermentation of biowastes. Moreover, Huang et al. (2016b) studied the effect of biosurfactants (e.g., surfactin, saponin, and rhamnolipid) on the fermentation of waste activated sludge for VFAs production. They found that the addition of plant-derived saponin inhibited the methanogenesis process but improved the solubility of organic matter, resulting in the accumulation of VFAs (i.e., 425.2 mg COD/g volatile suspended solids (VSS)). Dahiya and Mohan (2019) confirmed the benefits of some buffering/neutralizing agents (e.g., Na_2CO_3, NaOH, $Ca(OH)_2$, and $CaCO_3$) as additives to favor selective VFA production in acidogenic fermentation bioreactor of food waste. To further reduce the cost of additives, alternative buffering agents (e.g., cement asbestos wastes) deserve further study to facilitate VFA production on a large scale.

15.5.4 Bioaugmentation

Acid-producing bacterial communities play a decisive role in the acidogenesis of biowastes for VFA production. With a good understanding of bacterial communities, bioaugmentation employing beneficial microbes can help improve the fermentation processes. The rules governing dynamic succession of predominant species and their operational parameters are important, and they can be analyzed by a bioinformatic analysis of the metagenomics data (Zhang et al., 2019). The aforementioned information of predominant microbes in acidogenic bioreactors can lay the foundation for the implementation of bioaugmentation strategies to enhance VFA production. Bioaugmentation refers to a method to enhance the performance of microbial fermentation through adding preadapted consortium, pure cultures, or genetically engineered microbes (Herrero & Stuckey, 2015). A successful bioaugmentation strategy to increase butyric acid production using the bioaugmented mixed culture (i.e. *Clostridium butyricum*) was recently reported (Atasoy & Cetecioglu, 2020). Similarly, Reddy et al. (2018) applied an enriched mixed culture of *Clostridium kluyveri* to food waste fermentation for VFA production and obtained the highest yields for medium-chain VFAs (8.1 g/L of caproic acid) and short-chain VFA (8.9 g/L of butyric acid). Dams et al. (2018) successfully applied a mixed culture of *Clostridium acetobutylicum* in an acidogenic fermenter to enhance the production of *n*-caproic acid by improving the production of related precursors. More recently, Wang et al. (2020a) achieved an enhanced production of VFAs by adding a kind of sulfate-reducing bacteria to acidogenic fermentation of waste activated sludge under alkaline pH. It was known that the sulfate-reducing bacteria could compete with methanogens for electrons and accelerate the transformation of macromolecular organic polysaccharide and protein to VFAs (Tsui et al., 2018; Wang et al., 2020a). Another interesting example is the enhancement of dark fermentation under hyperthermophilic conditions (e.g., 70–80 °C) employing *Thermotoga neapolitana*, which further expanded the application scope of the bioaugmentation strategy (Okonkwo et al., 2020). Nevertheless, not all bioaugmentations can securely lead to positive outcomes due to species differences in the added microbes (Eder et al., 2020). The competition for nutrients between the indigenous microbes and bioaugmented species as well as the environmental stresses caused by altered operational conditions could result in the failure of bioaugmentation. To tackle this issue, multiple inoculation techniques for bioaugmentation could be a promising method.

15.5.5   Cofermentation

Cofermentation is a cost-effective and high-efficiency VFA production strategy as it can effectively enhance hydrolysis yield and acidogenesis yield (Wu et al., 2016). Hitherto, various cofermentation of different biowastes have been carried out, including waste activated sludge and tofu residue (Huang et al., 2019), food waste and excess sludge (Wu et al., 2016), macro- and microalgae (Xia et al., 2016), waste activated sludge and henna plant biomass (Huang et al., 2016a), wastewater and food waste (Li & Li, 2017), mushroom residue and sewage sludge (Fang et al., 2019), syngas and carbohydrate-rich synthetic wastewater (Liu et al., 2020b), as well as microalgae and rice residue (Sun et al., 2018). Through cofermentation, an appropriate C/N ratio (e.g. 25–26) can be achieved, which is essential to balance the nutrition in the fermenter (Chen et al., 2013; Lin et al., 2017). The lignocellulosic biomass (e.g., rice straw) usually has a relatively high C/N ratio such as 47–67 (Paul & Dutta, 2018), which is very unfavorable to microbial fermentation. Hence, cofermentation of lignocellulose biomass with other biowastes with low C/N ratio, such as food waste and sludge, to adjust the C/N ratio could significantly favor acidogenic fermentation and enhance VFA production (Fang et al., 2019). From a practical application point of view, cofermentation could simultaneously convert more kinds of biowastes into VFAs. Yet, the availability of biowastes should be taken into consideration when selecting the appropriate cosubstrate to conduct cofermentation.

15.6   Separation/recovery of VFAs from fermentation broth

One of the most challenging steps during acidogenic fermentation is the recovery/separation of the VFAs from fermentation broth (Outram & Zhang, 2018; Zacharof & Lovitt, 2013).

Hitherto, a series of separation methods have been investigated, including membrane-based separation, separation via ion exchange resins (adsorption), recovery via acidogenesis-electrodialysis integrated system, solvent extraction, gas stripping with absorption, etc. (Atasoy et al., 2018). Three of the separation or recovery methods, namely membrane separation, ion exchange resins, and electrodialysis, are reviewed in detail.

15.6.1 Membrane separation

Membrane separation processes of VFAs can be categorized into several types, involving pressure-driven reverse osmosis (RO), nanofiltration (NF), forward osmosis (FO), and supported liquid membrane (SLM). Among them, RO and NF are mostly reported. Bóna et al. (2020) investigated the separation of VFAs from model anaerobic effluents using RO, NF, FO, and SLM technologies. They found that RO and NF showed the highest average retention (84% at the applied pressure of 6 bar) and the highest permeance (6.5 L/m^{2}hbar), respectively. Separation performance of acetic acid from a xylose-glucose-acetic acid model solution by NF/RO membranes was evaluated by Zhou et al. (2013), who found that the separation factors of the RO membrane (i.e. 223.2 over xylose and 348.7 over glucose) were much higher than those of the NF membrane (i.e., 8.87 over xylose and 56.5 over glucose). Weng et al. (2010) found that the maximum separation factor of NF membrane (Desal-5 DK, GE-Osmonic) of acetic acid from sugars (i.e. xylose and arabinose) in dilute acid rice straw hydrolyzates was 49–52. Xiong et al. (2015) evaluated the ability of two commercially available NF membranes (i.e., Desal DK and Desal DL, GE-Osmonics) for carboxylic acid separation from the fermentation broth and achieved > 90% of sugar rejection, 0%–40% of most of the acid rejection, and 100% of rejection of butyric acid. Similarly, Zacharof et al. (2016) conducted NF of digested agricultural wastewater containing 21.08 mM of acetic acid and 15.81 mM of butyric acid with a range of NF membranes and identified the DK, DL (Osmonics, USA), and NF270 (Dow Chemicals, USA) as the promising candidates for carboxylic acids separation due to the fact that these membranes achieved retention ratios up to 75%.

Furthermore, the performance of RO and NF can be affected by a variety of factors including pH, temperature, pressure, membrane property, and feed concentration (Atasoy et al., 2018). For instance, the effect of pH on the retention of VFAs during the filtration of pretreated swine manure using three RO membranes was studied by Masse et al. (2008). It was observed that the retention of VFAs was reduced at pH < 5. The possible reason was due to the formation of unionized VFAs that can permeate more readily through the RO membranes. Aydin et al. (2018) investigated the influence of temperatures (i.e., 21, 30, and 38 °C) on VFA recovery from synthetic VFA mixture (1.25 g/L propionic, butyric, valeric, and caproic acids) through polytetrafluoroethylene (PTFE) and PTFE-trioctylamine membranes and achieved a higher separation factor at 38 °C compared to that of other temperatures. Both RO and NF are pressure-driven membrane processes, thereby being affected by the applied pressures. Wu et al. (2020) evaluated the performance of five pressure-driven membranes for acetic acid separation from fermentation broth and an RO membrane (BW30XFR) achieved the highest rejection (98.6% acetate rejection). A comprehensive review on various membrane processes for the recovery and purification of bio-based VFAs was recently published by Aktij et al. (2020), who highlighted several related challenges, including membrane fouling and contaminations, elevated cost for membrane separation of various components of VFAs, and lack of a cost-effective membrane method for large-scale application.

15.6.2 Separation *via* ion exchange resins (adsorption)

VFAs can also be recovered from the fermentation broth by the adsorption-based separation technique. Due to a remarkable capacity for VFAs and regeneration ability, the ion exchange resins were considered as a promising adsorbent for the recovery of VFAs (Reyhanitash et al., 2017). Eregowda et al. (2020) examined the removal of VFAs through adsorption on 11 anion exchange resins in batch systems and selected resins Amberlite IRA-67 and Dowex optipore L-493 for subsequent studies focusing on the adsorption kinetics, adsorption isotherms, and diffusion mechanism of the resins for VFAs. They achieved highly selective recovery rates (e.g., > 85% acetic acid and ~75% propionic acid). The acetic acid recovery efficiency of five anion-exchange resins (i.e., AmberLite IRN-78, Amberlyst A26, AmberLite IRA-67, DIAION WA10, and Dowex-66) was investigated by Wu et al. (2020), who found that IRN-78 was the best-performing resin for acetic acid separation. Also, IRN-78 had good reusability and was the better fit to a pseudo-second-order model. Moreover, ion exchange resins can also be used for the adsorptive removal of unsaturated fatty acids such as oleic, linoleic, and linolenic acids (Khedkar et al., 2021).

15.6.3 Electrodialysis

Electrodialysis as a mature membrane separation process can be utilized for recovery of VFAs from the fermentation broth. The separation mechanism of the traditional membrane electrodialysis has been well documented (Atasoy et al., 2018). Essentially, ion exchange membranes selectively transport anions or cations in individual cells under an applied direct-current potential (Sun et al., 2017). In the past decade, research was focused more on the use of bipolar membrane electrodialysis. Specifically, the feasibility of a bipolar membrane electrodialysis system scenario was validated for recovery of mixed VFAs from pig manure hydrolysate with a separation efficiency of 87% (Shi et al., 2018). Bak et al. (2019) investigated electrodialytic separation of VFAs from hydrogen-fermented food waste, achieved a maximum VFA purity of 96.2%, and demonstrated that VFA recovery was affected by the applied voltages and volume of the concentrate. A limitation of electrodialysis is the loss of nutrients from the fermentation broth as electrodialysis is not selective to VFAs alone. To overcome nutrient loss, Brown et al. (2020) studied the use of a PTFE membrane configuration coupled to electrodialysis for continuous extraction of VFAs from a mixed culture bioreactor, and achieved up to 98% of total VFA recovery. More recently, a continually fed 100 L food waste bioreactor by integrating filtration and electrodialysis was developed for continuous recovery and enhanced yields of VFAs, which was fit for downstream use with a concentration of 4 g/L (Jones et al., 2021).

15.7 Subsequent applications of VFAs

In addition to VFA utilization for production of biogas, biohydrogen, and electricity (Lee et al., 2014; Strazzera et al., 2018; Tsui et al., 2020), VFAs are the building blocks of valuable organic compounds such as alcohols, esters, ketones, aldehydes, and olefins (Tsui et al., 2019), which are demanded by many chemicals industries. VFAs can also be used as the substrates for various microbial fermentation processes of oleaginous yeasts for synthesis of microbial lipids that can be further converted into biodiesel. Besides, VFAs can also be used as carbon sources for polyhydroxyalkanoates (PHAs) production (Szacherska et al., 2021) as well as biological nitrogen and phosphorus removal in wastewater treatment process (Feng et al., 2021). Biodiesel produced from microbial lipids and PHAs produced from VFAs are typical applications of VFAs in production of bioenergy and biochemical, respectively. Hence, applications of VFAs for production of microbial lipids and PHAs are discussed next.

15.7.1   Microbial lipids for production of biodiesel

Previous studies have shown that biowastes-derived VFAs can be converted into microbial lipids via oleaginous yeasts and the resultant lipids can be further converted to biodiesel through a transesterification process (Llamas et al., 2020; Zhang et al., 2021). Biodiesel produced from nonedible microbial lipids avoids the competition with edible lipids such as soybean oil, palm oil, and rapeseed oil for food purposes, which therefore attracted great attention in the past decade. Compared to edible lipids, microbial lipids have a relatively low cost, which is beneficial to minimize the production costs of biodiesel. Moreover, the microbial lipids or single-cell oil synthesized from VFAs share similar main compositions with jatropha oil and soybean oil. Specifically, lipids accumulated in *Lipomyces starkeyi* contained about 46% of oleic acid (C18:1), 40.2% of palmitic acid (C16:0), 5.9% of stearic acid (C18:0), 4.5% of palmitoleic acid (C16:1), and 3.1 % of linoleic acid (C18:2), respectively (Liu et al., 2020c). These advantageous characteristics make microbial lipids be promising as a potential alternative to edible lipids for biodiesel production. Various VFA mixtures and hydrolysates derived from biowastes such as corn stover (Gong et al., 2016), food waste (Gao et al., 2017), vegetable waste (Chatterjee & Mohan, 2018), corncob (Liu et al., 2015), sludge (Liu et al., 2016), and waste office paper (Annamalai et al., 2020; Annamalai et al., 2018) have been investigated for microbial lipid production. The yield and composition of the obtained mixture of lipids can be significantly affected by the variations in carbon sources in terms of glucose, xylose, acetate, and other nutrient substances. For instance, cultivation of *Cryptococcus curvatus* using 30 g/L mixed VFAs as a source of carbon led to a lipid yield and content of 4.88–4.93 g/L and 45.1%–58%, respectively (Liu et al., 2017a; Liu et al., 2017b). In contrast, a yield of 1.4–5.8 g/L of lipids with a content in lipids of 22%–37.8% was achieved from a similar type of microorganism cultivated at a similar temperature using waste office paper hydrolysate as a substrate (Annamalai et al., 2018). Oleaginous microorganisms such as *Yarrowia lipolytica* (Kumar et al., 2021; Lazar et al., 2018), *Cryptococcus vishniaccii* (Deeba et al., 2016), *C. curvatus* (Chatterjee & Mohan, 2018), *Rhodotorula glutinis* (Liu et al., 2015), *L. starkeyi* (Di Fidio et al., 2020; Xavier et al., 2017), *Cryptococcus albidus* (Sathiyamoorthi et al., 2019), *Rhodosporidium toruloides* (Castañeda et al., 2018), and *Rhodosporidiobolus fluvialis* (Poontawee et al., 2018) have been studied. Hitherto, various strategies for enhancing lipid accumulation in the oleaginous yeasts have been developed to improve the technical and economic feasibility, including optimization of parameters (e.g. C/N ratio, pH, temperature, VFA concentration, and feeding methods) (Gorte et al., 2020; Leong et al., 2018), genetic and metabolic engineering approaches (Kamineni & Shaw, 2020; Wang et al., 2020b), and two-stage fermentation (Liu et al., 2020c).

15.7.2   PHAs for production of bioplastics

VFAs derived from biowastes can also be utilized to produce PHAs that are a family of environmentally friendly biodegradable thermoplastic and can be used as raw material for the production of bioplastics (Bhatia et al., 2021; Szacherska et al., 2021). PHAs production with biowastes-derived VFAs as suitable carbon sources would contribute simultaneously to waste reduction and resource recovery. As the PHAs-based bioplastics were regarded as a promising substitution of the conventional petrochemical-based plastics such as polyethylene, polystyrene, polypropylene, and polyethylene terephthalate, production of PHAs from biowastes-derived VFAs attracted global attention in the past decades (Tripathi et al., 2021). Hitherto, the technical feasibility of PHAs production from waste streams via PHAs-accumulating microbes has been confirmed; however, the large-scale application is still limited by the high production cost of PHAs, necessitating further efforts in increasing the conversion efficiency. Moreover, the unsatisfactory mechanical strength under thermal conditions is another technical limitation of the conventional technology of PHAs. Hence, a variety of studies have been conducted to reduce the costs of the biological process of PHAs synthesis while improving the material characteristics of the obtained PHAs (Blunt et al., 2018; Tarrahi et al., 2020). Recently, the next-generation industrial biotechnology was proposed to reduce the production cost and process complexity of PHAs production (Tan et al., 2020). The constituents and properties of PHAs polymer can be directly affected by the compositions of VFAs. The selection of an appropriate microbial producer is a critical factor for the profitable production of PHAs (Szacherska et al., 2021). Botturi et al. (2021) cultivated *Thauera* sp. Sel9 as pure strain using VFAs as a carbon source for the production of polyhydroxyalkanoated PHA-rich microbial biomass and achieved a PHAs content of 41%. Through optimization of the operational conditions such as the feeding of suitable VFA types, the PHAs contents in a range of 43.5%–76% were achieved during fermentation of mixed microbial consortia fed with fermented crude glycerol (Burniol-Figols et al., 2018a; Burniol-Figols et al., 2018b), waste stream (Perez-Zabaleta et al., 2021), sludge (Munir & Jamil, 2020; Tu et al., 2019), and organic fraction of municipal solid waste (Colombo et al., 2017). Furthermore, regulation of N and P contents in the fermentation broth is essential as excessive nutrients would inhibit the conversion of VFAs to PHAs but contribute to the growth of microorganisms (Albuquerque et al., 2007). The obtained PHAs can be manufactured into bioplastics films using various methods such as the conventional solution-casting method (Rech et al., 2020), extrusion, compression molding, and 3D printing (Frone et al., 2020). Yet, the unsatisfactory mechanical properties of bioplastics films have limited their wide applications. To reduce the brittleness of PHAs polymers, synthesis of copolymers such as poly (3-hydroxybutyrate-co-hydroxyvalerate), poly (3-hydroxybutyrate-co-hexanoate) and medium-chain length PHA copolymers has been investigated (Turco et al., 2021).

15.8   Conclusions and perspectives

Simultaneous organic waste management with the VFAs production has attracted great attention due to the fact that VFAs can be widely applied as platform chemicals in the industry. Numerous mixtures of VFAs have been produced successfully using pure or mixed cultures from various biowastes in acidogenic fermentation bioreactors. However, the acidogenic fermentation processes could be affected by a series of factors such as inocula, substrate properties, fermentation conditions (e.g. pH, temperature, and organic loading, etc.), bioreactors, and operation modes. To further increase the viability of acidogenic fermentation technology, several enhancing strategies have been developed for elevated VFA yield, including optimization of operating parameters, pretreatment of biowastes, additive strategy, bioaugmentation, and cofermentation. The downstream processing methods of VFAs directly affect the future development of waste management via acidogenic fermentation technology. Therefore, we have summarized the studies on separation and recovery of VFAs from the fermentation broth, focusing on membrane separation, separation via ion exchange resins (adsorption), and electrodialysis. Nevertheless, most of the current studies on membrane or leached-bed bioreactors are still lab-scale experiments, necessitating more efforts on scale-up practices. Moreover, subsequent novel application of recovered VFAs is a crucial step to close the loop of a circular economy. Hence, typical applications of VFAs in the production of bioenergy and biochemical, namely, microbial lipids for biodiesel production and synthesis of PHAs from VFAs, respectively, are discussed. Previous studies have confirmed the high value of VFAs produced from biowastes via acidogenic fermentation. Future studies should be conducted to further reduce the production cost and process complexity, contributing to successful commercial applications of biowastes-derived VFAs. Meanwhile, metabolic engineering and synthetic biology approaches can be applied to the related processes to develop engineered microorganisms with high production performance.

Acknowledgments

This work was funded by the National Research Foundation, Prime Minister's Office, Singapore under its Campus for Research Excellence and Technological Enterprise (CREATE) Program. For the copyrighted Fig. 15.2, the permission attained from the publisher for adaptation and reuse in the present work has been acknowledged.

References

Aktij, S., Zirehpour, A., Mollahosseini, A., Taherzadeh, M.J., Tiraferri, A., Rahimpour, A., 2020. Feasibility of membrane processes for the recovery and purification of bio-based volatile fatty acids: a comprehensive review. J. Ind. Eng. Chem. 81, 24–40.

Albuquerque, M., Eiroa, M., Torres, C., Nunes, B., Reis, M., 2007. Strategies for the development of a side stream process for polyhydroxyalkanoate (PHA) production from sugar cane molasses. J. Biotechnol. 130 (4), 411–421.

Annamalai, N., Sivakumar, N., Fernandez-Castane, A., Oleskowicz-Popiel, P., 2020. Production of microbial lipids utilizing volatile fatty acids derived from wastepaper: a biorefinery approach for biodiesel production. Fuel 276, 118087.

Annamalai, N., Sivakumar, N., Oleskowicz-Popiel, P., 2018. Enhanced production of microbial lipids from waste office paper by the oleaginous yeast *Cryptococcus curvatus*. Fuel 217, 420–426.

Arras, W., Hussain, A., Hausler, R., Guiot, S.R., 2019. Mesophilic, thermophilic and hyperthermophilic acidogenic fermentation of food waste in batch: effect of inoculum source. Waste Manag 87, 279–287.

Atasoy, M., Cetecioglu, Z., 2020. Butyric acid dominant volatile fatty acids production: bio-augmentation of mixed culture fermentation by *Clostridium butyricum*. J. Environ. Chem. Eng. 8 (6), 104496.

Atasoy, M., Eyice, O., Schnürer, A., Cetecioglu, Z., 2019. Volatile fatty acids production via mixed culture fermentation: revealing the link between pH, inoculum type and bacterial composition. Bioresour. Technol. 292, 121889.

Atasoy, M., Owusu-Agyeman, I., Plaza, E., Cetecioglu, Z., 2018. Bio-based volatile fatty acid production and recovery from waste streams: current status and future challenges. Bioresour. Technol. 268, 773–786.

Aydin, S., Yesil, H., Tugtas, A.E., 2018. Recovery of mixed volatile fatty acids from anaerobically fermented organic wastes by vapor permeation membrane contactors. Bioresour. Technol. 250, 548–555.

Bóna, Á., Bakonyi, P., Galambos, I., Bélafi-Bakó, K., Nemestóthy, N., 2020. Separation of volatile fatty acids from model anaerobic effluents using various membrane technologies. Membranes 10 (10), 252.

Bak, C., Yun, Y.-M., Kim, J.-H., Kang, S., 2019. Electrodialytic separation of volatile fatty acids from hydrogen fermented food wastes. Int. J. Hydrog. Energy 44 (6), 3356–3362.

Behera, S., Arora, R., Nandhagopal, N., Kumar, S., 2014. Importance of chemical pretreatment for bioconversion of lignocellulosic biomass. Renew. Sustain. Energy Rev. 36, 91–106.

Bhatia, S.K., Otari, S.V., Jeon, J.-M., Gurav, R., Choi, Y.-K., Bhatia, R.K., Pugazhendhi, A., Kumar, V., Banu, J.R., Yoon, J.-J., 2021. Biowaste-to-bioplastic (polyhydroxyalkanoates): conversion technologies, strategies, challenges, and perspective. Bioresour. Technol., 124733.

Bhatia, S.K., Yang, Y.-H., 2017. Microbial production of volatile fatty acids: current status and future perspectives. Rev. Environ. Sci. Biotechnol. 16 (2), 327–345.

Blunt, W., Levin, D.B., Cicek, N., 2018. Bioreactor operating strategies for improved polyhydroxyalkanoate (PHA) productivity. Polymers 10 (11), 1197.

Botturi, A., Battista, F., Andreolli, M., Faccenda, F., Fusco, S., Bolzonella, D., Lampis, S., Frison, N., 2021. Polyhydroxyalkanoated-rich microbial cells from bio-based volatile fatty acids as potential ingredient for aquaculture feed. Energies 14 (1), 38.

Burniol-Figols, A., Varrone, C., Daugaard, A.E., Le, S.B., Skiadas, I.V., Gavala, H.N., 2018a. Polyhydroxyalkanoates (PHA) production from fermented crude glycerol: study on the conversion of 1, 3-propanediol to PHA in mixed microbial consortia. Water Res. 128, 255–266.

Burniol-Figols, A., Varrone, C., Le, S.B., Daugaard, A.E., Skiadas, I.V., Gavala, H.N., 2018b. Combined polyhydroxyalkanoates (PHA) and 1,3-propanediol production from crude glycerol: selective conversion of volatile fatty acids into PHA by mixed microbial consortia. Water Res. 136, 180–191.

Cabrera, F., Serrano, A., Torres, Á., Rodriguez-Gutierrez, G., Jeison, D., Fermoso, F.G., 2019. The accumulation of volatile fatty acids and phenols through a pH-controlled fermentation of olive mill solid waste. Sci. Total Environ. 657, 1501–1507.

Castañeda, M.T., Nuñez, S., Garelli, F., Voget, C., De Battista, H., 2018. Comprehensive analysis of a metabolic model for lipid production in *Rhodosporidium toruloides*. J. Biotechnol. 280, 11–18.

Castilla-Archilla, J., Papirio, S., Lens, P.N.L., 2021. Two step process for volatile fatty acid production from brewery spent grain: hydrolysis and direct acidogenic fermentation using anaerobic granular sludge. Process Biochem. 100, 272–283.

Cavinato, C., Da Ros, C., Pavan, P., Bolzonella, D., 2017. Influence of temperature and hydraulic retention on the production of volatile fatty acids during anaerobic fermentation of cow manure and maize silage. Bioresour. Technol. 223, 59–64.

Chalmers Brown, R., Tuffou, R., Massanet Nicolau, J., Dinsdale, R., Guwy, A., 2020. Overcoming nutrient loss during volatile fatty acid recovery from fermentation media by addition of electrodialysis to a polytetrafluoroethylene membrane stack. Bioresour. Technol. 301, 122543.

Chang, S.-E., Saha, S., Kurade, M.B., Salama, E.-S., Chang, S.W., Jang, M., Jeon, B.-H., 2018. Improvement of acidogenic fermentation using an acclimatized microbiome. Int. J. Hydrog. Energy 43 (49), 22126–22134.

Chatterjee, S., Mohan, S.V., 2018. Microbial lipid production by *Cryptococcus curvatus* from vegetable waste hydrolysate. Bioresour. Technol. 254, 284–289.

Chen, H., Zeng, X., Zhou, Y., Yang, X., Lam, S.S., Wang, D., 2020. Influence of roxithromycin as antibiotic residue on volatile fatty acids recovery in anaerobic fermentation of waste activated sludge. J. Hazard. Mater. 394, 122570.

Chen, Y., Luo, J., Yan, Y., Feng, L., 2013. Enhanced production of short-chain fatty acid by co-fermentation of waste activated sludge and kitchen waste under alkaline conditions and its application to microbial fuel cells. Appl. Energy 102, 1197–1204.

Cho, H.U., Kim, Y.M., Park, J.M., 2018. Changes in microbial communities during volatile fatty acid production from cyanobacterial biomass harvested from a cyanobacterial bloom in a river. Chemosphere 202, 306–311.

Colombo, B., Favini, F., Scaglia, B., Sciarria, T.P., D'Imporzano, G., Pognani, M., Alekseeva, A., Eisele, G., Cosentino, C., Adani, F., 2017. Enhanced polyhydroxyalkanoate (PHA) production from the organic fraction of municipal solid waste by using mixed microbial culture. Biotechnol. Biofuels 10 (1), 1–15.

Da Ros, C., Conca, V., Eusebi, A.L., Frison, N., Fatone, F., 2020. Sieving of municipal wastewater and recovery of bio-based volatile fatty acids at pilot scale. Water Res. 174, 115633.

Dahiya, S., Mohan, S.V., 2019. Selective control of volatile fatty acids production from food waste by regulating biosystem buffering: a comprehensive study. Chem. Eng. J. 357, 787–801.

Dams, R.I., Viana, M.B., Guilherme, A.A., Silva, C.M., dos Santos, A.B., Angenent, L.T., Santaella, S.T., Leitão, R.C., 2018. Production of medium-chain carboxylic acids by anaerobic fermentation of glycerol using a bioaugmented open culture. Biomass Bioenergy 118, 1–7.

Deeba, F., Pruthi, V., Negi, Y.S., 2016. Converting paper mill sludge into neutral lipids by oleaginous yeast *Cryptococcus vishniaccii* for biodiesel production. Bioresour. Technol. 213, 96–102.

Di Fidio, N., Dragoni, F., Antonetti, C., De Bari, I., Galletti, A.M.R., Ragaglini, G., 2020. From paper mill waste to single cell oil: enzymatic hydrolysis to sugars and their fermentation into microbial oil by the yeast *Lipomyces starkeyi*. Bioresour. Technol. 315, 123790.

Duan, X., Chen, Y., Yan, Y., Feng, L., Chen, Y., Zhou, Q., 2019a. New method for algae comprehensive utilization: algae-derived biochar enhances algae anaerobic fermentation for short-chain fatty acids production. Bioresour. Technol. 289, 121637.

Duan, Y., Awasthi, S.K., Liu, T., Zhang, Z., Awasthi, M.K., 2019b. Response of bamboo biochar amendment on volatile fatty acids accumulation reduction and humification during chicken manure composting. Bioresour. Technol. 291, 121845.

Eder, A.S., Magrini, F.E., Spengler, A., da Silva, J.T., Beal, L.L., Paesi, S., 2020. Comparison of hydrogen and volatile fatty acid production by *Bacillus cereus*, *Enterococcus faecalis* and *Enterobacter aerogenes* singly, in co-cultures or in the bioaugmentation of microbial consortium from sugarcane vinasse. Environ. Technol. Innov. 18, 100638.

Eregowda, T., Rene, E.R., Rintala, J., Lens, P.N., 2020. Volatile fatty acid adsorption on anion exchange resins: kinetics and selective recovery of acetic acid. Sep. Sci. Technol. 55 (8), 1449–1461.

Esteban-Gutiérrez, M., Garcia-Aguirre, J., Irizar, I., Aymerich, E., 2018. From sewage sludge and agri-food waste to VFA: individual acid production potential and up-scaling. Waste Manag 77, 203–212.

Fang, W., Zhang, P., Zhang, T., Requeson, D.C., Poser, M., 2019. Upgrading volatile fatty acids production through anaerobic co-fermentation of mushroom residue and sewage sludge: performance evaluation and kinetic analysis. J. Environ. Manage. 241, 612–618.

Fang, W., Zhang, P., Zhang, X., Zhu, X., van Lier, J.B., Spanjers, H., 2018. White rot fungi pretreatment to advance volatile fatty acid production from solid-state fermentation of solid digestate: efficiency and mechanisms. Energy 162, 534–541.

Fang, W., Zhang, X., Zhang, P., Wan, J., Guo, H., Ghasimi, D.S., Morera, X.C., Zhang, T., 2020. Overview of key operation factors and strategies for improving fermentative volatile fatty acid production and product regulation from sewage sludge. J. Environ. Sci. 87, 93–111.

Feng, X.-C., Bao, X., Che, L., Wu, Q.-L., 2021. Enhance biological nitrogen and phosphorus removal in wastewater treatment process by adding food waste fermentation liquid as external carbon source. Biochem. Eng. J. 165, 107811.

Fernández-Domínguez, D., Astals, S., Peces, M., Frison, N., Bolzonella, D., Mata-Alvarez, J., Dosta, J., 2020. Volatile fatty acids production from biowaste at mechanical-biological treatment plants: focusing on fermentation temperature. Bioresour. Technol. 314, 123729.

Frone, A.N., Batalu, D., Chiulan, I., Oprea, M., Gabor, A.R., Nicolae, C.-A., Raditoiu, V., Trusca, R., Panaitescu, D.M., 2020. Morpho-structural, thermal and mechanical properties of PLA/PHB/cellulose biodegradable nanocomposites obtained by compression molding, extrusion, and 3D printing. Nanomaterials 10 (1), 51.

Gao, R.L., Li, Z.F., Zhou, X.Q., Cheng, S.K., Zheng, L., 2017. Oleaginous yeast *Yarrowia lipolytica* culture with synthetic and food waste-derived volatile fatty acids for lipid production. Biotechnol. Biofuels 10, 247.

Garcia-Aguirre, J., Aymerich, E., de Goñi, J.G.-M., Esteban-Gutiérrez, M., 2017. Selective VFA production potential from organic waste streams: assessing temperature and pH influence. Bioresour. Technol. 244, 1081–1088.

Garcia-Aguirre, J., Esteban-Gutiérrez, M., Irizar, I., González-Mtnez de Goñi, J., Aymerich, E., 2019. Continuous acidogenic fermentation: narrowing the gap between laboratory testing and industrial application. Bioresour. Technol. 282, 407–416.

Gong, Z., Zhou, W., Shen, H., Zhao, Z.K., Yang, Z., Yan, J., Zhao, M., 2016. Co-utilization of corn stover hydrolysates and biodiesel-derived glycerol by *Cryptococcus curvatus* for lipid production. Bioresour. Technol. 219, 552–558.

Gorte, O., Kugel, M., Ochsenreither, K., 2020. Optimization of carbon source efficiency for lipid production with the oleaginous yeast *Saitozyma podzolica* DSM 27192 applying automated continuous feeding. Biotechnol. Biofuels 13 (1), 1–17.

He, X., Yin, J., Liu, J., Chen, T., Shen, D., 2019. Characteristics of acidogenic fermentation for volatile fatty acid production from food waste at high concentrations of NaCl. Bioresour. Technol. 271, 244–250.

Herrero, M., Stuckey, D., 2015. Bioaugmentation and its application in wastewater treatment: a review. Chemosphere 140, 119–128.

Hu, Q., Jung, J., Chen, D., Leong, K., Song, S., Li, F., Mohan, B.C., Yao, Z., Prabhakar, A.K., Lin, X.H., 2020. Biochar industry to circular economy. Sci. Total Environ. 757, 143820.

Huang, J., Zhou, R., Chen, J., Han, W., Chen, Y., Wen, Y., Tang, J., 2016a. Volatile fatty acids produced by co-fermentation of waste activated sludge and henna plant biomass. Bioresour. Technol. 211, 80–86.

Huang, X., Mu, T., Shen, C., Lu, L., Liu, J., 2016b. Effects of bio-surfactants combined with alkaline conditions on volatile fatty acid production and microbial community in the anaerobic fermentation of waste activated sludge. Int. Biodeterior. Biodegradation 114, 24–30.

Huang, X., Zhao, J., Xu, Q., Li, X., Wang, D., Yang, Q., Liu, Y., Tao, Z., 2019. Enhanced volatile fatty acids production from waste activated sludge anaerobic fermentation by adding tofu residue. Bioresour. Technol. 274, 430–438.

Hussain, A., Filiatrault, M., Guiot, S.R., 2017. Acidogenic digestion of food waste in a thermophilic leach bed reactor: Effect of pH and leachate recirculation rate on hydrolysis and volatile fatty acid production. Bioresour. Technol. 245, 1–9.

Iglesias-Iglesias, R., Campanaro, S., Treu, L., Kennes, C., Veiga, M.C., 2019. Valorization of sewage sludge for volatile fatty acids production and role of microbiome on acidogenic fermentation. Bioresour. Technol. 291, 121817.

Islam, M.S., Zhang, C., Sui, K.-Y., Guo, C., Liu, C.-Z., 2017. Coproduction of hydrogen and volatile fatty acid via thermophilic fermentation of sweet sorghum stalk from co-culture of *Clostridium thermocellum* and *Clostridium thermosaccharolyticum*. Int. J. Hydrog. Energy 42 (2), 830–837.

Jankowska, E., Chwialkowska, J., Stodolny, M., Oleskowicz-Popiel, P., 2015. Effect of pH and retention time on volatile fatty acids production during mixed culture fermentation. Bioresour. Technol. 190, 274–280.

Jankowska, E., Chwialkowska, J., Stodolny, M., Oleskowicz-Popiel, P., 2017. Volatile fatty acids production during mixed culture fermentation—the impact of substrate complexity and pH. Chem. Eng. J. 326, 901–910.

Jiang, J., Zhang, Y., Li, K., Wang, Q., Gong, C., Li, M., 2013. Volatile fatty acids production from food waste: effects of pH, temperature, and organic loading rate. Bioresour. Technol. 143, 525–530.

Jiang, Y., Zhang, Y., Banks, C., Heaven, S., Longhurst, P., 2017. Investigation of the impact of trace elements on anaerobic volatile fatty acid degradation using a fractional factorial experimental design. Water Res. 125, 458–465.

Jin, Y., Lin, Y., Wang, P., Jin, R., Gao, M., Wang, Q., Chang, T.-C., Ma, H., 2019. Volatile fatty acids production from saccharification residue from food waste ethanol fermentation: effect of pH and microbial community. Bioresour. Technol. 292, 121957.

Jomnonkhaow, U., Uwineza, C., Mahboubi, A., Wainaina, S., Reungsang, A., Taherzadeh, M.J., 2020. Membrane bioreactor-assisted volatile fatty acids production and in situ recovery from cow manure. Bioresour. Technol. 321, 124456.

Jones, R.J., Fernández-Feito, R., Massanet-Nicolau, J., Dinsdale, R., Guwy, A., 2021. Continuous recovery and enhanced yields of volatile fatty acids from a continually-fed 100 L food waste bioreactor by filtration and electrodialysis. Waste Manag 122, 81–88.

Kamineni, A., Shaw, J., 2020. Engineering triacylglycerol production from sugars in oleaginous yeasts. Curr. Opin. Biotechnol. 62, 239–247.

Kandylis, P., Bekatorou, A., Pissaridi, K., Lappa, K., Dima, A., Kanellaki, M., Koutinas, A.A., 2016. Acidogenesis of cellulosic hydrolysates for new generation biofuels. Biomass Bioenergy 91, 210–216.

Kaur, G., Johnravindar, D., Wong, J.W.C., 2020. Enhanced volatile fatty acid degradation and methane production efficiency by biochar addition in food waste-sludge co-digestion: a step towards increased organic loading efficiency in co-digestion. Bioresour. Technol. 308, 123250.

Khan, M.A., Ngo, H.H., Guo, W., Liu, Y., Nghiem, L.D., Chang, S.W., Nguyen, D.D., Zhang, S., Luo, G., Jia, H., 2019. Optimization of hydraulic retention time and organic loading rate for volatile fatty acid production from low strength wastewater in an anaerobic membrane bioreactor. Bioresour. Technol. 271, 100–108.

Khedkar, M.A., Satpute, S.R., Bankar, S.B., Chavan, P.V., 2021. Adsorptive removal of unsaturated fatty acids using ion exchange resins. J. Chem. Eng. Data 66 (1), 308–321.

Kumar, L.R., Yellapu, S.K., Yan, S., Tyagi, R.D., Drogui, P., 2021. Elucidating the effect of impurities present in different crude glycerol sources on lipid and citric acid production by *Yarrowia lipolytica* SKY7. J. Chem. Technol. Biotechnol. 96 (1), 227–240.

Iaqa Kakar, F., Koupaie, E.H., Razavi, A.S., Hafez, H., Elbeshbishy, E., 2020. Effect of hydrothermal pretreatment on volatile fatty acids production from thickened waste activated sludge. BioEnergy Res 13 (2), 591–604.

Latif, M.A., Mehta, C.M., Batstone, D.J., 2017. Influence of low pH on continuous anaerobic digestion of waste activated sludge. Water Res. 113, 42–49.

Lazar, Z., Liu, N., Stephanopoulos, G., 2018. Holistic approaches in lipid production by *Yarrowia lipolytica*. Trends Biotechnol. 36 (11), 1157–1170.

Lee, W.S., Chua, A.S.M., Yeoh, H.K., Ngoh, G.C., 2014. A review of the production and applications of waste-derived volatile fatty acids. Chem. Eng. J. 235, 83–99.

Leong, W.-H., Lim, J.-W., Lam, M.-K., Uemura, Y., Ho, Y.-C., 2018. Third generation biofuels: a nutritional perspective in enhancing microbial lipid production. Renew. Sustain. Energy Rev. 91, 950–961.

Li, R.-H., Li, X.-Y., 2017. Recovery of phosphorus and volatile fatty acids from wastewater and food waste with an iron-flocculation sequencing batch reactor and acidogenic co-fermentation. Bioresour. Technol. 245, 615–624.

Li, X., Liu, G., Liu, S., Ma, K., Meng, L., 2018. The relationship between volatile fatty acids accumulation and microbial community succession triggered by excess sludge alkaline fermentation. J. Environ. Manage. 223, 85–91.

Lim, S.-J., Kim, B.J., Jeong, C.-M., Choi, J.-d.-r., Ahn, Y.H., Chang, H.N., 2008. Anaerobic organic acid production of food waste in once-a-day feeding and drawing-off bioreactor. Bioresour. Technol. 99 (16), 7866–7874.

Lin, R., Cheng, J., Murphy, J.D., 2017. Unexpectedly low biohydrogen yields in co-fermentation of acid pretreated cassava residue and swine manure. Energy Convers. Manag. 151, 553–561.

Liu, C., Ren, L., Yan, B., Luo, L., Zhang, J., Awasthi, M.K., 2020a. Electron transfer and mechanism of energy production among syntrophic bacteria during acidogenic fermentation: a review. Bioresour. Technol. 124637.

Liu, C., Wang, W., Sompong, O., Yang, Z., Zhang, S., Liu, G., Luo, G., 2020b. Microbial insights of enhanced anaerobic conversion of syngas into volatile fatty acids by co-fermentation with carbohydrate-rich synthetic wastewater. Biotechnol. Biofuels 13 (1), 1–14.

Liu, H., Xiao, H., Yin, B., Zu, Y., Liu, H., Fu, B., Ma, H., 2016. Enhanced volatile fatty acid production by a modified biological pretreatment in anaerobic fermentation of waste activated sludge. Chem. Eng. J. 284, 194–201.

Liu, J., Yuan, M., Liu, J.-N., Huang, X.-F., 2017a. Bioconversion of mixed volatile fatty acids into microbial lipids by *Cryptococcus curvatus* ATCC 20509. Bioresour. Technol. 241, 645–651.

Liu, J.N., Huang, X., Chen, R., Yuan, M., Liu, J., 2017b. Efficient bioconversion of high-content volatile fatty acids into microbial lipids by *Cryptococcus curvatus* ATCC 20509. Bioresour. Technol. 239, 394–401.

Liu, N., Jiang, J., Yan, F., Gao, Y., Meng, Y., Aihemaiti, A., Ju, T., 2018. Enhancement of volatile fatty acid production and biogas yield from food waste following sonication pretreatment. J. Environ. Manage. 217, 797–804.

Liu, W., Mao, W., Zhang, C., Lu, X., Xiao, X., Zhao, Z., Lin, J., 2020c. Co-fermentation of a sugar mixture for microbial lipid production in a two-stage fermentation mode under non-sterile conditions. Sustain. Energy Fuels 4 (5), 2380–2385.

Liu, X., Liu, H., Chen, Y., Du, G., Chen, J., 2008. Effects of organic matter and initial carbon-nitrogen ratio on the bioconversion of volatile fatty acids from sewage sludge. J. Chem. Technol. Biotechnol. 83 (7), 1049–1055.

Liu, Y., Wang, Y., Liu, H., Zhang, J., 2015. Enhanced lipid production with undetoxified corncob hydrolysate by *Rhodotorula glutinis* using a high cell density culture strategy. Bioresour. Technol. 180, 32–39.

Llamas, M., Magdalena, J.A., González-Fernández, C., Tomás-Pejó, E., 2020. Volatile fatty acids as novel building blocks for oil-based chemistry via oleaginous yeast fermentation. Biotechnol. Bioeng. 117 (1), 238–250.

Lu, J.-H., Chen, C., Huang, C., Zhuang, H., Leu, S.-Y., Lee, D.-J., 2020. Dark fermentation production of volatile fatty acids from glucose with biochar amended biological consortium. Bioresour. Technol. 303, 122921.

Lukitawesa, Eryildiz, B., Mahboubi, A., Millati, R., Taherzadeh, M.J., 2020. Semi-continuous production of volatile fatty acids from citrus waste using membrane bioreactors. Innov. Food Sci. Emerg. Technol. 67, 102545.

Lynam, J.G., Reza, M.T., Vasquez, V.R., Coronella, C.J., 2012. Effect of salt addition on hydrothermal carbonization of lignocellulosic biomass. Fuel 99, 271–273.

Ma, S., Hu, H., Wang, J., Liao, K., Ma, H., Ren, H., 2019. The characterization of dissolved organic matter in alkaline fermentation of sewage sludge with different pH for volatile fatty acids production. Water Res. 164, 114924.

Magdalena, J.A., Greses, S., González-Fernández, C., 2019. Impact of organic loading rate in volatile fatty acids production and population dynamics using microalgae biomass as substrate. Sci. Rep. 9 (1), 1–11.

Masse, L., Massé, D.I., Pellerin, Y., 2008. The effect of pH on the separation of manure nutrients with reverse osmosis membranes. J. Membr. Sci. 325 (2), 914–919.

Mockaitis, G., Bruant, G., Guiot, S.R., Peixoto, G., Foresti, E., Zaiat, M., 2020. Acidic and thermal pre-treatments for anaerobic digestion inoculum to improve hydrogen and volatile fatty acid production using xylose as the substrate. Renew. Energy 145, 1388–1398.

Moretto, G., Valentino, F., Pavan, P., Majone, M., Bolzonella, D., 2019. Optimization of urban waste fermentation for volatile fatty acids production. Waste Manag 92, 21–29.

Munir, S., Jamil, N., 2020. Polyhydroxyalkanoate (PHA) production in open mixed cultures using waste activated sludge as biomass. Arch. Microbiol. 202, 1907–1913.

Nguyen, D., Wu, Z., Shrestha, S., Lee, P.-H., Raskin, L., Khanal, S.K., 2019. Intermittent micro-aeration: new strategy to control volatile fatty acid accumulation in high organic loading anaerobic digestion. Water Res. 166, 115080.

Okonkwo, O., Papirio, S., Trably, E., Escudie, R., Lakaniemi, A.-M., Esposito, G., 2020. Enhancing thermophilic dark fermentative hydrogen production at high glucose concentrations via bioaugmentation with *Thermotoga neapolitana*. Int. J. Hydrog. Energy 45 (35), 17241–17249.

Outram, V., Zhang, Y., 2018. Solvent-free membrane extraction of volatile fatty acids from acidogenic fermentation. Bioresour. Technol. 270, 400–408.

Paul, S., Dutta, A., 2018. Challenges and opportunities of lignocellulosic biomass for anaerobic digestion. Resour. Conserv. Recycl. 130, 164–174.

Peng, L., Appels, L., Su, H., 2018. Combining microwave irradiation with sodium citrate addition improves the pre-treatment on anaerobic digestion of excess sewage sludge. J. Environ. Manage. 213, 271–278.

Perez-Zabaleta, M., Atasoy, M., Khatami, K., Eriksson, E., Cetecioglu, Z., 2021. Bio-based conversion of volatile fatty acids from waste streams to polyhydroxyalkanoates using mixed microbial cultures. Bioresour. Technol. 323, 124604.

Poontawee, R., Yongmanitchai, W., Limtong, S., 2018. Lipid production from a mixture of sugarcane top hydrolysate and biodiesel-derived crude glycerol by the oleaginous red yeast, *Rhodosporidiobolus fluvialis*. Process Biochem. 66, 150–161.

Rajapaksha, A.U., Chen, S.S., Tsang, D.C.W., Zhang, M., Vithanage, M., Mandal, S., Gao, B., Bolan, N.S., Ok, Y.S., 2016. Engineered/designer biochar for contaminant removal/immobilization from soil and water: potential and implication of biochar modification. Chemosphere 148, 276–291.

Rech, C.R., Brabes, K.C., Silva, B.E., Martines, M.A., Silveira, T.F., Alberton, J., Amadeu, C.A., Caon, T., Arruda, E.J., Martelli, S.M., 2020. Antimicrobial and physical-mechanical properties of polyhydroxybutyrate edible films containing essential oil mixtures. J. Polym. Environ. 29, 1202–1211.

Reddy, M.V., Hayashi, S., Choi, D., Cho, H., Chang, Y.-C., 2018. Short chain and medium chain fatty acids production using food waste under non-augmented and bio-augmented conditions. J. Clean. Prod. 176, 645–653.

Reyhanitash, E., Kersten, S.R., Schuur, B., 2017. Recovery of volatile fatty acids from fermented wastewater by adsorption. ACS Sustain. Chem. Eng. 5 (10), 9176–9184.

Saha, S., Jeon, B.-H., Kurade, M.B., Chatterjee, P.K., Chang, S.W., Markkandan, K., Salama, E.-S., Govindwar, S.P., Roh, H.-S., 2019. Microbial acclimatization to lipidic-waste facilitates the efficacy of acidogenic fermentation. Chem. Eng. J. 358, 188–196.

Sarkar, O., Kiran Katari, J., Chatterjee, S., Venkata Mohan, S., 2020a. Salinity induced acidogenic fermentation of food waste regulates biohydrogen production and volatile fatty acids profile. Fuel 276, 117794.

Sarkar, O., Rova, U., Christakopoulos, P., Matsakas, L., 2020b. Influence of initial uncontrolled pH on acidogenic fermentation of brewery spent grains to biohydrogen and volatile fatty acids production: optimization and scale-up. Bioresour. Technol. 319, 124233.

Sathiyamoorthi, E., Kumar, P., Kim, B.S., 2019. Lipid production by *Cryptococcus albidus* using biowastes hydrolysed by indigenous microbes. Bioprocess Biosyst. Eng. 42 (5), 687–696.

Sawatdeenarunat, C., Sung, S., Khanal, S.K., 2017. Enhanced volatile fatty acids production during anaerobic digestion of lignocellulosic biomass via micro-oxygenation. Bioresour. Technol. 237, 139–145.

Seo, C., Kim, W., Chang, H.N., Han, J.-I., Kim, Y.-C., 2016. Comprehensive study on volatile fatty acid production from *Ettlia* sp. residue with molecular analysis of the microbial community. Algal Res 17, 161–167.

Shi, L., Hu, Y., Xie, S., Wu, G., Hu, Z., Zhan, X., 2018. Recovery of nutrients and volatile fatty acids from pig manure hydrolysate using two-stage bipolar membrane electrodialysis. Chem. Eng. J. 334, 134–142.

Soomro, A.F., Abbasi, I.A., Ni, Z., Ying, L., Liu, J., 2020. Influence of temperature on enhancement of volatile fatty acids fermentation from organic fraction of municipal solid waste: Synergism between food and paper components. Bioresour. Technol. 304, 122980.

Sträuber, H., Schröder, M., Kleinsteuber, S., 2012. Metabolic and microbial community dynamics during the hydrolytic and acidogenic fermentation in a leach-bed process. Energy Sustain. Soc. 2 (1), 1–10.

Strazzera, G., Battista, F., Garcia, N.H., Frison, N., Bolzonella, D., 2018. Volatile fatty acids production from food wastes for biorefinery platforms: a review. J. Environ. Manage. 226, 278–288.

Sun, C., Xia, A., Liao, Q., Fu, Q., Huang, Y., Zhu, X., Wei, P., Lin, R., Murphy, J.D., 2018. Improving production of volatile fatty acids and hydrogen from microalgae and rice residue: effects of physicochemical characteristics and mix ratios. Appl. Energy 230, 1082–1092.

Sun, X., Lu, H., Wang, J., 2017. Recovery of citric acid from fermented liquid by bipolar membrane electrodialysis. J. Clean. Prod. 143, 250–256.

Szacherska, K., Oleskowicz-Popiel, P., Ciesielski, S., Mozejko-Ciesielska, J., 2021. Volatile fatty acids as carbon sources for polyhydroxyalkanoates production. Polymers 13 (3), 321.

Tampio, E.A., Blasco, L., Vainio, M.M., Kahala, M.M., Rasi, S.E., 2019. Volatile fatty acids (VFAs) and methane from food waste and cow slurry: comparison of biogas and VFA fermentation processes. Glob. Change Biol. Bioenergy 11 (1), 72–84.

Tan, D., Wang, Y., Tong, Y., Chen, G.-Q., 2020. Grand challenges for industrializing polyhydroxyalkanoates (PHAs). Trends Biotechnol. https://doi.org/10.1016/j.tibtech.2020.11.010.

Tang, J., Wang, X.C., Hu, Y., Zhang, Y., Li, Y., 2017. Effect of pH on lactic acid production from acidogenic fermentation of food waste with different types of inocula. Bioresour. Technol. 224, 544–552.

Tarrahi, R., Fathi, Z., Seydibeyoğlu, M.Ö., Doustkhah, E., Khataee, A., 2020. Polyhydroxyalkanoates (PHA): from production to nanoarchitecture. Int. J. Biol. Macromol. 146, 596–619.

Tripathi, A.D., Paul, V., Agarwal, A., Sharma, R., Hashempour-Baltork, F., Rashidi, L., Darani, K.K., 2021. Production of polyhydroxyalkanoates using dairy processing waste—a review. Bioresour. Technol. 326, 124735.

Tsui, T.H., Wong, J.W., 2019. A critical review: emerging bioeconomy and waste-to-energy technologies for sustainable municipal solid waste management. Waste Disposal Sustain. Energy 1 (3), 151–167.

Tsui, T.H., Ekama, G.A., Chen, G.H., 2018. Quantitative characterization and analysis of granule transformations: role of intermittent gas sparging in a super high-rate anaerobic system. Water Res. 139, 177–186.

Tsui, T.H., Wu, H., Song, B., Liu, S.S., Bhardwaj, A., Wong, J.W., 2020. Food waste leachate treatment using an upflow anaerobic sludge bed (UASB): effect of conductive material dosage under low and high organic loads. Bioresour. Technol. 304, 122738.

Tu, W., Zhang, D., Wang, H., 2019. Polyhydroxyalkanoates (PHA) production from fermented thermal-hydrolyzed sludge by mixed microbial cultures: the link between phosphorus and PHA yields. Waste Manag 96, 149–157.

Turco, R., Santagata, G., Corrado, I., Pezzella, C., Di Serio, M., 2021. In vivo and post-synthesis strategies to enhance the properties of PHB-based materials: a review. Front. Bioeng. Biotechnol. 8, 1454.

van Aarle, I.M., Perimenis, A., Lima-Ramos, J., de Hults, E., George, I.F., Gerin, P.A., 2015. Mixed inoculum origin and lignocellulosic substrate type both influence the production of volatile fatty acids during acidogenic fermentation. Biochem. Eng. J. 103, 242–249.

Wainaina, S., Awasthi, M.K., Horváth, I.S., Taherzadeh, M.J., 2020. Anaerobic digestion of food waste to volatile fatty acids and hydrogen at high organic loading rates in immersed membrane bioreactors. Renew. Energy 152, 1140–1148.

Wainaina, S., Lukitawesa, Kumar, Awasthi, M., Taherzadeh, M.J., 2019a. Bioengineering of anaerobic digestion for volatile fatty acids, hydrogen or methane production: a critical review. Bioengineered 10 (1), 437–458.

Wainaina, S., Parchami, M., Mahboubi, A., Horváth, I.S., Taherzadeh, M.J., 2019b. Food waste-derived volatile fatty acids platform using an immersed membrane bioreactor. Bioresour. Technol. 274, 329–334.

Wang, D., Liu, Y., Ngo, H.H., Zhang, C., Yang, Q., Peng, L., He, D., Zeng, G., Li, X., Ni, B.-J., 2017. Approach of describing dynamic production of volatile fatty acids from sludge alkaline fermentation. Bioresour. Technol. 238, 343–351.

Wang, D., Shuai, K., Xu, Q., Liu, X., Li, Y., Liu, Y., Wang, Q., Li, X., Zeng, G., Yang, Q., 2018. Enhanced short-chain fatty acids production from waste activated sludge by combining calcium peroxide with free ammonia pretreatment. Bioresour. Technol. 262, 114–123.

Wang, G., Wang, D., Huang, L., Song, Y., Chen, Z., Du, M., 2020a. Enhanced production of volatile fatty acids by adding a kind of sulfate reducing bacteria under alkaline pH. Colloids Surf. B 195, 111249.

Wang, J., Ledesma-Amaro, R., Wei, Y., Ji, B., Ji, X.-J., 2020b. Metabolic engineering for increased lipid accumulation in Yarrowia lipolytica. Bioresour. Technol., 123707.

Wang, S., Tao, X., Zhang, G., Zhang, P., Wang, H., Ye, J., Li, F., Zhang, Q., Nabi, M., 2019. Benefit of solid-liquid separation on volatile fatty acid production from grass clipping with ultrasound-calcium hydroxide pretreatment. Bioresour. Technol. 274, 97–104.

Wang, Y., Liu, X., Liu, Y., Wang, D., Xu, Q., Li, X., Yang, Q., Wang, Q., Ni, B.-J., Chen, H., 2020c. Enhancement of short-chain fatty acids production from microalgae by potassium ferrate addition: feasibility, mechanisms and implications. Bioresour. Technol. 318, 124266.

Weng, Y.-H., Wei, H.-J., Tsai, T.-Y., Lin, T.-H., Wei, T.-Y., Guo, G.-L., Huang, C.-P., 2010. Separation of furans and carboxylic acids from sugars in dilute acid rice straw hydrolyzates by nanofiltration. Bioresour. Technol. 101 (13), 4889–4894.

Wu, H., Valentino, L., Riggio, S., Holtzapple, M., Urgun-Demirtas, M., 2020. Performance characterization of nanofiltration, reverse osmosis, and ion exchange technologies for acetic acid separation. Sep. Purif. Technol. 118108. https://doi.org/10.1016/j.seppur.2020.118108.

Wu, Q.-L., Guo, W.-Q., Bao, X., Yin, R.-L., Feng, X.-C., Zheng, H.-S., Luo, H.-C., Ren, N.-Q., 2017. Enhancing sludge biodegradability and volatile fatty acid production by tetrakis hydroxymethyl phosphonium sulfate pretreatment. Bioresour. Technol. 239, 518–522.

Wu, Q.-L., Guo, W.-Q., Zheng, H.-S., Luo, H.-C., Feng, X.-C., Yin, R.-L., Ren, N.-Q., 2016. Enhancement of volatile fatty acid production by co-fermentation of food waste and excess sludge without pH control: the mechanism and microbial community analyses. Bioresour. Technol. 216, 653–660.

Xavier, M.C.A., Coradini, A.L.V., Deckmann, A.C., Franco, T.T., 2017. Lipid production from hemicellulose hydrolysate and acetic acid by *Lipomyces starkeyi* and the ability of yeast to metabolize inhibitors. Biochem. Eng. J. 118, 11–19.

Xia, A., Jacob, A., Tabassum, M.R., Herrmann, C., Murphy, J.D., 2016. Production of hydrogen, ethanol and volatile fatty acids through co-fermentation of macro-and micro-algae. Bioresour. Technol. 205, 118–125.

Xiang, C., Tian, D., Hu, J., Huang, M., Shen, F., Zhang, Y., Yang, G., Zeng, Y., Deng, S., 2021. Why can hydrothermally pretreating lignocellulose in low severities improve anaerobic digestion performances? Sci. Total Environ. 752, 141929.

Xie, J., Duan, X., Feng, L., Yan, Y., Wang, F., Dong, H., Jia, R., Zhou, Q., 2019. Influence of sulfadiazine on anaerobic fermentation of waste activated sludge for volatile fatty acids production: focusing on microbial responses. Chemosphere 219, 305–312.

Xiong, B., Richard, T.L., Kumar, M., 2015. Integrated acidogenic digestion and carboxylic acid separation by nanofiltration membranes for the lignocellulosic carboxylate platform. J. Membr. Sci. 489, 275–283.

Xiong, Z., Hussain, A., Lee, H.-S., 2019a. Food waste treatment with a leachate bed reactor: effects of inoculum to substrate ratio and reactor design. Bioresour. Technol. 285, 121350.

Xiong, Z., Hussain, A., Lee, J., Lee, H.-S., 2019b. Food waste fermentation in a leach bed reactor: reactor performance, and microbial ecology and dynamics. Bioresour. Technol. 274, 153–161.

Yin, J., Wang, K., Yang, Y., Shen, D., Wang, M., Mo, H., 2014. Improving production of volatile fatty acids from food waste fermentation by hydrothermal pretreatment. Bioresour. Technol. 171, 323–329.

Yin, J., Yu, X., Wang, K., Shen, D., 2016. Acidogenic fermentation of the main substrates of food waste to produce volatile fatty acids. Int. J. Hydrog. Energy 41 (46), 21713–21720.

Yu, L., Zhang, W., Liu, H., Wang, G., Liu, H., 2018. Evaluation of volatile fatty acids production and dewaterability of waste activated sludge with different thermo-chemical pretreatments. Int. Biodeterior. Biodegrad. 129, 170–178.

Zacharof, M.-P., Lovitt, R.W., 2013. Complex effluent streams as a potential source of volatile fatty acids. Waste Biomass Valoriz. 4 (3), 557–581.

Zacharof, M.-P., Mandale, S.J., Williams, P.M., Lovitt, R.W., 2016. Nanofiltration of treated digested agricultural wastewater for recovery of carboxylic acids. J. Clean. Prod. 112, 4749–4761.

Zhang, H., Wang, W., Wang, H., Ye, Q., 2017. Effect of e-beam irradiation and microwave heating on the fatty acid composition and volatile compound profile of grass carp surimi. Radiat. Phys. Chem. 130, 436–441.

Zhang, L., Loh, K.-C., Dai, Y., Tong, Y.W., 2020. Acidogenic fermentation of food waste for production of volatile fatty acids: bacterial community analysis and semi-continuous operation. Waste Manag 109, 75–84.

Zhang, L., Loh, K.-C., Kuroki, A., Dai, Y., Tong, Y.W., 2021. Microbial biodiesel production from industrial organic wastes by oleaginous microorganisms: current status and prospects. J. Hazard. Mater. 402, 123543.

Zhang, L., Loh, K.-C., Lim, J.W., Zhang, J., 2019. Bioinformatics analysis of metagenomics data of biogas-producing microbial communities in anaerobic digesters: a review. Renew. Sustain. Energy Rev. 100, 110–126.

Zhou, F., Wang, C., Wei, J., 2013. Separation of acetic acid from monosaccharides by NF and RO membranes: performance comparison. J. Membr. Sci. 429, 243–251.

Zhou, M., Yan, B., Wong, J.W., Zhang, Y., 2018. Enhanced volatile fatty acids production from anaerobic fermentation of food waste: a mini-review focusing on acidogenic metabolic pathways. Bioresour. Technol. 248, 68–78.

Zhuo, G., Yan, Y., Tan, X., Dai, X., Zhou, Q., 2012. Ultrasonic-pretreated waste activated sludge hydrolysis and volatile fatty acid accumulation under alkaline conditions: effect of temperature. J. Biotechnol. 159 (1), 27–31.

Functional microbial characteristics in acidogenic fermenters of organic wastes for production of volatile fatty acids

Le Zhang[a,b], Miao Yan[a,b], To-Hung Tsui[a,b], Jonathan T.E. Lee[a,b], Kai-Chee Loh[b,c], Yanjun Dai[b,d], Yen Wah Tong[a,b,c]

[a]NUS Environmental Research Institute, National University of Singapore, Singapore, [b]Energy and Environmental Sustainability for Megacities (E2S2) Phase II, Campus for Research Excellence and Technological Enterprise (CREATE), Singapore, [c]Department of Chemical and Biomolecular Engineering, National University of Singapore, Singapore, [d]School of Mechanical Engineering, Shanghai Jiao Tong University, Shanghai, China

16.1 Introduction

The sustainable management of solid organic wastes (e.g., food waste, lignocellulosic biomass waste, manure, crop residues, and sludge, etc.) has become a popular research topic in recent years (Antoniou et al., 2019; Nguyen et al., 2017; Zhang et al., 2019b). Currently, the most common waste treatment methods are incineration and landfill (Nabavi-Pelesaraei et al., 2017), which do not recover the significant resources (e.g., protein, carbohydrate, lipids, and lignocellulosic biomass) contained within the wastes, but could also cause secondary pollution. Specifically, leachate in the landfill can potentially pollute the soil and groundwater surrounding the site, posing risks to public health via contaminated drinking water (Gautam et al., 2019). Meanwhile, landfill is identified as a main source contributing greenhouse gas emission. To minimize its harmful climate effects, EU countries have imposed a landfill ban (Luo et al., 2020). As a consequence, incineration has been introduced in many countries for organics waste treatments and energy recovery. But it contributes contaminants (e.g., dioxins, furans, volatile heavy metals, and acidic gases) emission into atmosphere and is not recycling nutrients (Schuhmacher et al., 2019). Ergo, there is an impetus for adopting more environmentally beneficial and sustainable technologies (e.g., acidogenic fermentation, anaerobic digestion, gasification, and composting, etc.) for waste disposal. Among these technical options, acidogenic fermentation, a modified anaerobic digestion process with arrested methanogenesis, is considered a promising technology to achieve biowastes management and resource recovery simultaneously (Atasoy et al., 2018; Zhang et al., 2020b). The liquid end products of acidogenic fermentation, namely, volatile fatty acids (VFAs), are major building block for green energy, biofuels, and chemicals production (Llamas et al., 2020; Strazzera et al., 2018).

Nevertheless, several economic limitations need to be addressed before the acidogenic fermentation technology can be upscaled to industrial plants for the production of VFAs. To maintain a stable operation and efficient VFA production, previous studies have focused on process monitoring and the control of parameters such as pH (Atasoy et al., 2019), temperature (Soomro et al., 2020), solid retention time (SRT) (Calero et al., 2018), and organic loading rate (OLR) (Magdalena et al., 2019). The fermentation process, however, is mediated by a complex array of acidogenic microbes, including fermenting and acetogenic bacteria, which have a syntrophic relationship (Liu et al., 2020b). Fig. 16.1 shows a schematic of metabolic pathways and involved bacteria in acidogenic fermentation starting from biowastes.

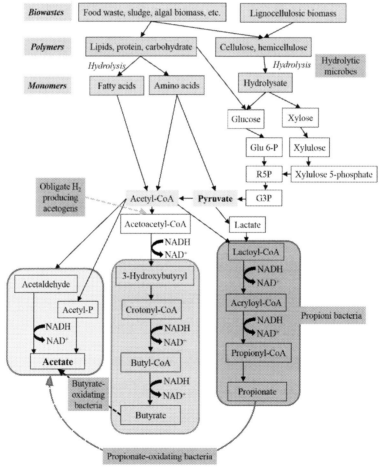

FIG. 16.1   A schematic of metabolic pathways and involved bacteria in acidogenic fermentation starting from biowastes. Adapted from Liu et al. (2020b) and Zhou et al. (2018) with permission/license granted by the publisher.

Hence, it is crucial to understand comprehensively the bacterial communities involved in terms of the taxonomic compositions, and relationships between operational conditions and community shifts, similarity, diversity, and major metabolic pathways, as well as interspecies electron transfer and other energetic mechanisms among syntrophic bacteria. Thanks to the rapid development of sequencing technology, substantial insights into microbial acidogenic fermentation have been achieved. For instance, bacterial communities in acidogenic fermenters fed with food waste (Jin et al., 2019b; Xiong et al., 2019), sludge (Iglesias-Iglesias et al., 2019; Xin et al., 2021), sludge + food waste (Moestedt et al., 2020b), and algal residue (Cho et al., 2018) have been studied. In addition, the effects of different fermentation conditions (e.g., temperature, water content, and batch size) on acidogenic bacterial communities have also been investigated. Nevertheless, many of the aforementioned studies are diverse and unconsolidated, which makes it difficult to identify the state-of-the-art.

Therefore, this chapter aims to fill in the gaps by summarizing the characteristics of acidogenic bacterial communities in terms of microbial community composition, shift, function, and metabolic interactions in various acidogenic fermenters for the production of VFAs. A comprehensive technical procedure of using bioinformatic approaches to analyze the metagenomics data of bacterial communities in acidogenic fermenters fed with various biowastes is presented. This is followed by a comprehensive overview of bioinformatic analyses of bacterial communities in acidogenic fermenters fed with various biowastes. Ultimately, the influence of various environmental factors on the bacterial communities in acidogenic fermenters has also been discussed.

16.2   Procedures for bacterial community analysis

Fig. 16.2 shows common procedures of bioinformatic analysis of bacterial communities in acidogenic fermenters. A prerequisite of bioinformatic analysis is DNA extraction. Most frequently used method for DNA extraction is usage of commercially available DNA extraction kits. The quality of the extracted DNA can be checked by measurements of 260/280 nm

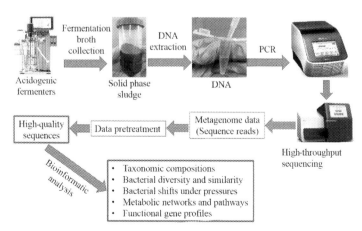

FIG. 16.2   Common procedures of bioinformatic analysis of bacterial communities in acidogenic fermenters.

and 260/230 nm ratios. The qualified DNA can be stored in Tris-ethylenediamine tetraacetic acid (EDTA) solution (TE buffer) at −20 °C prior to subsequent usage. The popular DNA sequencing platforms in the market include Illumina (Solexa) sequencing platform, Roche GS FLX 454 pyrosequencing platform, Applied Biosystems 3130xl Genetic Analyzer, and ABI SOLiDTM short-read DNA sequencing platform, etc.

The metagenome data (raw sequencing reads) obtained from high throughput sequencing should be filtered for subsequent data processing including assembly, binning, and annotation (Ju & Zhang, 2015; Yan et al., 2020). To compare patterns in microbiome variation, the relative abundance of microbes was achieved for operational taxonomic units clustering analysis, taxonomic composition analysis, alpha diversity analysis, and multivariate statistical analysis such as principal components analysis, principal coordinate analysis, nonmetric multidimensional scaling, redundancy analysis, and canonical correspondence analysis. Taxonomical assignment and functional analysis can provide high resolution of genome characteristic, function, and metabolic activity networks at gene levels. Zhang et al. (2019a) and Kumar & Chowdhary (2016) summarized the frequently used bioinformatics platforms and tools in microbial metagenomics-based studies, which provided a detailed introduction for beginners. A possible limitation is that the application of these reported bioinformatics software and platforms depend on microbial sampling environments (e.g. bioreactors, soil, and wastewater), thereby could not be totally applied to the bacterial communities in acidogenic fermenters. Hence, more efforts should be paid to the development of new powerful bioinformatics software and platforms for analyzing acidogenic fermentation microbes.

16.3   Microbial characteristics in acidogenic fermenters

Functional microbial community involved in acidogenic fermentation can be affected by a variety of factors, including substrate properties (Shen et al., 2017), pH, temperature (e.g., mesophilic or thermophilic), OLR (Tang et al., 2016), inoculum composition (Yin et al., 2016b), additives (Jin et al., 2019a), bioaugmentation (Chi et al., 2018), and microaerobic condition (Zhang et al., 2021). For the acidogenic fermenters with a specific waste type, the knowledge about dominant bacteria, bacterial function, effects of environmental factors, and potential application (e.g., bioaugmentation) of functional bacteria is beneficial to process optimization, which deserves deep investigation. Thus, microbial characteristics in acidogenic fermentation of food waste, lignocellulosic biomass waste, sludge, wastewater, algal residues, and mixed wastes are summarized.

16.3.1   Microbial characteristics in acidogenic fermentation of food waste

Food waste is usually rich in protein, carbohydrate, lipid, cellulose, and hemicellulose. These compositions can be degraded into VFAs by various hydrolytic, acidogenic, and acetogenic bacteria (Zhou et al., 2018). Table 16.1 shows microbial characteristics of acidogenic fermentation of food waste. Notably, almost all the surveyed studies directly used the mixed components (e.g., protein, carbohydrate, and lipid) as it is challenging to separate different fractions from heterogeneous food waste because different organic components are linked to distinct bacteria for degradation, leading to different VFA products. Thus, the microbial compositions

TABLE 16.1   Microbial characteristics of acidogenic fermentation of food waste.

Food waste types	Reactors	Inoculation sludge	Operating conditions	Main VFA components	Dominant microbes	Refs
Carbohydrate-rich food waste	Single-stage, 1 L CSTR reactors	-	25 °C; OLR 3 gVS/L/d; HRT 20–27 days	Isobutyric and caproic acids	Ruminococcus	(Greses et al., 2021)
Food waste from campus canteen in Singapore	1 L bioreactors	Sludge from wastewater treatment plant	35 °C and 55 °C; organic loading 22 gVS/L; pH 5–7	Acetic, propionic and butyric acids	Clostridia, Bacteroidia and Bacilli	(Zhang et al., 2020b)
Food waste in European Union	Immersed membrane bioreactors	Sludge from another acidogenic reactor	OLR 4, 6, 8, and 10 g VS/L/d; 37 °C; 150 rpm	Acetic and butyric acids	Firmicutes, Actinobacteria and Proteobacteria phyla dominated at 6 and 10 g VS/L/d; Clostridium and Lactobacillus enriched	(Wainaina et al., 2020)
Cheese whey	Sequencing batch reactor		30 °C; 80 rpm; OLR 6 g COD/L/d; SRT 5–15 d	Butyric and acetic acids	Phyla Firmicutes and Patescibacteria; Genus Lactobacillus and Megasphaera	(Lagoa-Costa et al., 2020)
Food waste from campus canteen in Canada	7.5 L leachate bed reactor	Sludge from wastewater treatment facility; treated at 75 °C for 15 min	Batch mode; 22 °C; 14 days; OLR 21.7 gVS/L	Butyric, acetic and propionic acids	pH 6: Bifidobacterium, Clostridium; pH 7–8: Bacteroides, Dysgonomonas	(Xiong et al., 2019)
Food waste from campus canteen in China	Lab-scale bioreactors	Sludge from domestic sewage treatment plant; treated in a water bath at 105 °C for 2 h	37 °C; 72 h; pH 5.5–6.5	Butyric, propionic and acetic acids	Vagococcus, Actinomyces, Bacteroides, and Fermentimonas	(Jin et al., 2019b)
Food waste from campus canteen in China	1 L bioreactors	Anaerobic granular sludge	30 °C; pH 6; NaCl 10 g/L	Acetic, propionic and butyric acids	Propionibacteria (high tolerant to NaCl)	(He et al., 2019a)
Fat, oil, and grease (FOG) from food wastes' leachate in Korea	250 mL serum bottles	Anaerobic sludge, supplemented with nutrition and treated at 100 °C for 1 h	37 °C; 150 rpm; Initial pH 6.86	C4 and C6 fatty acids	Firmicutes, Bacteroidetes, and Cloacimonetes	(Saha et al., 2019)

(continued)

TABLE 16.1 (Cont'd)

Food waste types	Reactors	Inoculation sludge	Operating conditions	Main VFA components	Dominant microbes	Refs
Food waste from campus canteen in China	Conical flasks	Anaerobic sludge from sewage treatment plant; treated at 105 °C for 2 h	37 °C; 120 h	Acetic acid (72%) and lactic acid	*Lactobacillus* and *Olsenella*	(Jin et al., 2019a)
Carbohydrate-rich wasted potato	Double-walled organic glass reactor	Sludge from wastewater treatment plant; incubated at 35 °C for 7 days	35 °C; OLR 6.7 gVS/L/d; pH 6–8	Acetic and butyric acid	Phylum: *Firmicutes* and *Bacteroidetes*; Genus: *Bacteroides*, *Lactobacillus*, *Lactococcus*, and *Clostridium*	(Li et al., 2019b)
Food waste	3 L continuous stirred tank reactor (CSTR)	Sludge from anther anaerobic digester	35 °C; 40 rpm; SRT 4 days; OLR 11.8 gVS/L/d	pH 3.2–4.5: lactic acid; pH 4.7–5.0: butyric acid	pH 3.2–4.5: *Lactobacillus*; pH 4.7–5.0: *Bifidobacterium*, *Lactobacillus*, and *Olsenella*	(Feng et al., 2018)
Vegetable wastes	4 L leaching bed reactor	Sludge from wastewater treatment plant	35 °C; inoculum-to-substrate ratio of 1:10	Acetic and butyric acids	*Proteobacteria*, *Firmicutes*, and *Bacteroidetes*	(Li et al., 2017)
Protein-rich substrates (tofu and egg white) in food waste	500 mL wide-mouthed bottles	Anaerobic granular sludge from an industrial reactor	30 °C; 120 rpm; pH 6.0; waste-to-microorganism ratio of 5:1	Acetic acid; mixed acids	Tofu: *Sporanaerobacter*, *Candidatus*; Egg white: *Aminobacterium* and *Bacteroides*	(Shen et al., 2017)

in previous studies using different substrates could be remarkably diverse. For instance, the main products during acidogenic fermentation of glucose, peptone, and glycerol are butyric, acetic, and propionic acids, respectively (Yin et al., 2016a; Tsui et al., 2020). Even so, some common microbes present in acidogenic fermenters fed with food waste were observed, including class *Clostridia*, class *Bacteroidia*, class *Bacilli*, phylum *Firmicutes*, phylum *Actinobacteria*, phylum *Proteobacteria*, genus *Lactobacillus*, genus *Ruminococcus*, genus *Bifidobacterium*, and genus *Olsenella*. (Greses et al., 2021; Wainaina et al., 2020; Xiong et al., 2019; Zhang et al., 2020b). Specifically, bacteria belonging to the class *Clostridia* can ferment carbohydrates to produce acetic and butyric acids (Łukajtis et al., 2018). Genus *Ruminococcus*, belonging to phylum *Firmicutes*, is related to polysaccharides hydrolysis and H_2 generation together with acetic acid through the reverse β-oxidation pathway (Greses et al., 2020; Kallscheuer et al., 2017). Genus *Bacteroides* as typical fermentation bacteria are capable of converting complex organic compounds into corresponding monomers (Zhang et al., 2018). Moreover, although the reported ranking of predominant bacterial species in previous studies was based on the relative abundance of each species rather than their absolute quantity, the information of dominant bacteria remains helpful to optimize operational conditions and develop process intensification methods such as bioaugmentation (Murali et al., 2021).

Previous studies have investigated microbial community shift in food waste acidogenic fermenters in response to various environmental changes such as redox potential (Yin et al., 2016b), temperature (Zhang et al., 2020b), pH (Feng et al., 2018; Jin et al., 2019b; Li et al., 2019b; Xiong et al., 2019), OLR (Tang et al., 2016), hydraulic retention time (HRT) (Domingos et al., 2017), high concentration of NaCl (He et al., 2019a), additives (e.g., nanoscale zerovalent iron (NZVI) (Jin et al., 2019a)), and leachate recirculation frequencies in leach bed reactors (Xu et al., 2014). These environmental factors could affect the enrichment of specific microbes, leading to changes in major acidogenic metabolic pathways. For example, seed sludge can directly shape the bacterial compositions in acidogenic fermenters, which further determines VFA yield (Moestedt et al., 2020b). Oxidation reduction potential (ORP) ranging from −100 to −200 mV contributes to higher acidogenic fermentation efficiency of food waste (Yin et al., 2016b). At these ORP levels, *Firmicutes* was one of the most common major phyla in acidogenic fermenters and involved in overall fermentation process. Arras et al. (2019) found that temperature changes (e.g., 35, 55, and 70 °C) clearly influenced the biotransformation pathways by shifting the bacterial population structures. During thermophilic acidogenic fermentation of canteen food waste, one of the major bacterial genera was identified as *Lactobacillus* (Zhang et al., 2020b). Genus *Lactobacillus*, an aerotolerant anaerobe or microaerophilic bacteria, capable of converting sugars into lactic acids, was also observed in other acidogenic fermenters fed with typical European and Chinese food waste (Feng et al., 2018; Wainaina et al., 2020). pH control was another common approach to achieve the targeted product spectrum because it can affect the microbial community structure, enzyme activity, and metabolic pathways (Tang et al., 2017). For instance, in the food waste acidogenic fermenters with a pH of 6.0, *Bifidobacterium* and *Clostridium* were the dominant bacteria, which shifted to *Dysgonomonas* and *Bacteroides* when the pH increased to 7.0–8.0 (Xiong et al., 2019). The presence of genus *Clostridium* (e.g., species *Clostridium autoethanogenum*) in the acidogenic fermenters of food waste contributed to the acidification of carbohydrates (Sarkar et al., 2016). In another food waste acidogenic fermenter, *Lactobacillus* was the abundant genus at pH 3.2–4.5, resulting in lactic acid fermentation. At the same time, *Bifidobacteria* was enriched significantly at pH 4.5, which led to an increase in acetic acid. When pH was 4.7–5.0, the dominant genera were still *Bifidobacterium*, *Lactobacillus*, and *Olsenella*; however, the metabolic pathways for butyric acid generation were activated, and the activities of lactic acid fermentation started to decrease (Feng et al., 2018).

Application of high OLR (e.g., 11 g VS (volatile solids)/L/d) had been a method in acidogenic fermentation to inhibit the methanogenesis through decreasing the pH by VFAs accumulation. However, there remain challenges such as potential process instability and trace amount of biogas production. To solve this issue, an inhibitor (i.e., 2-bromoethanesulfonate) is commonly utilized to inhibit methanogenesis in acidogenic fermenters. Regarding the use of increased OLR, Tang et al. (2016) investigated the effects of high OLR on the lactic acid production from food waste. They observed that lactic acid concentration had positive correlation with OLR less than 18 g total solids (TS)/L/d but negative correlation with OLR ranging from 18 to 22 g TS/L/d. Results of 454 high-throughput pyrosequencing indicated that genera *Lactobacillus* (43.6%) and *Weissella* (19.2%) were the predominant fermentative bacteria, which explained the high lactic acid in the reactors (Tang et al., 2016). Similarly, the performance of membrane bioreactors operated at OLRs of 4, 6, 8, and 10 g (volatile solids) VS/L/d and the microbial dynamics during the acidogenic fermentation of food waste to VFAs was investigated by Wainaina et al. (2020). They analyzed the bacterial populations using 16S rRNA gene amplicon sequencing and it was found that *Firmicutes* and *Actinobacteria* were predominant phyla at OLRs 4 and 8 g VS/L/d while phyla *Firmicutes*, *Actinobacteria*, and *Proteobacteria* dominated at 6 and 10 g VS/L/d. Additionally, the dominance of acetate, butyrate, and caproate was due to the presence of the genus *Clostridium* (Zhang et al., 2008), while the presence of genus *Lactobacillus* was correlated with the production of lactate (Wainaina et al., 2020). Furthermore, HRT is an essential operational parameter that can determine the contact time between microorganism and substrates (Bhatt et al., 2020). The relatively high HRT offers more time to hydrolytic, acidogenic, and acetogenic bacteria to grow and convert substrates into VFAs. In particular, long HRT contributes to improved hydrolysis of substrates and VS reduction, resulting in a higher concentration of VFAs (Ferrer et al., 2010). Hence, the HRT should also be optimized in an acidogenic fermenter with a specific waste type to promote acidification and enhance VFA production.

Besides, the effect of nanoscale NZVI on VFA production from food waste was investigated by Jin et al. (2019a). They observed the changes in microbial community structure in the NZVI amended bioreactor. Specifically, the addition of NZVI contributed to an increase in the abundance of genus *Olsenella* that could rapidly convert the VFAs into short-chain VFAs (Jin et al., 2019a). *Olsenella*, an obligate anaerobic and nonmotile bacterial genus, belongs to the family of *Coriobacteriaceae* and is capable of producing lactic acid and acetic acid anaerobically (Dewhirst et al., 2001; Feng et al., 2018). As a common food flavoring agent in food processing, salt (e.g., NaCl) is inevitably accumulated in food waste. He et al. (2019a) characterized the microbial communities during acidogenic fermentation of food waste with an NaCl concentration of 10–70 g/L for VFA production. At the end of fermentation, microbial community analysis confirmed the predominance of genera *Propionibacteria*, which showed a high tolerance to NaCl and contributed to the production of propionic acid (He et al., 2019a).

Upon elucidation of dominant bacteria in the acidogenic fermenters, bioaugmentation of the key players for acidogenic fermentation processes can be applied to enhance VFA yields. For instance, Chi et al. (2018) investigated the effects of bioaugmentation with *Clostridium tyrobutyricum* on the improvement of butyric acid production. They demonstrated that the bioaugmentation strategy enhanced butyric acid production up to 3.26-fold by significantly shifting the bacterial community. Reddy et al. (2018) observed the highest medium-chain fatty acids production in the bioaugmented reactor with a pure culture of *Clostridium kluyveri*, where they characterized microbial populations using denaturing gradient gel electrophoresis analysis. The predominant bacteria included class *Clostridia*, order *Sphingobacteriales*, family *Desulfobacteraceae*, and genus *Bacillus* in the bioaugmented reactors. Notably, *Clostridium cellulolyticum*, as mesophilic cellulolytic bacteria, play an essential role in the synthesis of glycosyl hydrolase enzymes and fermentation products such as acetate and ethanol (Desvaux, 2005; Williams et al., 2013). The order *Sphingobacteriales* is capable of fermenting glucose into isovaleric acid, isobutyric acid, and 2-methylbutyric acid (Allison, 1978). *Desulfobacteraceae*, as a family of *Proteobacteria*, is reported as sulfate-reducing bacteria that are key players of the carbon- and sulfur-cycles (Dörries et al., 2016; Tsui et al., 2018). Genus *Bacillus*, a member of the phylum *Firmicutes*, is one of the most predominant genera involved in hydrolysis and acidification of proteins, fats, and sugars (Jin et al., 2019b). However, a relatively high cost of isolation, cultivation, and identification of certain functional microbes could pose a technical limitation to bioaugmentation operations with pure microbial strains, especially for large-scale applications. To increase the economic feasibility, a low-cost alternative of pure microbial strains should be identified. Bioaugmentation of acclimatized microbial consortia could be a cost-effective method to enhance fermentation performance (Basak et al., 2021). For instance, Chang et al. (2018) successfully enhanced the hydrogenogenic acidogenic fermentation of food waste using acclimatized microbial consortia for coproduction of hydrogen and C4-C7 carboxylates. Furthermore, digestate recirculation positively affected fermentation performance on hydrolysis perspective (Gottardo et al., 2017; Wu et al., 2018).

16.3.2　Microbial characteristics in acidogenic fermentation of lignocellulosic biomass waste

Conversion of lignocellulosic biomass wastes into VFAs via acidogenic fermentation has attracted interests due to increasing feedstock product as well as high market demand and widespread application of VFAs. Hitherto, microbial characteristics of acidogenic fermenters of lignocellulosic biomass wastes have been investigated, involving hydrothermal pretreated olive mill solid waste, corn straw hydrolysate, banana waste, pretreated waste wood feedstock, duckweed (*Lemnaceae*), and rice straw (Table 16.2). Lignocellulosic biomass mainly consists of cellulose, hemicellulose, and lignin. Cellulose and hemicellulose as two primary components available to hydrolytic bacteria are hydrolyzed to corresponding monomers, that is, glucose, xylose, arabinose, galactose, mannose, etc., which are further converted into VFAs by acidogenic and acetogenic bacteria. Generally, frequently identified hydrolytic, acidogenic, and acetogenic microbes in acidogenic fermenters of lignocellulose biomass belong to phyla *Firmicutes*, *Bacteroidetes*, and *Proteobacteria*. These phyla are responsible for the consumption of the aforementioned monomers (Sun et al., 2019). Importantly, the dominant microbes in various acidogenic fermenters of lignocellulosic biomass wastes can be significantly different due to variations of waste type and operating conditions (e.g., pH and temperature), etc. For instance, in mesophilic acidogenic fermenters fed with hydrothermal pretreated olive mill solid waste, dominant microbes were phyla *Proteobacteria* and *Firmicutes*

TABLE 16.2 Microbial characteristics of acidogenic fermenters of lignocellulosic biomass wastes.

Waste types	Reactors	Inoculation sludge	Operating conditions	Main VFA components	Dominant microbes	Reference
Hydrothermal pretreated olive mill solid waste	120 mL glass bottles	Anaerobic sludge from pilot-scale mesophilic reactor treating sewage	35 °C; 180 rpm; initial pH 5.5	Acetic and butyric acids	Phyla Proteobacteria and Firmicutes; Genera *Enterobacter, Pseudomonas, Achromobacter*, etc.	(da Fonseca et al., 2021)
Corn straw hydrolysate	1 L CSTR reactors	Anaerobic and aerobic sludge from wastewater treatment plant	38 °C; HRT 5 days; OLR 8 g/L/d	Butyric, acetic, and propionic acids	Phylum Firmicutes, class Clostridia, family Clostridiaceae	(Li et al., 2020)
Banana waste	250 mL batch reactors	Microbiota obtained from natural fermentation of banana waste	Initial pH 5.5, 6.6, 7.5; 30–44 °C; Inoculum 5%–15%	Acetic acid, butyric acid, and lactic acid	Autochthonous acidogenic bacteria *Clostridium* and *Lactobacillus*	(Mazareli et al., 2020)
Pretreated waste wood feedstock	2.5 L glass bottles	Sludge from a municipal wastewater treatment plant	pH 6.0; 35 °C; Initial C/N ratio of 20–80	Acetic acid and propionic acid	Genera *Corynebacterium, Actinomyces* and *Bacteroides*	(Li et al., 2019a)
Duckweed (Lemnaceae)	300 mL batch reactors	Mesophilic seed: silage, rumen fluid, ad anaerobic wastewater sludge; Thermophilic sludge: compost	35 °C and 55 °C; acidic pH 5.3; basic pH 9.2; solid loading 25 g/L	Acetic acid	Acidic sample: *Veillonellaceae acidaminococcus*; basic sample: *Clostridiaceae alkaliphilus*	(Calicioglu et al., 2018)
Rice straw	250 mL flasks	Anaerobic sludge from sewage disposal plant	10% (v/v) inocula; optimized pH 9.58; 41.84 °C	Mixed VFAs	Members of *Ruminococcaceae, Bacteroidaceae, Porphyromonadaceae*, and *Lachnospiraceae*	(Park et al., 2015)
Rice straw	500 mL serum bottles	Enrichment cultures from peptone cellulose solution medium	35 °C; 140 rpm; 6 days; 5% (v/v) inocula	Butyric acid	*Bacteroidetes* and *Oscillospiraceae*	(Ai et al., 2013)

as well as genera *Enterobacter, Pseudomonas, Clostridium, Achromobacter, Ochrobactrum,* and *Peptoclostridium* (da Fonseca et al., 2021). In contrast, in another mesophilic acidogenic fermenter fed with pretreated waste wood feedstock, the predominant bacteria were genera *Corynebacterium, Actinomyces,* and *Bacteroides* (Li et al., 2019a). The dominant microbial species for acidogenic fermentation of olive mill solid waste depend on the contents of main components, that is, hemicelluloses, cellulose, lignin, and fat, which accounted for 13.54%, 8.19%, 33.80%, and 16.81%, respectively (da Fonseca et al., 2021). Genus *Pseudomonas* was reported to produce xylanases while genera *Achromobacter, Clostridium. Pseudomonas* were capable of producing cellulases, which jointly contributed to hydrolysis of lignocellulosic biomass (Cheng & Chang, 2011; Wierckx et al., 2011). In addition, the genus *Enterobacter* has been reported to contribute to the biological detoxification of furans from lignocellulose hydrolysates (Zhang et al., 2013). Even in two similar rice straw acidogenic fermenters reported by two different studies, the dominant microbes were nonidentical due to differences in inoculum and operation temperature (Ai et al., 2013; Park et al., 2015). Specifically, Park et al. (2015) analyzed microbial communities using the 454 pyrosequencing method. They confirmed the dominance of Family *Ruminococcaceae,* which contained bacterial species capable of hydrolyzing rice straw and fermenting xylose and glucose. However, in another rich straw acidogenic fermenter inoculated with mixed culture derived from peptone cellulose solution medium, the enriched bacteria Phyla were *Bacteroidetes* and *Oscillospiraceae,* which belong to cellulolytic and xylanolytic bacteria, and butyrate-producing bacteria (Ai et al., 2013).

Moreover, operating conditions could have a significant influence on the yield and composition of VFA. For instance, pH and temperature have been shown to significantly affect the microbial community structure for carboxylic acid production in an acidogenic fermenter for duckweed (Calicioglu et al., 2018). In the acidic fermenter for duckweed, *Veillonellaceae acidaminococcus* was enriched greatly, while some genera *Bacillaceae natronobacillus, Clostridiaceae natronincola,* and *Bacteroidaceae bacteroides* were not detected. In contrast, significant enrichments of *Clostridiaceae alkaliphilus* were observed. Additionally, the enriched bacterial genera included *Clostridiaceae thermoanaerobacterium, Tissierellaceae tepidimicrobium, Thermoanaerobacterales thermovenabulum,* and *Planococcaceae lysinibacillus* in the thermophilic acidogenic fermenter of duckweed, whereas in corresponding mesophilic reactors included *Prevotellaceae prevotella* and unidentified genera in the families of *Porphyromonadaceae* and *Enterobacteriaceae.* Indeed, the alteration of microbial community structure had close relationships with operating conditions (e.g., pH and temperature) and final end VFA products (Ziara et al., 2019). For instance, *Thermoanaerobacterium,* an enriched genus observed in an acidic thermophilic fermenter of duckweed (Calicioglu et al., 2018), can degrade cellulose, starch, and sucrose and favor a slightly acidic environment (Prasertsan et al., 2009). Furthermore, it was concluded that pH had a more significant impact than the temperature on both VFA production and the composition of microbial communities (Calicioglu et al., 2018).

16.3.3 Microbial characteristics in acidogenic fermentation of sludge

Table 16.3 shows the microbial characteristics of acidogenic fermenters of sludge. Several sludge types were summarized, including waste activated sludge (WAS) (with or without pretreatment), primary sludge, sewage sludge, excess sludge, and primary sedimentation sludge. Generally, to start the acidogenic fermentation of sludge, as shown in Table 16.3,

TABLE 16.3 Microbial characteristics of acidogenic fermenters of sludge.

Waste types	Reactors	Inoculation sludge	Operating conditions	Main VFA components	Dominant microbes	Reference
WAS	500 mL serum bottles	Mesophilic anaerobic sludge from fermenter treating WAS	35 °C; 80 rpm; SRT 25 days; WAS to inoculum ratio of 2:1 (w/w)	Acetic acid and *n*-butyric acid	Phyla *Proteobacteria, Bacteroidetes, Firmicutes,* and *Ignavibacteriae*	(She et al., 2020)
Primary sludge	250 mL batch reactors	-	35 °C; 200 rpm; 8 days; 0.05–0.30 g/gVS CaO$_2$	Acetic acid and propionic acid	*Alkaliphilus, Acinetobacter, Macellibacteroides, Arobacter* and *Zoogloea*	(Huang et al., 2019)
WAS and primary sludge	150 mL serum bottles	Three inocula (granular and slurry)	35 °C; pH 5, 8, 10, without adjustment; 120 rpm	Neutral pH: acetic acid; Acidic pH: acetic and butyric acids; Alkali pH: butyric acid	*Firmicutes* and *Chloroflexi*	(Atasoy et al., 2019)
Sewage sludge	2 L glass reactors	Anaerobic sludge from a brewery WWTP	35 °C; 150 rpm; substrate to inoculum ratio of 1, 2, 4, and 6 gVS/VS	Acetic acid and propionic acid	Phyla *Proteobacteria, Bacteroidetes* and *Firmicutes*	(Iglesias-Iglesias et al., 2019)
WAS	600 mL serum bottles	-	35 °C; 180 rpm; iron activated persulfate	Acetic, propionic, butyric, and valeric acids	Fe reactor: Phyla *Actinobacteria* and *Proteobacteria;* Fe/PS reactor: *Chloroflexi, Acidobacteria,* and *Saccharibacteria*	(Luo et al., 2019)
WAS after ferrate pretreatment	500 mL serum bottles	Sludge collected from an acidogenic reactor	35 °C; 160 rpm; 12 days; pH 10; ferrate dosage: 0.5 g/gVS	Acetic acid	Phyla *Firmicutes* and *Bacteroidetes;* Genera: *Macellibacteroides, Petrimonas, Proteiniclasticum,* and *Proteocatella*	(Li et al., 2018a)

Waste types	Reactors	Inoculation sludge	Operating conditions	Main VFA components	Dominant microbes	Reference
WAS	500 mL and 5.5 L reactors	Pure WAS taken from a wastewater treatment plant	35 °C; 550 rpm	Propionic acid	Classes Bacteroidetes, Actinobacteria, Alphaproteobacteria, Betaproteobacteria, Bacilli, and Clostridia	(Huang et al., 2018a)
WAS	1 L reactors	No inoculum	37 °C; 200 rpm; HRT 8 days; pH 10	Acetic acid	Phylum Firmicutes; Genera Natronincola and Desulfitispora	(Rao et al., 2018)
WAS	2 L reactors	Sludge taken from another anaerobic reactor	35 °C; 60 rpm; Operated for 15 days	Acetic acid	Proteiniborus, Lactobacillus and Anaerotruncus	(Xin et al., 2018a)
Primary sludge	250 mL reactors	-	35 °C; 200 rpm; 8 days; pH: 3, 5, 7, 9, 10, and 12, respectively	Acetic acid and propionic acid	Alkali treated: Firmicutes, Proteobacteria, Synergistates, and Actinobacteria; Acid treated: Firmicutes and Proteobacteria	(Huang et al., 2018b)
Primary sedimentation sludge	550 mL reactors	Fermentation sludge	37 °C; pH 5, 6, 8; HRT 4 days; OLR 4.3 g/L/d; iron-based chemically enhanced	-	pH 6: Bacteroidia and Erysipelotrichi; pH 8: Clostridia	(Lin & Li, 2018)

anaerobic sludge derived from a wastewater treatment plant or another acidogenic fermenter was commonly adopted as inoculation sludge. Generally, the microbial communities in various sludge acidogenic fermenters were generally similar at higher taxonomic levels (e.g., phylum) of microorganisms; however, the relative abundance of their microbial species could also be significantly affected by environmental factors such as additive and pH. For instance, in mesophilic acidogenic fermenters fed with sewage sludge, *Proteobacteria*, *Bacteroidetes*, and *Firmicutes* were identified as predominant bacterial phyla responsible for acidogenic fermentation of organic matter such as carbohydrates and amino acids (Iglesias-Iglesias et al., 2019). It was reported that an enrichment in bacterial phyla *Acidobacteria*, *Chloroflexi*, and *Saccharibacteria* was achieved after supplementation of iron-activated persulfate, which favored the WAS fermentation (Luo et al., 2019). In another mesophilic acidogenic digester fed with WAS after ferrate pretreatment at a dosage of 0.5 g/gVS, dominant microbial species belonged to phyla *Firmicutes* and *Bacteroidetes* as well as genera *Macellibacteroides*, *Proteiniclasticum*, and *Proteocatella* (Li et al., 2018a). On the one hand, the phyla *Firmicutes* and *Bacteroidetes* were considered the main functional players, and they were correlated well with the production of butyric and valeric acids (Cirne et al., 2012). On the other hand, *Macellibacteroides* were responsible for conversion of carbohydrate to acetic acid and butyric acid (She et al., 2020). Hence, the high abundance of genus *Macellibacteroides* had positive correlation with the yield of short-chain VFA. Additionally, the synergy between genera *Macellibacteroides* and *Proteiniclasticum* has been confirmed for the enhanced VFA production (He et al., 2019b). Furthermore, Wang et al. (2017) demonstrated that triclocarban, one typical antibacterial agent being used in soap, cosmetics, and shampoo, could increase the microbial community diversity in the acidogenic fermenter of WAS. Hu et al. (2018) observed that the addition of diclofenac, an antiinflammatory drug being used in human health care, could promote the relative abundance of acetogen. Moreover, Liu et al. (2019) found that supplementation of cationic polyacrylamide, a flocculation powder being utilized in wastewater pretreatment and sludge dewatering, in the acidogenic fermenter of WAS decreased microbial community diversity and altered activities of key enzymes responsible for VFA production. Huang et al. (2019) found that CaO_2 addition increased significantly abundance of *Alkaliphilus*, *Macellibacteroides*, *Acinetobacter*, *Zoogloea*, and *Arobacter*, thus successfully enhanced acidogenic fermentation of primary sludge for VFA production.

Hitherto, the effect of pH on VFA yield and microbial composition has been extensively investigated (Atasoy et al., 2019; Huang et al., 2018b; Lin & Li, 2018). For instance, the long-term effect of pH on microbial community in sludge acidogenic fermenters was studied by Yuan et al. (2015). They found that acidogenic bacterial genera of *Erysipelothrix*, *Paludibacter*, and *Tissierella* favored the production of short-chain VFA at a pH of 10.0. Similarly, Liu et al. (2014) investigated the effects of alkaline adjustment (pH 7.0–10.0) on microbial communities in acidogenic fermenters fed with ultrasonic-pretreated WAS at 20 °C. They observed that some bacteria belonging to *Proteobacteria*, *Firmicutes*, and *Bacteroidetes* eventually adapted to high alkaline environments, which contributed to biodegradation of proteins and carbohydrates and subsequent VFA production. Notably, although the optimal pH for acidogenic fermentation of excess sludge for VFA production was 10.0, a major bacteria *Pseudomonas* sp. capable of utilizing organic carbon was detected at all pH values between 5.0 and 12.0 (Jie et al., 2014). The maximum hydrolysis efficiency of a mixture of primary sludge and WAS was achieved at pH 9.9 by Chen et al. (2017). They found that phyla *Proteobacteria* and

Actinobacteria as well as *Firmicutes* were the most dominant population at a pH of 8.9 to 9.9, resulting in the maximum acidification efficiency. Based on the aforementioned studies, alkaline acidogenic fermentation seems a promising technology for efficient VFA production from sludge. In recent years, microbial community analysis has been reported in fermenters operating at alkaline conditions (Rao et al., 2018). For instance, microbial community succession during alkaline (pH = 10) fermentation of excess sludge was studied by Li et al. (2018b). They found that the dominant bacterial genera included *Clostridium, Amphibacillus, Bacillus,* and *Peptostreptococcaceae*. In another alkaline (pH = 9) fermentation of WAS with the addition of saponin (a biosurfactant), the dominant bacteria were identified as *Proteiniclasticum, Macellibacteroides, Fusibacter,* and *Petrimonas* (Huang et al., 2016). Additionally, in a fermenter operated at pH of 10.0, acidogenic bacteria were reported to be capable of assisting biodegradation of nonylphenol in WAS (Duan et al., 2018). Therefore, bioaugmentation of nonylphenol-degrading bacteria such as *Proteiniphilum acetatigenes* and *Propionibacterium acidipropionici* could be utilized for nonylphenol degradation in WAS.

16.3.4 Microbial characteristics in acidogenic fermentation of wastewater

Acidogenic fermentation of several kinds of wastewater has been investigated, including high strength wastewater (palm oil mill effluent), semisynthetic milk processing wastewater, sucrose-containing wastewater, molasses wastewater, swine wastewater, as well as synthetic wastewater containing lactate and sulfate. The fermenting bacteria varied in various wastewater types, as different components required corresponding microbes to degrade. Table 16.4 shows microbial characteristics in acidogenic fermenters of wastewater. For instance, in an acidogenic fermenter of palm oil mill effluent, the dominant microbes were genera *Lactobacillus, Olsenella, Aeriscardovia, Pseudoramibacter,* and *Atopobium* (Lim et al., 2020). In contrast, *Desulfobulbus, Desulfovibrio, Pseudomonas,* and *Clostridium* were identified as the dominant genus in another fermenter fed with synthetic wastewater containing lactate and sulfate (Zhao et al., 2008). It has been reported that acid-tolerant *Lactobacillus* and *Aeriscardovia* can form a highly efficient consortium that can favor the production of acetate and lactate during the fermentation process (Moestedt et al., 2020a). Genera *Olsenella* and *Atopobium* are recognized as lactic acid–producing bacteria (De Groof et al., 2020), while genus *Pseudoramibacter* is associated with chain elongation (Liu et al., 2020a). The predominance of these bacterial genera was positively correlated with a high concentration of *n*-caproate and *n*-caprylate (Liu et al., 2020a). Similarly, Yun and Cho (2016) found that concentration of butyrate and propionate were positively correlated with lactic acid bacteria, including *Sporolactobacillus, Bacillales,* and *Lactobacillus,* and that the close relationship between lactic acid bacteria and *Pseudoramibacter* (a chain-elongating bacteria) played a critical role in stable VFA production from molasses wastewater.

Similarly, microbial communities in acidogenic fermenters fed with wastewater can also be affected by various environmental factors such as temperature, inoculant, and HRT. Specifically, Liu et al. (2002) compared microbial communities between mesophilic and thermophilic fermenters of dairy wastewater. They found that an elevated temperature caused more significant and rapid microbial shift in the thermophilic fermenter. Acidogenic fermentation of semisynthetic milk processing wastewater was compared using pure and mixed cultures of *Clostridium aceticum, Clostridium butyricum,* and *P. acidipropionici*. It showed that the

TABLE 16.4 Microbial characteristics of acidogenic fermenters of wastewater.

Waste types	Reactors	Inoculation sludge	Operating conditions	Main VFA components	Dominant microbes	Reference
High strength wastewater (palm oil mill effluent)	12 L sequencing batch reactor	Anaerobic palm oil mill	29 °C; pH 4.8–5.5; HRT 5 days	Acetic, propionic, butyric, valeric, caproic, and heptanoic acids	Lactobacillus, Olsenella, Aeriscardovia, Pseudoramibacter, and Atopobium	(Lim et al., 2020)
Semisynthetic milk processing wastewater	100 mL serum bottles	Mixed cultures containing Clostridium aceticum, Clostridium butyricum, and Propionibacterium acidipropionici	35 °C; 120 rpm; Initial pH 10; VS content 2 g/L	Acetic, butyric and propionic acids	Clostridium aceticum, Clostridium butyricum, and Propionibacterium acidipropionici	(Atasoy et al., 2020)
Sucrose-containing wastewater	100 mL reactor	Sludge from a full-scale reactor treating starch-processing wastewater	30 °C; Flux of 200 mL/h; HRT 13 h	Acetic, propionic and butyric acids	-	(Pan et al., 2018)
Molasses wastewater	1 L CSTR reactors	Acclimated anaerobic sludge from wastewater treatment plant	35 °C; 150 rpm; HRT 24 h; Initial pH 7.0; OLR 19–35 gCOD/L/d	Butyric acid and acetic acid	Clostridiales, Lactobacillales, Pseudomonadaceae, and Micrococcineae	(Yun & Cho, 2016)
Swine wastewater	4 L CSTR reactors	Anaerobic seed sludge from wastewater treatment plant	40–60 °C; HRT 0.5–2.5 d	Acetate, propionate, n-butyrate, isobutyrate, n-valerate, isovalerate, n-caproate	Bacillus, Pseudomonas, Thermoactinomyces, and Thermomonas	(Kim et al., 2013)
Synthetic wastewater containing lactate and sulfate	1 L CSTR reactors	Seed sludge from moat sediment	35 °C; 200 rpm; Wastewater contained 4 gCOD/L and 2 g sulfate/L	Acetic acid	Desulfobulbus, Desulfovibrio, Pseudomonas, and Clostridium	(Zhao et al., 2008)

fermentation using mixed cultures has a faster pH regulation and acclimation than that of pure cultures (Atasoy et al., 2020). In addition, the effects of temperature and HRT on bacterial community structure in acidogenic fermenters for swine wastewater have been studied by Kim et al. (2013), who found that temperature had stronger influence on bacterial community than HRT. Further, Kim et al. (2013) claimed that the acidogenesis process was more dependent on the specific growth rate of acidogenic microbes and their contact time with feedstock than on the alteration of acidogenic microbes. Although the microbial communities in various wastewater acidogenic fermenters could be highly diversified, above-mentioned microbial information could provide guidance to design and optimize the related acidogenic fermenters for efficient VFA production from microbial perspective by bioaugmentation strategy.

16.3.5 Microbial characteristics in acidogenic fermentation of algal residues

Algal residue has been regarded as a promising biomass substrate for the production of biofuels through biorefinery processes (Chandra et al., 2019; Zhang et al., 2020a). Table 16.5 summarizes microbial characteristics of acidogenic fermentation of algal residues. In acidogenic fermenters fed with algal residue (e.g., *Ettlia* sp.), a significant positive correlation between *Sporanaerobacter acetigenes* (a dominant VFA producer) and VFA production was observed (Seo et al., 2016). Similarly, the correlation between bacterial community shifts and VFA profiles was statistically verified by Seon et al. (2014). They found that the dominant bacteria involved in hydrolysis and fermentation of alginate were *Bacteroides*-related microbes and *Clostridium* spp. A major limitation in acidogenic fermentation of microalgae was the relatively low hydrolysis rate due to recalcitrant algal cell wall that accounts for 11%–37% of total cell mass (Santos-Ballardo et al., 2016). Hence, several strategies such as additive (e.g., potassium ferrate) (Wang et al., 2020), alkali pretreatment, and thermal-alkali pretreatment (Cho et al., 2018) have been investigated to enhance cell wall disruption and promote acidification step. Specifically, the addition of potassium ferrate in an acidogenic fermenter for microalgae not only effectively destroyed the cell structure and surface morphology of microalgae, but also enriched hydrolytic and short-chain VFA-forming bacteria such as *Clostridium* sp. and *Terrisporobacter* sp. (Wang et al., 2020). Moreover, the alkali-pretreatment and thermal-alkaline pretreatment of cyanobacterial biomass increased the soluble protein concentration by 1.58 times and 1.81 times, respectively, which enhanced the subsequent acidification process. After alkaline pretreatment, bacterium *S. acetigenes*, belonging to protein-utilizing bacteria, was gradually acclimated to the protein component in algal biomass (Cho et al., 2018). It indicated the importance of the pretreatments and feedstock components in shaping acidogenic bacteria. Furthermore, the effects of different temperatures (e.g., 35–55 °C) (Gruhn et al., 2016) and OLR (Magdalena et al., 2019) on microbial community dynamics have been studied. For instance, Cho et al. (2015) investigated influence of temperatures from 35 °C to 55 °C on microbial community structure during acidogenic fermentation of microalgae. They found that the dominant bacterial communities at different operating temperatures belong to phyla *Firmicutes*, *Proteobacteria*, and *Bacteroidetes* for the efficient degradation of organic compounds (e.g., protein and carbohydrate) inside microalgal biomass. *Clostridium* sp. was one of the most abundant in an acidogenic fermenter fed with marine macroalgae (Kidanu et al., 2017). *Clostridium* sp. has been identified as one of the most effective producer of

TABLE 16.5 Microbial characteristics of acidogenic fermenters of algal residues.

Waste types	Reactors	Inoculation sludge	Operating conditions	Main VFA components	Dominant microbes	Reference
Thickened microalgae	600 mL serum bottles	Sludge from a municipal sewage treatment plant	20–25 °C; K_2FeO_4 dosage: 28.2–112.8 mg Fe(VI)/gVS; 120–300 r/min	Acetic acid	Enriched hydrolytic and VFA-forming bacteria	(Wang et al., 2020)
Enzyme-pretreated microalgae (*Chlorella vulgaris*) biomass	1 L CSTR reactors	Adapted anaerobic sludge from a previous anaerobic reactor	25 °C; 250 rpm; HRT 8 days	Butyric, acetic and propionic acids	Phyla *Firmicutes* and *Bacteroidetes*	(Magdalena et al., 2019)
Cyanobacterial biomass	1 L reactors	Sludge from an anaerobic digester fed with sewage sludge	35 °C; 150 rpm; pH 6.5; Thermal-alkaline pretreatment	Acetic, butyric, propionic, isovaleric, valeric, and isobutyric acid	Phyla *Firmicutes*, *Bacteroidetes*, *Chloroflexi*, and *Proteobacteria*	(Cho et al., 2018)
Air-dried macroalgae samples (*S. japonica*, *P. elliptica*, and *E. crinita*)	500 mL serum bottles	Sludge from wastewater treatment plant; heated at 100 °C for 15 min	35 °C; organic loading 35.0 g/L; 150 rpm; pH at 6.0–7.0	Acetic, butyric and propionic acids	*Clostridium* sp.	(Kidanu et al., 2017)
Lipid-extracted biomass of algae *Ettlia* sp.	–	Sludge from a mesophilic anaerobic digester	25–65 °C; pH 7; substrate to microorganism ratio of 0.5–2.5	Butyric, acetic, and propionic acids	Species *Sporanaerobacter acetigenes*	(Seo et al., 2016)
Algal biomass containing *Desmodesmus* sp., *Scenedesmus* sp., and *Chlamydomonas* sp.	1 L batch reactors	Anaerobic sludge from a mesophilic digester; heated at 100 °C for 2 h	35 °C; 45 °C; 55 °C; 150 rpm; Initial pH: 6.9	Acetic acid	Phyla *Proteobacteria*, *Bacteroidetes*, and *Firmicutes*	(Cho et al., 2015)
Alginate derived from algae (sodium alginate)	2 L reactors	Acidogenic microbes cultivated with glucose	35 °C; 120 rpm; pH 6–10; Alginate concentration 4–9 g/L	Acetic, butyric, and propionic acids	Phyla *Firmicutes*, *Actinobacteria*, and *Bacteroidetes*; Genus *Clostridium*	(Seon et al., 2014)

hydrogen and VFA, and they are capable of resisting unfavorable environmental conditions such as temperature shock (Atasoy & Cetecioglu, 2020; Chi et al., 2018). In addition, the effect of stepwise increased OLR from 3 to 15 g COD (chemical oxygen demand)/L/d on acidogenic bacteria during acidogenic fermentation of *Chlorella vulgaris* was assessed by Magdalena et al. (2019). They observed the dominance of *Firmicutes* (74%) and *Bacteroidetes* (20%) as well as the reduction of *Euryarchaeota* phylum (0.5%) at the OLR of 15 g COD/L/d and confirmed the feasibility of promoting VFA production by stepwise OLR increase.

16.3.6 Microbial characteristics in acidogenic cofermentation of mixed wastes

Compared to acidogenic fermentation of single feedstock, acidogenic cofermentation has been considered as a more attractive technology as it can not only simultaneously treat two or more kinds of wastes and reduce equipment investment, but also can achieve synergistic effects of microorganisms and balanced nutrients (Wu et al., 2016). Table 16.6 shows microbial characteristics during acidogenic cofermentation of mixed wastes. Common cofermentation scenarios included swine manure + corn silage, sludge + food waste, cheese whey + sludge, corn stover + WAS, WAS + corn stalk + livestock manure, and WAS + potato peel waste + food waste. On the one hand, cofermentation can efficiently enhance hydrolysis and acidogenesis yield. On the other hand, cofermentation can favor enrichment of hydrolytic and acidogenic bacteria, leading to a higher VFA production. Different wastes containing various organic matters may lead to microbial community shifts in different directions. For instance, microbial communities in acidogenic cofermentation of WAS with different organic wastes (e.g., potato peel waste and food waste) were analyzed by Ma et al. (2017). They observed that food waste (a representative of lipid and protein-rich waste) addition to WAS substantially enriched the class *Clostridia* and that potato peel waste (a representative of starch-rich waste) addition favored enrichment of class *Bacilli*. These results demonstrated that organic compositions in the cosubstrate significantly affected the bacterial communities and selectively enriched corresponding acidogenic bacteria. Li et al. (2018c) investigated the microbial communities during acidogenic cofermentation of sewage sludge and food waste. They found that cofermentation enhanced the hydrogen content and VFA yield by 30% and 8.4%, respectively. Meanwhile, they achieved an appropriate carbon/nitrogen (C/N) ratio of 15–23 in the cofermentation reactor, where VFA-producing bacteria such as genera *Veillonella* and *Clostridium* as well as orders *Bacteroidales* and *Lactobacillales* were dominant. Interestingly, in another cofermentation reactor fed with food waste and sewage sludge, these dominant bacteria such as *Lactobacillus* and *Clostridium* can be significantly affected by pH, leading to a diverse distribution of VFA products (Cheng et al., 2014).

Generally, agricultural wastes (e.g., corn stalk and corn silage) with a high C/N ratio (e.g., > 50) can be cofermented with the wastes (e.g., food waste, swine manure, and WAS) with low C/N ratios (e.g., < 15) (Cao et al., 2021; Xin et al., 2018b). For instance, Zhou et al. (2016) demonstrated the feasibility of cofermentation of WAS with corn stover for higher VFA production. They found that characteristic bacterial genera, with better hydrolysis and acidification abilities were enriched by the added contents of cellulose, hemicellulose, or the saccharification hydrolysates. Notably, when WAS is used as a substrate for cofermentation, the microbes originating from WAS may tend to dominate the bioreactor after long-term operation (Jankowska et al., 2018). Moreover, in an acidogenic cofermentation reactor fed

TABLE 16.6 Microbial characteristics in acidogenic fermentation of mixed wastes.

Waste types	Reactors	Inoculation sludge	Operating conditions	Main VFA components	Dominant microbes	Reference
Swine manure and corn silage	10 L anaerobic reactor	Sludge from another anaerobic reactor	37 °C; 150 rpm; C/N ratio 35; HRT 12 days; Initial pH 6.0–6.5; OLR 2.0–4.0 gVS/L/d	Acetic, n-butyric, and caproic acids	Hydrolytic bacteria: Clostridium, Terrisporobacter, Intestinibacter, and Turicibacter; Acidogenic bacteria: Acetobacter, Lactobacillus, Aeriscardovia, and Pseudomonas	(Cao et al., 2021)
Waste-activated sludge mixed with corn stalk and livestock manure	2 L anaerobic reactors	–	35 °C; 60 rpm; 10 days; C/N ratio 9–20	Acetic and propionic acids	Proteiniborus, Lactobacillus, Firmicutes, and Clostridium	(Xin et al., 2018b)
Sewage sludge and food waste	2 L reactors	Sludge collected from a food waste anaerobic digester	37 °C; pH 7.2–7.6; inoculum to substrate ratio of 10%	Acetic, butyric, propionic, lactic, and valeric acids	Orders Bacteroidales and Lactobacillales; Genera Veillonella and Clostridium	(Li et al., 2018c)
Cheese whey and sludge	4.5 L CSTR reactors	Anaerobic sludge from full-scale mesophilic digester of WWTP	35 °C; initial pH 5.2; OLR 1.65–32.9 gVS/L/d	Butyrate, valerate, caproate and heptate	Phyla Firmicutes, Cloacimonetes, Proteobacteria	(Jankowska et al., 2018)
WAS, potato peel waste, and food waste	500 mL serum bottles	Sludge from wastewater treatment plant; Heated at 105 °C for 2 h	35 °C; 120 rpm; substrate to microorganism ratio of 5.67; fermentation period: 12 days	Acetic, propionic, valeric, and butyric acids	Classes Bacilli and Clostridia; Phylum Firmicutes	(Ma et al., 2017)

Waste types	Reactors	Inoculation sludge	Operating conditions	Main VFA components	Dominant microbes	Reference
Food waste and excess sludge	500 mL reactors	Sludge from wastewater treatment plant	40 °C; OLR 9 gVS/L/d; SRT 7 d; food waste/excess sludge ratio: 5	Acetic, propionic and n-butyric acids	Hydrolytic and acidogenic bacteria: Clostridium, Sporanaerobacter, Tissierella and Bacillus	(Wu et al., 2016)
Cornstover and WAS	300 mL reactors	Sludge from municipal wastewater treatment plant	35 °C; 100 rpm; corn stover proportion of 50% and 35% in cosubstrates	Acetic and propionic acids	Acetic acid producing bacteria: genera Bacteroides, Proteiniclasticum and Fluviicola; Propionic acid producing bacteria: genera Mangroviflexus and Paludibacter	(Zhou et al., 2016)
Food waste and sewage sludge	3 L anaerobic reactor	Sludge from a digester treating wastewater	35 °C; 160–200 rpm; SRT 5–7 days; food waste/dewatered excess sludge 1:1; pH 5–11	Acetic, butyric, propionic, and valeric acids	Lactobacillus, Prevotella, Mitsuokella, Treponema, Clostridium, and Ureibacillus	(Cheng et al., 2014)

with swine manure and corn silage, a gradually increased OLR led to a decrease in the relative abundance of hydrolysis bacteria (e.g., some species belonging to *Clostridium*, *Terrisporobacter*, *Turicibacter*, and *Intestinibacter*) and an increase of relative abundance of acidogenic bacteria such as *Acetobacter*, *Lactobacillus*, *Pseudomonas*, and *Aeriscardovia* (Cao et al., 2021). The aforementioned information can be helpful for substrate selection and parameter control for acidogenic cofermentation for efficient VFA production.

16.4   Conclusions and perspectives

A substantial availability of biowastes provides promising substrates sources for VFAs production through acidogenic fermentation. However, an inadequate understanding of efficient microbiome in the acidogenic fermenters is now the bottleneck for the best utilization of biowastes. Therefore, this chapter comprehensively reviewed the microbial characteristics during acidogenic fermentation of various organic wastes, including food waste, lignocellulosic biomass waste, sludge, wastewater, manure, algal residues, and mixed organic wastes. The development of high-throughput sequencing of 16S rRNA amplicons enables deeper insights into microbial compositions, shifts, metabolism pathways, correlations, and functions. It is clear that the effects of operation parameters such as pH, OLR, HRT, temperature, inoculum, additives, and substrate pretreatment could influence microbial activities during acidogenic fermentation. Moreover, acidogenic cofermentation of two or more biowastes showed a great potential in practical application, given that it can develop more balanced element composition (e.g., C/N ratio) for microbial fermentation. In this regard, substrates should be carefully evaluated for cofermentation to achieve desired performance. Additionally, substrate pretreatments prior to acidogenic fermentation are recommended for the lignocellulosic wastes such as crop straw and algal residues.

Upon successfully identifying the functional microbes in various acidogenic fermenters, it is critical to apply the microbial information for practical uses of acidogenic fermentation enhancement thereby value-added chemicals production. Bioaugmentation is an effective approach through the application of recognized microbes with their ability on such purposes. Notwithstanding, the high cost of pure microbial strains for inoculation remains a big challenge for scale-up and long-term operations. To reduce the cost, an acclimatized microbial consortium obtained from the discharged digestate can be a low-cost alternative to pure microbial strains, and it presents advantages in the long-term bioaugmentation applications. Furthermore, a further elucidation of electron transfer mechanisms between syntrophic microbes (e.g., hydrolytic bacteria, acidogenic bacteria and homoacetogenic bacteria) would undoubtedly broaden the application scope of identified microbial resources for higher and more efficient VFA production.

Acknowledgments

This work was funded by the National Research Foundation, Prime Minister's Office, Singapore under its Campus for Research Excellence and Technological Enterprise (CREATE) Program.

References

Ai, B., Li, J., Song, J., Chi, X., Meng, J., Zhang, L., Ban, Q., 2013. Butyric acid fermentation from rice straw with undefined mixed culture: enrichment and selection of cellulolytic butyrate-producing microbial community. Int. J. Agric. Biol. 15 (6), 1075–1082.

Allison, M.J., 1978. Production of branched-chain volatile fatty acids by certain anaerobic bacteria. Appl. Environ. Microbiol. 35 (5), 872–877.

Antoniou, N., Monlau, F., Sambusiti, C., Ficara, E., Barakat, A., Zabaniotou, A., 2019. Contribution to circular economy options of mixed agricultural wastes management: coupling anaerobic digestion with gasification for enhanced energy and material recovery. J. Clean. Prod. 209, 505–514.

Arras, W., Hussain, A., Hausler, R., Guiot, S., 2019. Mesophilic, thermophilic and hyperthermophilic acidogenic fermentation of food waste in batch: effect of inoculum source. Waste Manage. 87, 279–287.

Atasoy, M., Cetecioglu, Z., 2020. Butyric acid dominant volatile fatty acids production: bio-augmentation of mixed culture fermentation by Clostridium butyricum. J. Environ. Chem. Eng. 8 (6), 104496.

Atasoy, M., Eyice, Ö., Cetecioglu, Z., 2020. Volatile fatty acid production from semi-synthetic milk processing wastewater under alkali pH: the pearls and pitfalls of microbial culture. Bioresour. Technol. 297, 122415.

Atasoy, M., Eyice, O., Schnürer, A., Cetecioglu, Z., 2019. Volatile fatty acids production via mixed culture fermentation: revealing the link between pH, inoculum type and bacterial composition. Bioresour. Technol. 292, 121889.

Atasoy, M., Owusu-Agyeman, I., Plaza, E., Cetecioglu, Z., 2018. Bio-based volatile fatty acid production and recovery from waste streams: current status and future challenges. Bioresour. Technol. 268, 773–786.

Basak, B., Patil, S.M., Saha, S., Kurade, M.B., Ha, G.-S., Govindwar, S.P., Lee, S.S., Chang, S.W., Chung, W.J., Jeon, B.-H., 2021. Rapid recovery of methane yield in organic overloaded-failed anaerobic digesters through bioaugmentation with acclimatized microbial consortium. Sci. Total Environ. 764, 144219.

Bhatt, A.H., Ren, Z., Tao, L., 2020. Value proposition of untapped wet wastes: carboxylic acid production through anaerobic digestion. iScience 23 (6), 101221.

Calero, R.R., Lagoa-Costa, B., Fernandez-Feal, M.M.d.C., Kennes, C., Veiga, M.C., 2018. Volatile fatty acids production from cheese whey: influence of pH, solid retention time and organic loading rate. J. Chem. Tech. Biotechnol. 93 (6), 1742–1747.

Calicioglu, O., Shreve, M.J., Richard, T.L., Brennan, R.A., 2018. Effect of pH and temperature on microbial community structure and carboxylic acid yield during the acidogenic digestion of duckweed. Biotechnol. Biofuels 11 (1), 275.

Cao, Q., Zhang, W., Lian, T., Wang, S., Dong, H., 2021. Short chain carboxylic acids production and dynamicity of microbial communities from co-digestion of swine manure and corn silage. Bioresour. Technol. 320, 124400.

Chandra, R., Iqbal, H.M., Vishal, G., Lee, H.-S., Nagra, S., 2019. Algal biorefinery: a sustainable approach to valorize algal-based biomass towards multiple product recovery. Bioresour. Technol. 278, 346–359.

Chang, S.-E., Saha, S., Kurade, M.B., Salama, E.-S., Chang, S.W., Jang, M., Jeon, B.-H., 2018. Improvement of acidogenic fermentation using an acclimatized microbiome. Int. J. Hydrog. Energy 43 (49), 22126–22134.

Chen, Y., Jiang, X., Xiao, K., Shen, N., Zeng, R.J., Zhou, Y., 2017. Enhanced volatile fatty acids (VFAs) production in a thermophilic fermenter with stepwise pH increase—investigation on dissolved organic matter transformation and microbial community shift. Water Res. 112, 261–268.

Cheng, C.-L., Chang, J.-S., 2011. Hydrolysis of lignocellulosic feedstock by novel cellulases originating from Pseudomonas sp. CL3 for fermentative hydrogen production. Bioresour. Technol. 102 (18), 8628–8634.

Cheng, W., Chen, H., Yan, S., Su, J., 2014. Illumina sequencing-based analyses of bacterial communities during short-chain fatty-acid production from food waste and sewage sludge fermentation at different pH values. World J. Microb. Biotechnol. 30 (9), 2387–2395.

Chi, X., Li, J., Wang, X., Zhang, Y., Leu, S.-Y., Wang, Y., 2018. Bioaugmentation with Clostridium tyrobutyricum to improve butyric acid production through direct rice straw bioconversion. Bioresour. Technol. 263, 562–568.

Cho, H.U., Kim, Y.M., Choi, Y.-N., Kim, H.G., Park, J.M., 2015. Influence of temperature on volatile fatty acid production and microbial community structure during anaerobic fermentation of microalgae. Bioresour. Technol. 191, 475–480.

Cho, H.U., Kim, Y.M., Park, J.M., 2018. Changes in microbial communities during volatile fatty acid production from cyanobacterial biomass harvested from a cyanobacterial bloom in a river. Chemosphere 202, 306–311.

Cirne, D.G., Bond, P., Pratt, S., Lant, P., Batstone, D.J., 2012. Microbial community analysis during continuous fermentation of thermally hydrolysed waste activated sludge. Water Sci. Technol. 65 (1), 7–14.

Dörries, M., Wöhlbrand, L., Kube, M., Reinhardt, R., Rabus, R., 2016. Genome and catabolic subproteomes of the marine, nutritionally versatile, sulfate-reducing bacterium *Desulfococcus multivorans* DSM 2059. BMC Genom. 17 (1), 1–20.

da Fonseca, Y.A., Silva, N.C.S., de Camargos, A.B., de Queiroz Silva, S., Wandurraga, H.J.L., Gurgel, L.V.A., Baêta, B.E.L., 2021. Influence of hydrothermal pretreatment conditions, typology of anaerobic digestion system, and microbial profile in the production of volatile fatty acids from olive mill solid waste. J. Environ. Chem. Eng. 9 (2), 105055.

De Groof, V., Coma, M., Arnot, T.C., Leak, D.J., Lanham, A.B., 2020. Adjusting organic load as a strategy to direct single-stage food waste fermentation from anaerobic digestion to chain elongation. Processes 8 (11), 1487.

Desvaux, M., 2005. *Clostridium cellulolyticum*: model organism of mesophilic cellulolytic clostridia. FEMS Microbiol. Rev. 29 (4), 741–764.

Dewhirst, F.E., Paster, B.J., Tzellas, N., Coleman, B., Downes, J., Spratt, D.A., Wade, W.G., 2001. Characterization of novel human oral isolates and cloned 16S rDNA sequences that fall in the family *Coriobacteriaceae*: description of *Olsenella* gen. nov., reclassification of *Lactobacillus uli* as *Olsenella uli* comb. nov. and description of *Olsenella profusa* sp. nov. Int. J. Syst. Evol. Microbiol. 51 (5), 1797–1804.

Domingos, J.M., Martinez, G.A., Scoma, A., Fraraccio, S., Kerckhof, F.-M., Boon, N., Reis, M.A., Fava, F., Bertin, L., 2017. Effect of operational parameters in the continuous anaerobic fermentation of cheese whey on titers, yields, productivities, and microbial community structures. ACS Sustain. Chem. Eng. 5 (2), 1400–1407.

Duan, X., Wang, X., Xie, J., Feng, L., Yan, Y., Wang, F., Zhou, Q., 2018. Acidogenic bacteria assisted biodegradation of nonylphenol in waste activated sludge during anaerobic fermentation for short-chain fatty acids production. Bioresour. Technol. 268, 692–699.

Feng, K., Li, H., Zheng, C., 2018. Shifting product spectrum by pH adjustment during long-term continuous anaerobic fermentation of food waste. Bioresour. Technol. 270, 180–188.

Ferrer, I., Vázquez, F., Font, X., 2010. Long term operation of a thermophilic anaerobic reactor: process stability and efficiency at decreasing sludge retention time. Bioresour. Technol. 101 (9), 2972–2980.

Gautam, P., Kumar, S., Lokhandwala, S., 2019. Advanced oxidation processes for treatment of leachate from hazardous waste landfill: a critical review. J. Clean. Prod. 237, 117639.

Gottardo, M., Micolucci, F., Bolzonella, D., Uellendahl, H., Pavan, P., 2017. Pilot scale fermentation coupled with anaerobic digestion of food waste-effect of dynamic digestate recirculation. Renew. Energy 114, 455–463.

Greses, S., Tomás-Pejó, E., Gónzalez-Fernández, C., 2020. Agroindustrial waste as a resource for volatile fatty acids production via anaerobic fermentation. Bioresour. Technol. 297, 122486.

Greses, S., Tomás-Pejó, E., González-Fernández, C., 2021. Short-chain fatty acids and hydrogen production in one single anaerobic fermentation stage using carbohydrate-rich food waste. J. Clean. Prod. 284, 124727.

Gruhn, M., Frigon, J.-C., Guiot, S.R., 2016. Acidogenic fermentation of Scenedesmus sp.-AMDD: comparison of volatile fatty acids yields between mesophilic and thermophilic conditions. Bioresour. Technol. 200, 624–630.

He, X., Yin, J., Liu, J., Chen, T., Shen, D., 2019a. Characteristics of acidogenic fermentation for volatile fatty acid production from food waste at high concentrations of NaCl. Bioresour. Technol. 271, 244–250.

He, Z.-W., Tang, C.-C., Liu, W.-Z., Ren, Y.-X., Guo, Z.-C., Zhou, A.-J., Wang, L., Yang, C.-X., Wang, A.-J., 2019b. Enhanced short-chain fatty acids production from waste activated sludge with alkaline followed by potassium ferrate treatment. Bioresour. Technol. 289, 121642.

Hu, J., Zhao, J., Wang, D., Li, X., Zhang, D., Xu, Q., Peng, L., Yang, Q., Zeng, G., 2018. Effect of diclofenac on the production of volatile fatty acids from anaerobic fermentation of waste activated sludge. Bioresour. Technol. 254, 7–15.

Huang, L., Chen, Z., Xiong, D., Wen, Q., Ji, Y., 2018a. Oriented acidification of wasted activated sludge (WAS) focused on odd-carbon volatile fatty acid (VFA): regulation strategy and microbial community dynamics. Water Res. 142, 256–266.

Huang, X., Dong, W., Wang, H., Feng, Y., 2018b. Role of acid/alkali-treatment in primary sludge anaerobic fermentation: insights into microbial community structure, functional shifts and metabolic output by high-throughput sequencing. Bioresour. Technol. 249, 943–952.

Huang, X., Dong, W., Wang, H., Sun, F., Feng, Y., 2019. Enhance primary sludge acidogenic fermentation with CaO_2 addition: investigation on soluble substrate generation, sludge dewaterability, and molecular biological characteristics. J. Clean. Prod. 228, 1526–1536.

Huang, X., Mu, T., Shen, C., Lu, L., Liu, J., 2016. Effects of bio-surfactants combined with alkaline conditions on volatile fatty acid production and microbial community in the anaerobic fermentation of waste activated sludge. Int. Biodeterior. Biodegrad. 114, 24–30.

Iglesias-Iglesias, R., Campanaro, S., Treu, L., Kennes, C., Veiga, M.C., 2019. Valorization of sewage sludge for volatile fatty acids production and role of microbiome on acidogenic fermentation. Bioresour. Technol. 291, 121817.

Jankowska, E., Duber, A., Chwialkowska, J., Stodolny, M., Oleskowicz-Popiel, P., 2018. Conversion of organic waste into volatile fatty acids–the influence of process operating parameters. Chem. Eng. J. 345, 395–403.

Jie, W., Peng, Y., Ren, N., Li, B., 2014. Volatile fatty acids (VFAs) accumulation and microbial community structure of excess sludge (ES) at different pHs. Bioresour. Technol. 152, 124–129.

Jin, Y., Gao, M., Li, H., Lin, Y., Wang, Q., Tu, M., Ma, H., 2019a. Impact of nanoscale zerovalent iron on volatile fatty acid production from food waste: key enzymes and microbial community. J. Chem. Tech. Biotechnol. 94 (10), 3201–3207.

Jin, Y., Lin, Y., Wang, P., Jin, R., Gao, M., Wang, Q., Chang, T.-C., Ma, H., 2019b. Volatile fatty acids production from saccharification residue from food waste ethanol fermentation: effect of pH and microbial community. Bioresour. Technol. 292, 121957.

Ju, F., Zhang, T., 2015. Experimental design and bioinformatics analysis for the application of metagenomics in environmental sciences and biotechnology. Environ. Sci. Technol. 49 (21), 12628–12640.

Kallscheuer, N., Polen, T., Bott, M., Marienhagen, J., 2017. Reversal of β-oxidative pathways for the microbial production of chemicals and polymer building blocks. Metab. Eng. 42, 33–42.

Kidanu, W.G., Trang, P.T., Yoon, H.H., 2017. Hydrogen and volatile fatty acids production from marine macroalgae by anaerobic fermentation. Biotechnol. Bioproc. E. 22 (5), 612–619.

Kim, W., Shin, S.G., Lim, J., Hwang, S., 2013. Effect of temperature and hydraulic retention time on volatile fatty acid production based on bacterial community structure in anaerobic acidogenesis using swine wastewater. Bioproc. Biosyst. Eng. 36 (6), 791–798.

Kumar, G.R., Chowdhary, N., 2016. Biotechnological and bioinformatics approaches for augmentation of biohydrogen production: a review. Renew. Sustain. Energy Rev. 56, 1194–1206.

Lagoa-Costa, B., Kennes, C., Veiga, M.C., 2020. Cheese whey fermentation into volatile fatty acids in an anaerobic sequencing batch reactor. Bioresour. Technol. 308, 123226.

Li, D., Yin, F., Ma, X., 2019a. Achieving valorization of fermented activated sludge using pretreated waste wood feedstock for volatile fatty acids accumulation. Bioresour. Technol. 290, 121791.

Li, L., He, J., Wang, M., Xin, X., Xu, J., Zhang, J., 2018a. Efficient volatile fatty acids production from waste activated sludge after ferrate pretreatment with alkaline environment and the responding microbial community shift. ACS Sustain. Chem. Eng. 6 (12), 16819–16827.

Li, X., Liu, G., Liu, S., Ma, K., Meng, L., 2018b. The relationship between volatile fatty acids accumulation and microbial community succession triggered by excess sludge alkaline fermentation. J. Environ. Manage. 223, 85–91.

Li, Y., Hua, D., Mu, H., Xu, H., Jin, F., Zhang, X., 2017. Conversion of vegetable wastes to organic acids in leaching bed reactor: performance and bacterial community analysis. J. Biosci. Bioeng. 124 (2), 195–203.

Li, Y., Xu, H., Hua, D., Zhao, B., Mu, H., Jin, F., Meng, G., Fang, X., 2020. Two-phase anaerobic digestion of lignocellulosic hydrolysate: focusing on the acidification with different inoculum to substrate ratios and inoculum sources. Sci. Total Environ. 699, 134226.

Li, Y., Zhang, X., Xu, H., Mu, H., Hua, D., Jin, F., Meng, G., 2019b. Acidogenic properties of carbohydrate-rich wasted potato and microbial community analysis: effect of pH. J. Biosci. Bioeng. 128 (1), 50–55.

Li, Z., Chen, Z., Ye, H., Wang, Y., Luo, W., Chang, J.-S., Li, Q., He, N., 2018c. Anaerobic co-digestion of sewage sludge and food waste for hydrogen and VFA production with microbial community analysis. Waste Manage. 78, 789–799.

Lim, J., Zhou, Y., Vadivelu, V., 2020. Enhanced volatile fatty acid production and microbial population analysis in anaerobic treatment of high strength wastewater. J. Water Process. Eng. 33, 101058.

Lin, L., Li, X.-y., 2018. Acidogenic fermentation of iron-enhanced primary sedimentation sludge under different pH conditions for production of volatile fatty acids. Chemosphere 194, 692–700.

Liu, B., Kleinsteuber, S., Centler, F., Harms, H., Sträuber, H., 2020a. Competition between butyrate fermenters and chain-elongating bacteria limits the efficiency of medium-chain carboxylate production. Front. Microbiol. 11, 336.

Liu, C., Ren, L., Yan, B., Luo, L., Zhang, J., Awasthi, M.K., 2020b. Electron transfer and mechanism of energy production among syntrophic bacteria during acidogenic fermentation: a review. Bioresour. Technol., 124637.

Liu, W.-T., Chan, O.-C., Fang, H.H., 2002. Microbial community dynamics during start-up of acidogenic anaerobic reactors. Water Res. 36 (13), 3203–3210.

Liu, X., Xu, Q., Wang, D., Wu, Y., Yang, Q., Liu, Y., Wang, Q., Li, X., Li, H., Zeng, G., 2019. Unveiling the mechanisms of how cationic polyacrylamide affects short-chain fatty acids accumulation during long-term anaerobic fermentation of waste activated sludge. Water Res. 155, 142–151.

Liu, Y., Li, X., Kang, X., Yuan, Y., Du, M., 2014. Short chain fatty acids accumulation and microbial community succession during ultrasonic-pretreated sludge anaerobic fermentation process: effect of alkaline adjustment. Int. Biodeterior. Biodegrad. 94, 128–133.

Llamas, M., Magdalena, J.A., González-Fernández, C., Tomás-Pejó, E., 2020. Volatile fatty acids as novel building blocks for oil-based chemistry via oleaginous yeast fermentation. Biotechnol. Bioeng. 117 (1), 238–250.

Łukajtis, R., Hołowacz, I., Kucharska, K., Glinka, M., Rybarczyk, P., Przyjazny, A., Kamiński, M., 2018. Hydrogen production from biomass using dark fermentation. Renew. Sustain. Energy Rev. 91, 665–694.

Luo, H., Zeng, Y., Cheng, Y., He, D., Pan, X., 2020. Recent advances in municipal landfill leachate: a review focusing on its characteristics, treatment, and toxicity assessment. Sci. Total Environ. 703, 135468.

Luo, J., Wu, L., Feng, Q., Fang, F., Cao, J., Zhang, Q., Su, Y., 2019. Synergistic effects of iron and persulfate on the efficient production of volatile fatty acids from waste activated sludge: understanding the roles of bioavailable substrates, microbial community & activities, and environmental factors. Biochem. Eng. J. 141, 71–79.

Ma, H., Liu, H., Zhang, L., Yang, M., Fu, B., Liu, H., 2017. Novel insight into the relationship between organic substrate composition and volatile fatty acids distribution in acidogenic co-fermentation. Biotechnol. Biofuels 10 (1), 1–15.

Magdalena, J.A., Greses, S., González-Fernández, C., 2019. Impact of organic loading rate in volatile fatty acids production and population dynamics using microalgae biomass as substrate. Sci. Rep. 9 (1), 1–11.

Mazareli, R.d.S., Villa-Montoya, A.C., Delforno, T.P., Centurion, V.B., de Oliveira, V.M., Silva, E.L., Varesche, M.B.A., 2020. Metagenomic analysis of autochthonous microbial biomass from banana waste: screening design of factors that affect hydrogen production. Biomass Bioenergy 138, 105573.

Moestedt, J., Müller, B., Nagavara Nagaraj, Y., Schnürer, A., 2020a. Acetate and lactate production during two-stage anaerobic digestion of food waste driven by *Lactobacillus* and *Aeriscardovia*. Front. Energy Res. 8, 105.

Moestedt, J., Westerholm, M., Isaksson, S., Schnürer, A., 2020b. Inoculum source determines acetate and lactate production during anaerobic digestion of sewage sludge and food waste. Bioengineering 7 (1), 3.

Murali, N., Srinivas, K., Ahring, B.K., 2021. Increasing the production of volatile fatty acids from corn stover using bioaugmentation of a mixed rumen culture with homoacetogenic bacteria. Microorganisms 9 (2), 337.

Nabavi-Pelesaraei, A., Bayat, R., Hosseinzadeh-Bandbafha, H., Afrasyabi, H., Chau, K.-W., 2017. Modeling of energy consumption and environmental life cycle assessment for incineration and landfill systems of municipal solid waste management—a case study in Tehran Metropolis of Iran. J. Clean. Prod. 148, 427–440.

Nguyen, D.D., Chang, S.W., Cha, J.H., Jeong, S.Y., Yoon, Y.S., Lee, S.J., Tran, M.C., Ngo, H.H., 2017. Dry semi-continuous anaerobic digestion of food waste in the mesophilic and thermophilic modes: new aspects of sustainable management and energy recovery in South Korea. Energy Convers. Manage. 135, 445–452.

Pan, X.-R., Li, W.-W., Huang, L., Liu, H.-Q., Wang, Y.-K., Geng, Y.-K., Lam, P.K.-S., Yu, H.-Q., 2018. Recovery of high-concentration volatile fatty acids from wastewater using an acidogenesis-electrodialysis integrated system. Bioresour. Technol. 260, 61–67.

Park, G.W., Seo, C., Jung, K., Chang, H.N., Kim, W., Kim, Y.-C., 2015. A comprehensive study on volatile fatty acids production from rice straw coupled with microbial community analysis. Bioproc. Biosyst. Eng. 38 (6), 1157–1166.

Prasertsan, P., Sompong, O., Birkeland, N.-K., 2009. Optimization and microbial community analysis for production of biohydrogen from palm oil mill effluent by thermophilic fermentative process. Int. J. Hydrog. Energy 34 (17), 7448–7459.

Rao, Y., Wan, J., Liu, Y., Angelidaki, I., Zhang, S., Zhang, Y., Luo, G., 2018. A novel process for volatile fatty acids production from syngas by integrating with mesophilic alkaline fermentation of waste activated sludge. Water Res. 139, 372–380.

Reddy, M.V., Hayashi, S., Choi, D., Cho, H., Chang, Y.-C., 2018. Short chain and medium chain fatty acids production using food waste under non-augmented and bio-augmented conditions. J. Clean. Prod. 176, 645–653.

Saha, S., Jeon, B.-H., Kurade, M.B., Chatterjee, P.K., Chang, S.W., Markkandan, K., Salama, E.-S., Govindwar, S.P., Roh, H.-S., 2019. Microbial acclimatization to lipidic-waste facilitates the efficacy of acidogenic fermentation. Chem. Eng. J. 358, 188–196.

Santos-Ballardo, D.U., Rossi, S., Reyes-Moreno, C., Valdez-Ortiz, A., 2016. Microalgae potential as a biogas source: current status, restraints and future trends. Crit. Rev. Environ. Sci. Technol. 15 (2), 243–264.

Sarkar, O., Kumar, A.N., Dahiya, S., Krishna, K.V., Yeruva, D.K., Mohan, S.V., 2016. Regulation of acidogenic metabolism towards enhanced short chain fatty acid biosynthesis from waste: metagenomic profiling. RSC Adv. 6 (22), 18641–18653.

Schuhmacher, M., Mari, M., Nadal, M., Domingo, J.L., 2019. Concentrations of dioxins and furans in breast milk of women living near a hazardous waste incinerator in Catalonia. Spain. Environ. Int. 125, 334–341.

Seo, C., Kim, W., Chang, H.N., Han, J.-I., Kim, Y.-C., 2016. Comprehensive study on volatile fatty acid production from *Ettlia* sp. residue with molecular analysis of the microbial community. Algal Res 17, 161–167.

Seon, J., Lee, T., Lee, S.C., Pham, H.D., Woo, H.C., Song, M., 2014. Bacterial community structure in maximum volatile fatty acids production from alginate in acidogenesis. Bioresour. Technol. 157, 22–27.

She, Y., Hong, J., Zhang, Q., Chen, B.-Y., Wei, W., Xin, X., 2020. Revealing microbial mechanism associated with volatile fatty acids production in anaerobic acidogenesis of waste activated sludge enhanced by freezing/thawing pretreatment. Bioresour. Technol. 302, 122869.

Shen, D., Yin, J., Yu, X., Wang, M., Long, Y., Shentu, J., Chen, T., 2017. Acidogenic fermentation characteristics of different types of protein-rich substrates in food waste to produce volatile fatty acids. Bioresour. Technol. 227, 125–132.

Soomro, A.F., Abbasi, I.A., Ni, Z., Ying, L., Liu, J., 2020. Influence of temperature on enhancement of volatile fatty acids fermentation from organic fraction of municipal solid waste: synergism between food and paper components. Bioresour. Technol. 304, 122980.

Strazzera, G., Battista, F., Garcia, N.H., Frison, N., Bolzonella, D., 2018. Volatile fatty acids production from food wastes for biorefinery platforms: a review. J. Environ. Manage. 226, 278–288.

Sun, J., Li, Z., Zhou, X., Wang, X., Liu, T., Cheng, S., 2019. Investigation on methane yield of wheat husk anaerobic digestion and its enhancement effect by liquid digestate pretreatment. Anaerobe 59, 92–99.

Tang, J., Wang, X., Hu, Y., Zhang, Y., Li, Y., 2016. Lactic acid fermentation from food waste with indigenous microbiota: effects of pH, temperature and high OLR. Waste Manage. 52, 278–285.

Tang, J., Wang, X.C., Hu, Y., Zhang, Y., Li, Y., 2017. Effect of pH on lactic acid production from acidogenic fermentation of food waste with different types of inocula. Bioresour. Technol. 224, 544–552.

Tsui, T.H., Ekama, G.A., Chen, G.H., 2018. Quantitative characterization and analysis of granule transformations: role of intermittent gas sparging in a super high-rate anaerobic system. Water Res. 139, 177–186.

Tsui, T.H., Wu, H., Song, B., Liu, S.S., Bhardwaj, A., Wong, J.W., 2020. Food waste leachate treatment using an upflow anaerobic sludge bed (UASB): effect of conductive material dosage under low and high organic loads. Bioresour. Technol. 304, 122738.

Wainaina, S., Awasthi, M.K., Horváth, I.S., Taherzadeh, M.J., 2020. Anaerobic digestion of food waste to volatile fatty acids and hydrogen at high organic loading rates in immersed membrane bioreactors. Renew. Energy 152, 1140–1148.

Wang, Y., Liu, X., Liu, Y., Wang, D., Xu, Q., Li, X., Yang, Q., Wang, Q., Ni, B.-J., Chen, H., 2020. Enhancement of short-chain fatty acids production from microalgae by potassium ferrate addition: feasibility, mechanisms and implications. Bioresour. Technol. 318, 124266.

Wang, Y., Wang, D., Liu, Y., Wang, Q., Chen, F., Yang, Q., Li, X., Zeng, G., Li, H., 2017. Triclocarban enhances short-chain fatty acids production from anaerobic fermentation of waste activated sludge. Water Res. 127, 150–161.

Wierckx, N., Koopman, F., Ruijssenaars, H.J., de Winde, J.H., 2011. Microbial degradation of furanic compounds: biochemistry, genetics, and impact. Appl. Microbiol. Biotechnol. 92 (6), 1095–1105.

Williams, K., Zheng, Y., McGarvey, J., Fan, Z., Zhang, R., 2013. Ethanol and volatile fatty acid production from lignocellulose by *Clostridium cellulolyticum*. ISRN Biotechnol. 2013, 137835.

Wu, C., Huang, Q., Yu, M., Ren, Y., Wang, Q., Sakai, K., 2018. Effects of digestate recirculation on a two-stage anaerobic digestion system, particularly focusing on metabolite correlation analysis. Bioresour. Technol. 251, 40–48.

Wu, Q.-L., Guo, W.-Q., Zheng, H.-S., Luo, H.-C., Feng, X.-C., Yin, R.-L., Ren, N.-Q., 2016. Enhancement of volatile fatty acid production by co-fermentation of food waste and excess sludge without pH control: the mechanism and microbial community analyses. Bioresour. Technol. 216, 653–660.

Xin, X., He, J., Li, L., Qiu, W., 2018a. Enzymes catalyzing pre-hydrolysis facilitated the anaerobic fermentation of waste activated sludge with acidogenic and microbiological perspectives. Bioresour. Technol. 250, 69–78.

Xin, X., He, J., Qiu, W., 2018b. Volatile fatty acid augmentation and microbial community responses in anaerobic co-fermentation process of waste-activated sludge mixed with corn stalk and livestock manure. Environ. Sci. Pollut. Res. 25 (5), 4846–4857.

Xin, X., She, Y., Hong, J., 2021. Insights into microbial interaction profiles contributing to volatile fatty acids production via acidogenic fermentation of waste activated sludge assisted by calcium oxide pretreatment. Bioresour. Technol. 320, 124287.

Xiong, Z., Hussain, A., Lee, J., Lee, H.-S., 2019. Food waste fermentation in a leach bed reactor: reactor performance, and microbial ecology and dynamics. Bioresour. Technol. 274, 153–161.

Xu, S., Selvam, A., Karthikeyan, O.P., Wong, J.W., 2014. Responses of microbial community and acidogenic intermediates to different water regimes in a hybrid solid anaerobic digestion system treating food waste. Bioresour. Technol. 168, 49–58.

Yan, M., Treu, L., Zhu, X., Tian, H., Basile, A., Fotidis, I.A., Campanaro, S., Angelidaki, I., 2020. Insights into ammonia adaptation and methanogenic precursor oxidation by genome-centric analysis. Environ. Sci. Technol. 54 (19), 12568–12582.

Yin, J., Yu, X., Wang, K., Shen, D., 2016a. Acidogenic fermentation of the main substrates of food waste to produce volatile fatty acids. Int. J. Hydrog. Energy 41 (46), 21713–21720.

Yin, J., Yu, X., Zhang, Y., Shen, D., Wang, M., Long, Y., Chen, T., 2016b. Enhancement of acidogenic fermentation for volatile fatty acid production from food waste: effect of redox potential and inoculum. Bioresour. Technol. 216, 996–1003.

Yuan, Y., Wang, S., Liu, Y., Li, B., Wang, B., Peng, Y., 2015. Long-term effect of pH on short-chain fatty acids accumulation and microbial community in sludge fermentation systems. Bioresour. Technol. 197, 56–63.

Yun, J., Cho, K.S., 2016. Effects of organic loading rate on hydrogen and volatile fatty acid production and microbial community during acidogenic hydrogenesis in a continuous stirred tank reactor using molasses wastewater. J. Appl. Microbiol. 121 (6), 1627–1636.

Zhang, B., He, P.-j., Ye, N.-f., Shao, L.-m., 2008. Enhanced isomer purity of lactic acid from the non-sterile fermentation of kitchen wastes. Bioresour. Technol. 99 (4), 855–862.

Zhang, D., Ong, Y.L., Li, Z., Wu, J.C., 2013. Biological detoxification of furfural and 5-hydroxyl methyl furfural in hydrolysate of oil palm empty fruit bunch by *Enterobacter* sp. FDS8. Biochem. Eng. J. 72, 77–82.

Zhang, L., Li, F., Kuroki, A., Loh, K.-C., Wang, C.-H., Dai, Y., Tong, Y.W., 2020a. Methane yield enhancement of mesophilic and thermophilic anaerobic co-digestion of algal biomass and food waste using algal biochar: semi-continuous operation and microbial community analysis. Bioresour. Technol. 302, 122892.

Zhang, L., Loh, K.-C., Dai, Y., Tong, Y.W., 2020b. Acidogenic fermentation of food waste for production of volatile fatty acids: bacterial community analysis and semi-continuous operation. Waste Manage. 109, 75–84.

Zhang, L., Loh, K.-C., Lim, J.W., Zhang, J., 2019a. Bioinformatics analysis of metagenomics data of biogas-producing microbial communities in anaerobic digesters: a review. Renew. Sustain. Energy Rev. 100, 110–126.

Zhang, L., Loh, K.-C., Zhang, J., 2019b. Enhanced biogas production from anaerobic digestion of solid organic wastes: current status and prospects. Bioresour. Technol. Rep. 5, 280–296.

Zhang, L., Zhang, J., Loh, K.-C., 2018. Activated carbon enhanced anaerobic digestion of food waste–laboratory-scale and pilot-scale operation. Waste Manage. 75, 270–279.

Zhang, Z., Tsapekos, P., Alvarado-Morales, M., Angelidaki, I., 2021. Impact of storage duration and micro-aerobic conditions on lactic acid production from food waste. Bioresour. Technol. 323, 124618.

Zhao, Y., Ren, N., Wang, A., 2008. Contributions of fermentative acidogenic bacteria and sulfate-reducing bacteria to lactate degradation and sulfate reduction. Chemosphere 72 (2), 233–242.

Zhou, A., Zhang, J., Wen, K., Liu, Z., Wang, G., Liu, W., Wang, A., Yue, X., 2016. What could the entire corn stover contribute to the enhancement of waste activated sludge acidification? Performance assessment and microbial community analysis. Biotechnol. Biofuels 9 (1), 1–14.

Zhou, M., Yan, B., Wong, J.W., Zhang, Y., 2018. Enhanced volatile fatty acids production from anaerobic fermentation of food waste: a mini-review focusing on acidogenic metabolic pathways. Bioresour. Technol. 248, 68–78.

Ziara, R.M., Miller, D.N., Subbiah, J., Dvorak, B.I., 2019. Lactate wastewater dark fermentation: the effect of temperature and initial pH on biohydrogen production and microbial community. Int. J. Hydrog. Energy 44 (2), 661–673.

Microbial fermentation for biodegradation and biotransformation of waste plastics into high value–added chemicals

Haojie Liu[a,1], Lijie Xu[a,1], Xinhui Bao[a], Jie Zhou[a,b], Xiujuan Qian[a], Weiliang Dong[a,b], Min Jiang[a,b]

[a]State Key Laboratory of Materials-Oriented Chemical Engineering, College of Biotechnology and Pharmaceutical Engineering, Nanjing Tech University, Nanjing, PR China, [b]Jiangsu National Synergetic Innovation Center for Advanced Materials, Nanjing Tech University, Nanjing, PR China
[1]These authors contributed equally to this work

17.1   Introduction

Since the 1950s, the average annual growth rate of global plastic production has remained at 8.5%, and the cumulative amount of waste plastics has exceeded 9 billion tons around the world. According to the British "Mirror" report, as of 2015, a total of 6.3 billion tons of waste plastic was generated globally. However, only 9% of these waste plastic was recycled, while 12% of it was incinerated, and the remaining 79% ended up by landfilling. In 2018, 360 million tons of waste plastics has been produced in a wide variety of applications around the world; however, only ~50 million tons of plastics were recycled (Geyer et al., 2017; Hundertmark et al., 2018). In addition, at least 8 million tons of waste plastics are discharged into the ocean every year, making up 80% of all marine debris from surface waters to deep-sea sediments. When plastics are exposed to sunlight or wave action, it will be degraded into microplastics (Jambeck et al., 2015; Lambert and Wagner, 2016). Over time, those microplastics will pollute marine ecosystem and full food chain, and finally gather in the human body (Jambeck et al., 2015; Lambert and Wagner, 2016).

The current waste plastics recycling mainly depend on physical and chemical methods (Evangelopoulos et al., 2019). For example, 45%–50% of waste plastics are recycled using physical and chemical methods in Europe (Geyer et al., 2017). However, physical recycling process is always complex and requires high economic and labor costs. Besides, limited by technical conditions, the plastics recycled from physical methods always suffered from poor color, heavy odor, and degraded quality; chemical recycling can recycle a high amount of thermal energy, but it makes high requirements on the purity and quality uniformity toward waste plastics. In addition, special chemical reagents are always needed during the chemical process. Currently, chemical recycling is mainly used for polyester plastics management (Ignatyev et al., 2014). Moreover, the process using either physical or chemical method converts plastics into relatively lower value plastic products, which is a downward recycle.

In recent years, biological technology using microorganisms or enzymes for biodegradation or biorefinery of low-value substrates have achieved remarkable progress, especially toward lignocellulose utilization. Plastics possess characteristics similar to lignocellulose in the terms of structure and molecular weight. Inspired by lignocellulose biorefinery, it seems feasible that biological recycling of waste plastics by using microorganisms or enzymes to first depolymerize plastics into small molecule oligomers or monomers, then convert these small molecule compounds into high-value products. As expected, biological recycling has attracted increasing attention due to its mild reaction condition, nonpollutants emission, and high product value. At present, the most recycled plastic using biological method is polyethylene terephthalate (PET), and a series of research work in the areas of mining and modification of PET depolymerase, optimization of PET depolymerization process, and biotransformation of PET depolymerization products into high-value chemicals has been carried out. In 2019, the French company Carbios built an industrial project for PET enzymatic recycling aiming to establish a complete industrial line contains PET waste biological depolymerization and monomer reuse (https://www.hbmedia.info/petplanet/). However, biological recycling of other waste plastics, such as polyethylene (PE) and polypropylene (PP), has not gotten remarkable achievements.

To deeply analyze the research progress toward all kinds of these synthetic plastics, this chapter comprehensively reviewed the development of mining and modification of various plastic depolymerizing microorganisms/enzymes, with emphasis on the analysis of the depolymerization mechanism and degradants distribution. In addition, the utilization of these degradants as substrate for high value–added chemicals synthesis through biotransformation is discussed in detail. Furthermore, the challenges and perspectives relying on the biological technologies are deeply analyzed and discussed, which might provide more valuable theoretical and technical support for the establishment of waste plastics recycling routes based on biological recycling (Fig. 17.1).

17.2   Classification of plastics

In market, the commonly used plastics are PET, polyurethane (PU), polyolefin (PE, PP), polybutene, polyvinyl chloride (PVC), polystyrene (PS), polyhydroxyalkanoates (PHA), etc. (Hee et al., 2019). Among them, PET is the most commonly produced polyester around the word and mainly used for packaging materials such as beverage bottles. The global PET

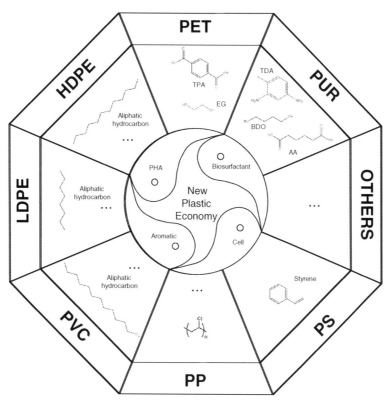

FIG. 17.1 **New plastic economy.**

production was as high as 33 million metric tons in 2015 (Geyer et al., 2017). Naturally, PET cannot be completely decomposed, the produced microplastics cause serious environmental problem, threatening the marine environment and human health (Nuelle et al., 2014; Lwanga et al., 2017).

PU is widely used in the medical, automotive, and industrial sectors. In 2015, the global production of PU reached 27 million metric tons. Many PU products are highly stable, resulting in the fact that only half of the produced PU has been generated as waste (Geyer et al., 2017). However, the stable characteristic of PU yet makes it difficult to be managed.

Polyolefin (PE, PP, PS, etc.) is the most produced and used synthetic plastic around the world. Take some examples, PE is mainly used in film, pipes, and injection products, and the output of PE accounts for about one-fourth of the total output of plastics. According to the difference toward polymerization method, molecular weight, and chain structure, PE can be divided into high-density polyethylene and low-density polyethylene. PP is mainly used in automobiles and electrical appliances, furniture, packaging materials and medical equipment, etc. PS is widely used in the optical industry such as optical glass and instruments.

According to different depolymerization mechanisms, the aforementioned plastics can be divided into hydrolyzed plastics and nonhydrolyzed plastics. Hydrolyzed plastics mainly refer to polymer plastics formed by polymerizing monomers via ester bonds, such as PET and PU. The degradation of hydrolyzed plastics mainly occurs by hydrolyzing the internal ester bonds, so that the biodegradation process is relatively simple. Currently, a variety of microorganisms have been screened and identified with the capability to degrade hydrolyzed plastics, and their degradation pathways are also relatively clear. Nonhydrolyzed plastics refer to polymer plastics made by polymerizing olefin monomers via C-C bond, such as PE and PS plastics, so the main chain chemical composition of nonhydrolyzed plastics is alkyl carbon. The C-C bond is highly inert and has a high reaction energy barrier, which makes it much difficult to be broken in nature (Wierckx et al., 2015).

17.3   Biodepolymerization and biotransformation of hydrolyzed plastics

17.3.1   Polyethylene terephthalate

17.3.1.1   *Structure and properties*

PET is a polymer compound formed by connecting terephthalic acid (TPA) with ethylene glycol (EG) through an ester bond (Webb et al., 2012). Traditionally, PET is mainly physically recycled for materials (Hopewell et al., 2009). However, lower than 21% of waste PET is physically recycled in the United States due to the degraded quality and high cost of recycled PET materials (e.g., \$1.3–1.5/kg waste PET, while virgin PET only needs \$1.1–1.3/kg) (Awaja and Pavel, 2005; Kuczenski and Geyer, 2010; Mar et al., 2014).

The use of biological methods for high-value utilization of waste PET plastics has achieved remarkable breakthroughs. Many microorganisms and enzymes have been isolated and identified as the candidate for PET biodepolymerization, once PET is biologically depolymerized into its monomers, such as TPA and EG, which can be further converted into more valuable biological compounds, such as polyhydroxyalkanoate (PHA) (Kenny et al., 2008), protocatechuic acid (PCA), gallic acid (GA), pyrogallol, catechol, muconic acid (MA), glycolic acid (GLA), and vanillic acid (VA) (Hee et al., 2019).

17.3.1.2   *Depolymerization of PET*

Crystallinity is one of the important factors affecting PET biodepolymerization. Higher crystallinity causes more difficulties in PET depolymerization. Therefore, amorphous or low-crystallinity films of PET plastics have always been used as model substrates to screen PET-degrading microorganisms and study their depolymerization characteristics. The currently reported PET degradation microorganisms are mainly *Fusarium solani* (Araújo et al., 2007), *Humicola Insolens* (Ronkvist et al., 2009), *Thermobifida fusca* (Then et al., 2016), *Saccharomonospora viridis* (Hu et al., 2008), and some actinomycetes, but most of these microorganisms can only degrade and modify the PET surface, their ability to degrade the actual PET waste plastic is very low. In 2016, Yoshid et al. isolated a PET-degrading strain of *Ideonella sakaiensis* 201-F6. After reacting at 30 °C for 6 weeks, a low crystallinity PET film was completely degraded (Yoshida et al., 2016), which is the best reported PET degradation bacteria till now.

Because PET plastic has a high glass transition temperature, the direct use of enzymes to depolymerize PET under high temperature (60–70 °C) conditions has become a research focus in recent years. Many types of enzymes have been identified with PET depolymerizing activity such as lipase, esterase, and cutinase. Among them, cutinase performance has the best PET depolymerization efficiency, especially, the cutinase TfH derived from *T. fusca* DSM43793 can lose 50% of the PET film with a crystallinity of 10% in 3 weeks under the catalyzed condition of 55 °C (Kleeberg et al., 2005). The cutinase HiC from *H. insolens* can almost completely degrade PET film with low crystallinity within 96 h at 70 °C. Besides, HiC, a fungal-derived polyester hydrolase, possesses the highest enzyme activity and the best thermal stability reported so far (Feder, 2013). Leaf-branch compost cutinase (LCC) derived from plant compost can catalyze the degradation of 25% amorphous PET film under the condition of 70 °C within 24 h (Table 17.1). This enzyme has certain homology with TfH polyester hydrolase (Sulaiman et al., 2014). In addition, lipases derived from *Thermomyces insolens, Candida antarctica, Aspergillus sp., Cladosporium, Cladosporioides*, and the esterase from *Melanocarpus albomyces, Penicillium citrinum* also possesses a certain depolymerizing effect on PET, but these enzymes can only increase the hydrophilicity of PET surface and cause PET surface morphology changing.

17.3.1.3 *Biotransformation of PET degradants*

Ethylene glycol

Ethylene glycol, referred to as EG, is one of the depolymerization products of PET (Fig. 17.1). In recent years, soil microorganisms such as *Pseudomonas putida* (*P. putida*) have been used to convert EG to high value–added GLA, glyoxylic acid, rhamnolipids, etc. In *P. putida* KT2440, EG is first converted to glycolaldehyde under the catalysis of periplasmic pyrroloquinoline quinone–dependent enzymes (PedE, PedH), followed by cytosolic aldehyde desorption. Hydrogenase PP_0545 and Pedi generate GLA, which is then converted into glyoxylic acid by the membrane-anchored glycolate oxidase GlcDEF. The produced glyoxylic acid can enter the Tricarboxylic acid (TCA) cycle via two pathways: (1) Condensed with succinic acid to form isocitrate catalyzed by isocitrate lyase (AceA) and (2) condensed with acetyl-CoA to form malic acid catalyzed by malate synthase (GlcB) (Muckschel et al., 2012). However, in pathway 1, the conversion of isocitrate to succinic acid is accompanied by the loss of two molecules of CO_2. Therefore, the two molecular weight carbon atoms provided by EG cannot enter the central metabolism for cell growth; in pathway 2, the lack of acetyl-CoA restricts the further conversion of glyoxylic acid conversion to malic acid. Therefore, *P. putida* KT2440 cannot use EG as the sole carbon source for cell growth. Interestingly, there is another EG metabolism pathway in *P. putida* JM37, where glyoxylic acid can be converted to tartrate semialdehyde (catalyzed by glyoxylate carboxylase (Gcl)), then to tartaric acid (catalyzed by hydroxypyruvate isomerase (hyi)), then glyceric acid is generated from tartaric acid catalyzed by ester semialdehyde reductase (glxR) and later is converted into 2-phosphoglycerate to enter into the glycolytic pathway, and finally enter into the TCA cycle (Muckschel et al., 2012). Therefore, *P. putida* JM37 can grow well on the medium with EG as the sole carbon source. Through a comprehensive investigation of the expression and transcription levels of each gene element in the EG metabolic pathway, it is found that Gcl and glxR are the key steps in EG metabolism. Once Gcl and glxR were simultaneously overexpressed, a fast cell growth was achieved by engineered *P. putida* grown on the medium with EG as the sole carbon source (Franden et al., 2018). In addition to the aforementioned pathways, Dragan Trifunović et al. found that propanediol dehydratase (PduCDE) and CoA-dependent propionaldehyde dehydrogenase (PduP) encoded by the pdu gene cluster in *Acetobacterium woodii* could also catalyze the degradation of EG to form acetaldehyde, which could further converted into acetyl-CoA and ethanol (Trifunovic et al., 2016) (Fig. 17.2).

TABLE 17.1 Strains for the valorization of depolymerization products of plastics.

Plastics	Depolymerization products	Strains	Metabolites	Yields	References
PET	TPA	*Pseudomonas putida* GO16, *Pseudomonas putida* GO19, *Pseudomonas frederiksbergensis* GO23	PHA	8.4 mg $^{-1}$ · h^{-1}	Kenny et al., 2008
	TPA	*Pseudomonas putida* GO16	PHA	108.8 mg $^{-1}$ $^{-1}$ · h^{-1}	Kenny et al., 2012
	TPA	*E. coli* PCA-1	PCA	0.81 g PCA per g TPA	
	TPA	*E. coli* GA-2b	GA	0.92 g GA per g TPA	
	TPA	*E. coli* MA-1	MA	0.85 g MA per g TPA	
	TPA	*E. coli* OMT-2	VA	0.45 g VA per g TPA	Kim et al., 2019
PET	EG	*Pseudomonas putida* KT2440	PHA	0.06 g PHA per g EG	Franden et al., 2018
	EG	*G. oxydans* KCCM 40109	GLA	0.98 g GLA per g EG	Kim et al., 2019
PU	AA	*Pseudomonas putida* KT2440	PHA	25% of the cell dry weight	Ackermann et al., 2021
PE	Paraffins from C8 to C32	*Pseudomonas aeruginosa* PAO1	PHA	25% of the cell dry weight	Guzik et al., 2014
PE	Oxidized hydrocarbons	*Ralstonia eutropha* H16	PHA	1.24 g $^{-1}$	Radecka et al., 2016
PS	Styrene	*Pseudomonas putida* CA-3	PHA	3.36 g $^{-1}$	Nikodinovic-Runic et al., 2011
PS	Styrene	*Pseudomonas putida* CA-3	PHA	62.5 mg PHA per g styrene	Ward et al., 2006
PS	Styrene	*Pseudomonas putida* CA-3	PHA	0.28 g PHA per g styrene	Goff et al., 2007
PP	Branched chain fatty alcohols and alkenes	*Yarrowia lipolytica* 78-003	Fatty acids	492 mg $^{-1}$ over 312 h	Mihreteab et al., 2019
PP	Oxidized PP fragments	*Ralstonia eutropha* H16	PHA	1.36 g $^{-1}$	Johnston et al., 2019

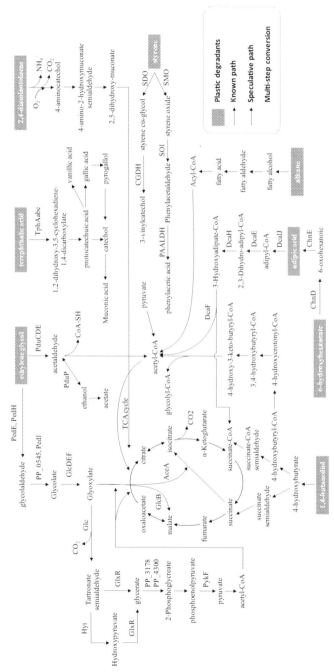

FIG. 17.2   The metabolism pathways of plastic degradants.

Bioconversion of EG to GLA: GLA is used as an exfoliant in cosmetics, which can be obtained through two-step conversion from EG. For example, *Gluconobacter oxydans* KCCM 40109 has been used to convert EG obtained from the PET hydrolysate into GLA; 65 reagent-grade samples with EG concentrations of 11.3, 28.6, and 67.6 mM were evaluated, after 12 h of biotransformation, the molar yields of 95.3, 99.7, and 89.4% were obtained to GLA (Hee et al., 2019).

Bioconversion of EG to polyhydroxyalkanoate (PHA): PHA is intracellular polyester, existing as a carbon source and energy storage substance, synthesized by many bacteria. PHA has the similar physical and chemical properties to synthetic plastics; besides, PHA possesses many excellent and unique properties, such as biodegradability, biocompatibility, optical activity, piezoelectricity, and gas barrier properties, which endow PHA a wide application in many areas. As reported, *P. putida* KT2440 could accumulate mclPHA under nitrogen limitation, but it could not efficiently utilize EG as a sole carbon source as mentioned before (Table 17.1). Through adaptive laboratory evolution, one mutant of *P. putida* KT2440 that could utilize EG as its sole carbon source was isolated. Comparative genomic analyses between the mutant and the wild-type strain revealed that a transcriptional regulator, GclR, plays a central role in repressing the glyoxylate carboligase pathway (Li et al., 2019). Based on this discovery, Franden demonstrated that through overexpression of the glyoxylate carboligase (Gcl) operon with the glycolate oxidase (GlcDEF) operon, the engineered *P. putida* KT2440 could utilize EG as a sole carbon source for cell growth and PHA accumulation (Franden et al., 2018).

Terephthalic acid

Conversion of TPA into high value–added aromatic compounds, such as PCA, GA, pyrogallol, catechol, MA, and VA, can undoubtedly improve the economics of PET recycling process (Hee et al., 2019). *Comamonas* sp. (Sasoh et al., 2006), *Delftia tsuruhatensis* (Toru et al., 2003), *Rhodococcus* (*Rhodococcus* sp.) (Choi et al., 2005), etc. can use TPA as the sole carbon source. The degradation pathway of TPA in microbial cells has been uncovered; TPA is catalyzed by 1,2-dioxygenase (TphAabc) and 1,2-Dihydroxy-3,5-cyclohexadiene-1,4-dicarboxylate (DCD) dehydrogenase to generate important intermediate PCA (Kim et al., 2019) (Fig. 17.2). PCA is a simple phenolic acid that can be further transformed into high value–added aromatic compounds such as GA, pyrogallol, MA, and VA, or entered into TCA cycle to synthesize bulk biological products such as rhamnolipids or PHA.

Bioconversion of TPA to PHA. In 2008, Kenny et al. first isolated three microorganisms, *P. putida* GO16, *P. putida* GO19, and *Pseudomonas frederiksbergensis* GO23, which could utilize TPA for not only growth but also accumulation of medium-chain length PHA (mclPHA). Subsequently, they used the TPA fraction from PET pyrolysis as the feedstock directly for mclPHA production, the maximal production rate of PHA reached approximately 8.4 mg/L/h by strains GO16 and GO19 (Kenny et al., 2008). Slow metabolism rate is one of the bottlenecks that restricts TPA fermentation. To solve this problem, Kenny et al., (2012) used TPA and glycerol as a cosubstrate for *P. putida* GO16 culture, the PHA productivity reached 108.8 mg/L/h.

Bioconversion of TPA to PCA: Two enzymes, TPA 1,2-dioxygenase and DCD dehydrogenase, are involved in converting TPA to PCA; TPA is first converted to DCD by TPA 1,2-dioxygenase, then converted to PCA by DCD dehydrogenase (Sasoh et al., 2006). Those enzymes

exist in one gene cluster and have been revealed in several bacteria, such as *Comamonas sp.*
E6 (Sasoh et al., 2006), *Delftia tsuruhatensis* T7 (Shigematsu et al., 2003), and *Rhodococcus sp.*
DK17 (Choi et al., 2005). TphAabc as TPA 1,2-dioxygenase and TphB as DCD dehydrogenase,
both from *Comamonas sp.* E6, were used for the biosynthetic route from TPA to PCA in *Escherichia coli* (Sasoh et al., 2006). A 2.8 mM PCA was produced at a molar yield of 81.4% after 3 h
(Kim et al., 2019).

Bioconversion of TPA to GA. GA is currently used in the pharmaceutical industry to produce
antibacterial agent, trimethoprim, antioxidant, and propyl gallate. Through overexpression
of PobA from *P. putida* KT2440, the engineered *E. coli* HBH-1 could produce 1.4 mM of GA
from PCA. Further, when double-mutant PobAMut (T294A/Y385F) was overexpressed, the
engineered strain HBH-2 could produce as high as 2.5 mM of GA from PCA, with a molar
yield of 74.3% (Kim et al., 2019).

Bioconversion of TPA to MA: Catechol synthesized from TPA via PCA can be converted to
MA, which is mainly used in antiultraviolet protective agent and military special products.
To construct an MA biosynthetic route from TPA, CatA, a catechol 1,2-dioxygenase from
P. putida KT2440, was selected for ring-cleavage, 4.5 mM of catechol was completely converted to MA after 10-min fermentation by *E. coli* strain CDO-1 overexpressing CatA. When
MA synthetic strain was cocultured with catechol synthetic strain CTL-1 (overexpressing
TphAabc, TphB, and AroY), which made it possible for MA synthesis directly from TPA, as
2.7 mM of MA could be obtained from 3.2 mM of TPA after 6 h fermentation, corresponding
with 85.4% of conversion yield (Hee et al., 2019).

Bioconversion of TPA to VA: VA is used as the direct precursor of vanillin in the pharmaceutical industry. PCA can be converted to VA by O-methyltransferase (OMT) *in vitro* or *in vivo* (Bai
et al., 2007). To supply a methyl group to this O-methylation reaction, S-adenosyl methionine
is usually used as a cosubstrate, and the adenosyl and methyl groups are supplied from ATP
and methionine, respectively. Most currently known OMTs are from eukaryotes. To produce
VA directly from TPA, the biosynthetic route from TPA to PCA was connected to the route
from PCA to VA, where HsOMT from *Homo sapiens* was heterologously expressed in engineered strain *E. coli* VA-1 (overexpressed TphAabc, TphB, and HsOMT). A 3.2 mM PCA was
converted to 1.0 mM VA at a molar yield of 29.4%. To relieve the genes expression pressure in
single strain, a coculture system with "from TPA to PCA" and "from PCA to VA" modules
was applied, where TphAabc and TphB were overexpressed in strain PCA-1 and HsOMTHis
was overexpressed in strain OMT-2His, after optimization of the inoculation ratio, highest VA
production of 1.4 mM VA at a molar yield of 41.6% was achieved (Hee et al., 2019).

Bioconversion of TPA to pyrogallol: Pyrogallol is another higher value chemical that can be
currently used as an antioxidant in the oil industry. On the basis of catechol synthesis from
TPA, the pyrogallol synthesis pathway was constructed by integrating the catechol hydroxylation module for catechol conversion to pyrogallol by PhKLMNOPQ, 2.6 mM pyrogallol can
be produced (Hee et al., 2019).

17.3.2 Polyurethane

17.3.2.1 *Structure and properties*

PU is a polymer with urethane bond repeating unit structure formed by the condensation of
three components of isocyanate, polyol, and chain extender, which is widely used in the

medical, automotive, and industrial sectors. PU is a semicrystalline thermosetting plastic, in which isocyanate constitutes its crystalline part, called the hard segment of PU, which determines the hardness and tensile strength of the PU plastic. Polyol and chain extender constitute its noncrystalline part, called the soft segment of PU, which determines its elasticity and elongation properties. PUs are the heterogeneous class of polymers regarding monomer composition, molar ratios, and additives used. This heterogeneity allows PU manufacturers to tailor the polymer to the individual applications and the requirements of the customers (Gama et al., 2018). Common monomers used in PU production are adipic acid (AA), 1,4-butanediol (BDO), EG, and the isocyanates 2,4-toluenediamine (2,4-TDA) (Fig. 17.1). Upon (enzymatic) depolymerization of PU, the 2,4-TDA monomers are released (Magnin et al., 2020). The resulting 2,4-TDA is toxic to humans and was classified as a carcinogen in animal studies. 2,4-TDA is water-soluble (38 mg/L at 25 °C) and, therefore, can easily contaminate soil or other environments (Romualdus et al., 2020).

17.3.2.2 Depolymerization of PU

Currently, most of the microorganisms with PU degradation ability reported till now are focused on polyester PU, while few studies reported on polyether PU degradation. *Fusarium sp., Curvularia sp., Cladosporium sp., Penicillium sp., Bacillus sp., Pseudomonas sp.*, etc. have been confirmed to be able to degrade PU plastics. Alvarezbarragan et al., (2016) isolated six strains of *Cladosporium sp.*, which could degrade 75%–85% of PU (Impranil DLN) within 2 weeks. *Aspergillus sp.* has been reported to have a polyester PU degradation ability, as *Aspergillus flavus* isolated from the waste dump soil could use polyester PU film as the sole carbon source, and it could degrade 60.6% of polyester PU within 30 days (Mathur and Prasad, 2012). *Aspergillus* sp. strain S45 separated from the dumping point of solid waste can degrade 20% of polyester PU film within 28 days (Osman et al., 2018). Moreover, *Aspergillus tubingensis* isolated from garbage dump, could degrade polyester PU film into fragments after 21 days catalysis with 2% glucose as a cocarbon source (Khan et al., 2017). However, most of the studies applied PU model as a substrate, few studies have tested realistic polyester PU waste plastics (Magnin et al., 2020).

Biological enzymes can depolymerize PU plastics by hydrolyzing the ester bond or urea bond. Commonly used enzymes include esterase, protease, and urease. In specific, esterase is currently reported as the most effective enzyme to degrade PU plastics, including PudA derived from *Comamonas acidovorans* TB-35 (Nomura et al., 1998; Akutsu et al., 1998); PueA, PueB, and PulA derived from *Pseudomonas* (Stern and Howard, 2000; Howard et al., 2001). Although these esterases can effectively depolymerize waterborne PU DLN, they almost have no effect on realistic PU waste. Except for esterase, some proteases have also been confirmed with PU degradation ability, after papain treatment of PU film at 37 °C for 1–6 months, Gel permeation chromatography (GPC) and FTIR analysis showed that the carbamate bond was broken to a certain extent (Phua et al., 1987). When α-chymotrypsin degrades PU, its average molecular weight can also be reduced by more than 30% after 10 days of reaction at 25 °C (Campiñez et al., 2013). In addition, urease was found to be able to break the urea bonds in PU so that to degrade PU plastic. Moreover, Juliane et al., (2017) found that the PET depolymerase of cutinase LCC, TfCut2, Tcur1278, and Tcur0390 also show a certain of polyester kind PU degradation ability. Under the condition of 70 °C, after 100 h of reaction, the degradation rate of PU plastic reached 0.3%~3.2%.

17.3.2.3 Biotransformation of PU degradants

2,4-toluenediamine

2,4-TDA is a kind of depolymerization of PU and other plastics. In 2020, Espinosa et al., (2020) isolated a strain of *Pseudomonas* TDA1 from the soil of a waste plastic dump that can use PU 2,4-diaminotoluene as sole carbon/nitrogen source to grow. Through genomics analysis, the degradation pathway of 2,4-diaminotoluene was initially proposed. The methyl group of 2,4-diaminotoluene is first oxidized, decarboxylated, and deaminated to form 4-aminocatechol. 4-aminocatechol may be converted into 5-amino-2-hydroxymuconic acid then further degraded through a metabolic pathway similar to catechin (Fig. 17.2). In the future, a comprehensive analysis of proteomics and transcriptomics of the relevant genes in the predicted metabolic pathways is necessary, with the investigation of the intermediates distribution and metabolic flux changes, so that to uncover the 2,4-diaminotoluene degradation pathway.

1,4-Butanediol

1,4-Butanediol is one of the main degrading products of PU. It was reported that *P. putida* KT2440 can grow in a medium with 1,4-butanediol as a sole carbon source, but the growth rate is very slow. Through adaptive evolution, the utilization efficiency of 1,4-butanediol by KT2440 was significantly improved (Li et al., 2020). Through analysis the genomics and proteomics between *P. putida* KT2440 and the mutant, Li et al., (2020) analyzed the biodegradation pathway of 1,4-butanediol. In detail, 1,4-butanediol is first oxidized to 4-hydroxybutyrate catalyzed by the highly expressed dehydrogenase encoded by the PP_2674-2680 ped gene cluster. The generated 4-hydroxybutyric acid can be metabolized by the following three pathways: (1) Conversion of succinate by the dehydrogenase encoded by the ped gene cluster; (2) conversion of amber Acyl-CoA catalyzed by acyl-CoA synthetase AcsA1 (PP_4487); and (3) conversion of glycolyl-CoA and acetyl-CoA through β-oxidation (LI et al., 2020) (Fig. 17.2). At present, only the third route has been confirmed, while the succinate synthetic route and succinyl-CoA synthetic route need to be further confirmed.

Adipic acid

AA is one of the degrading products of PU, and its metabolic pathway has been analyzed in *Acinetobacter* (PARKE et al., 2001). AA is first converted to adipyl-CoA under the catalysis of succinyl-CoA transferase (DcaIJ). Then, adipyl-CoA is converted to 2,3-dihydroadipyl-CoA under the catalysis of enoyl-CoA dehydratase (DcaE), which further form 3-hydroxyadipate-CoA catalyzed by 3-hydroxyacyl-CoA dehydrogenase (DcaA). Finally, acyl-CoA thiolase (DcaF) catalyzes 3-hydroxyadipate-CoA to generate succinyl-CoA and acetyl-CoA, which then enter the TCA cycle to maintain cell growth and metabolism (Fig. 17.2).

However, the way for valorization of PU waste via the biotechnology to production of value-added products is rough. Different from PET, which contains only TPA and EG in its main chain, PU polymers often have a more complex backbone, so that the degradants from PU plastic are diverse and unstable. Degradants derived from PU always contain amines, alcohols, acids, aromatics, and other residues (Magnin et al., 2019). In general, single strain always does not possess all substrate utilization ability, through mixed culture of multiple strains with different substrates utilization ability may do favor for mixed PU degradants utilization. For example, a mixed culture of three engineered *P. putida* KT2440 derivatives can utilize AA, BDO, and EG, respectively (Liu et al., 2021). In addition, the *Pseudomonas sp.* TDA1 was used to remove TDA. TDA inhibited the growth of mixed cultures and the utilization of mixed substrate. Reactive extraction of TDA was implemented before the full utilization of remaining PU monomers as carbon sources.

17.4   Biodepolymerization and biotransformation of nonhydrolyzed plastics

17.4.1   Polyethylene

Nonhydrolyzable plastics refer to polymer plastics formed by polymerization of olefin monomers, PE is the most used polyolefin plastics. The chemical composition of PE main chain is alkyl carbon, where C-C bond is inert with high energy barrier, which makes it difficult to be broken. This is also an important reason why PE plastic is difficult to be degraded in nature.

As early as the 1970s, Albertsson et al. carried out an experiment on microbial degradation of ^{14}C-labeled PE (average weight molecular of 300,000 Da) by using three different soil microbiotas (Albertsson et al., 1987). Through determining the release of $^{14}CO_2$, the microbial degradation ratio of PE was calculated to be in the range of 0.36%–0.39% after 2 years (Albertsson et al., 1987). When the ^{14}C-labeled PE was extracted with cyclohexane to get rid of its low molecular weight components (average weight molecular of 1000 Da), the microbial degradation ratio dropped to 0.16% (Albertsson, 1980). Therefore, it was concluded that the release of $^{14}CO_2$ was mainly derived from the microbial degradation of low-molecular-weight PE fraction, which was similar to the microbial degradation of straight-chain *n*-alkanes. Kawai et al. also pointed out that the upper limit of molecular weight for PE degradation by microorganisms was about 2000 Da (Kawai et al., 2004). Based on this conclusion, application of some physicochemical pretreatments, such as UV irradiation (Albertsson and Karlsson, 1990), chemical oxidizing agents (Brown et al., 1974), and thermal oxidation (Lee et al., 1991) to break long-chain PE into small moleculars, so that to facilitate microbial degradation process should be an efficient method (Hakkarainen and Albertsson, 2004).

Actually, researchers have isolated some microorganisms that can degrade PE plastics. For example, Han Qiuxia et al. isolated an *Aspergillus sp.* M6 from farmland soil that can use PE film as the sole carbon source, and the quality loss of PE film reached more than 20% after 30 days degradation (Han et al., 2009). Balasubramanian et al. isolated 15 high-density PE plastic-degrading bacteria from the plastic waste dump in Mannar Bay, India. Among them, *Arthrobacter sp.* GMB5 and *Pseudomonas sp.* GMB7 can degrade 12% and 15% of PE film within 30 days, respectively (Balasubramanian et al., 2010). Tribedi et al. isolated a low-density PE-degrading strain *Pseudomonas sp.* AKS2, which could degrade 4%–6% of PE within 45 days (Tribedi and Sil, 2013). Moreover, *Rhodococcus* sp. C208, isolated from waste mulch film, could degrade PE plastic film at a rate of 0.86% per week (Santo et al., 2013). Further, the laccase derived from *Rhodococcus ruber* C208 was cloned and expressed, after being treated by this laccase, the amount of carbonyl groups in PE plastic increases significantly, the average molecular weight (Mw) and average molecular weight (Mn) of the polymer were reduced by 20% and 15%, respectively, indicating the importance of laccase in PE degradation. Besides, laccase derived from *Trametes versicolor* can also accelerate the degradation rate of PE film under the mediation of 1-hydroxybenzotriazole (Fujisawa et al.,

2001). Interestingly, all the oxidation and fracture sites occurred in the amorphous region of the PE film.

On the other hand, alkane hydroxylases of the alkyl hydroxylase family can degrade hydrocarbons through their oxidation ends or subends (Rojo, 2010), which might show some potential in PE degradation. When the alkane hydroxylase (AH) encoding gene alkB derived from *Pseudomonas sp.* E4 was exogenously expressed in *E. coli* BL21, 20% of low-molecular-weight PE can be converted into CO_2 after 80 days culture at 37 °C (Gyung et al., 2012). Further integrating the AH catalytic system of *Pseudomonas aeruginosa* E7 (including alkane monooxygenase, red pigment, and red pigment reductase), the degradation rate of PE was increased to 30% (Jeon and Kim, 2015).

In terms of PE degradants utilization, Guzik et al., (2014) first used pyrolytic hydrocarbons of PE as substrate for PHA biological production, the result showed *P. aeruginosa* PAO-1 could accumulate PHA by 25% of dry cell weight from PE pyrolytic hydrocarbons. *Ralstonia eutropha* H16 (previously known as *Cupriavidus necator* or *Wausternia eutropha*) also exhibited PHA accumulation capability when supplied with nonoxygenated PE pyrolytic hydrocarbons in a nitrogen-rich tryptone soya broth growth medium (Johnston et al., 2017). In contrast to PE pyrolysis in the absence of air, pyrolysis in the presence of air would not only cleave the long chains of PE, but also introduce the carbonyl and hydroxyl groups into the backbone of pyrolytic hydrocarbons, which could improve the bioavailability of pyrolytic hydrocarbons for microbial fermentation (Radecka et al., 2016).

17.4.2 Polystyrene

There are relatively few reports on the degradation of PS plastic by microorganisms. In 2015, Eisaku et al. first isolated five strains of PS-degrading microorganisms from soil, including *Xanthomonas sp.*, *Sphingobacterium sp.*, and *Bacillus sp.* (Yang et al., 2015). In addition, Ji Rong et al. applied ^{14}C-labeled PS film pretreated by ozone oxidation to cultivate *Penicillium variabile* CCF3219, the PS material was almost completely mineralized into CO_2 and water within 16 weeks. In recent years, the use of insect gut microbial flora to degrade polyolefin plastics has developed rapidly. For example, Yang Jun et al. found that the larvae of *Tenebrio molitor Linnaeus* have a certain ability to degrade PS film, a PS-degrading bacterium *Exiguobacterium sp.* YT2 was further isolated from its intestine, strain YT2 could degrade 7.4% PS within 60 days (Yang et al., 2015). In future research, the insect gut microbial flora will provide an important source for screening of highly efficient polyolefin plastics degrading microorganisms. At present, only the nonheme hydroquinone peroxidase from the lignin decolorizing bacteria *Azotobacter beijerinckii* HM121 has been reported as PS depolymerase, which can degrade insoluble PS in a two-phase system (dichloromethane-water). In the presence of hydrogen and tetramethylhydroquinone, PS can be degraded into water-soluble small molecule products within 5 min (Jeon and Kim, 2015).

PS degradation mechanism has been uncovered in some strains. PS plastic can be directly used as a carbon source by some microorganisms through two different catabolic pathways. The degradation pathways mainly include ring opening and side chain oxidation (O'Leary et al., 2002). In 2005, Ward et al. first found that *P. putida* CA-3 could convert the metabolite of styrene, phenylacetic acid (PAA), into polyhydroxyalkanoate (PHA) under the nitrogen limitation (Ward et al., 2006). Their finding built the metabolic link between styrene degradation and

PHA accumulation in *P. putida* CA-3, and found a trail for the microbial valorization of PS waste into valuable chemicals (Nikodinovic-Runic et al., 2011). Ward et al. used styrene oil, the pyrolysis products of PS waste, as sole source for *P. putida* CA-3 cultivation, as a result 10% of PS was transformed into PHA (Ward et al., 2006). To improve the conversion rate, Goff et al. performed a batch fermentation of *P. putida* CA-3 grown on styrene oil in a stirred tank reactor with an optimized nitrogen feeding strategy. The highest yield of PHA was 0.28 g PHA per gram of styrene supplied with a nitrogen feeding rate of 1.5 mg/L/h (Goff et al., 2007).

17.4.3   Polypropylene

In 1993, the microbial degradation of PP was evaluated for the first time in a PE waste–polluted sandy soil. After 175 days of cultivation, 40% of PP material has been degraded. Further determination of the degradation products showed that 90% of them are aromatic esters derived from plasticizers, and only 10% of the degradation products were identified as hydrocarbons from PP polymer. This result shows that in addition to PP itself, plasticizers are also easily degraded by microorganisms (Cacciari et al., 1993). However, there are no enzymes reported to be able to degrade PP plastic, and little knowledge is available for the mechanism of microbial degradation of PP at present.

PP plastic products can obtain a variety of small molecular aliphatic hydrocarbons after pyrolysis treatment. The metabolic pathways of *aliphatic hydrocarbons* exist widely in nature.

17.4.4   Polyvinyl chloride

Although PVC can be depolymerized into hydrocarbons under a nitrogen flow at 300 °C (Yuan et al., 2014), there are no reports about microorganism that can utilize PVC pyrolysis products as a carbon source so far. Thus, the key enzymes involved in the microbial degradation of PVC are still unknown.

17.5   Conclusions and perspectives

The previously considered stubborn waste plastics can now be produced through microbial biotechnology to produce high-value products related to the circular economy. We believe that waste plastic can and should be established as a new second-generation carbon source for biotechnology. Obviously, it is similar to the development of super cellulose biotechnology: waste plastic, such as biomass, is just a carbon-rich polymer. In fact, considering that compared with biomass, the composition of plastics is relatively simple and certain, and their content is extremely high. Surprisingly, so far, this resource has hardly been regarded as a biotechnology raw material. Bio-based products for the production of bio-based fossil-based plastic monomers has been developed in parallel, such as ethylene, TPA, and EG, which will further promote the sustainability of plastics. Once successful, the renewal of plastics through biotechnology will increase the efficiency of resource utilization by adding value to a large amount of waste streams while also helping to reduce the burden of terrestrial resources required to provide food for the world's population. Therefore, through microbial biotechnology, the "value from waste plastic to waste plastic" workflow proposed in this chapter will be within the framework of a sustainable knowledge-based bioeconomy in various sectors including materials, chemistry, and environmental technology. It would help realize a new value chain and bring tangible benefits to the environment and society.

Waste plastic can and should be established as a new carbon source for microorganism cultivation. Compared with biomass, the composition of plastics is relatively simple, and their carbon content is extremely high. However, so far, plastic resource is rarely to be regarded as a biotechnology raw material. The lack of degradation enzyme components, low degradation efficiency, and difficulty for degradants utilization limit the development process of waste plastics biogradation. Further studies on the biological oxidation mechanism of nonhydrolyzable plastics, efficient heterologous expression of plastic depolymerase, construction of a multienzyme/mixed bacteria system for the depolymerization of mixed plastics, and engineering of high-value bio-refining pathways for plastic degradation products biotransformation, will contribute to boosting the waste plastic degradation, so as to mitigate the global plastic crisis.

References

Ackermann, Y.S., Li, W.-J., Blank, L.M., 2021. Engineering adipic acid metabolism in Pseudomonas putida. Metab. Eng. 67, 29–40.

Akutsu, Y., Nakajima-Kambe, T., Nomura, N., 1998. Purification and properties of a polyester polyurethane-degrading enzyme from *Comamonas acidovorans* TB-35. Appl. Environ. Microbiol. 64 (1), 62–67.

Albertsson, A.C., Andersson, S.O., Karlsson, S., 1987. The mechanism of biodegradation of polyethylene. Polym. Degrad. Stabil. 18, 73–87.

Albertsson, A.C., Karlsson, S., 1990. The influence of biotic and abiotic environments on the degradation of polyethylene. Prog. Polym. Sci. 15, 177–192.

Albertsson, A.C., 1980. Microbial and oxidative effects in degradation of polyethene. Appl. Polym. Sci. 25, 1655–1671.

Álvarez-Barragán, J., Domínguez-Malfavón, L., Vargas-Suárez, M., 2016. Biodegradative activities of selected environmental fungi on a polyester polyurethane varnish and polyether polyurethane foams. Appl. Environ. Microbiol. 82 (17), 5225–5235.

Araújo, R., Silva, C., O'neill, A., 2007. Tailoring cutinase activity towards polyethylene terephthalate and polyamide 6,6 fibers. J. Biotechnol. 128 (4), 849–857.

Awaja, F., Pavel, D., 2005. Recycling of PET. Eur. Polym. 41 (7), 1453–1477.

Bai, H.W., Shim, J.Y., Yu, J., Zhu, B.T., 2007. Biochemical and molecular modeling studies of the O-methylation of various endogenous and exogenous catechol substrates catalyzed by recombinant human soluble and membrane-bound catechol-O- methyltransferases. Chem. Res. Toxicol. 20 (10), 1409–1425.

Balasubramanian, V., Natarajan, K., Hemambika, B., 2010. High-density polyethylene (HDPE)-degrading potential bacteria from marine ecosystem of Gulf of Mannar. Lett. Appl. Microbiol. 51 (2), 205–211.

Brown, B.S., Mills, J., Hulse, J.M., 1974. Chemical and biological degradation of waste plastics. Nature 250, 161–163.

Cacciari, I., Quatrini, P., Zirletta, G., Mincione, E., Vinciguerra, V., Lupattelli, P., 1993. Isotactic polypropylene biodegradation byamicrobial community: physicochemical characterization of metabolites produced. Appl. Environ. Microbiol. 59, 3695–3700.

Campiñez, M.D., Aguilar-de-Leyva, Á., Ferris, C., 2013. Study of the properties of the new biodegradable polyurethane PU (TEG-HMDI) as matrix forming excipient for controlled drug delivery. Drug Dev. Ind. Pharm. 39 (11), 1758–1764.

Choi, K.Y., Kim, D., Sul, W.J., 2005. Molecular and biochemical analysis of phthalate and terephthalate degradation by Rhodococcus sp. strain DK17. FEMS Microbiol. Lett. 252 (2), 207–213.

Choi, K.Y., Kim, D., Sul, W.J., Chae, J.C., Zylstra, G.J., Kim, Y.M., Kim E., Molecular and biochemical analysis of phthalate and terephthalate degradation by *Rhodococcus* sp. Strain DK17. FEMS Microbiol. Lett. 252 (2), 207–213.

Espinosa, M.J.C., Blanco, A.C., Schmidgall, T., 2020. Toward biorecycling: isolation of a soil bacterium that grows on a polyurethane oligomer and monomer. Front. Microbiol 11, 404.

Evangelopoulos, P., Arato, S., Persson, H., 2019. Reduction of brominated flame retardants (BFRs) in plastics from waste electrical and electronic equipment (WEEE) by solvent extraction and the influence on their thermal decomposition. Waste Manage. (Oxford) 94, 165–171.

Feder, D., 2013. Humicola insolens cutinase; a novel catalyst for polymer synthesis reactions. PhD dissertations. Gradworks, Polytechnic Institute of New York University, New York.

Franden, M.A., Jayakody, L.N., Li, W.J., Wagner, N.J., Cleveland, N.S., Michener, W.E., Hauer, B., Blank, L.M., Wierckx, N., Klebensberger, J., Beckham, G.T., 2018. Engineering Pseudomonas putida KT2440 for efficient ethylene glycol utilization. Metab. Eng. 48, 197–207.

Fujisawa, M., Hirai, H., Nishida, T., 2001. Degradation of polyethylene and nylon-66 by the laccase-mediator system. Polym. Environ 9 (3), 103–108.

Gama, N., Ferreira, A., Barros-Timmons, A., 2018. Polyurethane foams: past, present, and future. Materials 11, 1841–1835.

Geyer, R., Jambeck, J.R., Law, K.L., 2017. Production, use, and fate of all plastics ever made. Sci. Adv. 3 (7), e1700782.

Goff, M., Ward, P.G., O'Connor, K.E., 2007. Improvement ofthe conversion of polystyrene to polyhydroxyalkanoate through the manipulation of the microbial aspect of the process: a nitrogen feeding strategy for bacterial cells in a stirred tank reactor. Biotechnology 132, 283–286.

Guzik, M.W., Kenny, S.T., Duane, G.F., Casey, E., Woods, T., Babu, R.P., 2014. Conversion of post consumer polyethylene to the biodegradable polymer polyhydroxyalkanoate. Appl. Microbiol. Biotechnol. 98, 4223–4232.

Gyung Yoon, M., Jeong Jeon, H., Nam Kim, M., 2012. Biodegradation of polyethylene by a soil bacterium and AlkB cloned recombinant cell. Bioremediat.Biodegrad. 3, 145.

Hakkarainen, M., Albertsson, A.C., 2004. Environmental degradation of polyethylene. In: Albertsson, A.C. (Ed.), Long Term Properties of Polyolefins. Springer, Berlin, pp. 177–200.

Han, Q., Wang, Q., Zhang, M., 2009. Study on the biodegradability of modified PE film. Plastics Ind 37 (10), 48–51.

Hee, T.K., Kim, J.K., Cha, H.G., Kang, M.J., Lee, H.S., Khang, T.U., Yun, E.J., Lee, D.-H., Song, B.K., Park, S.J., Joo, J.C., Kim, K.H., 2019. Biological valorization of poly(ethylene terephthalate) monomers for upcycling waste PET. ACS Sustain. Chem. Eng 7, 19396–19406.

Hopewell, J., Dvorak, R., Kosior, E., 2009. Plastics recycling: challenges and opportunities. Philos 364 (1526), 2115–2126.

Howard, G.T., Crother, B., Vicknair, J., 2001. Cloning, nucleotide sequencing and characterization of a polyurethanase gene (pueB) from *Pseudomonas chlororaphis*. Int. Biodeterior. Biodegrad. 47 (3), 141–149.

Hundertmark, T., Mayer, M., McNally, C., 2018. How plastics waste recycling could transform the chemical industry. McKinsey & Company, https://www.mckinsey.com/industries/chemicals/our-insights/how-plastics-waste-recycling-could-transform-the-chemical-industry#.

Hu, X., Osaki, S., Hayashi, M., 2008. Degradation of a terephthalate-containing polyester by thermophilic actinomycetes and bacillus species derived from composts. J. Polym. Environ. 16 (4), 103–108.

Ignatyev, I.A., Thielemans, W., Vander Beke, B., 2014. Recycling of polymers: a review. ChemSusChem 7 (6), 1579–1593.

Jambeck, J.R., Geyer, R., Wilcox, C., 2015. Plastic waste inputs from land into the ocean. Science 347 (6223), 768–771.

Jeon, H.J., Kim, M.N., 2015. Functional analysis of alkane hydroxylase system derived from *Pseudomonas aeruginosa* E7 for low molecular weight polyethylene biodegradation. Int. Biodeterior. Biodegrad. 103, 141–146.

Johnston, B., Jiang, G., Hill, D., Adamus, G., Kwiecieh, I., Zieba, M., 2017. The molecular level characterization of biodegradable polymers originated from polyethylene using non-oxygenated polyethylene wax as a carbon source for polyhydroxyalkanoate production. Bioengineering. 4, 73. doi:10.3390/bioengineering4030073.

Johnston, B., Radecka, I., Chiellini, E., Barsi, D., Ilieva, V.I., Sikorska, W., 2019. Mass spectrometry reveals molecular structure of polyhydroxyalkanoates attained by bioconversion of oxidized polypropylene waste fragments. Polymers 11, 1580.

Juliane, S., Ren, W., Thorsten, O., 2017. Degradation of polyester polyurethane by bacterial polyester hydrolases. Polymers 9 (12), 65.

Kawai, F., Watanabe, M., Shibata, M., Yokoyama, S., Sudate, Y., Hayashi, S., 2004. Comparative study on biodegradability of polyethylene wax by bacteria and fungi. Polym. Degrad. Stab. 86, 105–114.

Kenny, S.T., Runic, J.N., Kaminsky, W., Woods, T., Babu, R.P., Keely, C.M., Blau, W., O'Connor, K.E., 2008. Up-cycling of PET (polyethylene terephthalate) to the biodegradable plastic PHA (polyhydroxyalkanoate). Environ. Sci. Technol. 42 (20), 7696–7701.

Kenny, S.T., Runic, J.N., Kaminsky, W., 2012. Development of a bioprocess to convert PET derived terephthalic acid and biodiesel derived glycerol to medium chain length polyhydroxyalkanoate. Appl. Microbiol. Biotechnol. 95 (3), 623–633.

Khan, S., Nadir, S., Shah, Z.U., 2017. Biodegradation of polyester polyurethane by *Aspergillus tubingensis*. Environ. Pollut. 225 (1), 469–480.

Kim, H.T., Kim, J.K., Cha, H.G., 2019. Biological valorization of poly(ethylene terephthalate) monomers for upcycling waste PET. ACS Sustain. Chem. Eng. 7 (24), 19396–19406.

Kleeberg, I., Welzel, K., Vandenheuvel, J., 2005. Characterization of a new extracellular hydrolase from *Thermobifida fusca* degrading aliphatic-aromatic copolyesters. Biomacromolecules 6, 262–270.

Kuczenski, B., Geyer, R., 2010. Material flow analysis of polyethylene terephthalate in the US, 1996–2007. Resour. Conserv. Recycl. 54 (12), 1161–1169.

Lambert, S., Wagner, M., 2016. Characterisation of nanoplastics during the degradation of polystyrene. Chemosphere 145, 265–268.

Lee, B., Pometto, A.L., Fratzke, A., Bailey, T.B., 1991. Biodegradation of degradable plastic polyethylene by Phanerochaete and Streptomyces species. Appl. Environ. Microbiol. 57, 678–685.

Liu, J., He, J., Xue, R., Xu, B., Qian, X., Xin, F., Blank, L.M., Zhou, J., Wei, R., Dong, W., Jiang, M., 2021. Biodegradation and up-cycling of polyurethanes: progress, challenges, and prospects. Biotechnol. Adv. 48, 107730.

LI, W.-J., Narancic, T., Kenny, S.T., 2020. Unraveling 1,4-butanediol metabolism in *Pseudomonas putida* KT2440. Front. Microbiol. 11, 382.

Li, W.J., Jayakody, L.N., Franden, M.A., Wehrmann, M., Daun, T., Hauer, B., Blank, L.M., Beckham, G.T., Klebensberger, J., Wierckx, N., 2019. Laboratory evolution reveals the metabolic and regulatory basis of ethylene glycol metabolism by *Pseudomonas putida* KT2440. Environ. Microbiol. 21, 3669–3682.

Lwanga, E.H., Gertsen, H., Gooren, H., Peters, P., Salánki, T., van der Ploeg, M., Besseling, E., Koelmans, A.A., Geissen, 2017. Incorporation of microplastics from litter into burrows of *Lumbricus terrestris*. Environ. Pollut. 220, 523–531.

Magnin, A., Pollet, E., Phalip, V., Averous, L., 2020. Evaluation of biological degradation of polyurethanes. Biotechnol. Adv. 39, 107457.

Magnin, A., 2019. Enzymatic recycling of thermoplastic polyurethanes: synergistic effect of an esterase and an amidase and recovery of building blocks. Waste. Manage. 85, 141–150.

Mar Castro López, M., Pernas, A.I.A., Abad, M.J., 2014. Assessing changes on poly(ethylene terephthalate) properties after recycling: mechanical recycling in laboratory versus postconsumer recycled material. Mater. Chem. Phys. 147 (3), 884–894.

Mathur, G., Prasad, R., 2012. Degradation of polyurethane by *Aspergillus flavus* (ITCC 6051) isolated from soil. Appl. Biochem. Biotechnol 167 (6), 1595–1602.

Mihreteab, M., Stubblefield, B.A., Gilbert, E.S., 2019. Microbial bioconversion of thermally depolymerized polypropylene by Yarrowia lipolytica for fatty acid production. Appl. Microbiol. Biotechnol. 103, 7729–7740.

Muckschel, B., Simon, O., Klebensberger, J., Graf, N., Rosche, B., Altenbuchner, J., Pfannstiel, J., Huber, A., Hauer, B., 2012. Ethylene glycol metabolism by *Pseudomonas putida*. Appl. Environ. Microbiol. 78, 8531–8539.

Nikodinovic-Runic, J., Casey, E., Duane, G.F., Mitic, D., Hume, A.R., Kenny, S.T., 2011. Process analysis of the conversion of styrene to biomass and medium chain length polyhydroxyalkanoate in a two- phase bioreactor. Biotechnol. Bioeng. 108, 2447–2455.

Nomura, N., Shigeno-Akutsu, Y., Nakajima-Kambe, T., 1998. Cloning and sequence analysis of a polyurethane esterase of *Comamonas acidovorans* TB-35. J. Ferment. Bioeng 86 (4), 339–345.

Nuelle, M.T., Dekiff, J.H., Remy, D., Fries, E., 2014. A new analytical approach for monitoring microplastics in marine sediments. Environ. Pollut. 184, 161–169.

Osman, M., Satti, S.M., Luqman, A., 2018. Degradation of polyester polyurethane by *Aspergillus* sp. strain S45 isolated from soil. J. Polym. Environ. 26 (1), 301–310.

O'Leary, N.D., O'Connor, K.E., Goff, M., Dobson, A.D., 2002. Biochemistry, genetics and physiology of microbial styrene degradation. FEMS Microbiol. 26, 403–417.

Parke, D., Garcia, M.A., Ornston, L.N., 2001. Cloning and genetic characterization of dca genes required for beta-oxidation of straight-chain dicarboxylic acids in Acinetobacter sp. strain ADP1. Appl. Environ. Microbiol. 67 (10), 4817–4827.

Phua, S.K., Castillo, E., Anderson, J.M., 1987. Biodegradation of a polyurethane in vitro. J. Biomed. Mater. Res. 21 (2), 231–246.

Radecka, I., Irorere, V., Jiang, G., Hill, D., Williams, C., Adamus, G., 2016. Oxidized polyethylene wax as a potential carbon source for PHA production. Materials 9, 367.

Rojo, F., 2010. Enzymes for Aerobic Degradation of Alkanes. Springer, Berlin Heidelberg.

Romualdus, N.C.U., Li, W.-J., Tiso, T., Eberlein, C., Doeker, M., Heipieper, H.J., Jupke, A., Wierckx, N., Blank, L.M., 2020. Defined microbial mixed culture for utilization of polyurethane monomers. ACS Sustain. Chem. Eng. 8, 17466–17474.

Ronkvist, S.M., Xie, W., Lu, W., 2009. Cutinase-catalyzed hydrolysis of poly(ethylene terephthalate). Macromolecules 42 (14), 5128–5138.

Santo, M., Weitsman, R., Sivan, A., 2013. The role of the copper-binding enzyme-laccase-in the biodegradation of polyethylene by the actinomycete *Rhodococcus* ruber. Biodeter. Biodegrad. 84, 204–210.

Sasoh, M., Masai, E., Ishibashi, S., 2006. Characterization of the terephthalate degradation genes of *Comamonas* sp. Strain E6. Appl. Environ. Microbiol. 72 (3), 1825–1832.

Shigematsu, T., Yumihara, K., Ueda, Y., Morimura, S., Kida, K., 2003. Purification and gene cloning of the oxygenase component of the terephthalate 1,2-dioxygenase system from *Delftia tsuruhatensis*- strain T7. FEMS Microbiol. Lett. 220 (2), 255–260.

Stern, R.V., Howard, G.T., 2000. The polyester polyurethanase gene (pueA) from *Pseudomonas chlororaphis* encodes a lipase. FEMS Microbiol. Lett 185 (2), 163–168.

Sulaiman, S., You, D.J., Kanaya, E., 2014. Crystal structure and thermodynamic and kinetic stability of metagenome-derived LC-cutinase. Biochemistry 53 (11), 1858–1869.

Then, J., Wei, R., Oeser, T., 2016. A disulfide bridge in the calcium binding site of a polyester hydrolase increases its thermal stability and activity against polyethylene terephthalate. FEBS Open Bio 6 (5), 425–432.

Toru, S., Kazuyo, Y., Yutaka, U., 2003. Purification and gene cloning of the oxygenase component of the terephthalate 1,2-dioxygenase system from *Delftia tsuruhatensis* strain T7. FEMS Microbiol. Lett. 220 (2), 255–260.

Tribedi, P., Sil, A.K., 2013. Low-density polyethylene degradation by *Pseudomonas* sp, AKS2 biofilm. Environ. Sci. Pollut. Res. 20, 4146–4153.

Trifunovic, D., Schuchmann, K., Muller, V., 2016. Ethylene glycol metabolism in the acetogen *Acetobacterium woodii*. J. Bacteriol. 198 (7), 1058–1065.

Ward, P.G., Goff, M., Donner, M., Kaminsky, W., O'Connor, K.E., 2006. A two-step chemo-biotechnological conversion of polystyrene to a biodegradable thermoplastic. Environ. Sci. Technol. 40, 2433–2437.

Webb, H.K., Arnott, J., Crawford, R.J., 2012. Plastic degradation and its environmental implications with special reference to Poly(ethylene terephthalate). Polymers 5 (1), 1–18.

Wierckx, N., Prieto, M.A., Pomposiello, P., de Lorenzo, V., O'Connor, K., Blank, L.M., 2015. Plastic waste as a novel substrate for industrial biotechnology. Microb. Biotechnol. 8, 900–903.

Yang, Y., Chen, J., Wu, W.M., Zhao, J., Yang, J., 2015. Complete genome sequence of Bacillus sp. YP1, a polyethylene-degrading bacterium from waxworm's gut. Biotechnology 200, 77–78.

Yoshida, S., Hiraga, K., Takehana, T., 2016. A bacterium that degrades and assimilates poly(ethylene terephthalate). Science 353 (6278), 759.

Yuan, G., Chen, D., Yin, L., Wang, Z., Zhao, L., Wang, J.Y., 2014. High efficiency chlorine removal from polyvinyl chloride (PVC) pyrolysis with a gas-liquid fluidized bed reactor. Waste Manage. 34, 1045–1050.

Index

Printed in the United States
by Baker & Taylor Publisher Services